SCHAUM'S OUTLINE OF

THEORY AND PROBLEMS

of

ELECTRONICS TECHNOLOGY

•

MILTON KAUFMAN
President, Electronic Writers and Editors, Inc.

and

J. A. WILSON
Former Instructor
Kent State University

•

Consulting Editor:
PETER BROOKS
Adjunt Professor of Science and Electrical Technology
Rockland Community College

SCHAUM'S OUTLINE SERIES
McGRAW-HILL BOOK COMPANY

New York St. Louis San Francisco Auckland Bogotá Guatemala Hamburg Johannesburg Lisbon London Madrid Mexico Montreal New Delhi Panama Paris San Juan São Paulo Singapore Sydney Tokyo Toronto

MILTON KAUFMAN is President of Electronic Writers and Editors and has extensive experience as an instructor, working engineer, and writer. He received his BSEE at New York University. He has coauthored McGraw-Hill books on radio electronics and electricity, and also is editor of McGraw-Hill's *Handbook for Electronics Engineering Technicians, Handbook for Chemical Technicians, Handbook of Microcircuit Design and Application,* and *Handbook of Operational Amplifier Circuit Design.*

J. A. WILSON received his Bachelor's degree at Long Beach State College and his Master's degree at Kent State University. He has taught technology courses at Kent State and at Colorado State University and has been Director of CET Testing for the International Society of Certified Electronics Technicians (ISCET) and Technical Director for the National Electronics Service Dealers Association (NESDA). He is presently engaged in medical electronics research and free-lance writing.

Schaum's Outline of Theory and Problems of
ELECTRONICS TECHNOLOGY

3 4 5 6 7 8 9 10 11 12 13 14 15 16 17 18 19 20 SH SH 8 6 5 4

Sponsoring Editor, John Aliano
Editing Supervisor, John Fitzpatrick
Production Manager, Nick Monti

Library of Congress Cataloging in Publication Data

Kaufman, Milton.
Schaum's outline of theory and problems of electronics technology.

(Schaum's outline series)
Includes index.
1. Electronics. I. Wilson, J. A. II. Brooks, Peter, 1940- III. Title. IV. Title: Outline of theory and problems of electronics technology. V. Series.
TK7816.K38 621.381 81-14238
ISBN 0-07-070690-5 AACR2

Preface

This book is intended as a companion text to be used with any electronics technology textbook. It may also be used as an intermediate level electronics technology textbook by itself, with the instructor supplying derivations and adding to the text.

Chapter 1 presents most of the fundamentals in tabular form, because it is assumed this information is adequately covered in other sources. A knowledge of basic algebra and trigonometry and the ability to use a scientific calculator are also assumed. Chapters 2 through 7 have explanatory text which should serve as a brief review of the subject matter. Chapter 7 provides an introduction to the digital electronics field, including microprocessors, and should prove valuable for student reference. Chapter 8 covers single- and three-phase power supplies and uses tables to provide students with a convenient reference for work in this area.

As is customary with the Schaum's Outline Series, numerous worked-out problems are included in each chapter along with supplementary problems and answers. The problems chosen demonstrate the basic theory and were selected as representative of those the student might encounter in other treatments of the same material.

Thanks are due to the various companies, schools, teachers, and students for their thoughtful comments when reviewing the material and for the input they provided which helped organize the material to be covered. We would also like to thank the Schaum staff for its tireless efforts in bringing this book to fruition. We hope that it provides a needed support for students and teachers alike.

MILTON KAUFMAN
J. A. WILSON

Contents

Chapter 1

Review of Basic Electricity

1.1 VOLTAGE, CURRENT, AND RESISTANCE IN DC CIRCUITS

In order to have a good understanding of electronics, it is necessary to be familiar with basic electrical concepts and units of measure. These basic concepts and units will be reviewed in this chapter.

The fundamental (SI) units of measure in electricity are the volt (V), used to measure *voltage*, and the ampere (A), used to measure *current*. Time, and its unit of measure the second (s), along with voltage and current can be combined to form other useful quantities. The algebraic symbols for voltage and current are V and I, respectively.

Ohm's law defines the unit of *resistance* as 1 V divided by 1 A. The symbol for resistance is R and its unit of measure is the ohm (Ω). Ohm's law may be written as

$$R = \frac{V}{I} \tag{1-1a}$$

$$V = IR \tag{1-1b}$$

or

$$I = \frac{V}{R} \tag{1-1c}$$

The reciprocal of resistance is called *conductance* (G) and is measured in siemens (S). However, the old unit for conductance, the mho (℧), is still found in many books.

$$G = \frac{1}{R} \tag{1-2}$$

Other common units are the henry (H), for measuring *inductance* (L); the farad (F), for measuring *capacitance* (C); and the watt (W), for measuring *power* (P).

Table 1-1 shows the relationship of these quantities to the fundamental units.

Table 1-1

Units	Abbreviation	Dimensions
Ohms	Ω	Volts / Ampere
Siemens (or mhos)	S (℧)	Amperes / Volt
Henries	H	Volt-seconds / Ampere
Farads	F	Ampere-seconds / Volt
Watts	W	Volt-amperes

Voltage sources are usually marked with + and − signs on their terminals, although sometimes only the positive (+) terminal is indicated. *Conventional current* is presumed to flow from the positive terminal of the source, through the external circuit, and back to the negative terminal of the source. Conventional current is also presumed to flow from the negative to the positive terminal *within* the source.

Electron current is presumed to flow in the opposite direction—that is, from negative to positive *external* to the source and positive to negative *within* the source. These conventions are shown in Fig. 1-1. In this book, conventional current will be used unless otherwise stated.

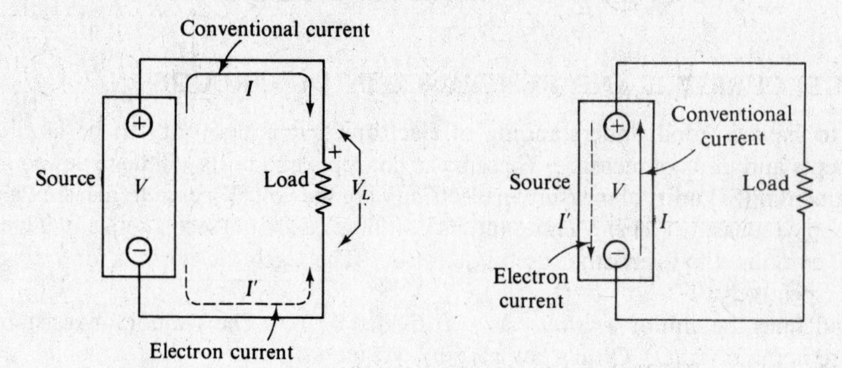

Fig. 1-1

The symbol V is used in this book to mean both a voltage drop and a voltage source or rise, such as that produced by a battery or generator.

The important characteristics and equations for series and parallel circuits are listed in Table 1-2, where n is an integer (1, 2, 3, etc.), R_{eq} is the equivalent resistance, and G_{eq} is the equivalent conductance.

Table 1-2

Series circuits (See Fig. 1-2.)	Parallel circuits (See Fig. 1-3.)
Current I is the same through all the resistors.	Voltage V is the same across all the resistors.
Total voltage $= V_T$	Total current $= I_T$
Voltage across $R_n = V_n$ $V_T = V_1 + V_2 + V_3 + \cdots + V_n$ (1-3)	Current through $R_n = I_n$ $I_T = I_1 + I_2 + I_3 + \cdots + I_n$ (1-6)
$R_{eq} = \frac{V_T}{I} = R_1 + R_2 + R_3 + \cdots + R_n$ (1-4)	$G_{eq} = \frac{1}{R_{eq}} = \frac{I_T}{V} = G_1 + G_2 + G_3 + \cdots + G_n$ (1-7a) $\frac{1}{R_{eq}} = \frac{1}{R_1} + \frac{1}{R_2} + \frac{1}{R_3} + \cdots + \frac{1}{R_n}$ (1-7b)
$V_n = \frac{R_n}{R_{eq}} V_T$ (1-5) (Voltage divider rule)	$I_n = \frac{G_n}{G_{eq}} I_T = \frac{R_{eq}}{R_n} I_T$ (1-8) (Current divider rule)
	Special case: two resistors $R_{eq} = \frac{R_1 R_2}{R_1 + R_2}$ (1-7c) $I_1 = \frac{R_2}{R_1 + R_2} I_T$ (1-8a) $I_2 = \frac{R_1}{R_1 + R_2} I_T$ (1-8b)

Remember that for a series circuit the highest voltage drop is across the *highest* value of resistance and for a parallel circuit the highest current is through the *lowest* value of resistance.

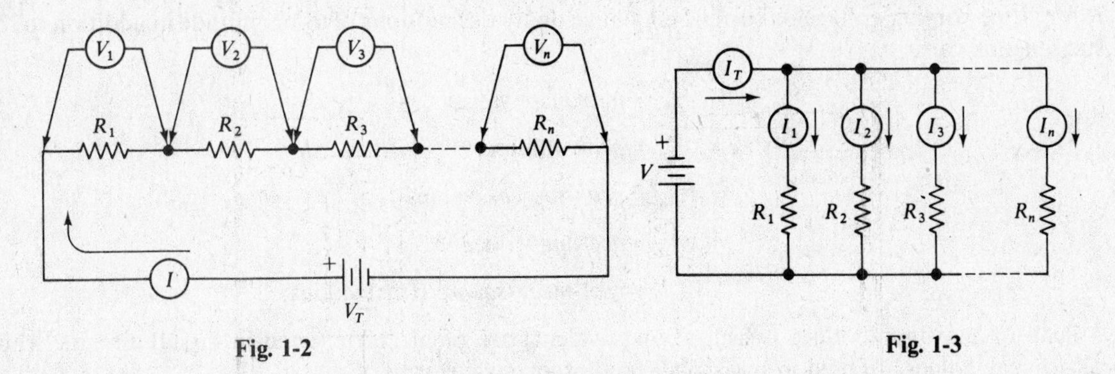

Fig. 1-2

Fig. 1-3

1.2 VOLTAGE AND CURRENT IN AC CIRCUITS

Although it is common practice to think of dc voltages and currents as steady quantities (not dependent on time), it is well to define a dc voltage or current as one which is *unidirectional* (acting in one direction only).

An ac or alternating voltage or current is then defined as *bidirectional.* In other words, its polarity changes with time. Usually, ac voltages are repetitive or *periodic.* That is, they repeat the same waveshape at a specified rate or *frequency* f. Another way of describing a periodic waveform is by specifying its *period* T, defined as the length of time between identical points on the waveform. The period and the frequency are the reciprocals of each other:

$$T = \frac{1}{f} \tag{1-9}$$

When f is measured in cycles per second or *hertz* (Hz), the period is measured in seconds.

Examples of dc waveforms are given in Fig. 1-4, while typical ac waveforms are shown in Fig. 1-5.

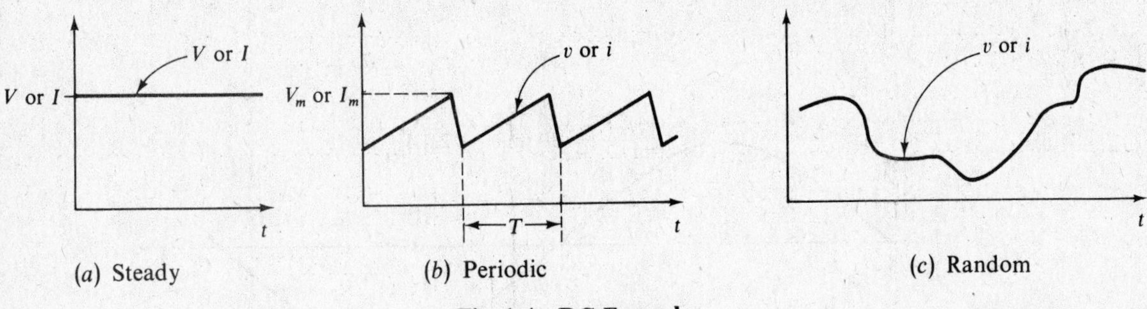

(*a*) Steady (*b*) Periodic (*c*) Random

Fig. 1-4 DC Examples

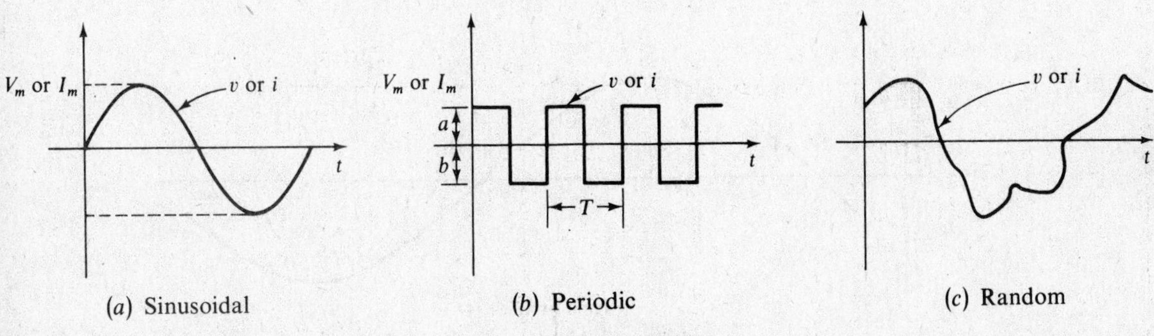

(*a*) Sinusoidal (*b*) Periodic (*c*) Random

Fig. 1-5 AC Examples

In this text we denote any time-varying value with a lowercase letter and any value that does not vary with time by a capital letter. The time-varying values are known as *instantaneous* values.

Any time-varying voltage or current may have many descriptions of its magnitude in addition to the instantaneous value:

$$V_m, I_m = \textit{maximum} \text{ values}$$

$$V_p, I_p = \textit{peak} \text{ values}$$

$$V_{p\text{-}p}, I_{p\text{-}p} = \textit{peak-to-peak} \text{ values}$$

$$V_o, I_o = \textit{average} \text{ values}$$

$$V, I = \textit{root-mean-square} \text{ (rms) values}$$

Peak or maximum values (which are equivalent) are often encountered in digital circuits, while peak-to-peak values are used in measuring oscilloscope waveforms.

Average values of polarity symmetrical waveforms are taken as half-period (half-cycle) averages since full-period averages would be zero.

The rms value represents the *effective* value and is the equivalent steady dc voltage that would produce the same heat in a resistor as the given voltage or current. Most ac voltmeters and ammeters are calibrated to read the rms value but give a true reading only when the waveform is sinusoidal.

For sinusoidal voltages and currents such as shown in Fig. 1-6, the instantaneous values v and i are given by

$$v = V_m \sin(\omega t + \phi) = V_m \sin(2\pi f t + \phi) \tag{1-10a}$$

and

$$i = I_m \sin(\omega t + \phi') = I_m \sin(2\pi f t + \phi') \tag{1-10b}$$

where V_m, I_m = maximum values, V and A respectively
f = linear frequency, Hz
ω = angular frequency, rad/s
t = time, s
ϕ, ϕ' = phase angles, rad

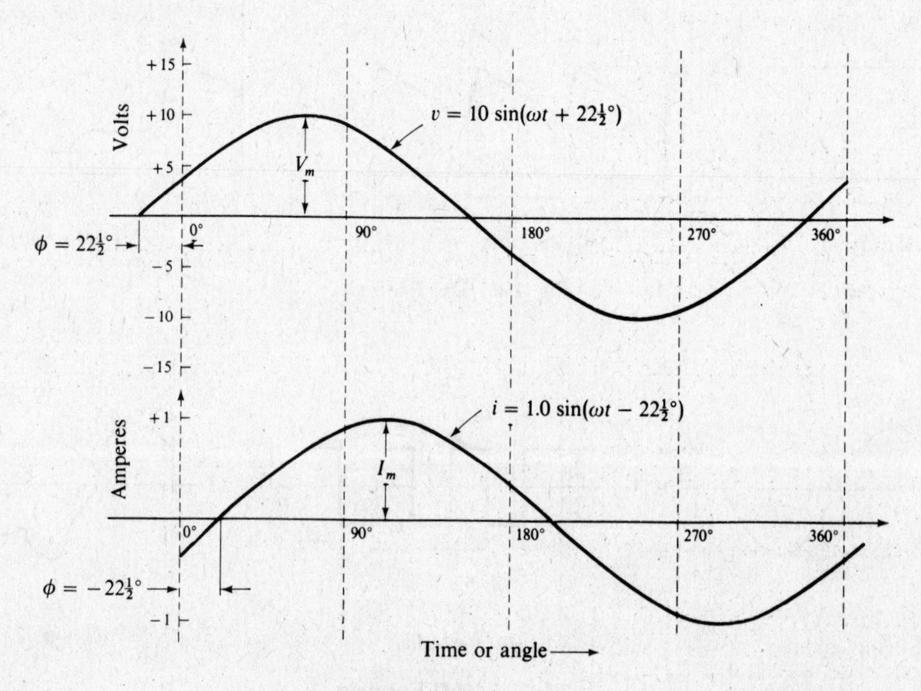

Fig. 1-6

The angular frequency ω is related to the linear frequency f by

$$\omega = 2\pi f \tag{1-11}$$

The phase angle and ωt (or $2\pi ft$) should be measured in identical units for proper results. When ω is used, the phase angle should be measured in rad and when f is used the phase angle should be measured in °.

The conversion from rad to ° is easily obtained with most scientific calculators or by using

$$\pi \text{ rad} = 180° \tag{1-11a}$$

$$1 \text{ rad} = 57.29577951° \tag{1-11b}$$

or

$$1° = 1.745329252 \times 10^{-2} \text{ rad} \tag{1-11c}$$

The voltage and current phase angles ϕ and ϕ' are not usually equal and each can be measured between $x = 0$ and the point where the wave crosses the x axis. The x axis itself can be labeled in angle units (° or rad) or time units (s). If labeled in time units, the phase angle is more properly called a phase time.

The relation between peak, rms, and half-cycle average values is shown in Fig. 1-7 and by the following equations:

$$I = \frac{I_m}{\sqrt{2}} = 0.707 I_m \tag{1-12a}$$

$$V = \frac{V_m}{\sqrt{2}} = 0.707 V_m \tag{1-12b}$$

$$I_o = \frac{2}{\pi} I_m = 0.636 I_m \tag{1-13a}$$

$$V_o = \frac{2}{\pi} V_m = 0.636 V_m \tag{1-13b}$$

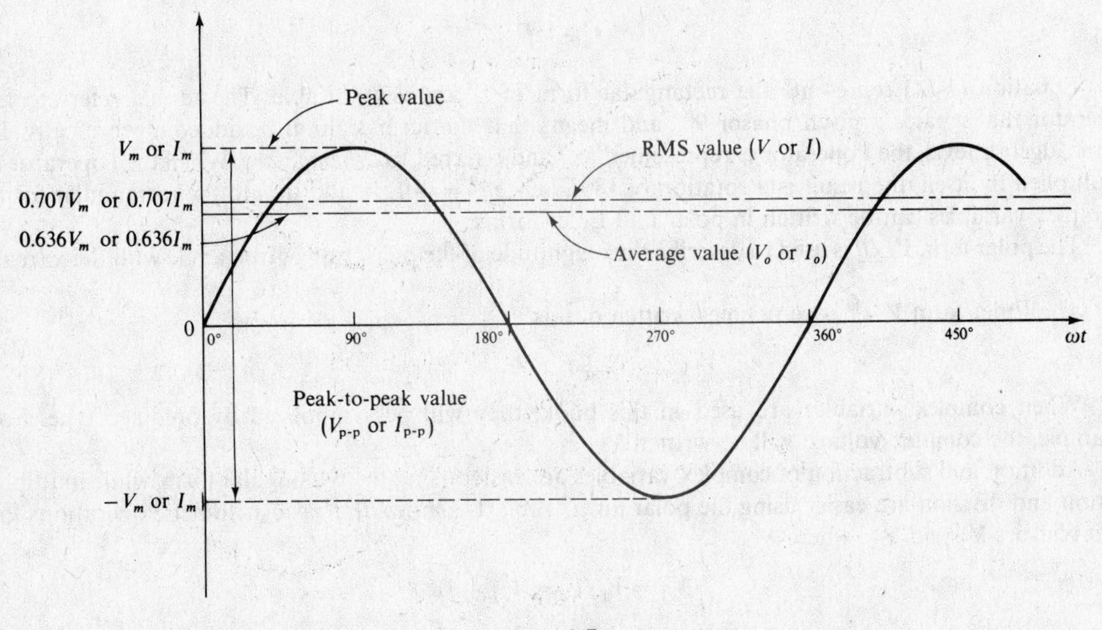

Fig. 1-7

1.3 COMPLEX VARIABLES AND PHASORS

For sine waves, we find it convenient to describe circuit relationships by means of vector diagrams. Since there are no true vectors involved, these are usually called *phasor*, or *rotor*, diagrams. The phasors are presumed to be rotating in the standard counterclockwise direction as they generate a sine wave (see Fig. 1-8). Either the voltage or the current can be used as a reference, in which case it is shown at the zero, or three o'clock, position.

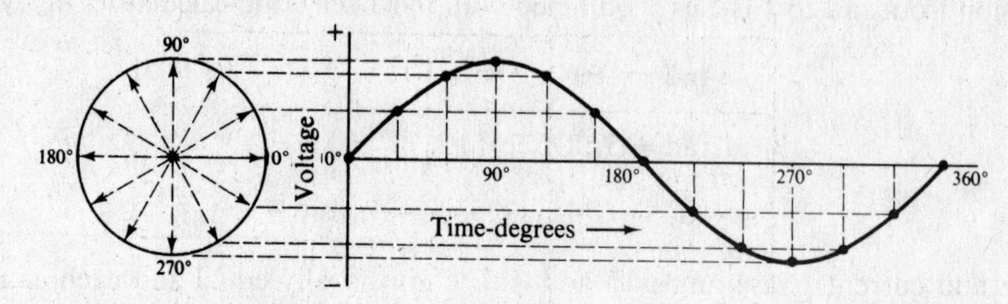

Fig. 1-8

When phasors are at right angles to each other (in *quadrature*), we can represent their projections on the real and quadrature axes by complex variables.

For example, if two voltages differ in phase by 90°, we may represent their sum by writing

$$V_T = V_1 + jV_2 \tag{1-14}$$

This is shown graphically in Fig. 1-9.

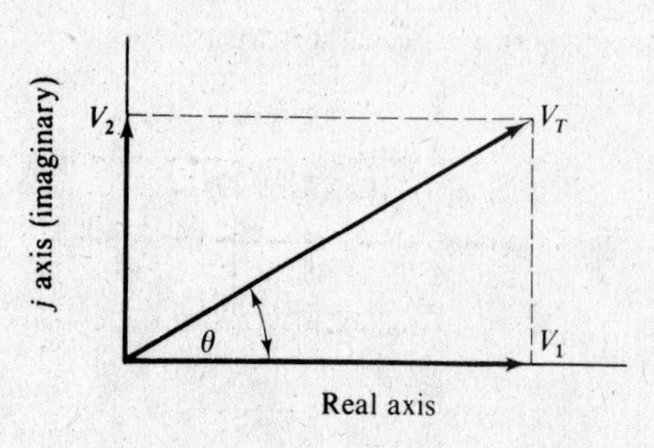

Fig. 1-9

Equation (*1-14*) represents the rectangular form of a complex variable. The term j refers to an operator that rotates a given phasor 90° and means that the terms cannot be added algebraically. In some algebra texts, the j operator is represented as i and referred to as *imaginary*. When the j operator is multiplied by itself, the result is a rotation of 180°, hence $j^2 = -1$. In addition to the rectangular form, complex variables can be written in polar and Euler forms.

The polar form $V_T\underline{/\theta}$ is used to describe the magnitude of the phasor at a given angle with the x (real) axis.

The Euler form $V_T\epsilon^{j\theta}$ is sometimes written out as

$$V_T\epsilon^{j\theta} = V_T \cos\theta + jV_T \sin\theta$$

When complex variables are used in this book, they will be symbolized by boldface type. For example, the complex voltage will be written **V**.

Addition and subtraction of complex variables are easier using the rectangular form, while multiplication and division are easier using the polar form. Table 1-3 shows the basic arithmetic operations for two phasors $\mathbf{V}_1$ and $\mathbf{V}_2$, where

$$\mathbf{V}_1 = V_1\underline{/\theta_1} = V_{r1} + jV_{x1}$$
$$\mathbf{V}_2 = V_2\underline{/-\theta_2} = V_{r2} - jV_{x2}$$

Table 1-3

Operation	Result
$\mathbf{V}_1 + \mathbf{V}_2$ or $\mathbf{V}_2 + \mathbf{V}_1$	$(V_{r1} + V_{r2}) + j(V_{x1} - V_{x2})$
$\mathbf{V}_1 - \mathbf{V}_2$	$(V_{r1} - V_{r2}) + j(V_{x1} + V_{x2})$
$\mathbf{V}_2 - \mathbf{V}_1$	$(V_{r2} - V_{r1}) - j(V_{x1} + V_{x2})$
$(\mathbf{V}_1)(\mathbf{V}_2)$ or $(\mathbf{V}_2)(\mathbf{V}_1)$	$(V_1)(V_2)\underline{/(\theta_1 - \theta_2)}$
$\mathbf{V}_1 \div \mathbf{V}_2$	$\frac{V_1}{V_2}\underline{/(\theta_1 + \theta_2)}$
$\mathbf{V}_2 \div \mathbf{V}_1$	$\frac{V_2}{V_1}\underline{/-(\theta_1 + \theta_2)}$

Another definition involving complex variables that we find useful is the complex *conjugate*. Two complex numbers are said to be conjugates if their real values are equal in magnitude and polarity and their imaginary values are equal in magnitude but opposite in polarity.

The notation for the complex conjugate of **A** is **A***. For example, if $\mathbf{Z} = R + jX$, then $\mathbf{Z}^* = R - jX$.

Table 1-4 shows some forms of conjugate numbers.

Table 1-4

Complex number	Conjugate
$\mathbf{Z}$	$\mathbf{Z}^*$
$\pm A \pm jB$	$\pm A \mp jB$
$Z\underline{/\pm\theta}$	$Z\underline{/\mp\theta}$
$Z\epsilon^{\pm j\theta}$	$Z\epsilon^{\mp j\theta}$

1.4 IMPEDANCE AND ADMITTANCE

The opposition that a circuit offers to alternating current is called *impedance* (Z), and the reciprocal of impedance is called *admittance* (Y), since it is a measure of how well the circuit "admits" current. Z, like R, is measured in Ω; Y, like G, is measured in S.

The impedance of any circuit can be found by modifying Ohm's law to read

$$z = \frac{v}{i} \tag{1-15}$$

The magnitude of the impedance Z can be found from any of the following forms

$$Z = \frac{V}{I} = \frac{V_o}{I_o} = \frac{V_m}{I_m} = \frac{V_{p\text{-}p}}{I_{p\text{-}p}} \tag{1-15a}$$

While resistors offer equal opposition to both direct and alternating current, the opposition that inductive and capacitive components offer to alternating current is frequency dependent and is termed *reactance* (X). The reciprocal of reactance is called *susceptance* (B).

A list of the characteristics and relationships involving impedance, reactance, admittance and susceptance is shown in Table 1-5.

Table 1-5

For sinusoidal waves

Series circuits (See Fig. 1-10.)		Parallel circuits (See Fig. 1-11.)	
$Z = \frac{1}{Y} = \frac{V}{I}$	(1-15b)	$Y = \frac{1}{Z} = \frac{I}{V}$	(1-21)
$X_L = \omega L = 2\pi fL$	(1-16)	$B_L = \frac{1}{X_L} = \frac{1}{\omega L} = \frac{1}{2\pi fL}$	(1-22)
$X_C = \frac{1}{\omega C} = \frac{1}{2\pi fC}$	(1-17)	$B_C = \frac{1}{X_C} = \omega C = 2\pi fC$	(1-23)
$X_T = X_L - X_C$	(1-18a)	$B_T = B_C - B_L$	(1-24a)
$= \omega L - \frac{1}{\omega C} = \frac{\omega^2 LC - 1}{\omega C}$	(1-18b)	$= \omega C - \frac{1}{\omega L} = \frac{\omega^2 LC - 1}{\omega L}$	(1-24b)
$Z^2 = R^2 + X_T^2$	(1-19)	$Y^2 = G^2 + B_T^2$	(1-25)
$\phi = \arctan \frac{X_T}{R}$	(1-20)	$\phi' = \arctan \frac{B_T}{G}$	(1-26)
where X_L = inductive reactance X_C = capacitive reactance X_T = total reactance		B_L = inductive susceptance B_C = capacitive susceptance B_T = total susceptance	

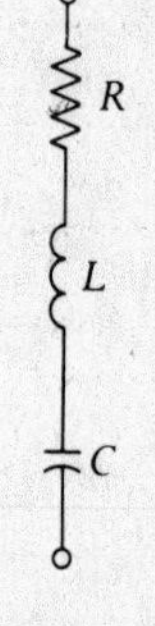

Fig. 1-10

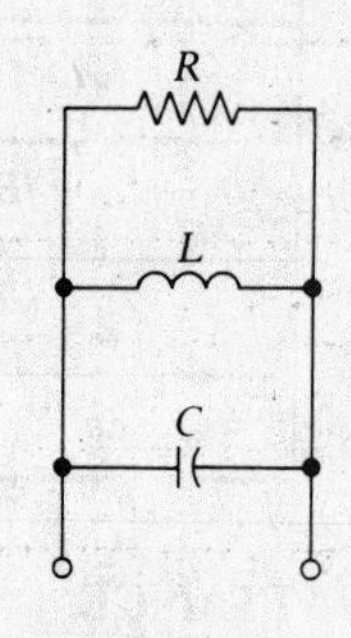

Fig. 1-11

When the frequency is zero (dc) a capacitor acts like an open circuit and an inductor acts like a short circuit.

Practically, capacitors exhibit leakage current and inductors have resistance associated with them, but for most cases both can be treated as if they were ideal.

As the frequency increases, the inductor causes the current to increasingly *lag* the voltage, while a capacitor causes the current to increasingly *lead* the voltage.

A resistor in an ac circuit maintains the same phase between the voltage across it and the current through it. This is described as an in-step or *in-phase* relationship.

Impedance and admittance triangles and phasor diagrams are helpful in understanding the phase relationships in ac circuits and are shown in Figs. 1-12 and 1-13. From these figures we note that

$$Z^2 = R^2 + X_T^2 \tag{1-19}$$

and

$$Y^2 = G^2 + B_T^2 \tag{1-25}$$

X_L R X_C

Z ϕ R $X_L - X_C = X_T$

R $-\phi$ Z $X_C - X_L = X_T$

When $X_L > X_C$, $\mathbf{Z} = R + jX_T$ When $X_C > X_L$, $\mathbf{Z} = R - jX_T$

R, X_L, and X_C phasors

Fig. 1-12 Impedance triangles

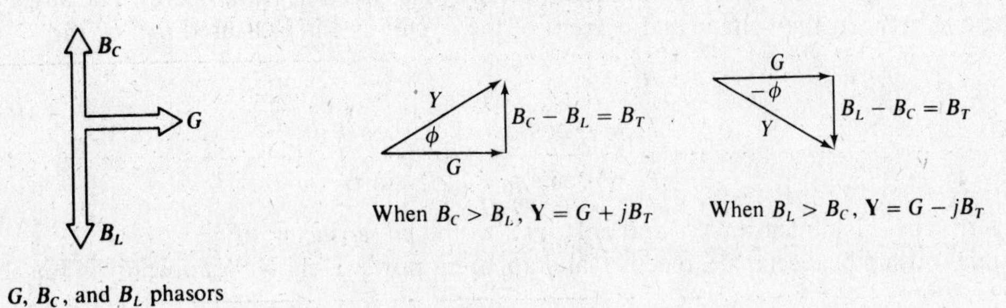

G, B_C, and B_L phasors

Fig. 1-13 Admittance triangles

We may also use complex variable notation for Z and Y as shown in Table 1-6.

Table 1-6

Rectangular	Polar	Euler
$\mathbf{Z}_L = R + jX_L$	$Z_L\underline{/\phi}$	$Z_L\epsilon^{j\phi}$
$\mathbf{Z}_C = R - jX_C$	$Z_C\underline{/-\phi}$	$Z_C\epsilon^{-j\phi}$
$\mathbf{Z}_T = R + j(X_L - X_C) = R \pm jX$	$Z_T\underline{/\pm\theta}$	$Z_T\epsilon^{\pm j\theta}$
$\mathbf{Y}_C = G + jB_C$	$Y_C\underline{/\phi'}$	$Y_C\epsilon^{j\phi'}$
$\mathbf{Y}_L = G - jB_L$	$Y_L\underline{/-\phi'}$	$Y_L\epsilon^{-j\phi'}$
$\mathbf{Y}_T = G + j(B_C - B_L) = G \pm jB_T$	$Y_T\underline{/\pm\phi'}$	$Y_T\epsilon^{\pm j\phi'}$

1.5 POWER IN DC AND AC CIRCUITS

The power associated with any circuit is defined as the total voltage across the circuit multiplied by the total current through the circuit. For dc circuits this reduces to

$$P = VI \tag{1-27a}$$

$$= I^2R \tag{1-27b}$$

$$= \frac{V^2}{R} \tag{1-27c}$$

For ac circuits the instantaneous power p is given by

$$p = vi \tag{1-27d}$$

and for sinusoidal waveforms the phasor notation becomes

$$\mathbf{P} = \mathbf{VI} \tag{1-27e}$$

$$= \mathbf{I}^2\mathbf{Z} \tag{1-27f}$$

$$= \frac{\mathbf{V}^2}{\mathbf{Z}} \tag{1-27g}$$

The real component of p is called true, or real, power P_r, is measured in W, and is the power actually dissipated by the circuit resistance.

The j, or imaginary, component of p is called the reactive, or imaginary, power P_q, is measured in var (voltamperes reactive), and is the energy stored by the reactance in the circuit.

The magnitude of p is called the apparent power P, measured in voltamperes. The angle of p is the phase angle ϕ between the voltage and current of the circuit and is measured in ° or rad.

$$P = V_m I_m \tag{1-28a}$$

$$P_r = P \cos \phi = V_m I_m \cos \phi \tag{1-28b}$$

$$P_q = P \sin \phi = V_m I_m \sin \phi \tag{1-28c}$$

The term $\cos \phi$ is dimensionless and is known as the *power factor PF*.

The relationship between true, reactive, and apparent power is shown graphically in Fig. 1-14 and is given by

$$P^2 = P_r^2 + P_q^2 \tag{1-29}$$

For the case of in-phase sinusoidal voltage and current (purely resistive or resonant circuit) the following relationships are useful:

$$p = V_m I_m \sin^2 \omega t = \left(\frac{V_m I_m}{2}\right)(1 - \cos 2\omega t) \tag{1-30a}$$

$$P_o = VI = \frac{V_m I_m}{2} \tag{1-30b}$$

$$P_p = P = V_m I_m = 2P_o = P_r \tag{1-30c}$$

where P_o = average power (full period)
P_p = peak power

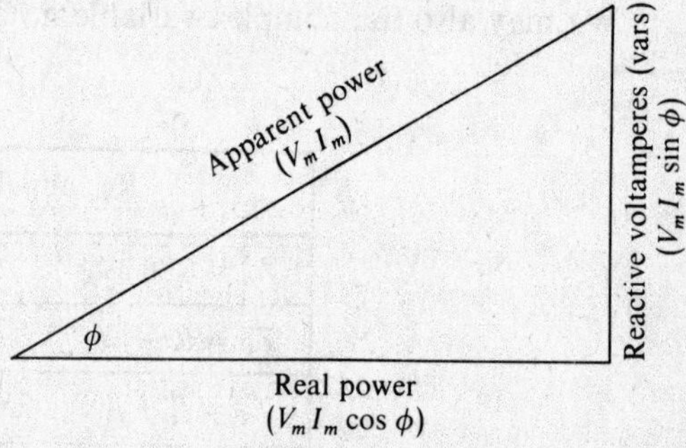

Fig. 1-14

Note that multiplying the rms voltage by the rms current results in the average power, although in high fidelity terminology this product is mistakenly referred to as the rms power.

1.6 DECIBELS

There are many applications in electronics where powers, voltages, or currents are compared. For example, in the amplifier shown in Fig. 1-15*a*, the output signal power P_2 is higher than the input signal power P_1, and we define the decibel (dB) as

$$\text{dB} = 10 \log \frac{P_2}{P_1} \tag{1-34a}$$

If the input and output impedances are the same, we can replace P_2/P_1 by either $(V_2/V_1)^2$ or $(I_2/I_1)^2$, so

$$\text{dB} = 10 \log \frac{P_2}{P_1} = 10 \log \left(\frac{V_2}{V_1}\right)^2 = 20 \log \frac{V_2}{V_1} \tag{1-34b}$$

Also

$$\text{dB} = 10 \log \left(\frac{I_2}{I_1}\right)^2 = 20 \log \frac{I_2}{I_1} \tag{1-34c}$$

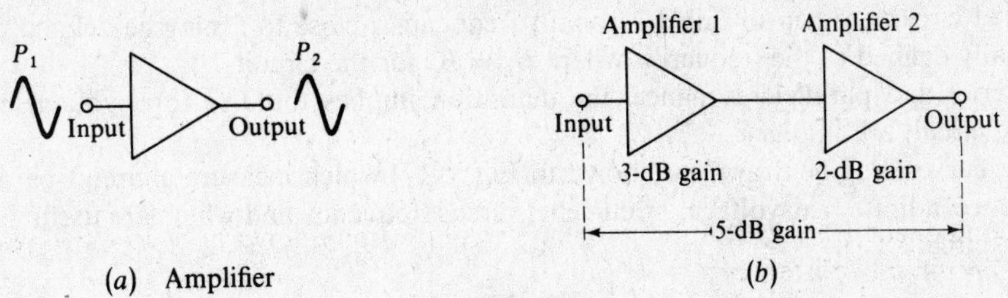

(a) Amplifier (b)

Fig. 1-15

Note that the dB "gain" will be negative if the output is lower than the input.

The symbol "log" is used to mean $\log_{10}$ (the common or Briggsian logarithm), and a subscript is not used unless a different base is required.

The use of logarithms to the base ϵ (natural or Naperian logarithms) is extensive in electronics. Power, voltage, and current ratios expressed with Naperian logarithms are called *nepers.*

Using the symbol ln to mean Naperian logarithms:

$$\text{Nepers} = \tfrac{1}{2} \ln \frac{P_2}{P_1} \tag{1-35a}$$

Again if the input and output impedances are the same, the power ratios can be replaced by the voltage or current ratios squared and

$$\text{Nepers} = \tfrac{1}{2} \ln \frac{P_2}{P_1} = \tfrac{1}{2} \ln \left(\frac{V_2}{V_1}\right)^2 = \ln \frac{V_2}{V_1} \tag{1-35b}$$

Also
$$\text{Nepers} = \tfrac{1}{2} \ln \left(\frac{I_2}{I_1}\right)^2 = \ln \frac{I_2}{I_1} \tag{1-35c}$$

The use of nepers is common outside of the United States, in which decibels are normally used.

An immediate advantage of expressing amplifier gain in decibels or nepers is seen in Fig. 1-15*b*. The output of amplifier 1 feeds the input of amplifier 2. When amplifiers are connected in this manner, they are said to be *cascaded.* The gain of the two-stage system is obtained by simply adding the decibel or neper gain of each stage. This is *not* possible when the gains are expressed as power, voltage, or current ratios.

1.7 SERIES AND PARALLEL RESONANCE

Resonance is an important special condition of ac circuits that depends upon frequency, component values and circuit configuration.

For a series RLC circuit such as is shown in Fig. 1-16*a*, resonance occurs at a single frequency $f_r(f_o)$ where the inductive and capacitive reactances are equal ($X_L = X_C$).

For a parallel circuit as shown in Fig. 1-16*b*, resonance occurs at the frequency $f_r(f_o)$ where the inductive and capacitive susceptances are equal ($B_L = B_C$).

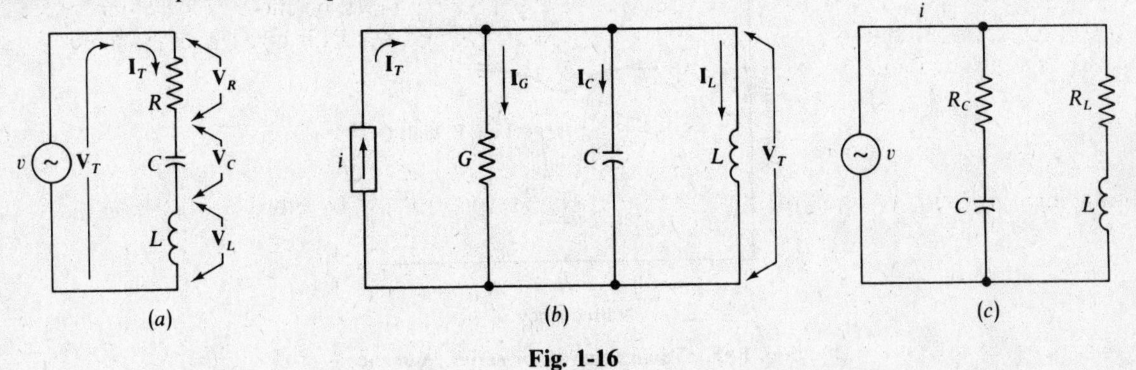

Fig. 1-16

For parallel circuits, other so-called resonant frequencies close to f_r may be defined, but for most work f_r is usually defined at the frequency where $B_L = B_C$ for the circuit.

In both series and parallel resonance, the definition implies that the total voltage and the total current for the circuit are in phase.

Resonance curves may be drawn (as shown in Fig. 1-17) which measure a circuit parameter (gain, power, impedance, admittance, voltage, or current) versus frequency and which are useful in comparing tuning or filter circuits.

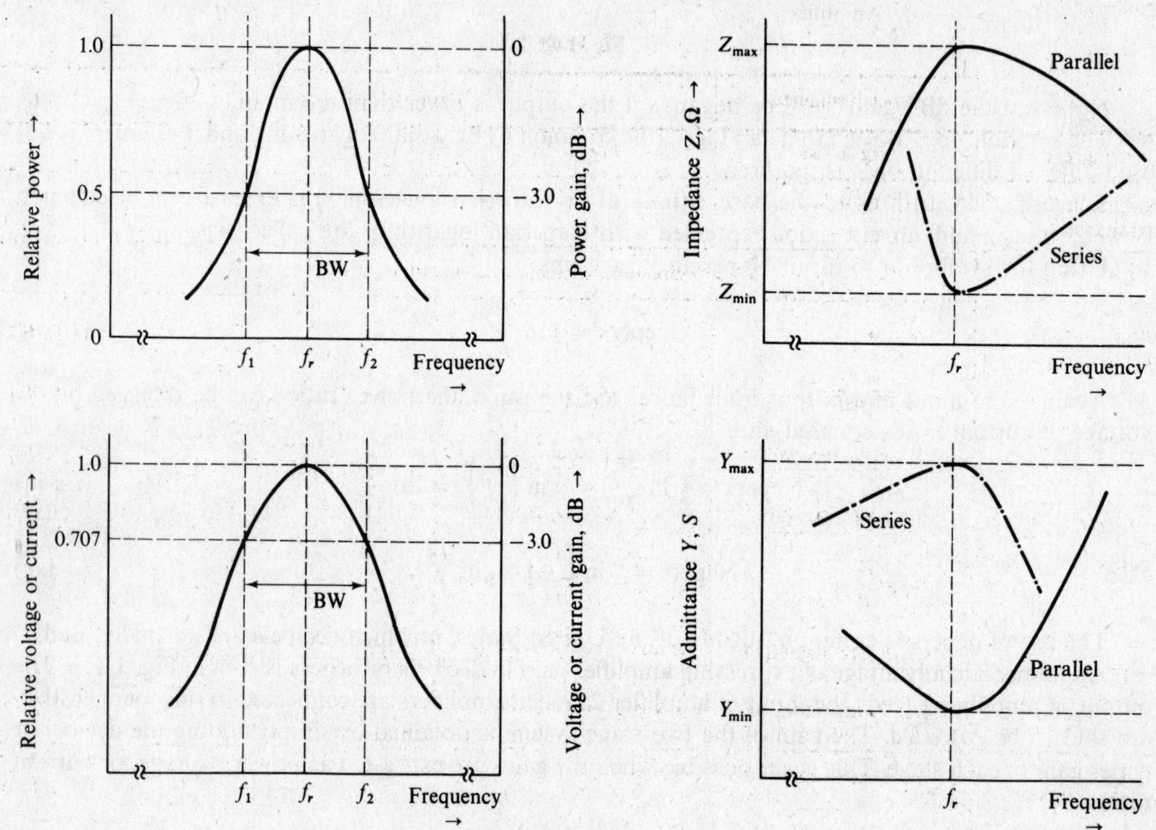

Fig. 1-17 Resonance curves

The shape of the resonance curve is affected by component values and is a convenient index for determining how sharply a circuit will tune. The *quality factor Q* is used to compare the resonance curves as shown in Fig. 1-18 and is defined as the ratio of stored to dissipated energy of the circuit. The lower the resistance in a circuit, the higher the Q. The higher the Q of a circuit, the more sharply it will tune.

Another useful measure for comparing resonance curves is the bandwidth BW. *Bandwidth* is defined as the distance on the frequency axis between the half-power points.

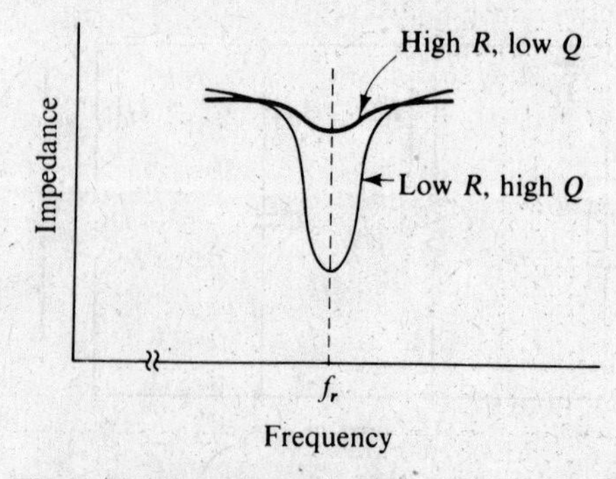

Fig. 1-18 Tuning curves for series resonance

The half-power points are sometimes known as the -3 dB points and occur where the power is half the resonant power or where the current or voltage is 0.707 times the resonant current or voltage. The bandwidth and half-power points are shown in Fig. 1-17.

A list of commonly encountered resonance relationships is compiled in Table 1-7.

By combining series and parallel resonant circuits, we can create passive networks called *filters*. Adjusting the number of resonant circuits and their values allows us to "shape" a frequency response curve.

Table 1-7

	Series resonance	Parallel resonance (anti-resonance)	
Diagram	Fig. 1-16*a*	Fig. 1-16*b*	Fig. 1-16*c*
Condition for resonance	$X_L = X_C$	$B_L = B_C$	
Circuit characteristics	Total circuit current and total circuit voltage are in phase		
	$Z = R$	$Y = G$	$B_C = \dfrac{X_C}{R_C^2 + X_C^2} = \dfrac{1}{X_C}\left(\dfrac{Q_C^2}{1 + Q_C^2}\right)$ (1-39)
	$V_R = V_T\underline{/\theta}$	$I_R = I_T\underline{/\phi}$	$B_L = \dfrac{X_L}{R_L^2 + X_L^2} = \dfrac{1}{X_L}\left(\dfrac{Q_L^2}{1 + Q_L^2}\right)$ (1-40)
	$V_L = QV_T\underline{/\theta + 90^\circ}$	$I_L = QI_T\underline{/\phi - 90^\circ}$	$G = \dfrac{R_C}{R_C^2 + X_C^2} + \dfrac{R_L}{R_L^2 + X_L^2}$ (1-41a)
	$V_C = QV_T\underline{/\theta - 90^\circ}$	$I_C = QI_T\underline{/\phi + 90^\circ}$	$= \dfrac{1}{R_C}\left(\dfrac{1}{1 + Q_C^2}\right) + \dfrac{1}{R_L}\left(\dfrac{1}{1 + Q_L^2}\right)$ (1-41b)
	$I_T = \dfrac{V_T}{R}\underline{/\theta}$	$V_T = \dfrac{I_T}{G}\underline{/\phi}$	For Q_C and Q_L both > 5,
	Y and I_T are maximum	Z and V_T are maximum	$B_C = \dfrac{1}{X_C}$ $B_L = \dfrac{1}{X_L}$
	Z and V_T are minimum	Y and I_T are minimum	$G = \dfrac{1}{Q_C^2 R_C} + \dfrac{1}{Q_L^2 R_L}$ (1-41c)
Resonant frequency:			$f_r = f_0 = \dfrac{1}{2\pi\sqrt{LC}}\sqrt{\dfrac{R_L^2 C - L}{R_C^2 C - L}}$ (1-42a)
Angular	$\omega_r = \omega_0 = \dfrac{1}{\sqrt{LC}}$ (1-36a)		$= \dfrac{1}{2\pi\sqrt{LC}}\left(\dfrac{Q_L}{Q_C}\right)\sqrt{\dfrac{1 + Q_C^2}{1 + Q_L^2}}$ (1-42b)
Linear	$f_r = f_0 = \dfrac{1}{2\pi\sqrt{LC}}$ (1-36b)		If $R_L = R_C$ or $Q_L = Q_C$ or both Q_L and $Q_C > 5$, $f_r = f_0 = \dfrac{1}{2\pi\sqrt{LC}}$
Quality factor	$Q = \dfrac{f_r}{BW}$ (1-37a)		
	$Q = \dfrac{X_L}{R} = \dfrac{X_C}{R}$ (1-37b)	$Q = \dfrac{B_C}{G} = \dfrac{B_L}{G}$ (1-37e)	If Q_L and Q_C are both > 5,
	$= \dfrac{\omega_0 L}{R} = \dfrac{1}{\omega_0 CR}$ (1-37c)	$= \dfrac{\omega_0 C}{G} = \dfrac{1}{\omega_0 LR}$ (1-37f)	$Q = \dfrac{1}{G}\sqrt{\dfrac{C}{L}}$ with
	$= \dfrac{1}{R}\sqrt{\dfrac{L}{C}}$ (1-37d)	$= \dfrac{1}{G}\sqrt{\dfrac{L}{C}}$ (1-37g)	$G = \dfrac{1}{Q_C^2 R_C} + \dfrac{1}{Q_L^2 R_L}$ (1-41d)
			$Q_C = \dfrac{1}{\omega_0 CR_C}$ (1-37h)
			$Q_L = \dfrac{\omega_0 L}{R_L}$ (1-37i)
Bandwidth Linear	$BW = f_2 - f_1 = \dfrac{f_r}{Q}$ (1-38)		

Solved Problems

1.1 What value of resistance must be connected in parallel with a 39-kΩ resistor to obtain an equivalent resistance of 21.3 kΩ?

Since all the values are given in kilohms, we may drop the factor of 1×10^{-3} in our equation if we remember that our answer will be in kilohms.

Using Eq. (*1-7c*), we have

$$R_{eq} = \frac{R_1 R_2}{R_1 + R_2}$$

$$21.3 = \frac{39R_2}{39 + R_2}$$

Cross multiplying,

$$830.7 + 21.3R_2 = 39R_2$$

$$830.7 = 17.7R_2$$

$$R_2 = 46.9 \text{ k}\Omega$$

The closest *preferred* value of resistance is 47 kΩ. A preferred value is one that is manufactured in large quantities. These values are available as "off the shelf" items. Most carbon-composition resistors are manufactured in preferred values.

1.2 What is the value of current through R_1 in Fig. 1-19?

Using the current divider rule (*1-8a*),

$$I_1 = \frac{R_2}{R_1 + R_2} I_T$$

$$= \left(\frac{150}{50 + 150}\right)(100 \times 10^{-3}) = 75 \times 10^{-3} \text{ A} = 75 \text{ mA}$$

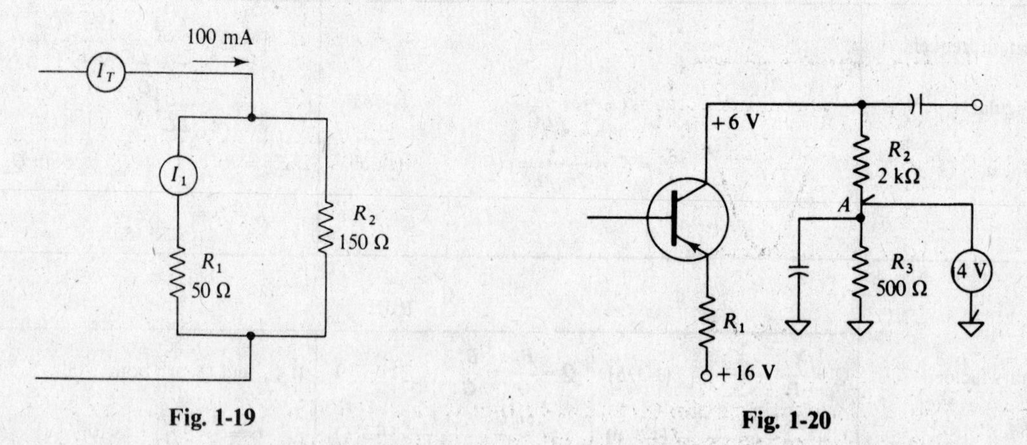

Fig. 1-19

Fig. 1-20

1.3 Figure 1-20 shows a transistor circuit. The manufacturer of the circuit shows the collector voltage to be +6 V. A technician measures the voltage at point *A* and finds the value to be +4 V. Is the circuit operating correctly?

The voltage divider method can be used to find the voltage across R_3, which is the voltage at point *A*:

$$V_A = \frac{R_3}{R_2 + R_3} V_T = \left(\frac{500}{2000 + 500}\right)6 = \frac{3000}{2500} = 1.2 \text{ V} \qquad (1\text{-}5)$$

The circuit is not operating properly. The voltage at point *A* should be 1.2 V instead of 4 V

1.4 To produce the oscilloscope trace shown in Fig. 1-21, the sweep is set so that it requires 0.2 ms (milliseconds) for the beam to go from A to B. What is the frequency of the wave?

The relationship between the period T of a wave and its frequency is

$$T = \frac{1}{f} \tag{1-9}$$

Therefore

$$f = \frac{1}{T}$$

The display is set for a period of 0.2 ms or 0.2×10^{-3} s. Thus

$$f = \frac{1}{T} = \frac{1}{0.2 \times 10^{-3}} = 5 \times 10^3 \text{ Hz} = 5 \text{ kHz}$$

1.5 The vertical *sensitivity* of the oscilloscope in Fig. 1-21 is set to 10 V per inch (V/in) to produce the 5-kHz (kilohertz) waveform shown. Write an equation for the instantaneous sine wave voltage.

From Fig. 1-21, the peak of the wave is 2 in high. Therefore, the peak voltage is

$$V_m = 2 \not{\text{in}} \times \frac{10 \text{ V}}{\not{\text{in}}} = 20 \text{ V}$$

The value of ω is

$$\omega = 2\pi f \tag{1-11}$$
$$= 2\pi(5 \times 10^3) = 3.14 \times 10^4$$

Knowing the value of V_m and ω, the equation can now be written (since we are considering only one waveform, we take $\phi = 0$):

$$v = V_m \sin(\omega t \pm \phi) \tag{1-10a}$$
$$= 20 \sin(3.14 \times 10^4)t$$

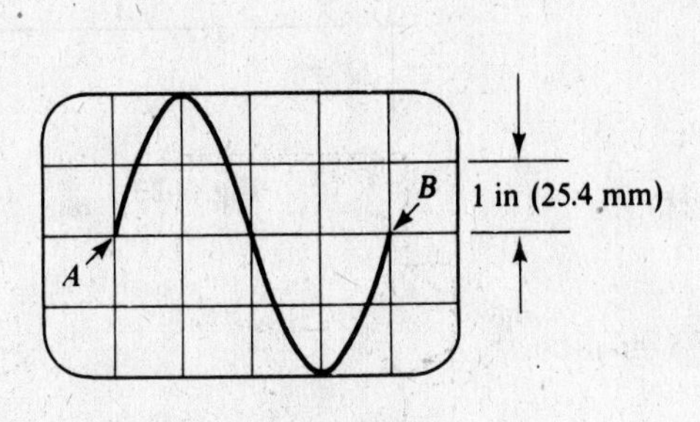

Fig. 1-21

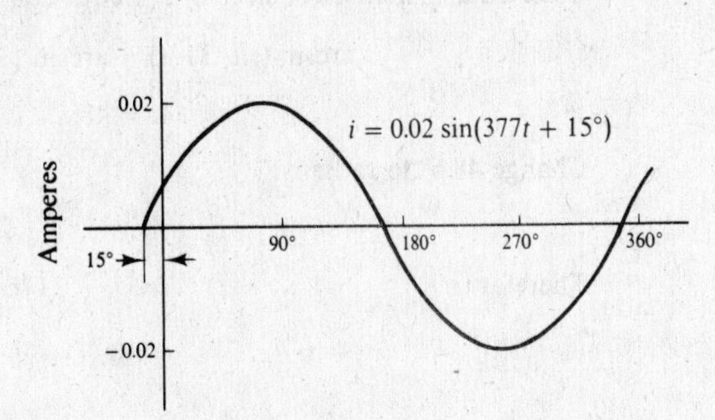

Fig. 1-22

1.6 The equation for a certain sine wave current is $i = 0.02 \sin(377t + 15°)$. Sketch the waveform. Find (*a*) the rms value of the current, (*b*) the average value of the current, (*c*) the instantaneous value of the current, and (*d*) the frequency of the alternating current when $t = 1$ ms.

The waveform is shown in Fig. 1-22.

(*a*) The peak value of current is 0.02 A. The rms value is then

$$I = 0.707 I_m = 0.707 \times 0.02 = 0.0141 \text{ A} = 14.1 \text{ mA (milliamperes)}$$

(*b*) The average value of current is

$$I_o = 0.636 I_m = 0.636 \times 0.02$$
$$= 0.0127 \text{ A} = 12.7 \text{ mA} \tag{1-13a}$$

(*c*) When $t = 1$ ms $(1 \times 10^{-3}$ s),

$$i = 0.02 \sin (377t + 15°) = 0.02 \sin [377(1 \times 10^{-3}) + 15°] = 0.02 \sin (0.377 + 15°)$$

In this equation the value 0.377 is in radians. There are two possibilities. The radians can be converted to degrees and the values added, or the degrees can be converted to radians and the values added. In this example the radians will be converted to degrees:

$$\pi \text{ rad} = 180°$$

So 0.377 rad = 21.6°. Therefore,

$$i = 0.02 \sin 21.6° + 15° = 0.02 \sin 36.6° = 0.0119 \text{ A} = 11.9 \text{ mA}$$

(*d*) From the given equation, $\omega = 377$. And since

$$\omega = 2\pi f$$

$$2\pi f = 377$$

$$f = \frac{377}{2\pi} = 60 \text{ H}$$

1.7 A *thyristor* is a semiconductor component that can be used as a fast-acting switch. An SCR (silicon controlled rectifier) is an example of a thyristor. In one application, an SCR holds an ac voltage to 0 V for a time (t_a), then switches the voltage on. This is shown in Fig. 1-23. The effect is to lower the rms value of the wave by removing the part shown by the shaded area. Calculate the value of t_a in Fig. 1-23 such that the voltage is 90 V when the thyristor switch is closed.

The equation for the voltage is $v = 120 \sin 314t$. The problem is to find the value of t such that $v = 90$ V.

$$v = 120 \sin 314t$$

$$90 = 120 \sin 314t$$

$$\sin 314t = \tfrac{90}{120} = 0.75$$

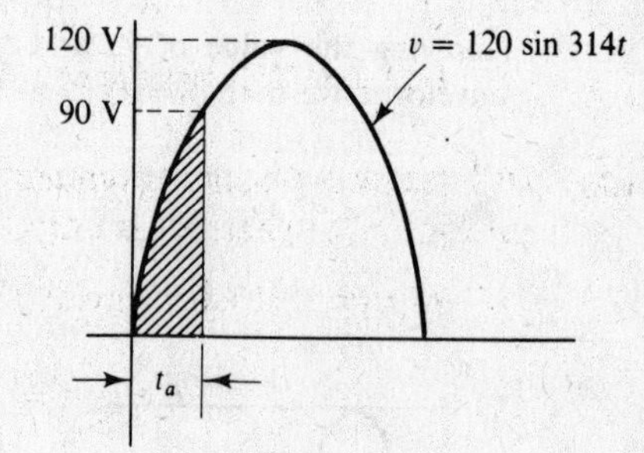

Fig. 1-23

Take an arcsin on both sides of the equation:

$$\arcsin (\sin 314t) = \arcsin (0.75)$$

$$314t = 48.6°$$

Change 48.6° to radians:

$$48.6° = 0.848 \text{ rad}$$

Therefore

$$314t = 0.848$$

$$t = 0.0027 \text{ s} = 2.7 \text{ ms}$$

1.8 In the circuit shown in Fig. 1-24, does the generator current lead or lag the generator voltage?

This question can be answered by finding the equivalent circuit impedance as seen by the generator. Impedances in complex form combine in the same way as series and parallel resistances:

$$\mathbf{Z}_{eq} = \frac{\mathbf{Z}_1\mathbf{Z}_2}{\mathbf{Z}_1 + \mathbf{Z}_2}$$

From Fig. 1-24 we see that Z_1 is the series combination of R_1 and X_L, so

$$\mathbf{Z}_1 = R_1 + jX_L = 15 + j25$$

Similarly, Z_2 is the series combination of R_2 and X_C, so

$$\mathbf{Z}_2 = R_2 - jX_C = 20 - j30$$

Therefore

$$\frac{\mathbf{Z}_1\mathbf{Z}_2}{\mathbf{Z}_1 + \mathbf{Z}_2} = \frac{(15 + j25)(20 - j30)}{15 + j25 + 20 - j30} = \frac{1050 + j50}{35 - j5}$$

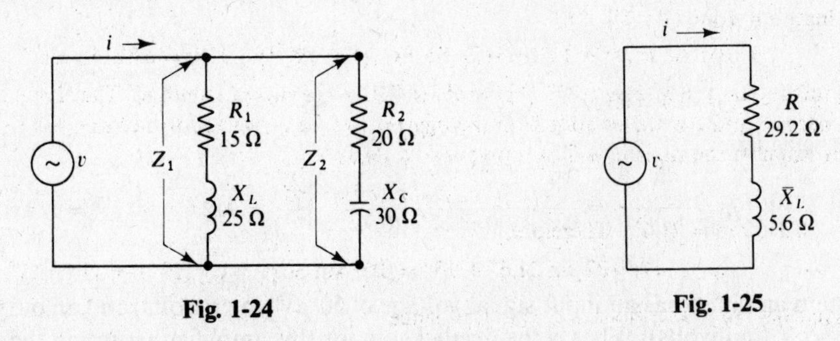

Fig. 1-24 Fig. 1-25

Division is more readily accomplished by converting these values to polar form:

$$R \pm jX = \sqrt{R^2 + X^2}\,\underline{/\arctan(\pm X/R)}$$

$$1050 + j50 = 1051\,\underline{/+2.73^\circ}$$

$$35 - j5 = 35.36\,\underline{/-8.13^\circ}$$

$$\mathbf{Z}_{eq} = \frac{1051\,\underline{/+2.73^\circ}}{35.36\,\underline{/-8.13^\circ}} = 29.7\,\underline{/10.86^\circ}$$

The positive angle indicates that the equivalent impedance is an *RL* circuit. This means that the generator current is lagging the generator voltage. The equivalent *RL* circuit can be drawn by converting the equivalent polar impedance to rectangular form:

$$Z\,\underline{/\pm\phi} = Z\cos\phi \pm jZ\sin\phi$$

$$29.7\,\underline{/10.86^\circ} = 29.7\cos 10.86^\circ + j29.7\sin 10.86^\circ = 29.2 + j5.6$$

Figure 1-25 shows the equivalent circuit.

1.9 What is the resonant frequency of the parallel-tuned circuit in Fig. 1-26 when R_L is adjusted to 50 Ω?

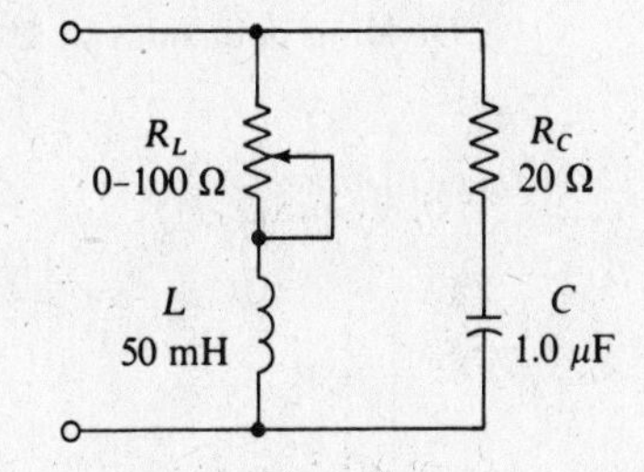

Fig. 1-26

In a parallel-tuned circuit, the resonant frequency is

$$f_r = \frac{1}{2\pi\sqrt{LC}}\sqrt{\frac{R_L^2C - L}{R_C^2C - L}} \qquad (1\text{-}42a)$$

Substituting from Fig. 1-26 and using $R_L = 50\ \Omega$,

$$f_r = \frac{1}{2\pi\sqrt{(50\times10^{-3})(1\times10^{-6})}}\sqrt{\frac{(50)^2(1\times10^{-6}) - (50\times10^{-3})}{(20)^2(1\times10^{-6}) - (50\times10^{-3})}} = (711.76)(0.9786) = 696.53\ \text{Hz}$$

In a series-tuned circuit, the equation for the resonant frequency is $f_r = \frac{1}{2}\pi\sqrt{LC}$. Resistance in a series-tuned circuit affects the *Q*, or selectivity, but not the resonant frequency.

If R_L in Fig. 1-26 were tuned to equal R_C (20 Ω in this case), f_r would be the same as in the series resonant case:

$$f_r = \frac{1}{2\pi\sqrt{LC}} = \frac{1}{2\pi\sqrt{(50\times10^{-3})(1\times10^{-6})}} = 711.76\ \text{Hz}$$

1.10 What is the *Q* of a resonant circuit tuned to 3.2 kHz if the bandwidth *BW* at the −3 dB points is 800 Hz?

$$Q = \frac{f_r}{BW} = \frac{3.2\times10^3}{800} = 4 \qquad \text{(no units)} \qquad (1\text{-}37a)$$

1.11 What value of capacitance is needed to series-tune a 0.9-mH coil to a frequency of 500 kHz?

$$f_r = \frac{1}{2\pi\sqrt{LC}} \qquad (1\text{-}36b)$$

Squaring both sides,

$$f_r^2 = \frac{1}{4\pi^2 LC}$$

Rearranging terms,

$$C = \frac{1}{4\pi^2 f_r^2 L} = \frac{1}{(4\pi^2)(500 \times 10^3)^2(0.9 \times 10^{-3})} = 112.6 \times 10^{-12} = 112.6 \text{ pF}$$

1.12 A certain amplifier has an input signal voltage of 50 μV (microvolts) and an output signal voltage of 200 mV (millivolts). What is the decibel gain for this amplifier assuming the input and output impedances are equal?

$$\text{dB} = 20 \log \frac{V_2}{V_1} \qquad (1\text{-}34b)$$

$$= 20 \log \frac{200 \times 10^{-3}}{50 \times 10^{-6}} = 20 \log (4 \times 10^3) = 20(\log 4 + 3) = 20(\log 4) + 60 = 72.04$$

1.13 A certain amplifier has a voltage gain of 15 dB. If the input signal voltage is 0.8 V, what is the output signal voltage assuming constant impedance?

$$\text{dB} = 20 \log \frac{V_2}{V_1} \qquad (1\text{-}34b)$$

$$15 = 20 \log \frac{V_2}{0.8}$$

Dividing both sides by 20 and taking antilogs,*

$$\text{antilog} \frac{15}{20} = \text{antilog} \left(\log \frac{V_2}{0.8} \right)$$

$$\text{antilog } 0.75 = \text{antilog} \left(\log \frac{V_2}{0.8} \right)$$

$$10^{0.75} = \frac{V_2}{0.8}$$

$$5.62 = \frac{V_2}{0.8}$$

$$V_2 = 4.50 \text{ V}$$

* Remember, if $L = \log N$, then $N = \text{antilog } L = 10^L$ and antilog $(\log M) = M$.

Supplementary Problems

1.14 For the circuit shown in Fig. 1-27, find the frequency f, the capacitive reactance X_C, and the phase angle ϕ between the voltage and the current.

1.15 How long does it take the voltage $v = 13 \sin 2000t$ to reach an instantaneous value of 6 V after passing through zero and going in a positive direction?

1.16 Derive an equation for finding the rms value of voltage when the average value is known.

1.17 Write the value of total impedance $\mathbf{Z}_T$ for the circuit shown in Fig. 1-28. Give the impedance in rectangular and polar form.

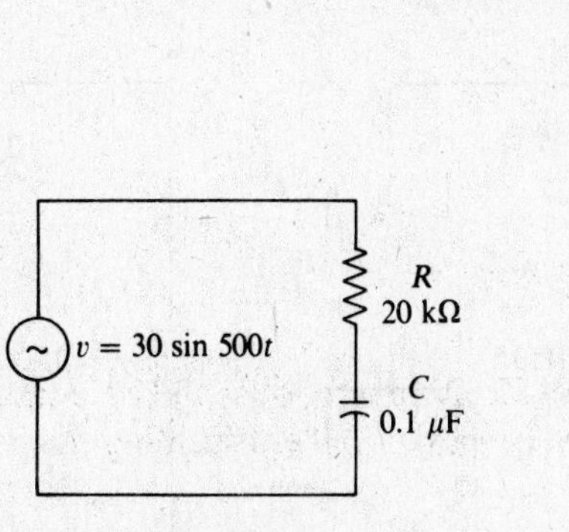

Fig. 1-27

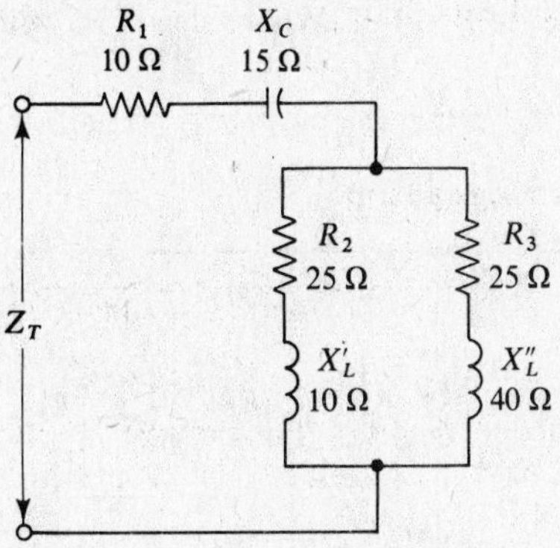

Fig. 1-28

1.18 A sine wave voltage is sometimes used to *synchronize* a *relaxation oscillator*. A relaxation oscillator is a circuit that produces nonsinusoidal ac voltages. When an oscillator is synchronized, its frequency is set by an input signal.

A 5000-Hz signal is used to synchronize a certain oscillator. The SYNC signal has a peak value of 1.5 V, and synchronizing occurs when the signal reaches -1.11 V immediately after passing through zero. Figure 1-29 shows the condition. Find the time t required to reach the sync voltage after $t = 0$.

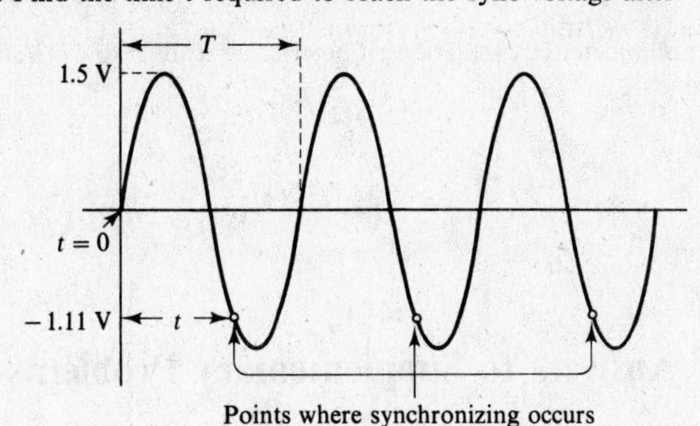

Fig. 1-29

1.19 The standard direction of rotation for phasors is ____________.

1.20 Figure 1-30 shows a circuit with the rated values of the resistors and the actual current measured by a milliammeter. Is the circuit operating correctly? What is the value of V?

1.21 What is the equation for instantaneous current in a circuit with an applied voltage of $v = 15 \sin 377t$ across an impedance of $\mathbf{Z} = 3 + j4$? *Hint*: From Eq. (*1-15*), $\mathbf{Z} = v/i$. Write $\mathbf{Z}$ in polar form. Divide the magnitude Z into V_m to get I_m [Eq. (*1-15a*)] and subtract the phase angle ϕ from ωt.

1.22 What is the value of current I in the circuit shown in Fig. 1-31?

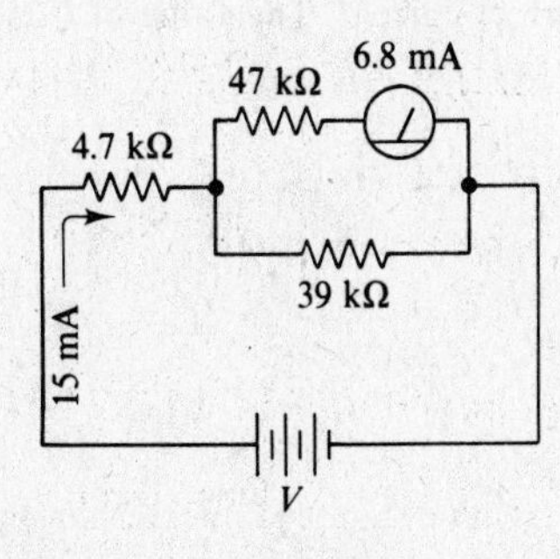

Fig. 1-30

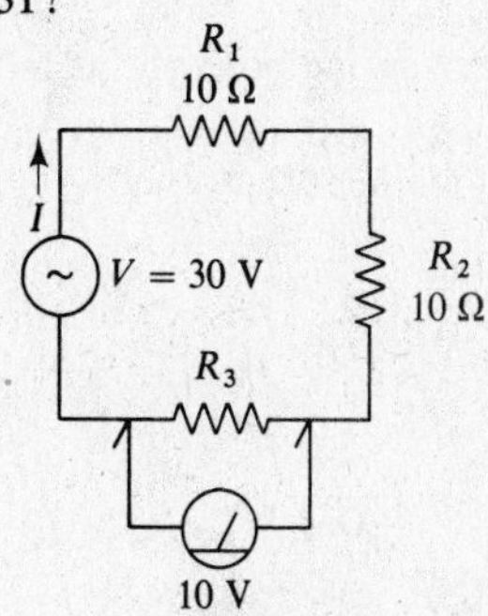

Fig. 1-31

1.23 Refer to Fig. 1-32. What value of R_3 will produce a current I_3 through it of 2 A?

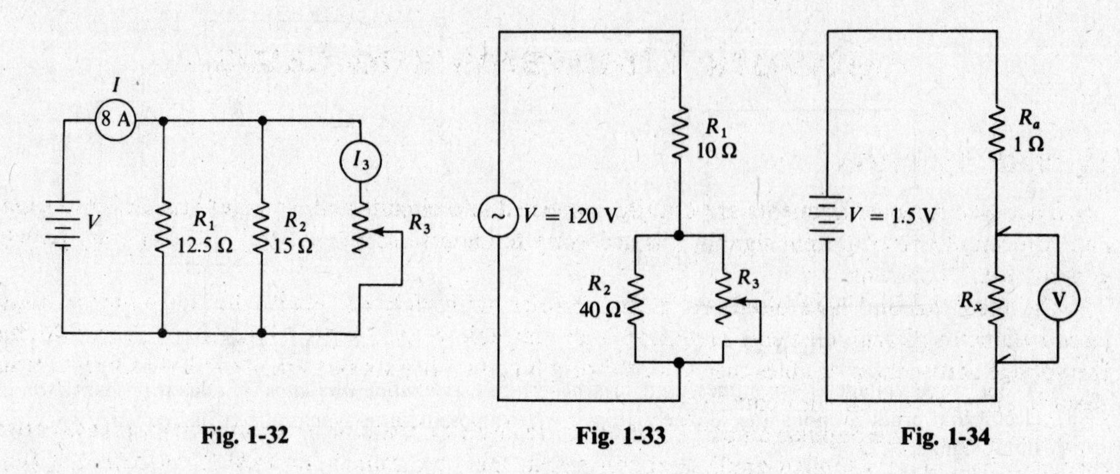

Fig. 1-32 **Fig. 1-33** **Fig. 1-34**

1.24 Refer to Fig. 1-33. To what value must R_3 be adjusted so that the total circuit dissipates 900 W?

1.25 Figure 1-34 shows an ohmmeter circuit. The voltmeter scale reads 0 to 1.5 V. What is the resistance of R_x when the voltmeter reads 1.3 V?

Answers to Supplementary Problems

1.14 $f = 79.6$ Hz, $X_C = 20\,000\ \Omega$, $\phi = 45°$.

1.15 $t = 0.000\,24$ s $= 240\ \mu$s.

1.16 $V = 1.11 V_o$.

1.17 $\mathbf{Z}_T = 24.75 - j4.75\ \Omega$ (rectangular form); $\mathbf{Z}_T = 25.2\underline{/-10.86°}\ \Omega$ (polar form).

1.18 $t = 126.5\ \mu$s.

1.19 Counterclockwise.

1.20 The circuit is working properly, as indicated by the correct current. The value of V is slightly over 390 V.

1.21 $i = 3 \sin (377t - 36.87°)$.

1.22 1 A.

1.23 20.45 Ω.

1.24 7.06 Ω.

1.25 6.5 Ω.

Chapter 2

Network Theorems and Laws

2.1 INTRODUCTION

Since electronic components are usually connected into circuits with voltages and currents applied, and with input and/or output signals, it is necessary to understand the network theorems and laws that govern circuit behavior.

The theorems and laws discussed in this chapter apply only to circuits that have linear, bilateral circuit elements. Circuit elements are linear when they follow Ohm's law relationships so that doubling the voltage across them doubles the current through them. They are bilateral if current can pass through them in either direction with equal ease.

Since none of the popular amplifying devices are linear or bilateral, it might seem that these network theorems have a very limited application. However, there are some additional techniques that can be brought into play such as the load line technique described in Chap. 4 that allow us to use diodes and other nonlinear unilateral devices in conjunction with the theorems and laws developed here.

2.2 DUALITY

Whenever we have a network theorem or law involving voltage, current, and some form of impedance in a series circuit, we may by *duality* write an equivalent theorem or law for a parallel circuit simply by writing voltage for current, current for voltage, admittance for impedance, and parallel for series. Table 2-1 lists the various duals. Note that power, frequency, and time are not listed as they are not affected by the transformation.

Table 2-1

Series	Parallel
Voltage	Current
Impedance	Admittance
Resistance	Conductance
Reactance	Susceptance
Inductance	Capacitance

When using Table 2-1, locate the word in the theorem you wish to find the dual of and replace it with the matching word from the opposite column.

Example 2.1 Find the dual for a voltage generator in the series form feeding a load resistor. See Fig. 2-1*a*.

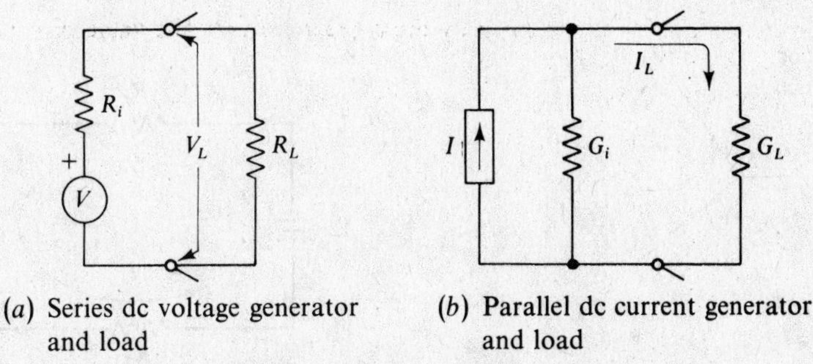

(*a*) Series dc voltage generator and load

(*b*) Parallel dc current generator and load

Fig. 2-1 Parts (*a*) and (*b*) are duals

When an external resistor is connected to a series voltage generator, the voltage across the load (see Fig. 2-1*a*) is given by the voltage divider rule as

$$V_L = \left(\frac{R_L}{R_i + R_L}\right)V \tag{2-1a}$$

Using Table 2-1, the equivalent dual would be a current generator in the parallel form feeding a load conductance as shown in Fig. 1-1*b*.

The load current (see Fig. 2-1*b*) would then be given by

$$I_L = \left(\frac{G_L}{G_i + G_L}\right)I \tag{2-1b}$$

which is the same result we would obtain using the current divider rule.

Example 2.2 In a parallel ac circuit, the current through a capacitor is given by $I_C = \omega CV$. Write the equivalent dual statement.

Again using Table 2-1, the equivalent statement would be that in a series ac circuit, the voltage across an inductor is given by $V_L = \omega LI$.

Once we establish a law for either series or parallel circuits, duality allows us to easily write the equivalent form.

A very important set of dual circuits are the Thevenin and Norton generators discussed in Sec. 2.7.

2.3 KIRCHHOFF'S CURRENT AND VOLTAGE LAWS

Kirchhoff's current and voltage laws are valuable tools for analyzing circuits—especially circuits with two or more sources. Statements of each law are given here:

- *Current law*—The algebraic sum of the currents at any node (junction) is zero. (*Or*, The sum of the currents entering a node equals the sum of the currents leaving that node.)
- *Voltage law*—The algebraic sum of the voltages around any closed circuit path (loop or mesh) is zero. (*Or*, The sum of the voltage rises in any closed circuit loop equals the sum of the voltage drops in that loop.)

Conventional notation assigns a positive value to the currents entering a node, so in Fig. 2-2 the current law would yield

$$I_A + I_B - I_C + I_D = 0$$

or

$$I_A + I_B + I_D = I_C$$

For sources, a voltage drop occurs when you go from positive to negative, and a voltage rise occurs when you go from negative to positive. For circuit elements such as resistors, a voltage drop occurs when you travel the loop in the direction of the assumed current. A voltage rise occurs when you travel the loop in a direction opposite to the assumed current.

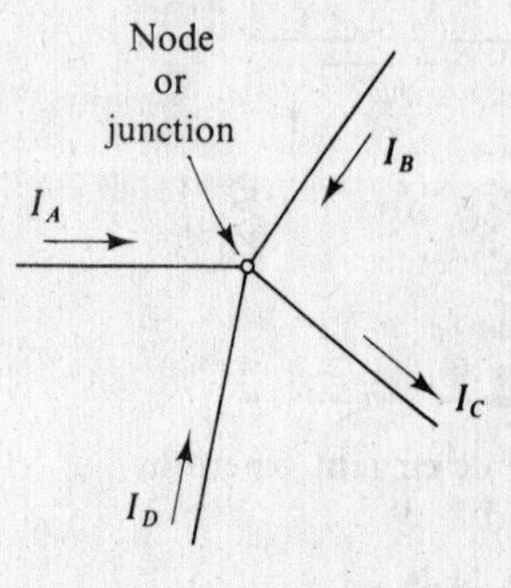

Fig. 2-2

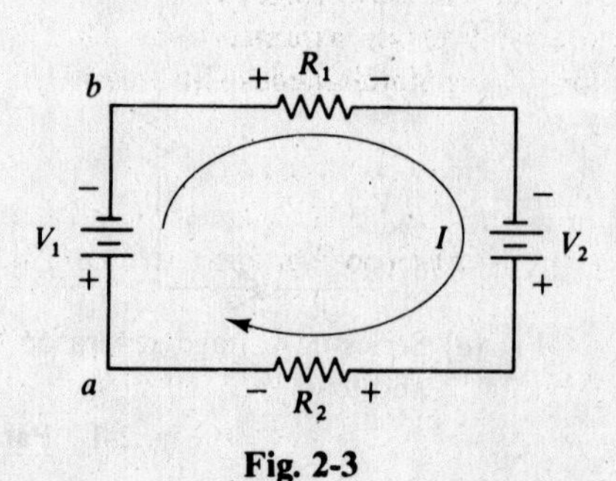

Fig. 2-3

Figure 2-3 shows how this works. The assumed current I is clockwise, and we normally travel the loop in the direction of the assumed current. By Ohm's law the voltage drop across R_1 is IR_1 and across R_2 is IR_2. Remember that these are voltage drops, so they are negative.

Starting at point a and traveling clockwise with the current, the voltage equation is

$$-V_1 - IR_1 + V_2 - IR_2 = 0$$

Source V_1 is negative because the positive side is entered first. If we traveled the loop in the direction opposite to the assumed current, we would write, starting at point b,

$$V_1 + IR_2 - V_2 + IR_1 = 0$$

This is the same equation as before except that all the signs are reversed, and it gives the same result.

Example 2.3 Find I_3 in the circuit shown in Fig. 2-4 by using Kirchhoff's laws.

The first step is to assume a direction of current for each branch, and label the assumed currents. It is not necessary to attempt to find the actual current directions. If the direction assumed is wrong, the solved value will be negative. If the assumed direction is correct, the solved value will be positive. Write as many equations as needed to include all unknown currents in the circuit.

The second step is to write voltage equations for each loop. It is necessary to write enough voltage equations to include all the unknown currents.

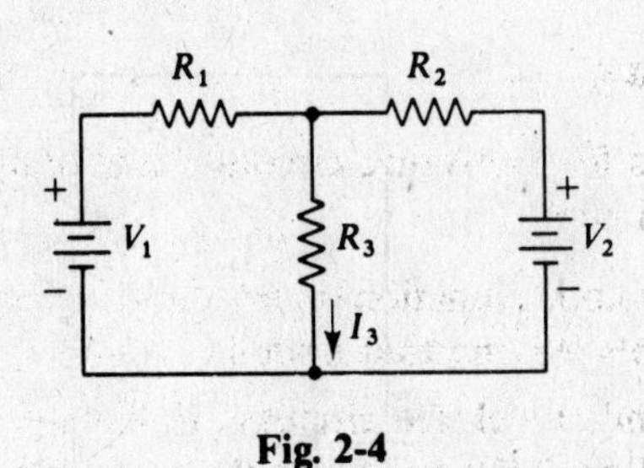

Fig. 2-4

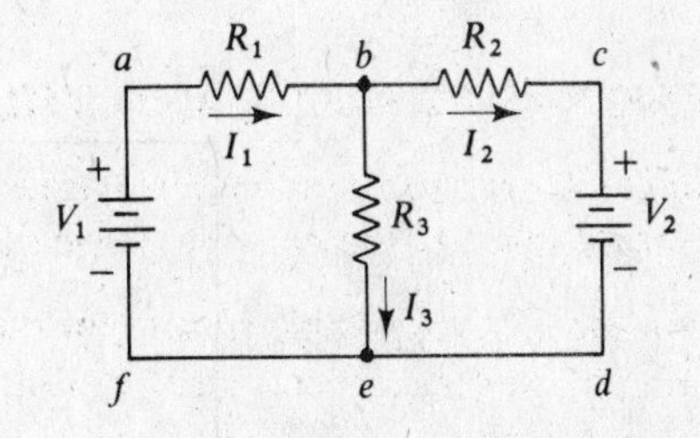

Fig. 2-5

Figure 2-5 shows the circuit redrawn with the assumed currents. The Kirchhoff's equations related to the simplified circuit shown in Fig. 2-5 are the following:

- Current equation for node b:

$$I_1 - I_2 - I_3 = 0$$

- Voltage equation for loop *abefa* (clockwise from point a):

$$-I_1 R_1 - I_3 R_3 + V_1 = 0$$

- Voltage equation for loop *bcdeb* (clockwise from point b):

$$-I_2 R_2 - V_2 + I_3 R_3 = 0$$

In the last equation, the voltage across R_3 is written as $+I_3 R_3$ because in going around loop *bcdeb* you are traveling against the direction assumed for I_3. If the three equations are solved by determinants, or by any other method used for solving simultaneous equations, the value for I_3 can be found. By determinants:

$$I_3 = \frac{N_3}{\Delta}$$

The equations are rewritten so that each unknown is in the same column:

$$\begin{aligned} I_1 \quad - I_2 \quad - I_3 \quad &= 0 \\ -I_1 R_1 + 0 \quad - I_3 R_3 &= -V_1 \\ 0 \quad - I_2 R_2 + I_3 R_3 &= V_2 \end{aligned}$$

Since the unknowns are I_1, I_2, and I_3, we may solve by the method of determinants and obtain I_3 in the form

$$I_3 = \frac{\begin{vmatrix} 1 & -1 & 0 \\ -R_1 & 0 & -V_1 \\ 0 & -R_2 & V_2 \end{vmatrix}}{\Delta} = \frac{-V_1 R_2 - V_2 R_1}{\Delta}$$

where
$$\Delta = \begin{vmatrix} 1 & -1 & -1 \\ -R_1 & 0 & -R_3 \\ 0 & -R_2 & R_3 \end{vmatrix} = -R_2 R_3 - R_1 R_3 - R_1 R_2$$

So
$$I_3 = \frac{V_1 R_2 + V_2 R_1}{R_1 R_2 + R_1 R_3 + R_2 R_3}$$

A knowledge of Kirchhoff's laws can simplify the understanding of some basic relationships in circuits. Literature describing circuit actions often assumes that the reader understands these laws and their applications. For example, the equation for currents in a bipolar transistor as shown in Fig. 2-6 is

$$I_E = I_B + I_C$$

This equation is based on Kirchhoff's current law. A transistor can be considered a junction of currents, and the sum of the currents leaving the junction $(I_B + I_C)$ equals the sum of the currents entering (I_E). Two examples, shown in Fig. 2-7, will show how Kirchhoff's current law is used.

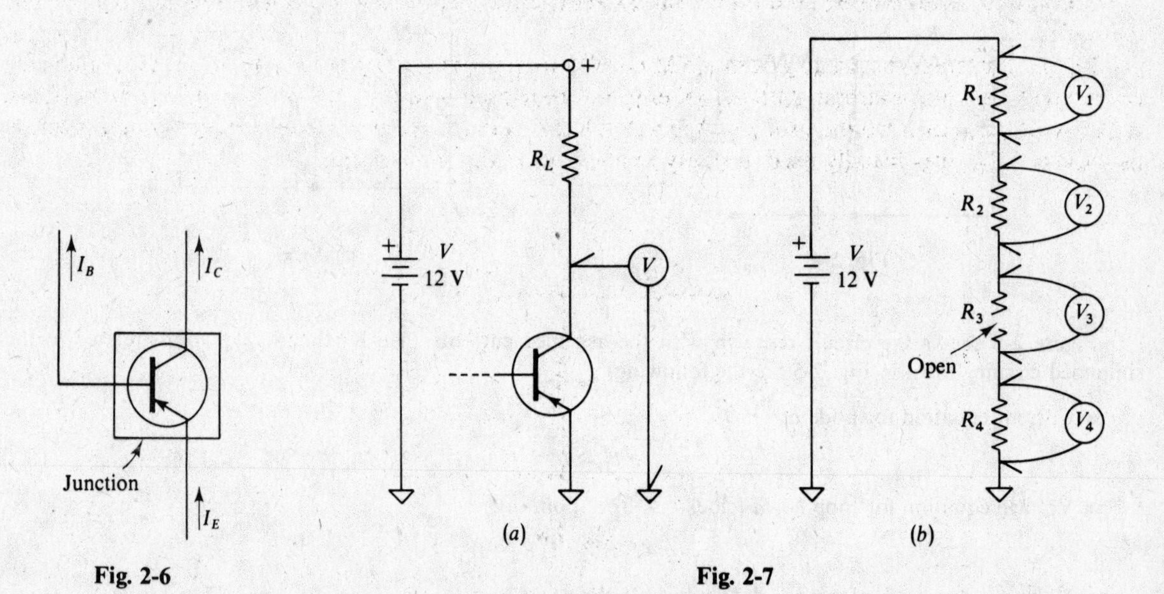

Fig. 2-6

Fig. 2-7

In Fig. 2-7*a*, the transistor collector voltage V_C is being measured. If the voltmeter reads 12 V, then there can be no voltage drop across R_L. This can be expressed in terms of Kirchhoff's laws.

The voltage from collector to ground is equal to the applied voltage minus the voltage drop across R_L. This is obtained by writing a Kirchhoff's voltage equation for the applied voltage V, the voltage drop V_L across R_L, and the voltage V_C across the transistor:

$$V = V_L + V_C$$

Solving for V_C:

$$V_C = V - V_L$$

So, if $V_C = V$; there is no voltage drop across V_L. This means the circuit is open at some point, or the transistor is cut off.

If the voltage reading is 0 V, then it follows that $V_L = V$. In other words, all the voltage is dropped across the load resistor, and there is no drop across the transistor. This indicates that the transistor is *saturated* (the current cannot be increased with an increase in base current).

In Fig. 2-7*b*, there are four resistors connected in series, and resistor R_3 is open. A voltmeter can be used to find the open resistor. Measurements V_1, V_2, and V_4 will be 0 V. The circuit is open, so there is no current through the resistors and no voltage drop across them. Voltage reading V_3 will be 12 V. This is apparent when a voltage equation is written for the circuit:

$$V = V_1 + V_2 + V_3 + V_4$$

If $V_1 = V_2 = V_4 = 0$ V, then

$$V = V_3$$

2.4 MAXWELL'S LOOP EQUATIONS AND NODAL ANALYSIS

Maxwell's Loop Equations

Maxwell's loop equations and nodal analysis can be thought of as two simplifications of the solution of network problems by Kirchhoff's laws. These methods may reduce the number of equations needed to solve a problem.

The two methods will be used on the same circuit (Fig. 2-4) that was previously used for Kirchhoff's methods.

Figure 2-8 shows the assumptions for Maxwell's loop (or mesh) equations. A current is assumed for each closed loop in the circuit. Although there are three closed-loop currents (I_1, I_2, and I_3), we need use only two of them to solve the problem. We will select I_1 and I_2, the inner loop currents for our solution as these are the ones usually used, but any two will yield the same result.

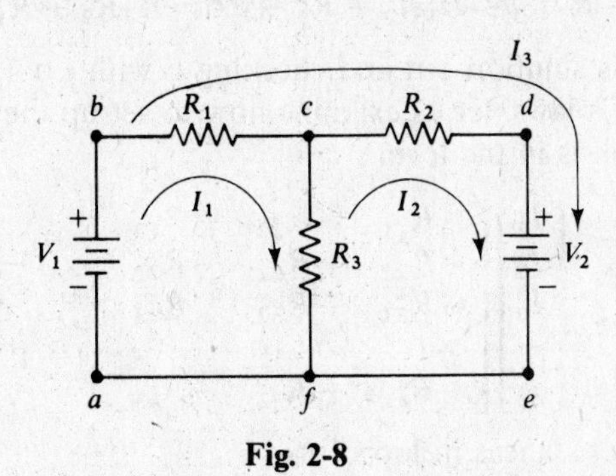

Fig. 2-8

Normal procedure calls for drawing each assumed loop current in the clockwise direction. If we are incorrect in this assumption, the solved current will have a minus sign associated with it.

There are two important considerations in the assumptions for a Maxwell solution. First, there are two voltage drops across R_3: One drop, $I_1 R_3$, due to current I_1, and a second drop, $I_2 R_3$, due to current I_2. With our direction of currents, the two voltages are of opposite sign.

Second, there are two currents through R_3, and the actual current is their algebraic sum. Again notice that, for our assumed current directions, I_1 and I_2 oppose each other through R_3.

In our case, if we go from point *c* to point *f*, we have $I_1 - I_2$, and traveling in the other direction (*f* to *c*) we have $I_2 - I_1$.

We now write Kirchhoff's voltage equation for each loop, traveling in the same direction as the assumed currents (clockwise) and writing voltage drops as negative and voltage rises as positive.

Starting at point *a*, for the first current loop, we have

$$V_1 - I_1 R_1 - (I_1 - I_2)R_3 = 0$$

For loop two, starting at point *d*, we obtain

$$-V_2 - (I_2 - I_1)R_3 - I_2R_2 = 0$$

Expanding these two equations, we have

$$V_1 - I_1R_1 - I_1R_3 + I_2R_3 = 0$$

and

$$-V_2 - I_2R_3 + I_1R_3 - I_2R_2 = 0$$

Rearranging and grouping terms:

$$V_1 = (R_1 + R_3)I_1 - R_3I_2 \tag{2-3}$$

$$-V_2 = -R_3I_1 + (R_2 + R_3)I_2 \tag{2-4}$$

We may solve this set of equations by substitution or by the method of simultaneous equations. However, for more than two loops a matrix solution is useful. Let us solve the two loop equation by matrices and then extend the method for any number of current loops.

Writing Eqs. (*2-3*) and (*2-4*) in matrix form,

$$\begin{bmatrix} V_1 \\ -V_2 \end{bmatrix} = \begin{bmatrix} I_1 \\ I_2 \end{bmatrix} \begin{bmatrix} R_1 + R_3 & -R_3 \\ -R_3 & R_2 + R_3 \end{bmatrix}$$

The solutions are then

$$I_1 = \frac{\begin{vmatrix} V_1 & -R_3 \\ -V_2 & R_2 + R_3 \end{vmatrix}}{\Delta} \quad \text{and} \quad I_2 = \frac{\begin{vmatrix} R_1 + R_3 & V_1 \\ -R_3 & -V_2 \end{vmatrix}}{\Delta}$$

where

$$\Delta = \begin{vmatrix} R_1 + R_3 & -R_3 \\ -R_3 & R_2 + R_3 \end{vmatrix} = (R_1 + R_3)(R_2 + R_3) - (-R_3)(-R_3)$$

$$= R_1R_2 + R_2R_3 + R_1R_3 + R_3^2 - R_3^2 = R_1R_2 + R_1R_3 + R_2R_3$$

Before actually writing this solution out and checking it with Eq. (*2-2*) that we obtained from the direct application of Kirchhoff's laws, let us examine how to set up the general matrix.

The general matrix solution is in the form

$$\begin{bmatrix} \Phi_1 \\ \Phi_2 \\ \Phi_3 \\ \vdots \\ \Phi_n \end{bmatrix} = \begin{bmatrix} I_1 \\ I_2 \\ I_3 \\ \vdots \\ I_n \end{bmatrix} \begin{bmatrix} R_{11} & -R_{12} & -R_{13} & \cdots & -R_{1n} \\ -R_{21} & R_{22} & -R_{23} & \cdots & -R_{2n} \\ -R_{31} & -R_{32} & R_{33} & \cdots & -R_{3n} \\ \cdots & \cdots & \cdots & \cdots & \cdots \\ -R_{n1} & -R_{n2} & -R_{n3} & \cdots & R_{nn} \end{bmatrix}$$

where R_{11} = sum of all the resistances in loop 1
R_{22} = sum of all the resistances in loop 2, etc.
R_{12} = sum of all the resistances that are common to loops 1 and 2
R_{23} = sum of all the resistances that are common to loops 2 and 3, etc.
Φ_1 = sum of all the voltage rises in loop 1, traveling in the clockwise direction
Φ_2 = sum of all the voltage rises in loop 2, traveling in the clockwise direction, etc.

Since the matrix is symmetrical, $R_{12} = R_{21}$, $R_{13} = R_{31}$, $R_{23} = R_{32}$, etc. These resistances are called *mutual resistances* and are inserted into the matrix with minus signs because the currents are in opposite directions through the mutual resistance.

For our case, in the circuit shown in Fig. 2-8, $\Phi_1 = V_1$ and $\Phi_2 = -V_2$; since in the direction of I_2, V_2 is a voltage drop,

$$R_{11} = R_1 + R_3 \quad \text{and} \quad R_{22} = R_2 + R_3$$

The only mutual resistance in our circuit is R_3 (the resistor common to loops 1 and 2), so

$$R_{12} = R_{21} = R_3$$

Knowing the general matrix form and the rules that define the quantities in the matrices, we could have written the matrix solution by inspection:

$$\begin{bmatrix} V_1 \\ -V_2 \end{bmatrix} = \begin{bmatrix} I_1 \\ I_2 \end{bmatrix} \begin{bmatrix} R_1 + R_3 & -R_3 \\ -R_3 & R_2 + R_3 \end{bmatrix}$$

For ac circuits we use the same matrix form, but substitute ac voltages and currents and complex impedances for the dc voltages, currents, and resistances.

To check our solution, we note that I_3 in Fig. 2-4 is the same as $I_1 - I_2$ in Fig. 2-8.

Writing out our result,

$$\Delta = \begin{vmatrix} R_1 + R_3 & -R_3 \\ -R_3 & R_2 + R_3 \end{vmatrix} = (R_1 + R_3)(R_2 + R_3) - (-R_3)^2$$

$$= R_1R_2 + R_1R_3 + R_2R_3 + \cancel{R_3^2} - \cancel{R_3^2} = R_1R_2 + R_1R_3 + R_2R_3$$

Then $$I_1 = \frac{\begin{vmatrix} V_1 & -R_3 \\ -V_2 & R_2 + R_3 \end{vmatrix}}{\Delta} = \frac{V_1(R_2 + R_3) - (-V_2)(-R_3)}{\Delta} = \frac{V_1R_2 + V_1R_3 - V_2R_3}{R_1R_2 + R_1R_3 + R_2R_3}$$

and $$I_2 = \frac{\begin{vmatrix} R_1 + R_3 & V_1 \\ -R_3 & -V_2 \end{vmatrix}}{\Delta} = \frac{-V_2(R_1 + R_3) - (V_1)(-R_3)}{\Delta} = \frac{V_1R_3 - V_2R_1 - V_2R_3}{R_1R_2 + R_1R_3 + R_2R_3}$$

For $I_1 - I_2$ we may subtract directly since both currents have the same denominator:

$$I_1 - I_2 = \frac{(V_1R_2 + \cancel{V_1R_3} - \cancel{V_2R_3}) - (\cancel{V_1R_3} - V_2R_1 - \cancel{V_2R_3})}{R_1R_2 + R_1R_3 + R_2R_3} = \frac{V_1R_2 + V_2R_1}{R_1R_2 + R_1R_3 + R_2R_3}$$

This is the same result as Eq. (*2-2*) previously obtained.

Example 2.4 Write the loop matrix solution for the circuit shown in Fig. 2-9.

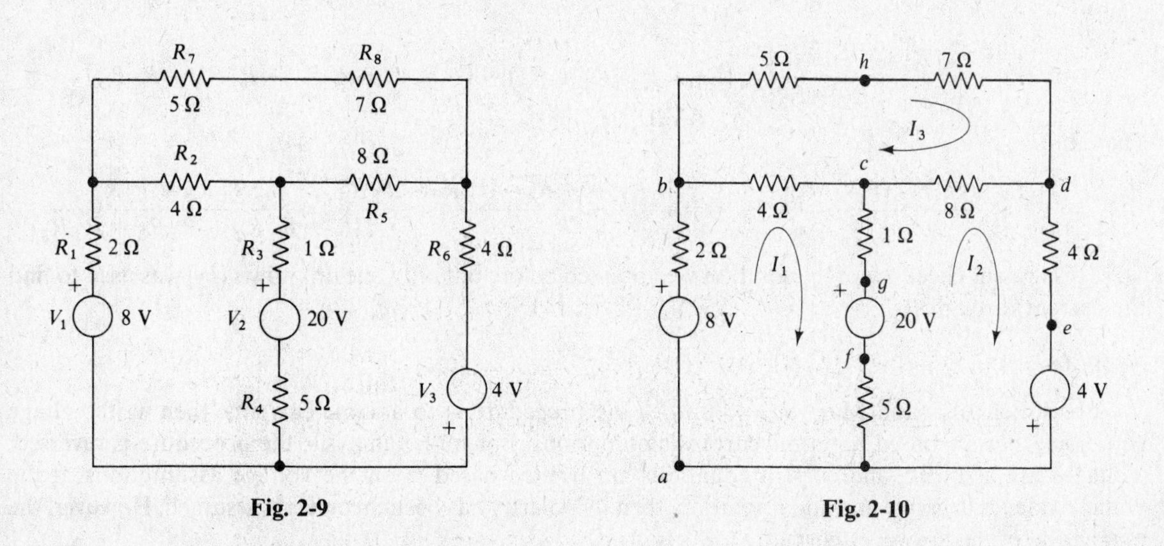

Fig. 2-9

Fig. 2-10

Taking the clockwise direction for the assumed currents as in Fig. 2-10, we may then write the matrix solution for the three unknown currents (I_1, I_2, I_3) as

$$\begin{bmatrix} \Phi_1 \\ \Phi_2 \\ \Phi_3 \end{bmatrix} = \begin{bmatrix} I_1 \\ I_2 \\ I_3 \end{bmatrix} \begin{bmatrix} R_{11} & -R_{12} & -R_{13} \\ -R_{21} & R_{22} & -R_{23} \\ -R_{31} & -R_{32} & R_{33} \end{bmatrix}$$

From Fig. 2-10, we find that

$$\Phi_1 = 8 - 20 = -12\ \text{V} \qquad R_{11} = 2 + 4 + 1 + 5 = 12\ \Omega \qquad R_{12} = R_{21} = 1 + 5 = 6\ \Omega$$

$$\Phi_2 = 4 + 20 = 24\ \text{V} \qquad R_{22} = 5 + 1 + 8 + 4 = 18\ \Omega \qquad R_{13} = R_{31} = 4\ \Omega$$

$$\Phi_3 = \text{none} = 0\ \text{V} \qquad R_{33} = 5 + 7 + 8 + 4 = 24\ \Omega \qquad R_{23} = R_{32} = 8\ \Omega$$

Substituting yields

$$\begin{bmatrix} -12 \\ 24 \\ 0 \end{bmatrix} = \begin{bmatrix} I_1 \\ I_2 \\ I_3 \end{bmatrix} \begin{bmatrix} 12 & -6 & -4 \\ -6 & 18 & -8 \\ -4 & -8 & 24 \end{bmatrix}$$

While it is still possible to solve this numerically without too much difficulty to obtain $I_1 = -0.0667^A$, $I_2 = 1.53^A$, and $I_3 = 0.500^A$, for matrices with more than three unknowns a computer or calculator method is commonly used.

In Example 2.3 there were only two equations to solve, but not much time was saved because it was necessary to solve for both unknowns. If the circuit were rearranged as shown in Fig. 2-11 *before* the equations were written, then only one unknown (I_2) would have to be solved to find the current through R_3.

For the circuit shown in Fig. 2-11, the voltage rise for loop 1 = $V_1 - V_2$, the voltage rise for loop 2 = V_2, and

$$R_{11} = R_1 + R_2 \qquad R_{22} = R_2 + R_3 \qquad R_{12} = R_{21} = R_2$$

So
$$\begin{bmatrix} V_1 - V_2 \\ V_2 \end{bmatrix} = \begin{bmatrix} I_1 \\ I_2 \end{bmatrix} \begin{bmatrix} R_1 + R_2 & -R_2 \\ -R_2 & R_2 + R_3 \end{bmatrix}$$

and I_2, the current we want, can be written

$$I_2 = \frac{\begin{vmatrix} R_1 + R_2 & V_1 - V_2 \\ -R_2 & V_2 \end{vmatrix}}{\Delta}$$

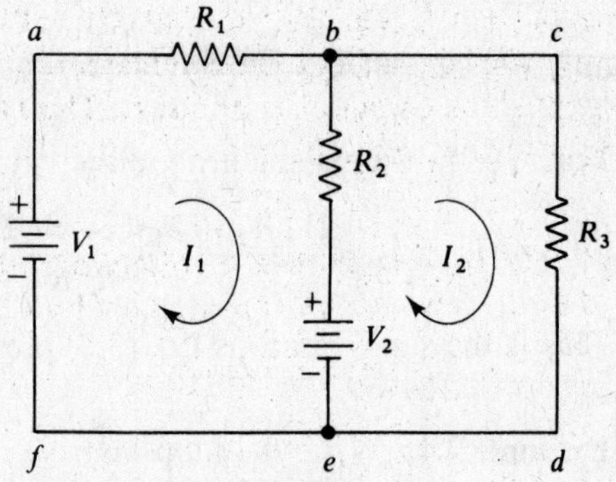

Fig. 2-11

where

$$\Delta = \begin{vmatrix} R_1 + R_2 & -R_2 \\ -R_2 & R_2 + R_3 \end{vmatrix} = (R_1 + R_2)(R_2 + R_3) - (-R_2)^2 = R_1R_2 + R_1R_3 + R_2R_3$$

Therefore

$$I_2 = \frac{V_2(R_1 + R_2) - (V_1 - V_2)(-R_2)}{\Delta} = \frac{V_2R_1 + \cancel{V_2R_2} + V_1R_2 - \cancel{V_2R_2}}{\Delta} = \frac{V_1R_2 + V_2R_1}{R_1R_2 + R_1R_3 + R_2R_3}$$

This answer checks with the solution we obtained before but only one unknown (I_2) was used to find the current through R_3.

Nodal Analysis

For Kirchhoff's and Maxwell's solutions, the procedure is to assume currents, then write voltage equations that are based upon the current assumptions. For nodal analysis, the procedure is reversed. Voltages are assumed, and current equations are written based upon the voltage assumptions. If the voltage value is negative in a nodal solution, then its polarity has been incorrectly assumed. However, the magnitude of the answer is correct.

Figure 2-12 shows the assumptions made for a solution by nodal analysis. A common point—called a *reference node*—is chosen first. This is a point that can be considered to be 0 V in the circuit. An obvious choice would be ground, or common, if such a point exists. There are no universally accepted rules for choosing a reference node, but experience in solving problems with nodal analysis will show that the point should be chosen in such a way that the least number of unknown voltage points results. Point e in Fig. 2-12 has been chosen for this discussion.

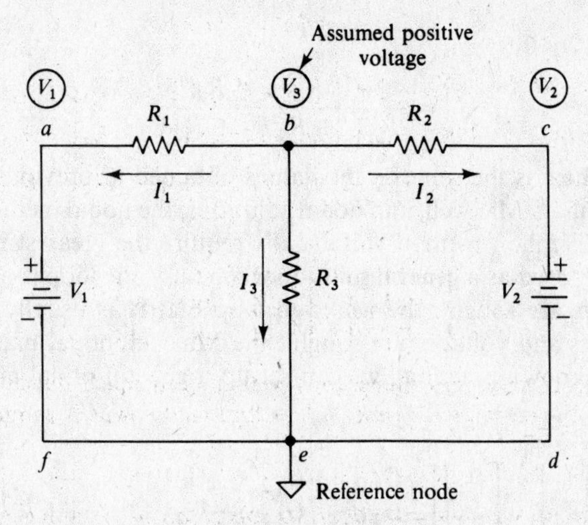

Fig. 2-12

In Fig. 2-12 if point e (or d or f) is chosen, then the voltages at points a and c are known. These values are V_1 and V_2, respectively. Only the voltage at point b is unknown.

After choosing the reference node, the next step is to write a current equation for each of the nodes where the voltage is unknown. The currents are written in the same way as in the Kirchhoff's law solution. At node b:

$$I_1 + I_2 + I_3 = 0$$

Current I_1 is equal to the voltage across R_1 divided by the value of R_1. The voltage across R_1 is equal to $V_3 - V_1$. The current through R_1 by Ohm's law is

$$I_1 = \frac{V_3 - V_1}{R_1}$$

The voltage across R_3 is V_3. By Ohm's law, the current through R_3 is

$$I_3 = \frac{V_3}{R_3}$$

The voltage across R_2 is $V_3 - V_2$, and

$$I_2 = \frac{V_3 - V_2}{R_2}$$

The values just determined for I_1, I_2, and I_3 are substituted into the junction current equation:

$$I_1 + I_2 + I_3 = 0$$

$$\frac{V_3 - V_1}{R_1} + \frac{V_3 - V_2}{R_2} + \frac{V_3}{R_3} = 0$$

The only unknown is V_3, so this equation can be solved. Both sides of the equation are multiplied by $R_1 R_2 R_3$ to give

$$R_2 R_3(V_3 - V_1) + R_1 R_2 V_3 + R_1 R_3(V_3 - V_2) = 0$$

Expanding,

$$R_2 R_3 V_3 - R_2 R_3 V_1 + R_1 R_2 R_3 + R_1 R_3 V_3 - R_1 R_3 V_2 = 0$$

Solving for V_3:

$$V_3 = \frac{R_2 R_3 V_1 + R_1 R_3 V_2}{R_1 R_2 + R_1 R_3 + R_2 R_3}$$

But $I_3 = V_3/R_3$, so

$$I_3 = \frac{V_1 R_2 + V_2 R_1}{R_1 R_2 + R_1 R_3 + R_2 R_3}$$

The value obtained for I_3 here is the same as the values obtained in previous solutions.

In comparing the Kirchhoff, Maxwell, and nodal solutions, the nodal method will usually require the fewest equations and the Kirchhoff method will usually require the greatest number of equations. The Kirchhoff method is usually used as a general solution when only one loop is present. For problems with many loops, where currents are sought, the Maxwell loop matrix is usually employed. For problems where there are many nodes and voltages are sought, the Maxwell nodal matrix is used.

For more than one unknown potential, we can set up a general nodal matrix which would be the dual of the mesh matrix:

$$\begin{bmatrix} I_1 \\ I_2 \\ I_3 \\ \vdots \\ I_n \end{bmatrix} = \begin{bmatrix} \Phi_1 \\ \Phi_2 \\ \Phi_3 \\ \vdots \\ \Phi_n \end{bmatrix} \begin{bmatrix} G_{11} & -G_{12} & -G_{13} & \cdots & -G_{1n} \\ -G_{21} & G_{22} & -G_{23} & \cdots & -G_{2n} \\ -G_{31} & -G_{32} & G_{33} & \cdots & -G_{3n} \\ \cdots & \cdots & \cdots & \cdots & \cdots \\ -G_{n1} & -G_{n2} & -G_{n3} & \cdots & G_{nn} \end{bmatrix}$$

where I_1 = sum of the currents entering node 1
I_2 = sum of the currents entering node 2, etc.
G_{11} = sum of the conductances attached to node 1
G_{22} = sum of the conductances attached to node 2, etc.
$G_{12} = G_{21}$ = sum of the conductances connecting node 1 directly to node 2
$G_{23} = G_{32}$ = sum of the conductances connecting node 2 directly to node 3, etc.
Φ_1 = voltage at node 1
Φ_2 = voltage at node 2, etc.

Note that the mutual conductances are again symmetrical and inserted into the matrix with a minus sign.

Example 2.5 Write the nodal matrix solution for the circuit shown in Fig. 2-13*a*.

Assigning Φ_1 to point *b*, Φ_2 to point *c*, and Φ_3 to point *d* as in Fig. 2-13*b*, the nodal matrix solution becomes

$$\begin{bmatrix} I_1 \\ I_2 \\ I_3 \end{bmatrix} = \begin{bmatrix} \Phi_1 \\ \Phi_2 \\ \Phi_3 \end{bmatrix} \begin{bmatrix} G_{11} & -G_{12} & -G_{13} \\ -G_{21} & G_{22} & -G_{23} \\ -G_{31} & -G_{32} & G_{33} \end{bmatrix}$$

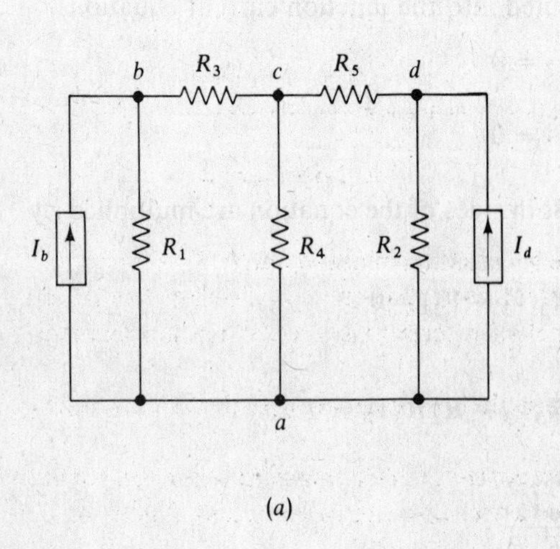

(*a*)

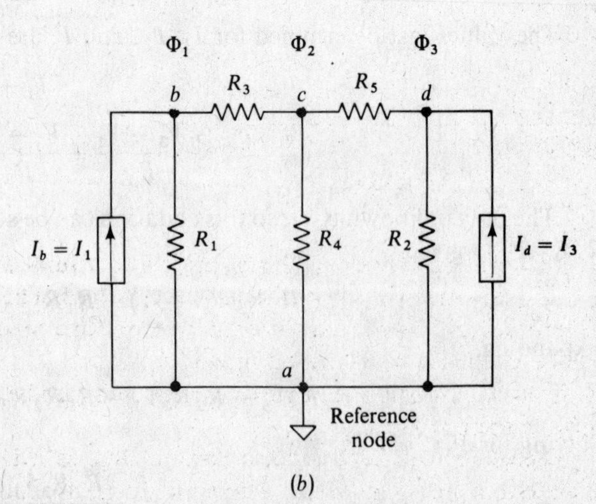

(*b*)

Fig. 2-13

From Fig. 2-13*b*, we see that

$$I_1 = I_b \qquad G_{11} = \frac{1}{R_1} + \frac{1}{R_3} \qquad G_{12} = G_{21} = \frac{1}{R_3}$$

$$I_2 = \text{none} = 0 \qquad G_{22} = \frac{1}{R_3} + \frac{1}{R_4} + \frac{1}{R_5} \qquad G_{13} = G_{31} = \text{none} = 0$$

$$I_3 = I_d \qquad G_{33} = \frac{1}{R_5} + \frac{1}{R_2} \qquad G_{23} = G_{32} = \frac{1}{R_5}$$

By substitution we obtain

$$\begin{bmatrix} I_b \\ 0 \\ I_d \end{bmatrix} = \begin{bmatrix} \Phi_1 \\ \Phi_2 \\ \Phi_3 \end{bmatrix} \begin{bmatrix} \frac{1}{R_1} + \frac{1}{R_3} & -\frac{1}{R_3} & 0 \\ -\frac{1}{R_3} & \frac{1}{R_3} + \frac{1}{R_4} + \frac{1}{R_5} & -\frac{1}{R_5} \\ 0 & -\frac{1}{R_5} & \frac{1}{R_5} + \frac{1}{R_2} \end{bmatrix}$$

2.5 Δ-Y AND Y-Δ TRANSFORMS

There are some circuits that cannot be solved as series-parallel combinations. A good example is shown in Fig. 2-14. This circuit, called a *bridged T*, is a form of the Wheatstone bridge. It is not possible to find the resistance between terminals *a* and *b* in this circuit by treating the resistors as series and parallel combinations.

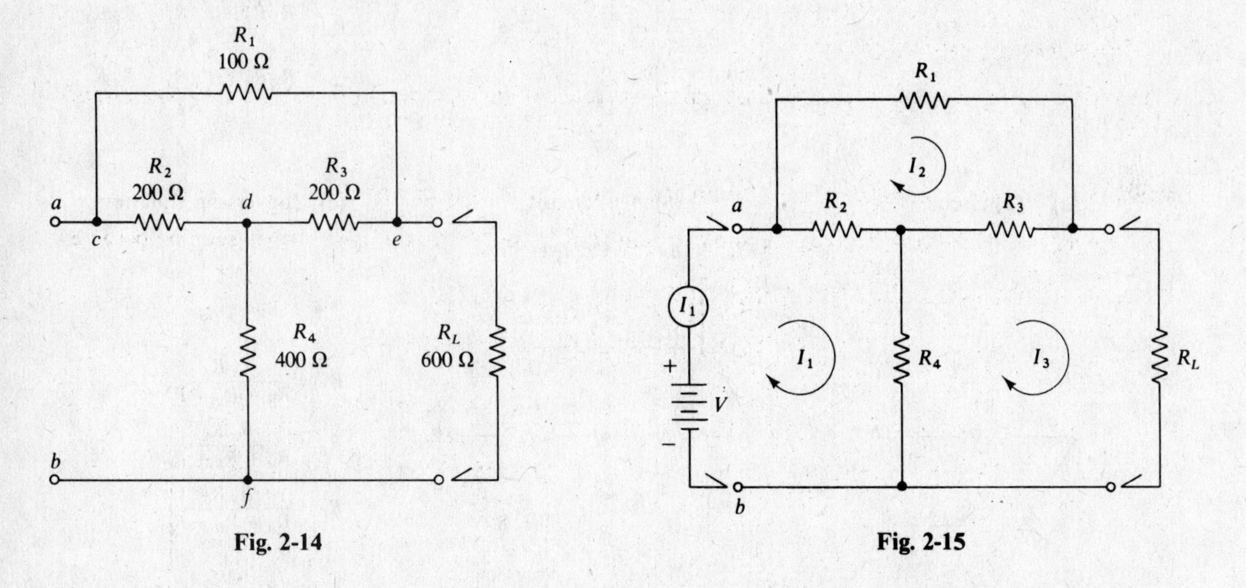

Fig. 2-14

Fig. 2-15

One method of finding the resistance is to assume a voltage V across terminals *a* and *b*. This is shown in Fig. 2-15. Any voltage value can be assumed. The amount of battery current I_1 flowing from the battery is found by Maxwell's mesh method. The resistance between terminals *a* and *b* is the assumed voltage V divided by the resulting current I.

It does not matter what voltage value is chosen because the ratio of voltage to current will always be equal to the resistance.

There are two mathematical conversions, called Δ-Y *transforms* or Y-Δ *transforms*, that make it possible to simplify the circuit shown in Fig. 2-14. Δ and Y networks also are called π and *T networks*. Figure 2-16 shows that they are three-terminal networks.

Δ circuit Y circuit

Fig. 2-16

Y or T circuit Δ or π circuit

Fig. 2-17

As shown in Fig. 2-17, resistors R_2, R_3, and R_4 from Fig. 2-14 form a Y network that can be converted to a Δ; and resistors R_1, R_2, and R_3 form a Δ network that can be converted to a Y.

Whenever there is a Y or a Δ network within a circuit, there is a good chance that the network can be reduced in complexity by making a Y-Δ or a Δ-Y transform. It often saves time to determine if this is a possibility before starting a solution for currents or voltages in a complex circuit.

Figure 2-18 shows the method of converting a Y configuration to a Δ, and Fig. 2-19 shows the method for converting a Δ configuration to a Y. If either of these conversions is made in the circuit shown in Fig. 2-14, the problem will be simplified to the point where the resistance between terminals a and b can be treated as a series-parallel combination.

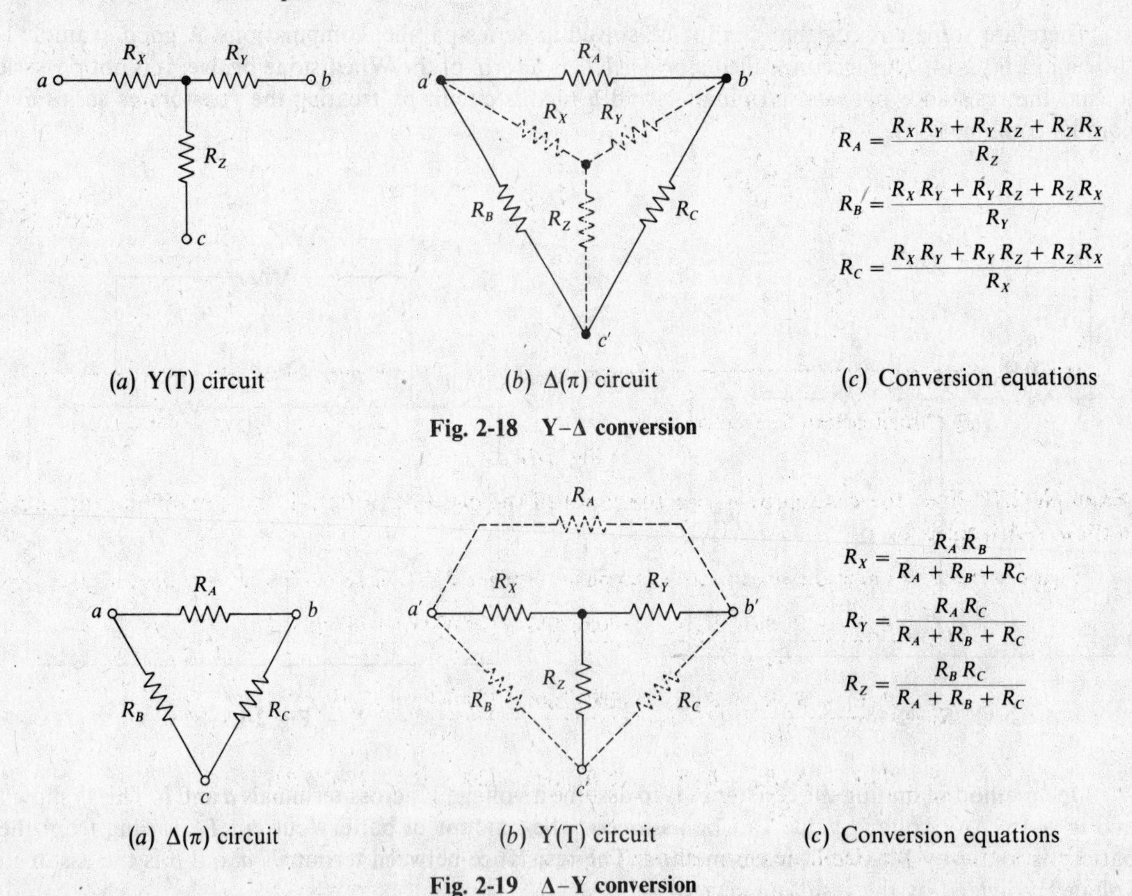

(*a*) Y(T) circuit (*b*) Δ(π) circuit (*c*) Conversion equations

Fig. 2-18 Y–Δ conversion

(*a*) Δ(π) circuit (*b*) Y(T) circuit (*c*) Conversion equations

Fig. 2-19 Δ–Y conversion

In practical circuits, the identifying letters for the resistances will most likely be different from the ones given in the illustrations, so it is better to learn the transforms by configuration rather than by equation. Note that each resistor in the equivalent Y (T) (Fig. 2-19) is equal to the product of the adjacent resistors in the Δ (π) divided by the sum of the resistors in the Δ.

Each resistor in the equivalent Δ (π) (Fig. 2-18) is equal to the sum of the products of the resistors in the Y (T) $(R_X R_Y + R_Y R_Z + R_X R_Z)$ divided by the opposite resistor. In the Y circuit shown in Fig. 2-16, R_Z is opposite to R_A, R_Y is opposite to R_B, and R_X is opposite to R_C.

Example 2.6 Find the resistance R_{ab} at the input of the circuit in Fig. 2-14 by simplifying that circuit with a Δ-Y transform.

Figure 2-20 shows how the circuit illustrated in Fig. 2-14 appears after being simplified with a Δ-Y transform. The calculations for the equivalent Y circuit are given here:

$$R_X = \frac{R_1 R_2}{R_1 + R_2 + R_3} = \frac{(100)(200)}{100 + 200 + 200} = 40\ \Omega$$

$$R_Y = \frac{R_1 R_3}{R_1 + R_2 + R_3} = \frac{(100)(200)}{100 + 200 + 200} = 40\ \Omega$$

$$R_Z = \frac{R_2 R_3}{R_1 + R_2 + R_3} = \frac{(200)(200)}{100 + 200 + 200} = 80\ \Omega$$

The resistance of the simplified circuit shown in Fig. 2-20 can be readily found.

$$R_{ab} = R_X + \frac{(R_Y + R_L)(R_Z + R_4)}{(R_Y + R_L) + (R_Z + R_4)} = 40 + \frac{(40 + 600)(80 + 400)}{(40 + 600) + (80 + 400)} = 314.3\ \Omega$$

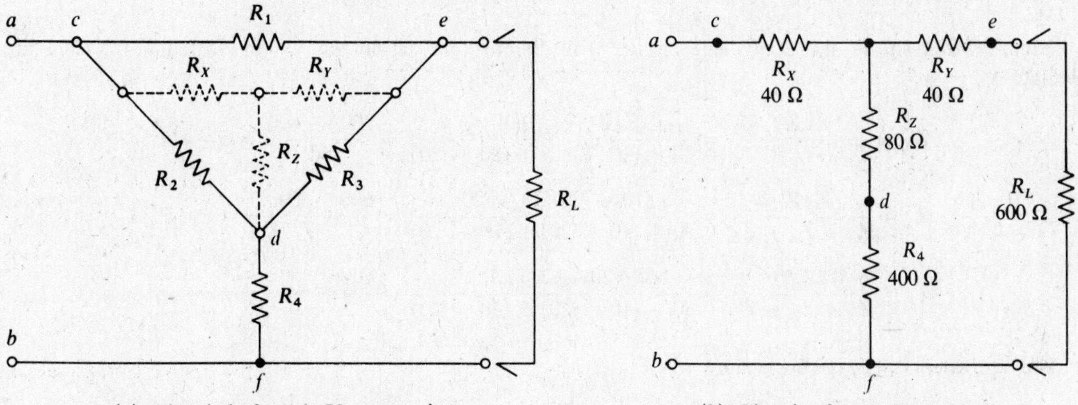

(*a*) Circuit before Δ–Y conversion

(*b*) Circuit after Δ–Y conversion

Fig. 2-20

Example 2.7 Find the resistance R_{ab} at the input of the circuit in Fig. 2-14 by simplifying the circuit with a Y-Δ transform.

Figure 2-21*a* shows how the circuit can be simplified using a Y-Δ transform. The calculations are given here:

$$R_A = \frac{R_2 R_3 + R_3 R_4 + R_2 R_4}{R_4} = \frac{(200)(200) + (200)(400) + (400)(200)}{400} = 500\ \Omega$$

$$R_B = \frac{R_2 R_3 + R_3 R_4 + R_2 R_4}{R_3} = \frac{(200)(200) + (200)(400) + (400)(200)}{200} = 1000\ \Omega$$

$$R_C = \frac{R_2 R_3 + R_3 R_4 + R_2 R_4}{R_2} = \frac{(220)(200) + (200)(400) + (400)(200)}{200} = 1000\ \Omega$$

The resistance of the simplified circuit shown in Fig. 2-21*b* can readily be found:

$$R_{ab} = \frac{R_b[R_1 R_A/(R_1 + R_A) + R_C R_L/(R_C + R_L)}{R_b + R_1 R_A/(R_1 + R_A) + R_C R_L/(R_C + R_L)}$$

$$= \frac{1000[(100)(500)/(100 + 500) + (1000)(600)/(1000 + 600)]}{1000 + (100)(500)/(100 + 500) + (1000)(600)/(1000 + 600)} = 314.3\ \Omega$$

This is the same answer that was obtained when a Δ-Y transform was used.

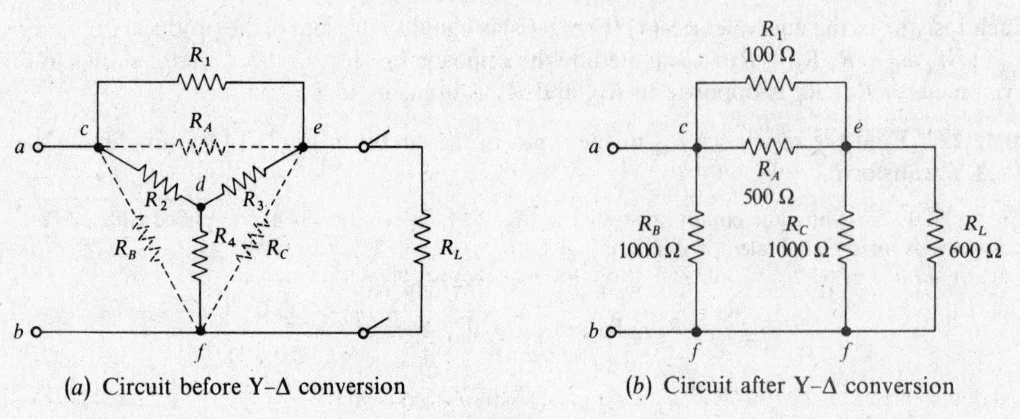

(*a*) Circuit before Y–Δ conversion (*b*) Circuit after Y–Δ conversion

Fig. 2-21

The Δ-Y transform procedure is the same for ac circuits as for dc circuits; however, complex variables must be used for the ac impedances that replace the resistors.

Example 2.8 Convert the Δ circuit in Fig. 2-22*a* to an equivalent Y.

Figure 2-22*b* shows the equivalent circuit with the original circuit shown in dotted lines. Here are the calculations:

$$\mathbf{Z}_X = \frac{\mathbf{Z}_A\mathbf{Z}_B}{\mathbf{Z}_A + \mathbf{Z}_B + \mathbf{Z}_C} = \frac{(10 - j20)(30)}{30 + (10 - j20) + (20 + j20)} = \frac{300 - j600}{60} = 5 - j10$$

$$\mathbf{Z}_Y = \frac{\mathbf{Z}_A\mathbf{Z}_C}{\mathbf{Z}_A + \mathbf{Z}_B + \mathbf{Z}_C} = \frac{(10 - j20)(20 + j20)}{30 + (10 - j20) + (20 + j20)} = \frac{600 - j200}{60} = 10 - j3.33$$

$$\mathbf{Z}_Z = \frac{\mathbf{Z}_B\mathbf{Z}_C}{\mathbf{Z}_A + \mathbf{Z}_B + \mathbf{Z}_C} = \frac{(20 + j20)(30)}{30 + (10 - j20) + (20 + j20)} = \frac{600 + j600}{60} = 10 + j10$$

The equivalent Y (T) circuit is shown in Fig. 2-22*c*.

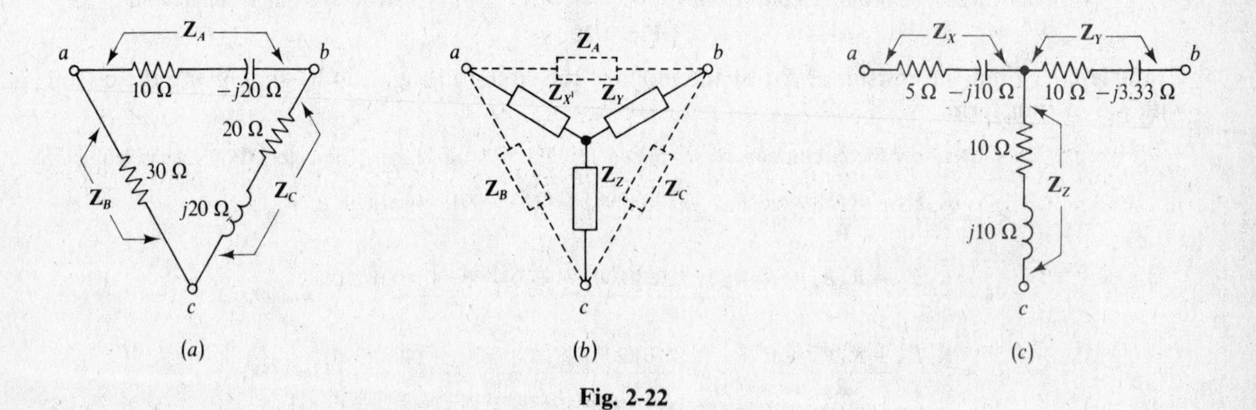

Fig. 2-22

2.6 SUPERPOSITION THEOREM

The superposition theorem states that any circuit with two or more sources can be analyzed by taking one source at a time. When analyzing the circuit, all other voltage sources must be short-circuited and all other current sources open-circuited. However, the internal impedances of the sources must be left in the circuit.

Example 2.9 Figure 2-23 shows a simple circuit with two voltage sources and their associated internal resistances feeding a load resistor, R_3. Using the superposition theorem, find I_3, the current through R_3.

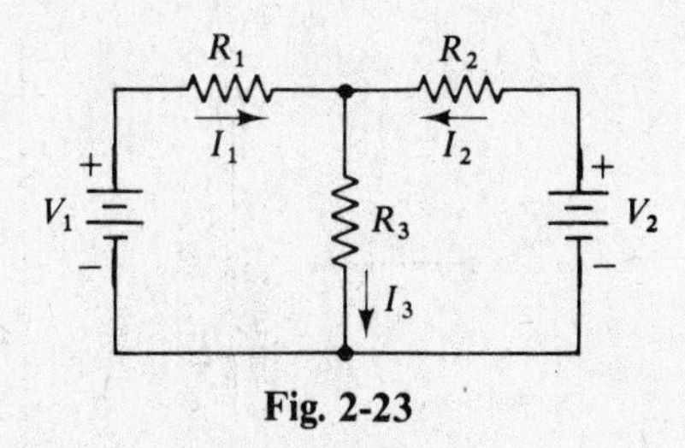

Fig. 2-23

Step 1 (Fig. 2-24*a*). Find the current I'_3 through R_3 that is due to V_1 with V_2 shorted out but with its internal resistance remaining in the circuit.

Start by finding I_1 in Fig. 2-24*a*. This is the total current supplied by V_1:

$$I_1 = \frac{V_1}{R_1 + [R_2 R_3/(R_2 + R_3)]} = \frac{V_1(R_2 + R_3)}{R_1 R_2 + R_1 R_3 + R_2 R_3}$$

Using the current divider method, the part of this current I_1 that flows through R_3 with V_1 shorted is

$$I'_3 = I_1\left(\frac{R_2}{R_2 + R_3}\right) = \frac{V_1(R_2 + R_3)}{R_1 R_2 + R_1 R_3 + R_2 R_3}\left(\frac{R_2}{R_2 + R_3}\right) = \frac{V_1 R_2}{R_1 R_2 + R_1 R_3 + R_2 R_3}\downarrow$$

The arrow indicates that the current due to V_1 flows down through R_3.

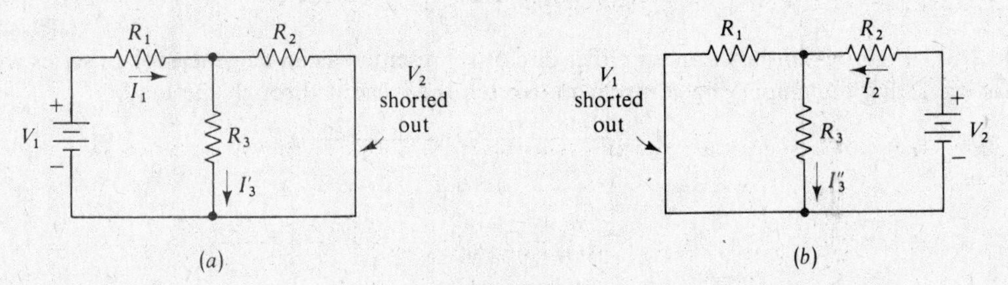

Fig. 2-24

Step 2 (Fig. 2-24*b*). Find the current I''_3 through R_3 that is due to V_2 with V_1 shorted out. Start by finding I_2, which is the total current supplied by V_2:

$$I_2 = \frac{V_2}{R_2 + [R_1 R_3/(R_1 + R_3)]} = \frac{V_2(R_1 + R_3)}{R_1 R_2 + R_1 R_3 + R_2 R_3}$$

By the current divider method, the part of this current I'_2 that flows through R_3 with V_1 shorted is

$$I''_3 = I_2\left(\frac{R_1}{R_1 + R_3}\right) = \frac{V_2(R_1 + R_3)}{R_1 R_2 + R_1 R_3 + R_2 R_3}\left(\frac{R_1}{R_1 + R_3}\right) = \frac{V_2 R_1}{R_1 R_2 + R_1 R_3 + R_2 R_3}\downarrow$$

The arrow indicates that the current due to V_2 flows down through R_3.

Since both currents I'_3 and I''_3 are in the same direction, they are added to find the actual current I_3. This current is shown in Fig. 2-23. (If the currents had been in opposite directions, they would have been subtracted and the direction of the larger value would be the resultant direction.)

$$I_3 = I'_3 + I''_3 = \frac{V_1 R_2}{R_1 R_2 + R_1 R_3 + R_2 R_3} + \frac{V_2 R_1}{R_1 R_2 + R_1 R_3 + R_2 R_3} = \frac{V_1 R_2 + V_2 R_1}{R_1 R_2 + R_1 R_3 + R_2 R_3}\downarrow$$

This checks with Eq. (*2-2*) obtained by previous methods.

Every amplifier circuit has at least two voltages applied. The dc voltage used for operating the tube, transistor, or FET, is one and the signal voltage is the other. An amplifier can be considered to be the generator of an ac signal in the circuit. Thus, the load resistor is supplied by an ac and a dc generator, and the actual current through the load resistor can be determined by considering each source separately.

An equation in the form $v = V_{dc} + V_m \sin(\omega t \pm \phi)$ represents an ac voltage superimposed on a dc voltage. This is shown in Fig. 2-25. The equation for an ac current superimposed upon a dc current is in the general form $i = I_{dc} + I_m \sin(\omega t \pm \phi)$. An example will show how an ac and dc voltage in a circuit results in a current with the above type of equation.

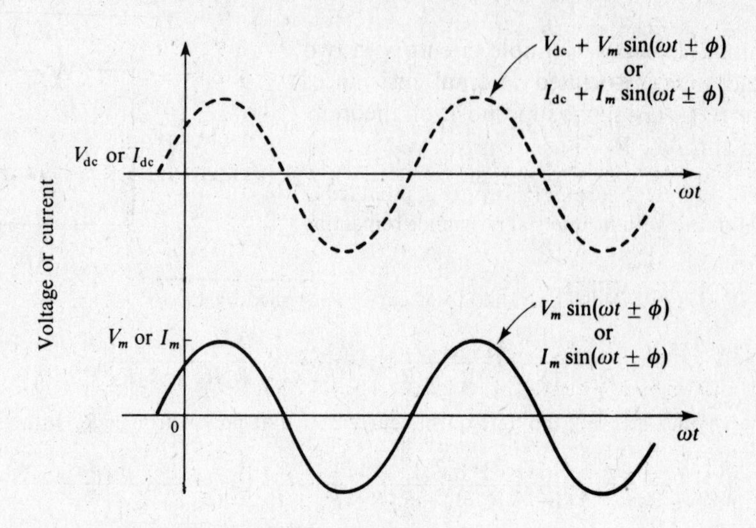

Fig. 2-25

Example 2.10 Figure 2-26 shows an amplifier circuit represented as an ac generator in series with a dc source for operating the amplifying component. Sketch the current through the load.

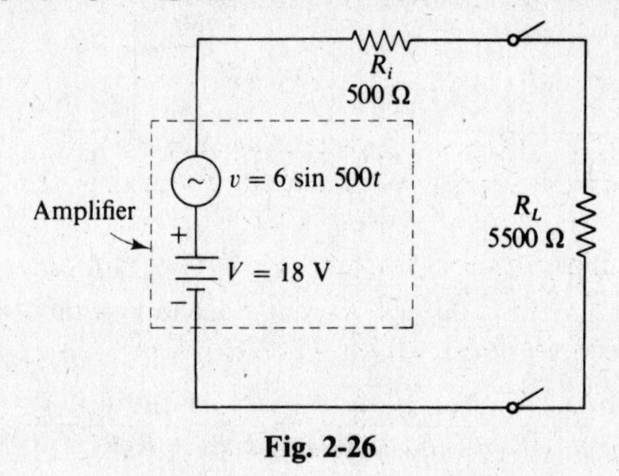

Fig. 2-26

The superposition theorem makes it possible to find the load current by considering the generators one at a time.

In Fig. 2-27*a*, the ac source has been shorted. The current due to the dc source is

$$I' = \frac{V}{R_i + R_L} = \frac{18}{500 + 5500} = 3.0 \text{ mA}$$

In Fig. 2-27*b*, the dc source has been shorted. The current due to the ac source is

$$i' = \frac{v}{R_i + R_L} = \frac{6 \sin 500t}{500 + 5500} = \frac{6}{500 + 5500} \sin 500t = 1.0 \sin 500t \text{ mA}$$

The equation for ac current is in the form $i = I_M \sin \omega t$. The peak value of current is 1.0 mA, and the value of ω is 500. The frequency can be determined from the value of ω as follows:

$$\omega = 2\pi f = 500$$

$$f = \frac{500}{2\pi} = 79.6 \text{ Hz}$$

The time T for one cycle is $1/f = 1/79.6 = 0.01256$ s, or 12.56 ms. Figure 2-27*c* shows the ac current waveform superimposed upon the dc current.

The equation for the waveform is

$$i = I_{dc} + I_m \sin \omega t = 3.0 + 1.0 \sin 500t$$

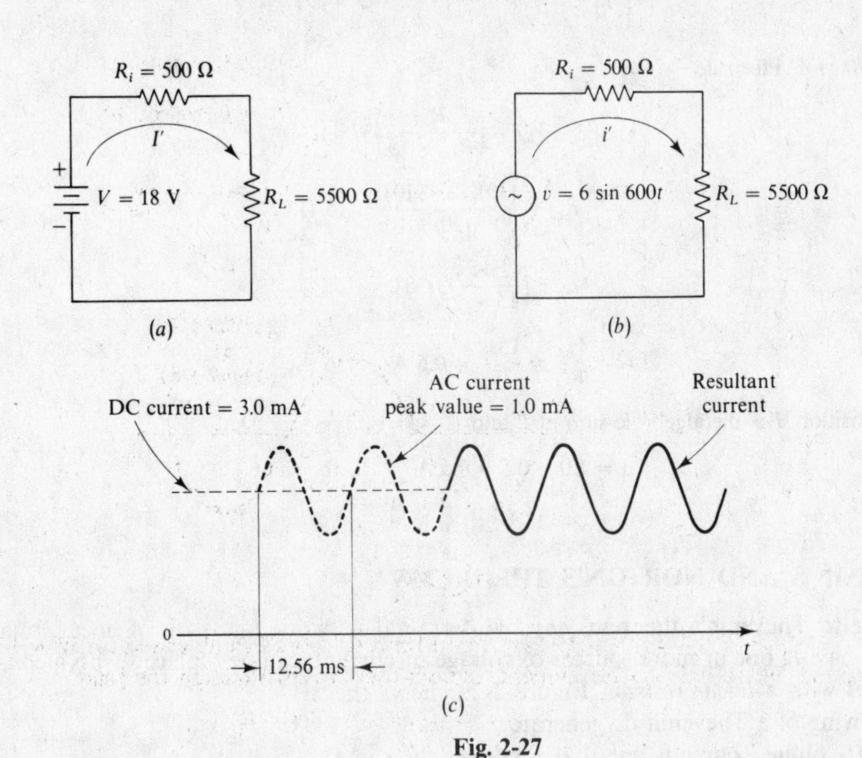

Fig. 2-27

When troubleshooting a circuit, it is a good idea to consider each source separately. The dc voltages for operating the amplifying component are measured first. Then the ac signal is traced or injected to check the ac circuitry.

We may also apply the principle of superposition to a circuit with both voltage and current sources.

Example 2.11 Find the current I through the 20-Ω resistor in the circuit shown in Fig. 2-28*a*.

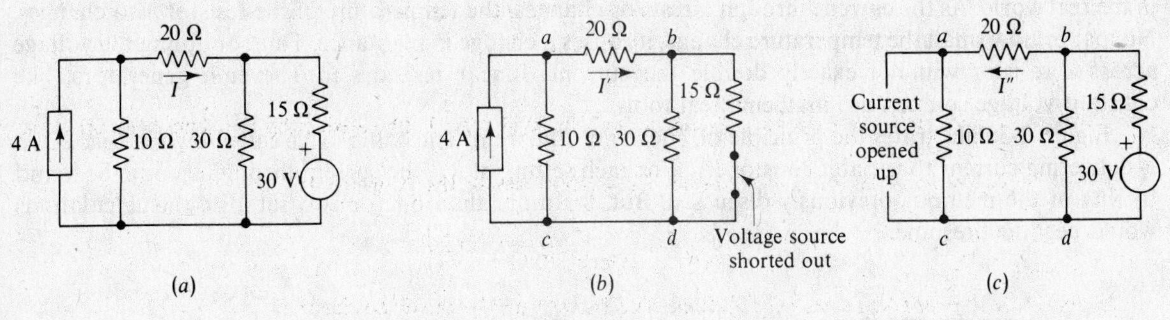

Fig. 2-28

First short the voltage source and calculate the current through the 20-Ω resistor due to the 4-A source. From Fig. 2-28*b*, we see that the 4-A current divides at point a according to the current divider rule; therefore

$$I' = \left(\frac{10}{10 + R'}\right)4.0$$

where

$$R' = 20 + \frac{(15)(30)}{15 + 30} = 20 + 10 = 30\ \Omega$$

Therefore

$$I' = \left(\frac{10}{10 + 30}\right)4.0 = 1.0\ \text{A} \qquad \text{to the right}$$

From Fig. 2-28*c*, with the voltage source replaced and the current source opened, $V_{bd}/R'' = I''$ with $R'' = (20 + 10) = 30\ \Omega$.

By the voltage divider rule,

$$V_{bd} = \left(\frac{R_{bd}}{R_{bd} + 15}\right)30.0$$

with $$R_{bd} = \frac{(30)(20 + 10)}{30 + 20 + 10} = 15\ \Omega$$

So $$V_{bd} = \left(\frac{15}{15 + 15}\right)30.0 = 15.0\ \text{V}$$

and $$I'' = \frac{V_{bd}}{R''} = \frac{15.0}{30} = 0.5\ \text{A} \quad \text{to the left}$$

By superposition, I is the algebraic sum of I' and I'', so

$$I = 1.0 - 0.5 = 0.5\ \text{A} \quad \text{to the right}$$

2.7 THEVENIN'S AND NORTON'S THEOREMS

According to Thevenin's theorem, *any* two-terminal network made up of linear, bilateral circuit elements, and having one or more sources of voltage or current, can be replaced with a *constant-voltage source* in series with a *linear resistor*. Figure 2-29 shows the schematic drawing of a Thevenin dc generator.

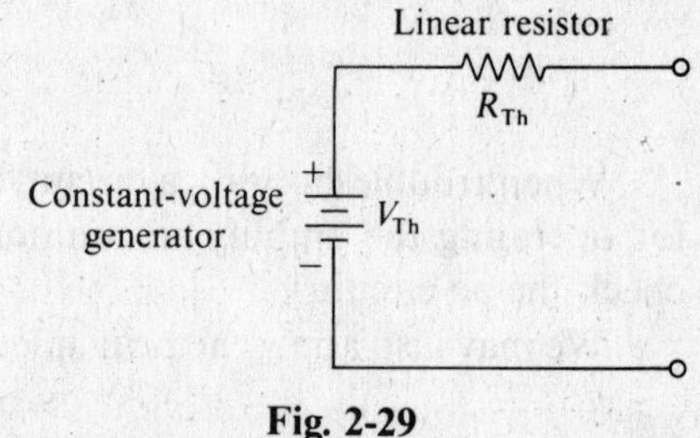

Fig. 2-29

A constant-voltage source is one that produces the same voltage regardless of the amount of load current it is delivering. This means that a constant-voltage source has no internal resistance. There is no such thing as a constant-voltage source in the real world. It is a mathematical tool that simplifies the solution of certain types of network problems.

The current through a linear resistor doubles whenever the voltage across it is doubled. There are no truly linear resistors in the real world. As the current through a resistor changes, the temperature of the resistor also changes. No matter how small the temperature change, it causes a change in resistance. Thus, doubling the voltage across a resistor will not exactly double the current. Linear resistors in Thevenin generators, like constant-voltage sources, are mathematical tools.

Figure 2-30 illustrates the principle of Thevenin's theorem. The complex circuit shown in Fig. 2-30*a* is delivering current to variable resistor R_L. For each setting of R_L, the current through it could be found by any of the methods previously discussed. But, for more than one or two settings, the calculations would become tiresome.

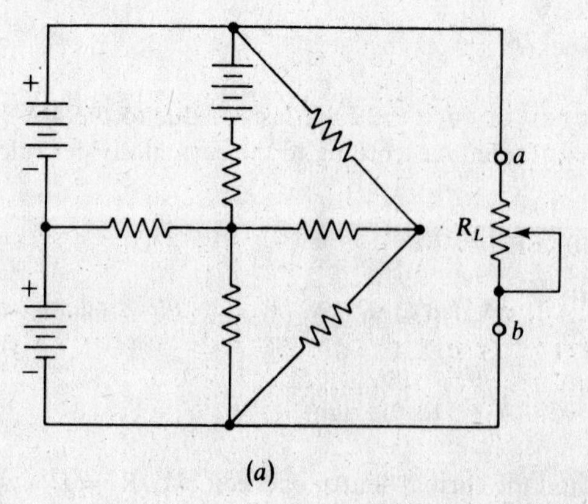

(*a*)

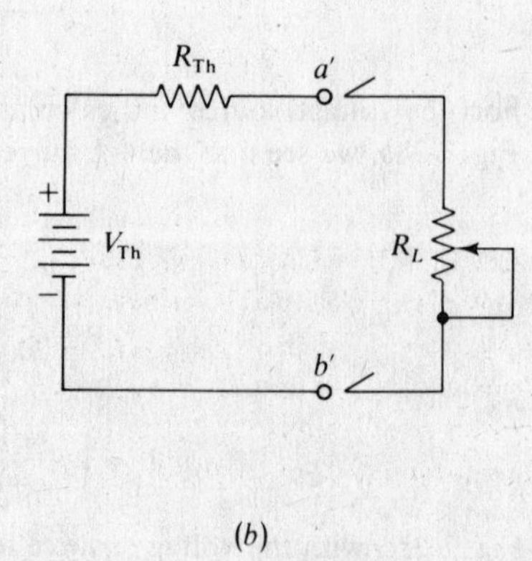

(*b*)

Fig. 2-30

A better way is to replace the circuit with the Thevenin generator shown in Fig. 2-30*b*. Now the calculation of current is quite simple:

$$I = \frac{V_{\text{Th}}}{R_{\text{Th}} + R_L}$$

A *constant-current source* and a *linear resistor* can be used to replace any two-terminal network that has linear, bilateral components and one or more sources of voltage. This equivalent circuit is called a *Norton generator* and is illustrated in Fig. 2-31. It may be considered the dual of the Thevenin generator.

A constant-current generator delivers the same amount of current regardless of the size of load resistance. Like the Thevenin constant-voltage generator, the Norton generator is a mathematical tool used for the solution of network problems. No such generator really exists, although certain "crowbar" power supplies closely approach the ideal for limited current ranges.

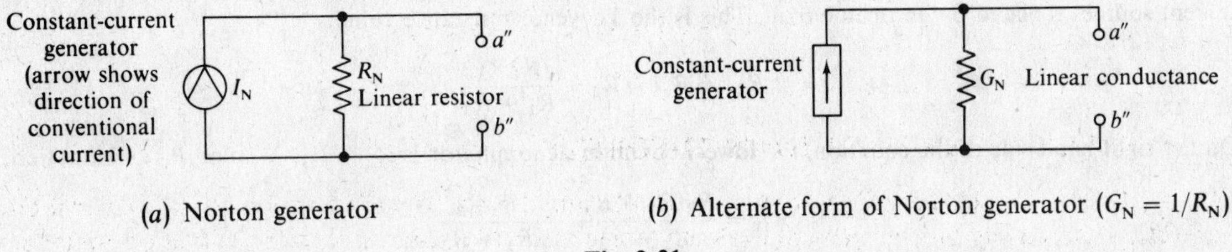

(*a*) Norton generator (*b*) Alternate form of Norton generator ($G_N = 1/R_N$)

Fig. 2-31

The linear shunt resistor in the Norton equivalent circuit is also a mathematical tool like the linear series resistor in the Thevenin equivalent circuit and, in fact, for a given circuit, has the same numerical value ($R_{\text{Th}} = R_N$).

The Thevenin equivalent is usually preferred if the circuit is to be analyzed in terms of voltages and impedances such as a vacuum tube or an FET (field-effect transistor) amplifier.

The Norton equivalent circuit is preferred when the circuit is to be analyzed in terms of currents and admittances such as a bipolar transistor amplifier circuit.

The Norton generator could be used to replace the complex circuit shown in Fig. 2-30*a*. Since the Thevenin generator and the Norton generator for this circuit are duals, it follows that for any circuit the equivalent Thevenin generator can be converted to a Norton generator and the equivalent Norton generator can be converted to a Thevenin generator.

Example 2.12 Figure 2-32 shows a load resistor R_L connected to a circuit made up of R_1, R_2, R_3, and V. Find the equivalent generator as "seen" by the load resistor.

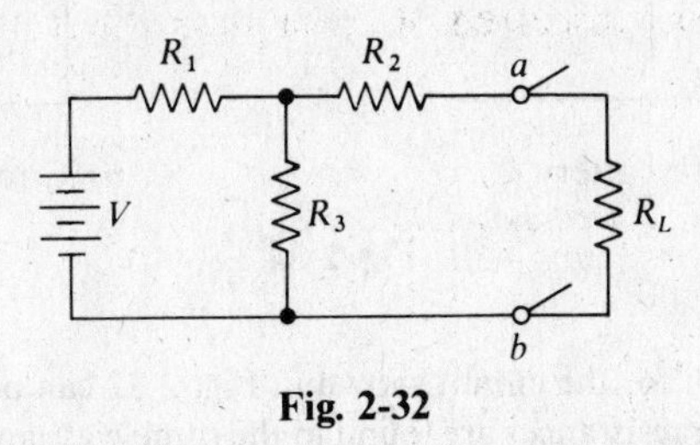

Fig. 2-32

This circuit will be replaced first with a Thevenin generator, then with a Norton generator. Figure 2-33 shows the procedure described in the following steps.

Step 1 (see Fig. 2-33*a*). Remove the load resistor R_L and find the voltage V_{ab} across terminals *a* and *b*. Note this important point: When R_L is removed, V_{ab} is equal to the voltage V_3 across R_3 because terminals *a* and *b* are open-circuited and there is no current through R_2, and hence no voltage drop across it.

By the voltage divider method,

$$V_{\text{Th}} = V_3 = V\left(\frac{R_3}{R_1 + R_3}\right) = \frac{VR_3}{R_1 + R_3}$$

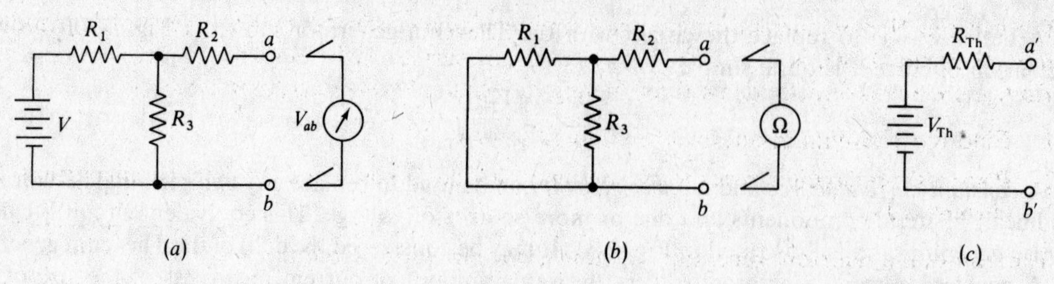

Fig. 2-33

Since this is the open-circuit voltage, it is also the Thevenin voltage.

Step 2 (see Fig. 2-33*b*). Find the resistance R_{ab} with all voltage sources replaced by a short circuit and all current sources replaced by an open circuit. This is the Thevenin resistance value:

$$R_{\text{Th}} = R_{ab} = R_2 + \frac{R_1 R_3}{R_1 + R_3}$$

On the right-hand side of the equation, the lowest common denominator is $R_1 + R_3$. Writing R_2 as a fraction:

$$R_{\text{Th}} = \frac{R_2(R_1 + R_3)}{R_1 + R_3} + \frac{R_1 R_3}{R_1 + R_3}$$

Since the fractions have the same denominator, they can be added. After simplifying the numerator, this gives

$$R_{\text{Th}} = \frac{R_1 R_2 + R_1 R_3 + R_2 R_3}{R_1 + R_3}$$

Step 3 (see Fig. 2-33*c*). Draw the equivalent Thevenin generator.

When shorting out the batteries or opening current sources, it is important to leave any internal resistances that are known. Thus, if the circuit shown in Fig. 2-34 were being Thevenized, the Thevenin resistance would be calculated with the internal resistance in place, i.e., the battery (with its internal resistance) must be replaced before the Thevenin resistance is measured.

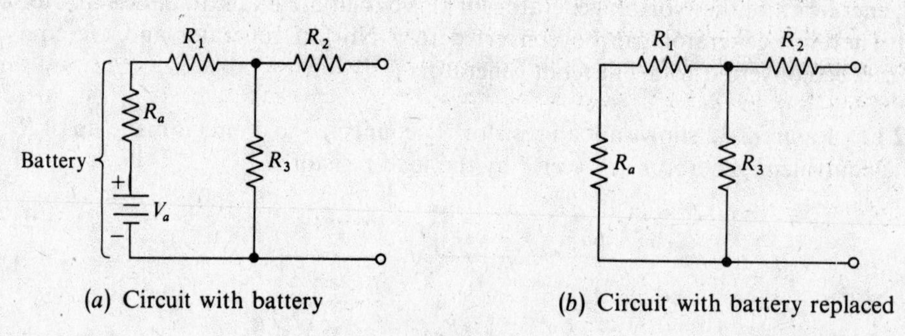

(*a*) Circuit with battery (*b*) Circuit with battery replaced

Fig. 2-34

The Norton equivalent generator for the circuit shown in Fig. 2-32 can be obtained directly from the Thevenin generator. The Thevenin and Norton resistances are found in the same way, and therefore, they have the same value:

$$R_{\text{N}} = R_{\text{Th}} = \frac{R_1 R_2 + R_1 R_3 + R_2 R_3}{R_1 + R_3}$$

The Norton generator current is the short-circuit current between the output terminals of the two-terminal network. Since the Thevenin generator shown in Fig. 2-33 is equivalent to the original circuit, it follows that the short-circuit current is the same for both. Thus, to find I_{N}, simply short-circuit terminals $a'b'$ on the Thevenin circuit and find the resulting current:

$$I_{\text{N}} = \frac{V_{\text{Th}}}{R_{\text{Th}}} = \frac{VR_3/\cancel{(R_1 + R_3)}}{(R_1 R_2 + R_1 R_3 + R_2 R_3)/\cancel{(R_1 + R_3)}} = \frac{VR_3}{R_1 R_2 + R_1 R_3 + R_2 R_3}$$

The value of Norton current can also be found by placing a short circuit between points a and b in the original circuit shown in Fig. 2-32. The short-circuit current is the Norton generator current.

Thus, there are two methods used for finding the value of Norton generator current:

1. Find the Thevenin generator, then find I_N from V_{Th}/R_{Th}.
2. Place a short circuit across the two terminals in question and find the short-circuit current. This current is I_N.

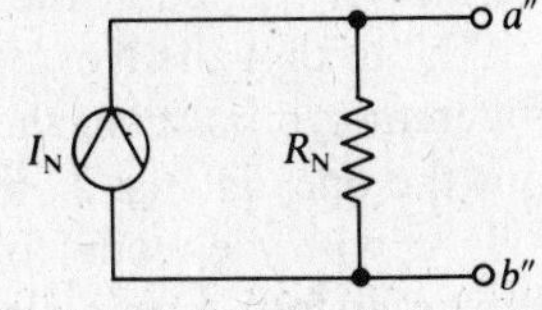

Fig. 2-35

The equivalent Norton generator is shown in Fig. 2-35. Note the direction of the current arrow in the generator. It indicates the direction of conventional current flow from the generator. This arrow must be shown in such a way that the current from the Norton generator will flow through terminals a'' and b'' in the same direction as current would flow in the original circuit when a short circuit is connected across the terminals.

Measuring the open-circuit voltage and the resistance looking back in at the circuit as in Fig. 2-33a and b suggests that the approximate value of the Thevenin voltage and current can be found in the lab using a voltmeter and an ohmmeter. This is true provided the voltmeter is a high-impedance type that draws very little current. Somewhat greater accuracy may be obtained by measuring V_{Th} directly across R_3. This eliminates the possibility of the voltmeter current introducing a drop across R_2 which would subtract from the true V_{Th} voltage across R_3. It is also important to remember that an ohmmeter reading must be taken with *no voltage present in the circuit*.

The ability to convert any two-terminal active network into a Thevenin or Norton equivalent generator makes it possible to work network problems that have nonlinear and unilateral components. The first step is to remove the nonlinear component, and then Thevenize or Nortonize the remaining two-terminal circuit.

As an example, suppose that R_L in Fig. 2-30 is actually a diode. The first step in the solution would be to remove the diode and calculate the Thevenin (or Norton) equivalent generator for the active network. Once the Thevenin or Norton generator has been determined, the diode can then be attached to the equivalent circuit and a solution obtained. Another mathematical tool called the *load line* will be needed before the complete solution can be obtained. Basically, a load line makes it possible to determine the voltage across and current through a nonlinear unilateral device when it is attached to a Thevenin generator. The method of plotting load lines will be discussed in Chap. 3.

Another application of Thevenin's and Norton's theorems is in making equivalent models of amplifying components. Figure 2-36 shows three examples of such models.

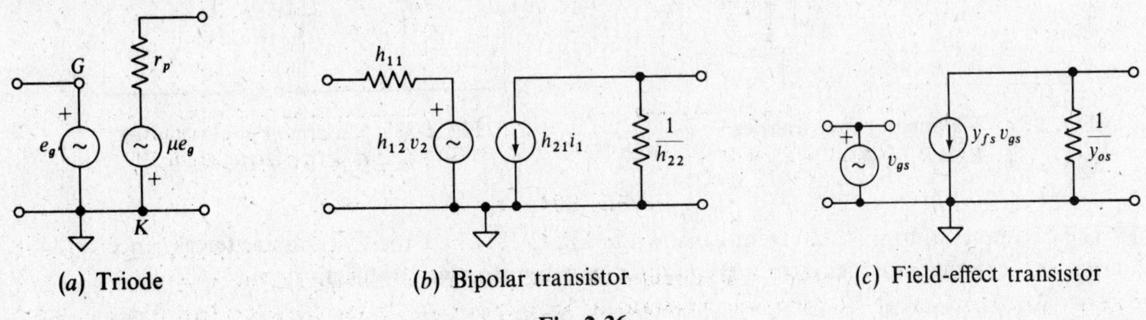

(a) Triode (b) Bipolar transistor (c) Field-effect transistor

Fig. 2-36

The model for the triode contains a Thevenin's generator with a voltage of μe_g, and a Thevenin's resistance r_p. This, of course, is the plate resistance the tube offers to the flow of dc current, and μ is the tube amplification factor. So, μe_g is the input signal voltage times the amplification factor. The models shown in Fig. 2-36 are considered to be four-terminal networks although two of the terminals are the same.

The model for the bipolar transistor has a Norton generator delivering a current of $h_{21}i_1$ to the output terminals. The Norton generator resistor is $1/h_{22}$. The input is a Thevenin generator with a voltage of $h_{12}V_2$ and a resistance of h_{11}. The four h terms (h_{11}, h_{12}, h_{21}, and h_{22}) are called *hybrid parameters*. These parameters are characteristics of the transistor, and their meaning will be described in Chap. 3.

The model for the field-effect transistor also uses a Norton generator, but a Thevenin generator could be used as well. The input signal voltage is designated v_{gs}, and this signal voltage is multiplied by the transconductance y_{fs} to obtain the output current. The Norton admittance for the FET is y_{os}, a parameter related to the field-effect transistor's operation.

Note that the Norton's generators in the bipolar transistor and field-effect transistor models have a different symbol than the ones that were used in Figs. 2-31*a* and 2-35. Both types of symbols are in use, but the ones in Fig. 2-36 are more popular.

As with the model for the tube (triode), the bipolar transistor and the field-effect transistor are both shown as four-terminal networks.

2.8 MAXIMUM POWER TRANSFER THEOREM

In dc circuits the maximum possible power that can be delivered to a load occurs when the load resistance equals the internal resistance of the dc generator. This is a statement of the maximum power transfer theorem.

With an ac generator having an internal impedance of $\mathbf{Z}_i$, the maximum power is delivered to the load impedance $\mathbf{Z}_L$ when the load impedance is the conjugate of the internal impedance.

When conjugate impedances are connected in series, as shown in Fig. 2-37, the reactive terms cancel:

$$\mathbf{Z}_i = R_i + jX_{Li} \qquad \mathbf{Z}_L = R_L - jX_{CL}$$

$$\mathbf{Z}_i = R_i - jX_{ci} \qquad \mathbf{Z}_L = R_L + jX_{LL}$$

The circuit then consists only of the internal and external (load) resistance values, which by the definition of conjugate are equal. The circuit is then the same as the dc case and maximum power transfer occurs.

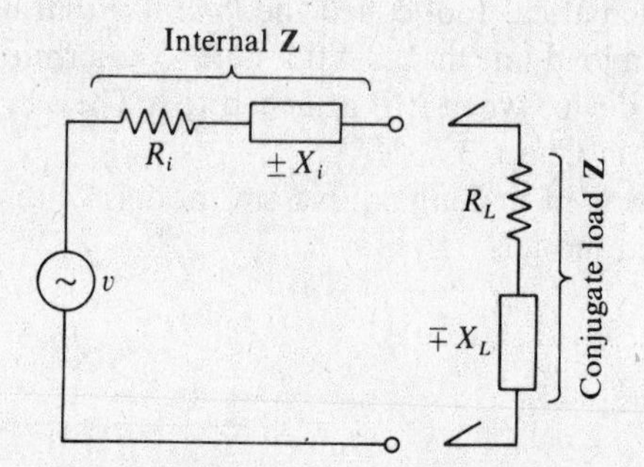

Fig. 2-37 Maximum power transfer: $R_i = R_L$, $X_i = -X_L$

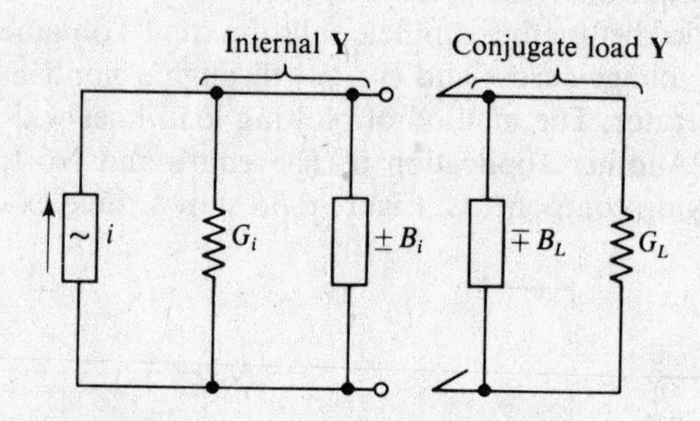

Fig. 2-38 Maximum power transfer: $G_i = G_L$, $B_i = -B_L$

The current dual to this statement is shown in Fig. 2-38, where the susceptance terms cancel and the internal and external conductances are equal for maximum power transfer:

$$\mathbf{Y}_i = G_i + jB_{Ci} \qquad \mathbf{Y}_L = G_L - jB_{LL}$$

$$\mathbf{Y}_i = G_i - jB_{Li} \qquad \mathbf{Y}_L = G_L + jB_{CL}$$

Example 2.13 Find the maximum power that can be delivered to R_L for the circuit shown in Fig. 2-39.

In Fig. 2-39, the power P_L dissipated in the load resistor R_L can be written as

$$P_L = \frac{V_L^2}{R_L}$$

where [from (2-1)] $$V_L = \left(\frac{R_L}{R_i + R_L}\right)V$$

Therefore $$P_L = \frac{[R_L/(R_i + R_L)]^2 V^2}{R_L} = \frac{R_L}{(R_i + R_L)^2} V^2 \qquad (2\text{-}5)$$

Substituting known values from Fig. 2-39,

$$P_L = \frac{R_L}{(5 + R_L)^2}(10)^2 = \frac{100R_L}{(5 + R_L)^2}$$

R_i 5 Ω; $V = 10$ V; R_L 0–20 Ω

Fig. 2-39

Values of R_L can be assigned between 0 and 20 Ω, and the resulting power calculated. Table 2-2 shows a number of examples.

Table 2-2

R_L, Ω	0	2.5	5	7.5	10	15	20
P_L, W	0	4.44	5.00	4.80	4.44	3.75	3.20

A graph of the power dissipated by R_L vs. the value of R_L is shown in Fig. 2-40. It peaks at $R_L = 5$, which is the value of internal resistance of the battery. Regardless of the different values of R_i, V, and R_L, the graph will always peak (show maximum power) at a point where $R_L = R_i$.

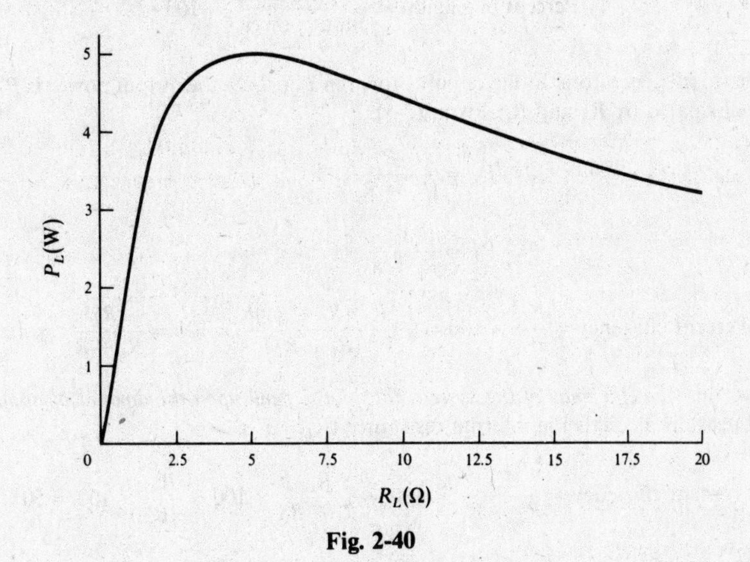

Fig. 2-40

Example 2.14 For the ac generator of Fig. 2-41, what is the maximum power that can be delivered to a load?

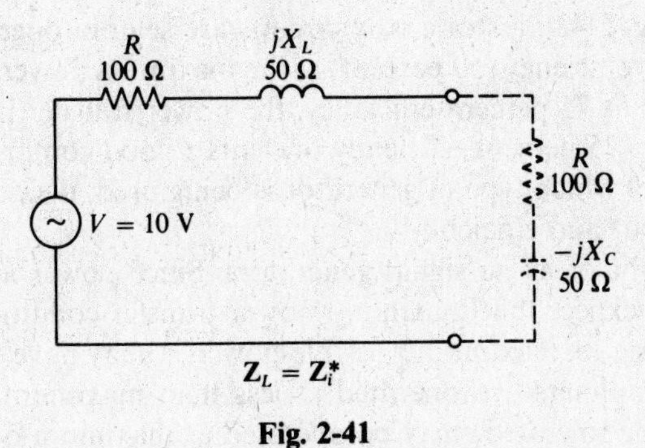

Fig. 2-41

Maximum power transfer will occur when the load impedance is $R - jX_C = 100 - j50$. This is shown in dotted lines in Fig. 2-41. The total circuit impedance, including the internal and load impedances, is

$$\mathbf{Z}_T = \mathbf{Z}_i + \mathbf{Z}_L = (100 + j50) + (100 - j50) = 200\ \Omega$$

Note that, by making the internal impedance and load impedance conjugates, there is no phase angle between the voltage and current, so the power can be obtained directly from the rms value of voltage V.

Solving for the power in the load P_L, we make use of Eq. (2-5):

$$P_L = \frac{R_L}{(R_i + R_L)^2} V^2$$

With $R_L = R_i$ at peak power,

$$P_L = \frac{R_i}{(2R_i)^2} V^2 = \frac{100}{[(2)(100)]^2} 10^2 = 0.25\ \text{W}$$

Example 2.15 Show by a graph how the percent efficiency of the circuit shown in Fig. 2-39 is related to the amount of power transferred.

The efficiency rating of any type of electric generator is a measure of how well that generator is able to deliver power to a load. Mathematically:

$$\text{Percent efficiency} = \frac{\text{output power}}{\text{input power}} \times 100 \qquad (2\text{-}6)$$

This equation applies to all generators. In the circuit shown in Fig. 2-39, the output power is P_L. The input power is the total power P_T dissipated by R_L and R_i. From (2-5),

$$P_L = \frac{R_L}{(R_i + R_L)^2} V^2$$

and

$$P_T = \frac{V^2}{R_i + R_L} \qquad (2\text{-}7)$$

$$\text{Percent efficiency} = \frac{P_L}{P_T} \times 100 = \frac{V^2 R_L/(R_i + R_L)^2}{V^2/(R_i + R_L)} \times 100 = \frac{R_L}{R_i + R_L} \times 100 \qquad (2\text{-}8)$$

Equation (2-8) shows that *the efficiency of the system does not depend upon the amount of applied voltage.*

When the load resistance equals the internal resistance ($R_L = R_i$),

$$\text{Percent efficiency} = \frac{R_i}{R_L + R_i} = 100 = \frac{R_i}{R_i + R_i} \times 100 = \frac{R_i}{2R_i} \times 100 = 50\%$$

The maximum power condition *always* results in an efficiency of only 50 percent.

Figure 2-42 shows a graph of the percent efficiency for the circuit illustrated in Fig. 2-39. The maximum power curve for the same circuit is shown on the same graph.

The curves shown in Fig. 2-42 illustrate why circuits are seldom operated under maximum power transfer conditions. The poor efficiency (50 percent) at the maximum power point cannot be tolerated for most applications. Note that at 75 percent efficiency, the power transfer is only slightly reduced.

For extended battery life, 75 percent efficiency presents a good compromise between output power and battery life. Regardless of which type of generator is being used, the designer is always faced with a tradeoff between power output and efficiency.

Amplifiers can be thought of as ac signal generators. Since power amplifiers are used to deliver power to a load, you might expect that maximum power transfer conditions are used. Unfortunately, when an amplifier is delivering its maximum possible power, it may have too much *distortion* for some applications. Thus, audio amplifiers are operated at less than maximum power, but servo amplifiers (where more distortion can be tolerated) may be operated at maximum output power.

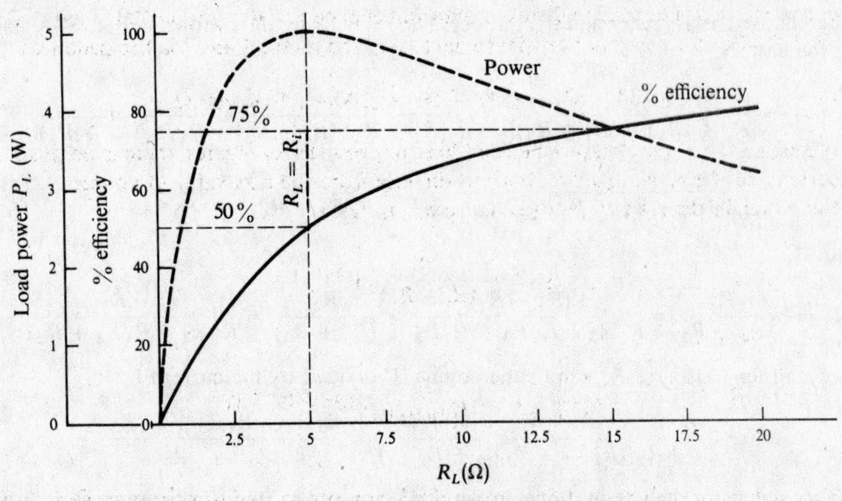

Fig. 2-42

2.9 RECIPROCITY THEOREM

The reciprocity theorem states that the voltage V of a voltage source in one part of the circuit divided by a current I measured by an ideal ammeter in any other part remains the same when the voltage source and ammeter are interchanged. This ratio is constant and is called the *transfer resistance*. According to the reciprocity theorem, as long as we are considering only one voltage source at a time, if the voltage source is interchanged with the ammeter, the transfer resistance will not be changed.

The term "ideal ammeter" means an ammeter with zero internal resistance. If the voltage source has an internal resistance as in the case of a Thevenin generator, the complete generator must be interchanged with the ammeter.

Example 2.16 In the circuit shown in Fig. 2-43, find the transfer resistance between points a and b and c and d. Then show that, when the battery and ammeter are interchanged, the transfer resistance will be the same value.

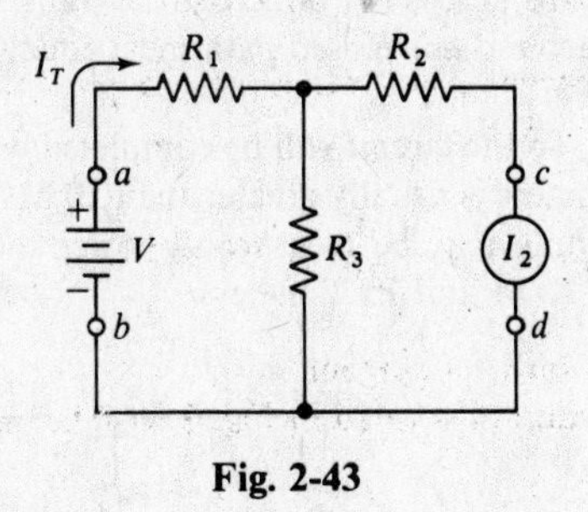

Fig. 2-43

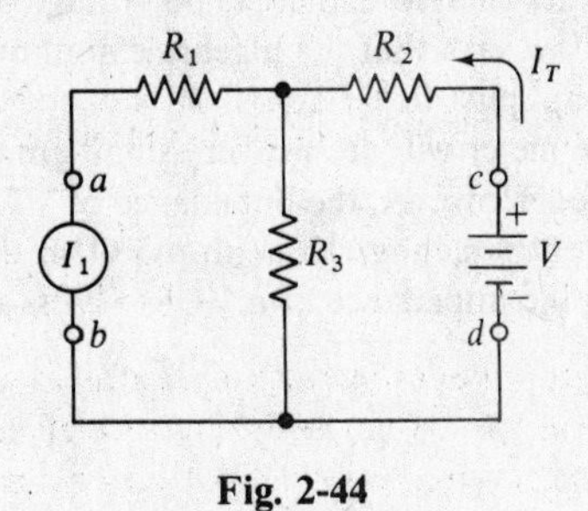

Fig. 2-44

Step 1. Find the value of I_T:

$$I_T = \frac{V}{R_1 + [R_2 R_3/(R_2 + R_3)]} = \frac{V(R_2 + R_3)}{R_1 R_2 + R_1 R_3 + R_2 R_3}$$

Step 2. Using the current divider method, find the part of I_T that flows through R_2:

$$I_2 = I_T \frac{R_3}{R_2 + R_3} = \frac{V(R_2 + R_3)}{R_1 R_2 + R_1 R_3 + R_2 R_3} \times \frac{R_3}{(R_2 + R_3)} = \frac{V R_3}{R_1 R_2 + R_1 R_3 + R_2 R_3}$$

Step 3. The transfer resistance R_X is the value of voltage V divided by the value of current I_2:

$$R_X = \frac{V}{V R_3/(R_1 R_2 + R_1 R_3 + R_2 R_3)} = \frac{R_1 R_2 + R_1 R_3 + R_2 R_3}{R_3}$$

Step 4. Interchange the ammeter and the battery. The revised circuit is shown in Fig. 2-44. Calculate the total current I_T from the battery:

$$I_T = \frac{V}{R_2 + [R_1R_3/(R_1+R_3)]} = \frac{V}{R_2(R_1+R_3)/(R_1+R_3) + R_1R_3/(R_1+R_3)}$$

$$= \frac{V}{[R_2(R_1+R_3)+R_1R_3]/(R_1+R_3)} = \frac{V(R_1+R_3)}{R_1R_2+R_1R_3+R_2R_3}$$

Step 5. Find I_1:

$$I_1 = I_T\frac{R_3}{R_1+R_3} = \frac{V(R_1+R_3)}{R_1R_2+R_1R_3+R_2R_3} \times \frac{R_3}{(R_1+R_3)} = \frac{VR_3}{R_1R_2+R_1R_3+R_2R_3}$$

Step 6. The transfer resistance R_X equals the voltage V divided by the current I_1:

$$R_X = \frac{V}{VR_3/(R_1R_2+R_1R_3+R_2R_3)} = \frac{R_1R_2+R_1R_3+R_2R_3}{R_3}$$

The transfer resistance for the circuit shown in Fig. 2-43, as given in Step 3, is the same as the transfer resistance for the circuit shown in Fig. 2-44, as given in Step 6. Thus, the transfer resistance between two sets of terminals is not affected by interchanging the battery and the point where the current is being measured.

The reciprocity theorem is used in deriving equations in some network theorems such as transforms and four-terminal networks. In a few isolated cases, it can be used to simplify the solution of network problems.

The network theorems and laws discussed in this chapter are the basic ones encountered in electronics literature. Additional theorems and laws will be included in later chapters as they are needed.

Solved Problems

2.1 In the circuit shown in Fig. 2-45, the emitter stabilization resistor R_3 is open. A high-impedance voltmeter connected between point a and common will read ____ V.

Approximately 20 V. Remember that there is no voltage drop across a resistor unless there is current through it. Therefore, there is no voltage drop across R_2, or across R_1, or across the transistor. Kirchhoff's voltage law says that the algebraic sum of the voltages around *any* closed path must equal zero. The entire 20-V drop must occur across R_3.

The meter will draw a very small amount of current, so the circuit will be completed when the meter is connected. However, the impedance of VTVM or FET meters is usually greater than 10 MΩ (megohms), so the current is negligible. With a VOM, the voltage reading may be measurably lower than 20 V. This is because the impedance of a VOM is less than that of VTVM and FET meters.

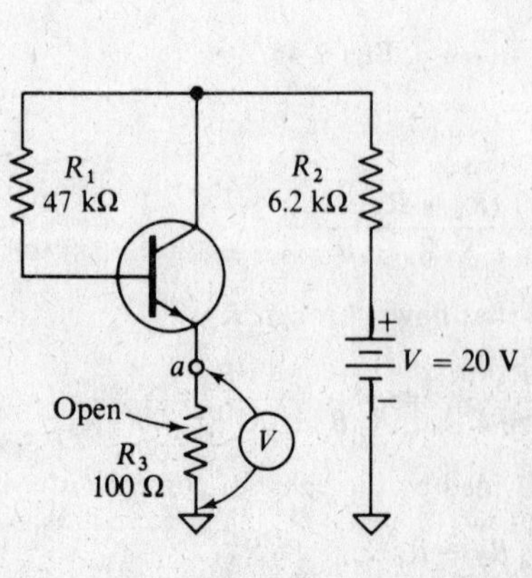

Fig. 2-45

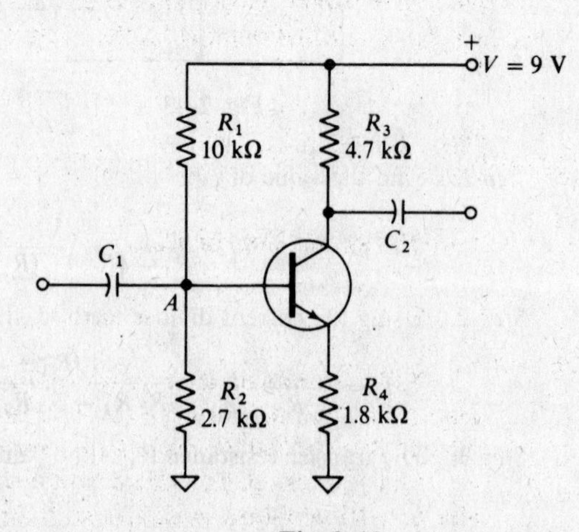

Fig. 2-46

2.2 A technician is troubleshooting the circuit shown in Fig. 2-46. What is the approximate transistor base voltage he or she should measure at point A? (*Hint:* Use the voltage divider method.)

Voltages are measured with respect to ground, or common. The voltage across R_2 is required here:

$$\text{Voltage across } R_2 = V_2 = \left(\frac{R_2}{R_1 + R_2}\right)V = \left(\frac{2.7 \times 10^3}{(10 \times 10^3) + (2.7 \times 10^3)}\right)9 = 1.91 \text{ V} \quad \text{approximately}$$

This value will be slightly off because there are two currents through R_1: the current through R_2 and the current from the base of the transistor. However, the base current is normally very small, so the calculated voltage is reasonably close.

Before making any circuit voltage measurement, the most logical starting point is to measure the 9-V power-supply voltage.

2.3 Figure 2-47 shows a meter movement with a maximum scale deflection of I_m and a meter resistance of R_M. It is desired to use this meter movement to measure a current I that is larger than I_m, so a shunt path R_{Sh} is provided. This shunt path assures that the current through the meter will never exceed I_m. (When measuring current I, the meter should deflect to full scale.)

Derive an equation for finding the shunt resistance when the following are known: I_m, R_M, and I.

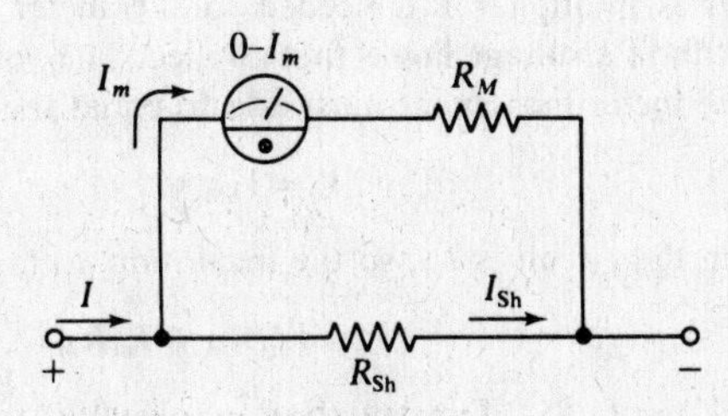

Fig. 2-47

Equations of this type are easy to derive using basic network theory.

In the circuit shown in Fig. 2-47, the *full-scale* current reading on the meter is I_m. The meter resistance is R_M. The *maximum* value of current to be measured is I. Since this current is greater than the maximum (full-scale) value of current that can be read on the meter, a parallel path, or shunt, is provided for the excess current to flow. This shunt is marked R_{Sh}.

According to Kirchhoff's current law, the shunt current I_{Sh} equals the current I to be read minus the current I_m through the meter movement:

$$I_{Sh} = I - I_m$$

The voltage across the shunt must equal the voltage across the meter because the voltages across all parts of a parallel circuit are equal. The maximum voltage across the meter movement is $I_m R_M$, and the voltage across the shunt is $I_{Sh} R_{Sh}$. Setting these voltages equal:

$$I_m R_M = I_{Sh} R_{Sh}$$

But, $I_{Sh} = I - I_m$, so

$$I_m R_M = (I - I_m) R_{Sh}$$

Solving this equation for R_{Sh}:

$$R_{Sh} = \frac{I_m R_M}{I - I_m} \tag{2-9}$$

The quantities on the right-hand side of the equation are known and the value of shunt resistance R_{Sh} required so that the current I can be read on the meter can be determined. Of course, the meter scale will have to be modified for reading the new current values.

An important point is demonstrated here. It is not necessary to memorize many equations in order to be able to work problems in electronics. A knowledge of network theorems and laws and a basic knowledge of algebra make it possible to derive equations as they are needed.

2.4 Figure 2-48 shows a meter movement with a maximum scale deflection of I_m and a meter resistance of R_M. It is desired to use this meter movement to measure a maximum value of voltage V. (The meter should deflect to full scale when measuring V.) A series multiplier resistor R_{SM} is connected in series with the meter to limit the current through the meter movement so that it will never exceed I_m.

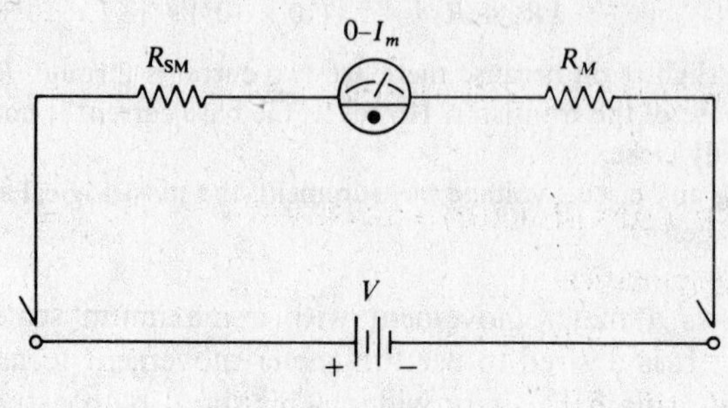

Fig. 2-48

Derive an equation for finding R_{SM} when the following are known: I_m, R_M, and V.

To calculate the series multiplier R_{SM} needed so the meter multiplier combination can measure a *full-scale* voltage V, Kirchhoff's voltage law is first applied. The voltage V_{SM} across the series multiplier plus the voltage V_M across the meter movement must add to equal the applied voltage V:

$$V = V_{SM} + V_M$$

The maximum current in the circuit is I_m, so the maximum voltage V that can be read is

$$V = I_m R_{SM} + I_m R_M$$

since $V_{SM} = I_m R_{SM}$ and $V_M = I_m R_M$. This equation is solved for the unknown value of the series multiplier resistance:

$$V = I_m(R_{SM} + R_M)$$

Rearranging terms,

$$R_{SM} + R_M = \frac{V}{I_m}$$

$$R_{SM} = \frac{V}{I_m} - R_M = \frac{V - I_m R_M}{I_m} \tag{2-10}$$

In many cases the meter resistance is so small compared to the resistance of R_{SM} that it can be disregarded. Then, the equation becomes

$$R_{SM} = \frac{V}{I_m} \tag{2-10a}$$

2.5 Using the equations derived in Prob. 2.3 and 2.4, find the resistances required to convert a 50-μA meter movement to a 0- to 50-mA current meter and a 0- to 5-V voltmeter. Assume the meter resistance is 200 Ω.

The value of shunt resistance needed to convert the 50-μA meter movement to a 0- to 50-mA meter movement is determined by converting the full-scale values (50 μA and 50 mA) to amperes, then applying Eq. (*1-9*) for R_{Sh}:

$$R_{Sh} = \frac{I_m R_M}{I - I_m} = \frac{(50 \times 10^{-6})(200)}{(30 \times 10^{-3}) - (50 \times 10^{-6})} = \frac{1 \times 10^{-2}}{4.995 \times 10^{-2}} \approx 2.00 \times 10^{-1} \approx 0.2\ \Omega$$

Values of shunt resistance this small are manufactured in flat metal strips.

The value of series multiplier R_{SM} is calculated with the meter current expressed in amperes (*2-10*):

$$R_{SM} = \frac{V}{I_m} - R_M = \frac{5}{50 \times 10^{-6}} - 200 = 100\,000 - 200 = 99\,800\ \Omega$$

2.6 Find the internal resistance of the battery shown in Fig. 2-49 if the following measurements are known:

Milliammeter reading with SW open $= 15 \text{ mA} = I_1$

Milliammeter reading with SW closed $= 30 \text{ mA} = I_2$

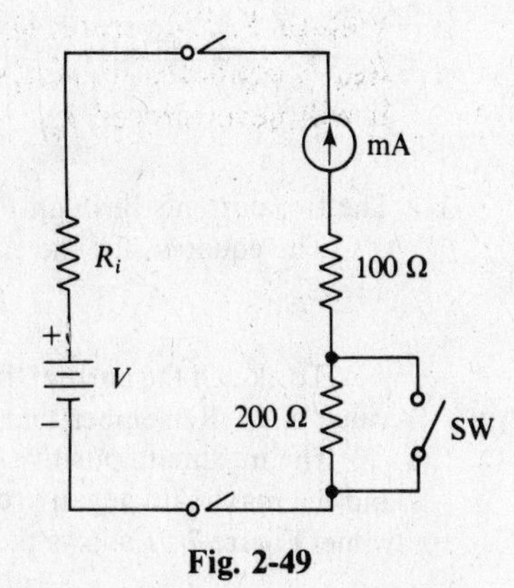

Fig. 2-49

With SW open:

$$V = I_1 R = 0.015\,(R_i + 300) = 0.015R_i + 4.5$$

With SW closed:

$$V = I_2 R = 0.03\,(R_i + 100) = 0.03R_i + 3$$

There are two equations for V:

$$V = 0.015R_i + 4.5$$

$$V = 0.03R_i + 3$$

Since both quantities on the right-hand side of the equation equal V, they must be equal to each other:

$$0.015R_i + 4.5 = 0.03R_i + 3$$

$$0.015R_i = 1.5$$

$$R_i = 100\ \Omega$$

2.7 Sketch the waveform of the current through R_3 in the circuit shown in Fig. 2-50.

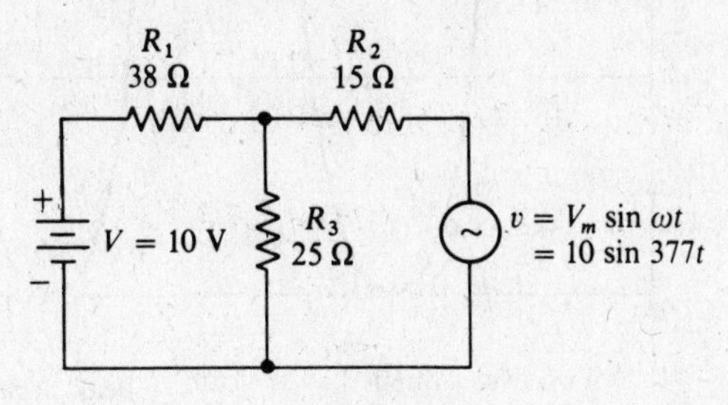

Fig. 2-50

The circuit has an ac generator and a dc battery, so there are two kinds of current (ac and dc) through R_3. The superposition method makes it possible to determine each current separately.

With battery V shorted out, the ac generator "sees" a series parallel circuit with a resistance of

$$R_{eq} = R_2 + \frac{R_1 R_3}{R_1 + R_3} = 15 + \frac{38 \times 25}{38 + 25} = 15 + 15.1 = 30.1\ \Omega$$

The ac voltage across R_3 can be determined by the voltage divider rule:

$$v_3 = \frac{vR_{eq}}{R_{eq} + R_2} = 10 \sin 377t \frac{30.1}{30.1 + 15} \approx 6.67 \sin 377t$$

The ac current through R_3 is equal to the voltage v_3 across R_3 divided by the resistance of R_3:

$$i_3 = \frac{v_3}{R_3} = \frac{6.67 \sin 377t}{25} = 0.267 \sin 377t \text{ A} = 267 \sin 377t \text{ mA}$$

The current through R_3 that is due to V will be called I_3. To find this current, with the ac generator shorted, first find the total current delivered by the battery. Then, by the current divider method, find the part of the current through R_3.

The resistance seen by the battery is

$$R = R_1 + \frac{R_2 R_3}{R_2 + R_3} = 38 + \frac{15 \times 25}{15 + 25} = 47.4\ \Omega$$

The battery current is

$$I = \frac{V}{R} = \frac{10}{47.4} = 0.211 \text{ A} = 211 \text{ mA}$$

By the current divider method:

$$I_3 = I\left(\frac{R_2}{R_2 + R_3}\right) = (0.211)\left(\frac{15}{15 + 25}\right) = 0.079 \text{ A} = 79 \text{ mA}$$

The two currents through R_3 have now been found.

The equation for the current waveform can now be written

$$i = I_{dc} + I_m \sin \omega t \pm \phi = 0.079 + 0.267 \sin 377t$$

To sketch the current, first draw the dc current as a base line. The ac current will fluctuate around that value of dc. Remember that the ac current is sinusoidal.

The maximum positive current occurs when the positive peak of the ac waveform adds to the dc value, and the maximum negative current occurs when the negative peak of the ac waveform subtracts from the dc value. Figure 2-51 shows the result.

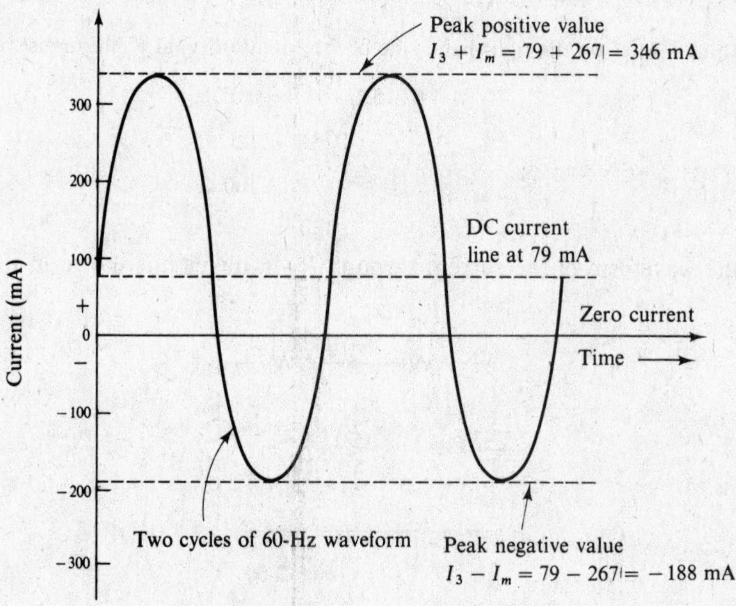

Fig. 2-51

2.8 What is the value of resistance for R_L such that it will receive the maximum possible amount of power from the circuit shown in Fig. 2-52? What is the value of that maximum power?

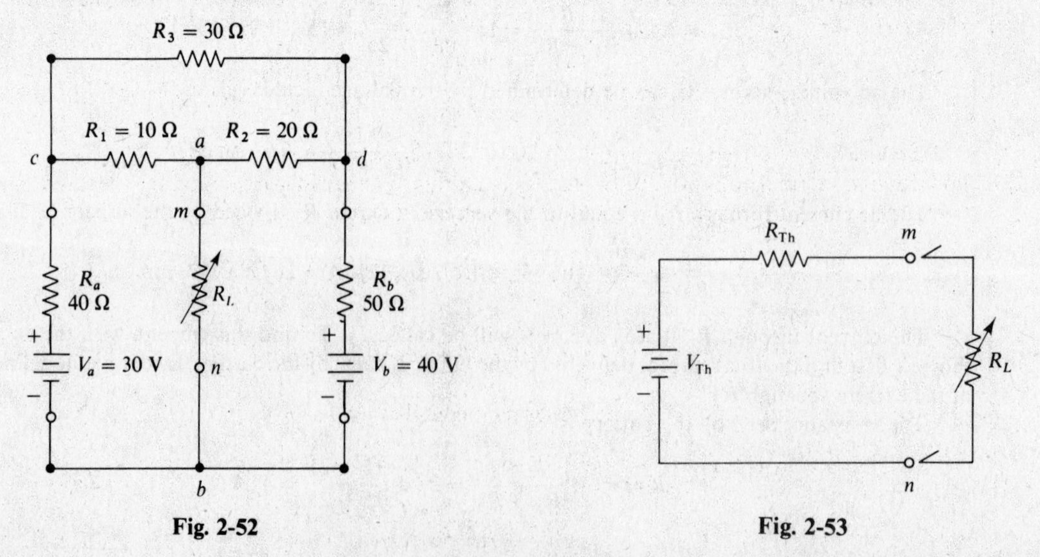

Fig. 2-52

Fig. 2-53

Consider terminals *mn*. If the variable resistor is disconnected, then an equivalent Thevenin generator can be made for the circuit. Figure 2-53 shows the Thevenin generator with R_L connected. Note that R_{Th} can be considered to be the internal resistance of the battery, so the maximum power transfer theorem will apply.

Step 1. Find the value of Thevenin resistance R_{Th}.

Figure 2-54 shows the circuit redrawn with R_L removed. The Thevenin resistance is equal to the resistance between points *m* and *n* with R_L removed. Note that the short circuits are shown in dotted lines around the batteries only. The internal resistances remain in the circuit. (It must be understood that the short circuit is only for the purpose of making calculations. Batteries are *never* short-circuited for the purpose of making measurements.)

To simplify the calculation, the Δ circuit (R_1, R_2, and R_3) can be converted to the Y (R_X, R_Y, and R_Z):

$$R_X = \frac{R_1 R_3}{R_1 + R_2 + R_3} = \frac{(10)(30)}{10 + 20 + 30} = 5\ \Omega$$

$$R_Y = \frac{R_1 R_2}{R_1 + R_2 + R_3} = \frac{(10)(20)}{10 + 20 + 30} = 3.3\ \Omega$$

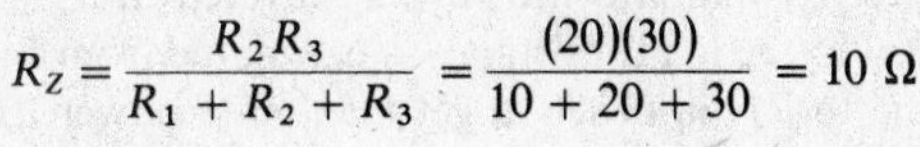

$$R_Z = \frac{R_2 R_3}{R_1 + R_2 + R_3} = \frac{(20)(30)}{10 + 20 + 30} = 10\ \Omega$$

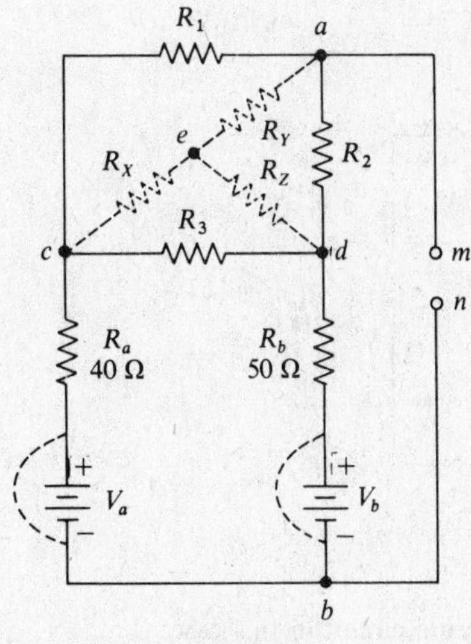

Fig. 2-54

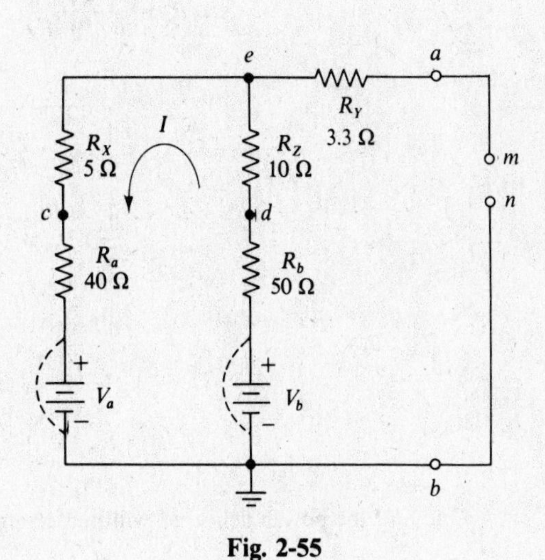

Fig. 2-55

The circuit is redrawn in Fig. 2-55 with the equivalent Y. The Thevenin resistance is equal to the resistance from *m* to *n* with the batteries shorted:

$$R_{mn} = R_{Th} = R_Y + \frac{(R_X + R_a)(R_Z + R_b)}{R_X + R_a + R_Z + R_b} = 3.3 + \frac{(5 + 40)(10 + 50)}{5 + 40 + 10 + 50} = 29.014\ \Omega$$

Step 2. Find the Thevenin voltage V_{mn} in Fig. 2-55 by selecting point *b* as the reference node. A Maxwell nodal equation can then be written for node *e*. Since terminals *m* and *n* are open-circuited, there is no current from *e* to *m* and no voltage drop across R_Y. Therefore the voltage assigned to node $e = V_{mn}$.

$$\text{Current entering } e \text{ from branch } e\text{-}c\text{-}b = \frac{V_{mn} - V_a}{R_a + R_X}$$

$$\text{Current entering } e \text{ from branch } e\text{-}d\text{-}b = \frac{V_{mn} - V_b}{R_b + R_Z}$$

$$\text{Current entering } e \text{ from branch } e\text{-}m = 0$$

So we may write Maxwell's nodal equation:

$$\frac{V_{mn} - V_a}{R_a + R_X} + \frac{V_{mn} - V_b}{R_b + R_Z} + 0 = 0$$

Substituting values:

$$\frac{V_{mn} - 30}{5 + 40} + \frac{V_{mn} - 40}{10 + 50} = 0$$

Rearranging terms:

$$\frac{V_{mn}}{45} + \frac{V_{mn}}{60} = \frac{30}{45} + \frac{40}{60}$$

Grouping fractions and canceling:

$$\frac{60V_{mn} + 45V_{mn}}{(45)(60)} = \frac{(30)(60) + (40)(45)}{(45)(60)}$$

So

$$105V_{mn} = 3600$$

$$V_{mn} = 34.29 \text{ V}$$

Step 3. Match R_L to R_{Th} and find the power delivered to R_L.

The power delivered to the load resistance is calculated from Fig. 2-56 (observe that the load resistance R_L has been made equal to R_{Th} in order to get maximum power transfer):

$$P_T = \frac{V_{\text{Th}}^2}{R_{\text{Th}} + R_L} = \frac{(34.29)^2}{29.01 + 29.01} = \frac{1175.80}{58.02} = 20.3 \text{ W}$$

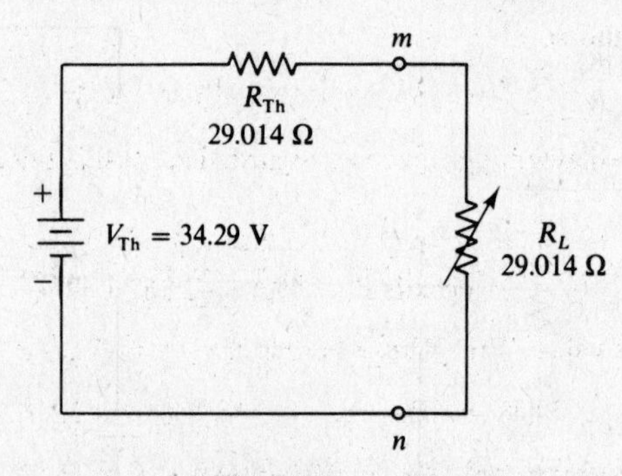

Fig. 2-56

One-half the power delivered will be delivered to the external circuit in this case:

$$P_m = \frac{P_T}{2} = \frac{20.3}{2} = 10.15 \text{ W}$$

where P_m equals maximum power.

2.9 Convert the two-terminal circuit shown in Fig. 2-57 to an equivalent Thevenin generator, and find the true power delivered to a 25-Ω resistor.

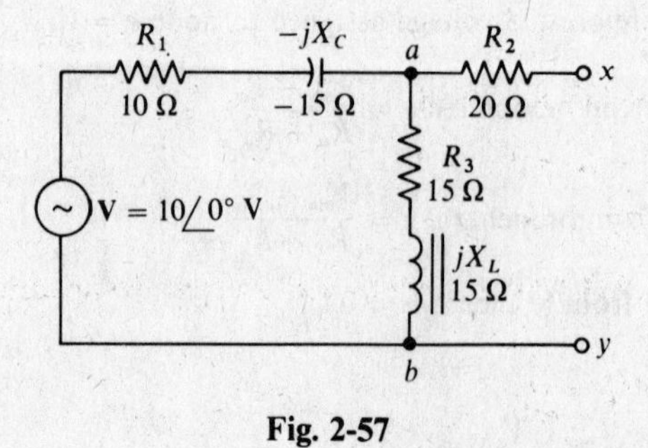

Fig. 2-57

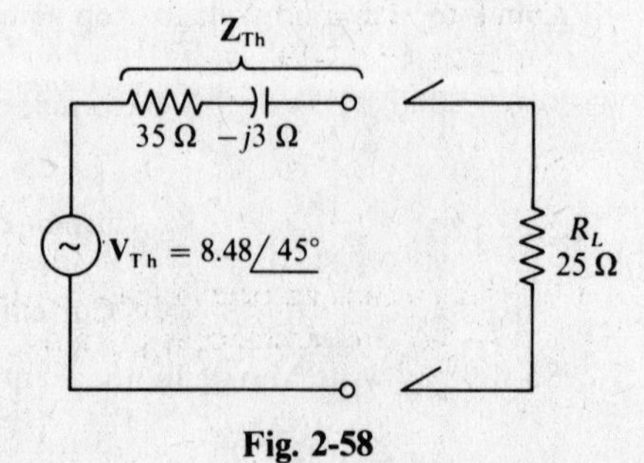

Fig. 2-58

With terminals x and y open, the voltage across terminals a and b can be found (this is the Thevenin voltage):

$$V_{\mathrm{Th}} = \mathbf{V}_{ab} = V\left(\frac{R_3 + jX_L}{(R_1 - jX_C) + (R_3 + jX_L)}\right) = 10\,/0^\circ\left(\frac{15 + j15}{10 - \cancel{j15} + 15 + \cancel{j15}}\right)$$

$$= \frac{(10\,/0^\circ)(21.2\,/45^\circ)}{25} = \frac{212\,/45^\circ}{25} = 8.48\,/45^\circ \text{ V}$$

The generator voltage is given as **V**, which is the symbol for rms voltage. Therefore, the Thevenin voltage, which is obtained by using the generator voltage in an equation, is also an rms voltage.

The Thevenin impedance is obtained by shorting the generator and calculating the impedance from terminal x to terminal y:

$$Z_{\mathrm{Th}} = \mathbf{Z}_{xy} = R_2 + \frac{(R_1 - jX_C)(R_3 + jX_L)}{R_1 - jX_C + R_3 + jX_L}$$

$$= 20 + \frac{(10 - j15)(15 + j15)}{10 - \cancel{j15} + 15 + \cancel{j15}} = 20 + \frac{(18.03\,/-56.3^\circ)(21.2\,/45^\circ)}{25}$$

$$= 20 + 15.3\,/-11.3^\circ = 20 + (15 - j3) = 35 - j3\ \Omega$$

Figure 2-58 shows the Thevenin generator circuit.

To find the true power delivered, we find the circuit current by dividing $\mathbf{V}_{\mathrm{Th}}$ by the circuit **Z**.

$$\text{Circuit } \mathbf{Z} = \mathbf{Z}_{\mathrm{Th}} + R_L$$

But $\mathbf{Z}_{\mathrm{Th}} = 35 - j3$, and so

$$\mathbf{Z} = 35 - j3 + 25 = 60 - j3 = 60.1\,/-2.86^\circ$$

[This angle (-2.86°) is small enough to be disregarded in most applications. It will *not* be disregarded in this example.]

$$\text{Circuit current} = \mathbf{I} = \frac{\mathbf{V}_{\mathrm{Th}}}{\text{circuit } \mathbf{Z}} = \frac{8.48\,/45^\circ}{60.1\,/-2.86^\circ} = 0.141\,/47.9^\circ \text{ A} = 141\,/47.9^\circ \text{ mA}$$

In this case, **I** is an rms value. The voltage across R_L is found next:

$$\mathbf{V}_L = \mathbf{I}R_L = (0.141\,/47.9^\circ)(25) = 3.53\,/47.9^\circ \text{ V}$$

True power delivered to a resistor in an ac circuit is given as

$$P = VI$$

where V = rms voltage across the resistor
I = rms current through the resistor

For our circuit, $V = 3.53$ V and $I = 141$ mA. Therefore

$$P = (3.53)(141 \times 10^{-3}) = 0.498 \text{ W} = 498 \text{ mW}$$

2.10 What values of load components are needed to obtain maximum power from the generator shown in Fig. 2-59?

In Sec. 2.8, the maximum power transfer theorem states that for ac circuits maximum power is transferred when the load impedance Z_L is the complex conjugate of the generator impedance Z_i. Therefore $\mathbf{Z}_L = \mathbf{Z}_i^*$.

From Fig. 2-59,

$$\mathbf{Z}_i = R_i + jX_i = 1.5 + j3$$

So

$$\mathbf{Z}_i^* = R_i - jX_i = 1.5 - j3$$

and

$$\mathbf{Z}_L = \mathbf{Z}_i^* = 1.5 - j3$$

This means we need a load consisting of a 1.5-Ω resistor in series with a capacitive reactance of 3 Ω. Capacitive reactance is given as

$$X_C = \frac{1}{\omega C}$$

Solving for C,

$$C = \frac{1}{\omega X_C}$$

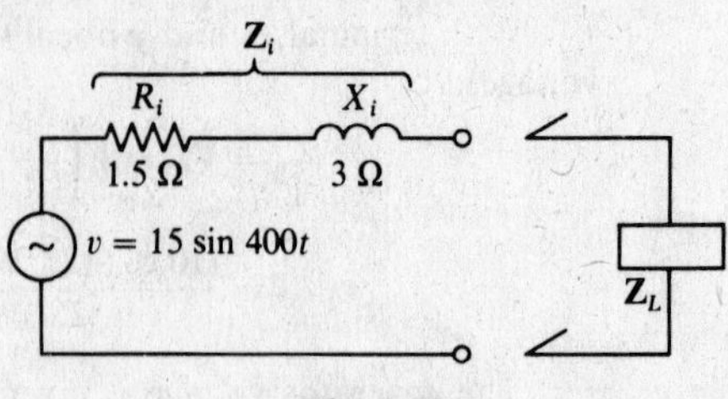

Fig. 2-59

The given voltage is in the form

$$v = V_m \sin \omega t = 15 \sin 400t$$

and so $\omega = 400$ rad/s. Substituting $X_C = 3$ and $\omega = 400$, we obtain

$$C = \frac{1}{(400)(3)} = 8.33 \times 10^{-4} = 833 \times 10^{-6} = 833\ \mu\text{F}$$

2.11 What is the voltage at point a in the circuit shown in Fig. 2-60?

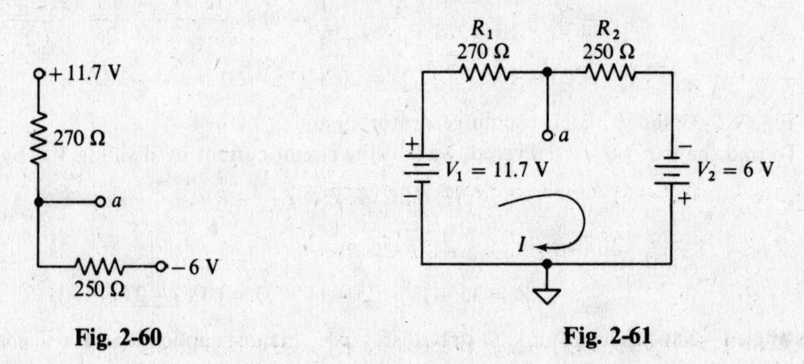

Fig. 2-60

Fig. 2-61

Voltages on schematics are usually given with respect to the ground, or the common point. The best way to solve problems like this is to redraw the circuit and assign callouts as shown in Fig. 2-61. With the assumed loop current shown, the value of current I is found:

$$V_1 - IR_1 - IR_2 + V_2 = 0$$

$$I = \frac{V_1 + V_2}{R_1 + R_2} = \frac{11.7 + 6}{270 + 250} = 0.034\ \text{A} = 34\ \text{mA}$$

The voltage drop across the 270-Ω resistor is

$$V_{R1} = IR_1 = (0.034)(270) = 9.18 \approx 9.2\ \text{V}$$

The voltage at point a is the voltage of V_1 minus the voltage drop across R_1:

$$V_a = V_1 - V_{R1} = 11.7 - 9.2 = 2.5\ \text{V}$$

The answer can be checked by taking the algebraic sum of the voltage across R_2 and V_2.

$$V_{R2} = IR_2 = 0.034 \times 250 = 8.5\ \text{V}$$

$$V_a = V_{R2} - V_2 = 8.5 - 6 = 2.5\ \text{V}$$

2.12 In the circuit shown in Fig. 2-62, find the transfer resistance between terminals a and b and c and d.

To find the transfer resistance, the current through R_4 must be determined. This is assumed current I_1 in the illustration.

To solve for I_1, write the Maxwell mesh matrix

$$\begin{bmatrix} \Phi_1 \\ \Phi_2 \end{bmatrix} = \begin{bmatrix} I_1 \\ I_2 \end{bmatrix} \begin{bmatrix} R_{11} & -R_{12} \\ -R_{21} & R_{22} \end{bmatrix}$$

From Fig. 2-62,

$$\Phi_1 = -2 \text{ V} \qquad R_{11} = 5 + 4 + 3 = 12\ \Omega$$

$$\Phi_2 = +2 - 3 = -1 \text{ V} \qquad R_{22} = 3 + 3 = 6\ \Omega$$

$$R_{12} = R_{21} = 3\ \Omega$$

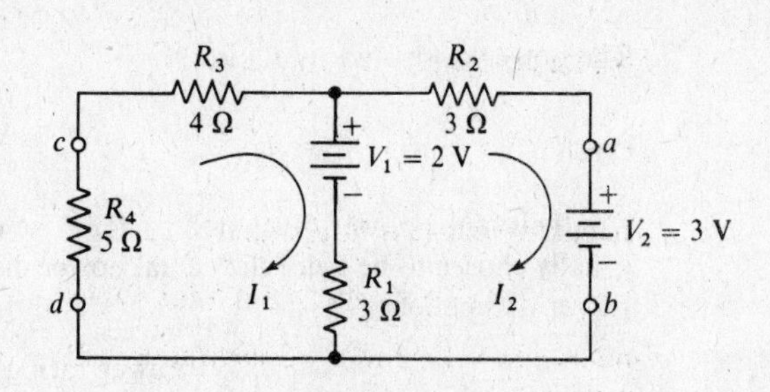

Fig. 2-62

Substituting,

$$\begin{bmatrix} -2 \\ -1 \end{bmatrix} = \begin{bmatrix} I_1 \\ I_2 \end{bmatrix} \begin{bmatrix} 12 & -3 \\ -3 & 6 \end{bmatrix}$$

Therefore

$$I_1 = \frac{\begin{vmatrix} -2 & -3 \\ -1 & 6 \end{vmatrix}}{\Delta}$$

where

$$\Delta = \begin{vmatrix} 12 & -3 \\ -3 & 6 \end{vmatrix} = 72 - 9 = 61$$

and

$$I_1 = -\frac{12 - 3}{61} = -\frac{15}{61} = -0.238 \text{ A}$$

The negative sign means the wrong direction was chosen for I_1.

$$\text{Transfer resistance } R_X = \frac{V_2}{I_1} = \frac{3}{0.238} = 12.6\ \Omega$$

2.13 What should be the power rating of resistor R_4 in the circuit shown in Fig. 2-63?

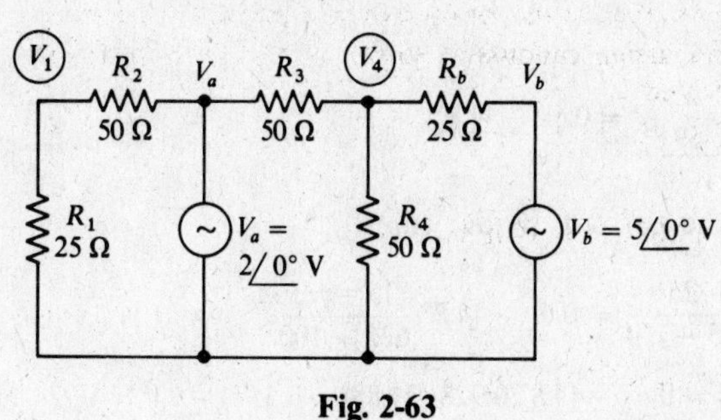

Fig. 2-63

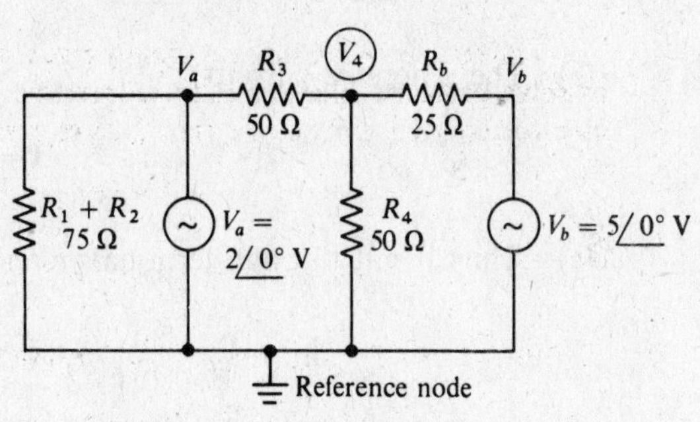

Fig. 2-64

The rms values of voltage are given in polar form, and the generators are shown to be in phase. Nodal analysis can be used to find the voltage across R_4. Assumed voltages for the problem are shown in circles, and the equations *could* be written with these assumptions. However, the problem can be simplified greatly by redrawing the circuit. Voltage V_1 in Fig. 2-63 is not pertinent to the problem. By redrawing and relabeling the problem as shown in Fig. 2-64, the voltage across R_4 is more readily found. Note that branch $R_1 + R_2$ does not enter into the calculation.

By the nodal method:

$$\frac{V_a - V_4}{R_3} - \frac{V_4}{R_4} = \frac{V_4 - V_b}{R_b}$$

$$\frac{2 - V_4}{50} - \frac{V_4}{50} = \frac{V_4 - 5}{25}$$

Multiply both sides of the equation by 50:

$$(2 - V_4) - V_4 = 2V_4 - 10$$

$$4V_4 = 12$$

$$V_4 = 3 \text{ V}$$

The power dissipated by R_4 is

$$P_4 = \frac{(V_4)^2}{R_4} = \frac{9}{50} = 0.18 \text{ W}$$

This is the power dissipated by R_4. The question asks what should be the power *rating*. The rating is usually chosen to be twice the actual power dissipated in normal operation. This provides a *safety factor* for power dissipation:

$$\text{Power rating} = 2P_4 = 2 \times 0.18 = 0.36 \text{ W}$$

The next step is to determine the type of resistor to be used (carbon composition, wire-wound, etc.). The type of resistor to be used is probably not manufactured in 0.36 W sizes, so the next larger size can be chosen. For example, if the standard sizes are 0.25 and 0.5 W, the 0.5-W size would normally be chosen.

2.14 Find the following for the circuit shown in Fig. 2-65.

(*a*) The impedance as "seen" by the generator:

$$\mathbf{Z} = jX_L + \frac{(R)(-jX_C)}{R - jX_C}$$

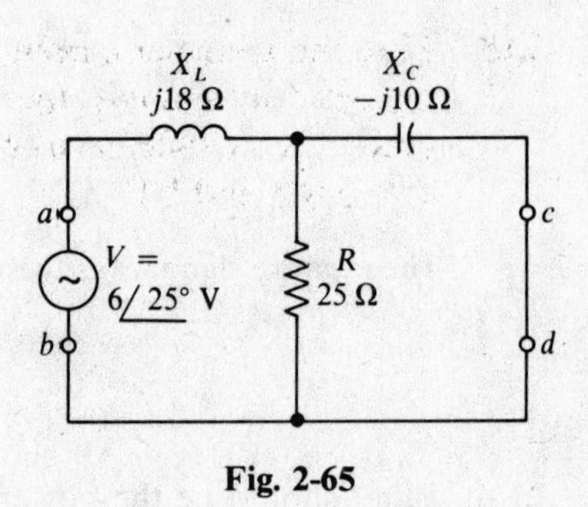

Fig. 2-65

Substitute values:

$$\mathbf{Z} = j18 + \frac{(25)(-j10)}{25 - j10} = j18 + \left[\frac{(25)(-j10)}{25 - j10}\right]\left[\frac{25 + j10}{25 + j10}\right]$$

$$= j18 + \frac{2500 - j6250}{725} = j18 + 3.45 - j8.62$$

$$= 3.45 + j9.38 = 10\underline{/69.8^\circ}$$

(*b*) The generator current:

$$\mathbf{I} = \frac{\mathbf{V}}{\mathbf{Z}} = \frac{6\underline{/25^\circ}}{10\underline{/69.8^\circ}} = 0.6\underline{/-44.8^\circ}$$

(*c*) The current through terminals c and d, by the current divider rule:

$$\mathbf{I}_{c\text{-}d} = \mathbf{I}\frac{R}{R - jX_C} = 0.6\underline{/-44.8^\circ}\frac{25}{25 - j10} = 0.6\underline{/-44.8^\circ}\frac{625 + j250}{625 + 100}$$

$$= 0.6\underline{/-44.8^\circ}(0.862 + j0.345) = 0.6\underline{/-44.8^\circ}(0.928\underline{/21.8^\circ}) = 0.557\underline{/-23^\circ}$$

(*d*) What is the transfer impedance between terminals a and b and c and d?

$$\mathbf{Z}_X = \frac{\mathbf{V}}{\mathbf{I}_{c\text{-}d}} = \frac{6\underline{/25^\circ}}{0.557\underline{/-23^\circ}} = 10.77\underline{/48^\circ}$$

(*e*) Interchange **V** and the short circuit between terminals c and d. Show that the transfer impedance is the same. (See Fig. 2-66).

Z as seen by the generator:

$$\mathbf{Z} = \mathbf{Z}_{c\text{-}d} = -j10 + \frac{(25)(j18)}{25 + j18} = -j10 + \left(\frac{j450}{25 + j18}\right)\left(\frac{25 - j18}{25 - j18}\right)$$

$$= 8.535 + j1.85 = 8.73\underline{/12.23^\circ}$$

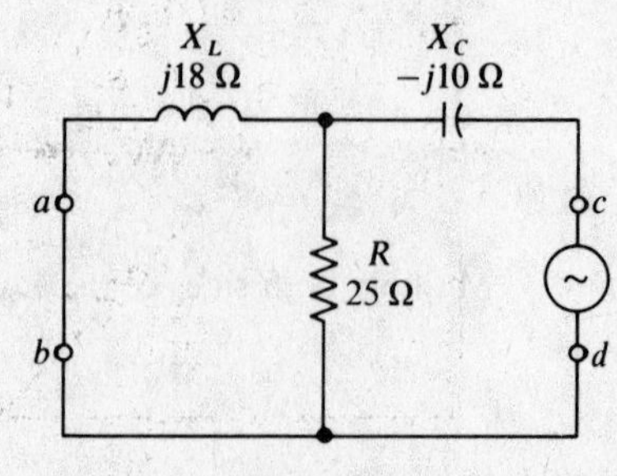

Fig. 2-66

Generator current:

$$\mathbf{I} = \frac{\mathbf{V}}{\mathbf{Z}} = \frac{6\underline{/25^\circ}}{8.73\underline{/12.23^\circ}} = 0.687\underline{/12.77^\circ}$$

Current through terminals c and d, by the current divider rule:

$$\mathbf{I}'_{c\text{-}d} = \mathbf{I}\left(\frac{R}{R + jX_L}\right) = 0.687\,\underline{/12.77^\circ}\left(\frac{25}{25 + j18}\right) = 0.687\,\underline{/12.77^\circ}(0.659 - j0.474)$$

$$= 0.687\,\underline{/12.77^\circ}(0.812\,\underline{/-35.73^\circ}) = 0.558\,\underline{/-22.96^\circ}$$

Transfer impedance:

$$\mathbf{Z}'_X = \frac{\mathbf{V}}{\mathbf{I}'_{c\text{-}d}} = \frac{6\,\underline{/25^\circ}}{0.558\,\underline{/-22.96^\circ}} = 10.75\,\underline{/47.96^\circ}$$

This value is substantially the same as the value $\mathbf{Z}_X = 10.77\,\underline{/48^\circ}$ that was found before.

(*f*) Which network theorem has been demonstrated in this problem?

The reciprocity theorem as applied to ac circuits.

2.15 Find the ammeter current in the circuit shown in Fig. 2-67. (*Hint:* Interchange the battery and ammeter to simplify the circuit. This is permitted by the reciprocity theorem. The ammeter current is the difference between the currents through R_1 and R_3 in the modified circuit shown in Fig. 2-68*a* and *b*.)

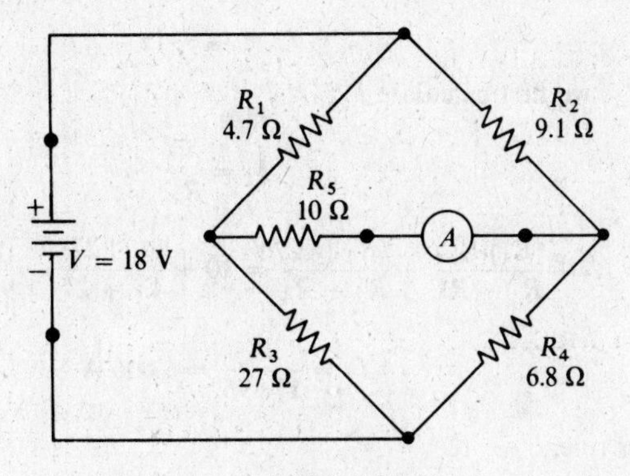

Fig. 2-67

We may solve for I_1 and I_2 by Maxwell's mesh matrix:

$$\begin{bmatrix} \Phi_1 \\ \Phi_2 \\ \Phi_3 \end{bmatrix} = \begin{bmatrix} I_T \\ I_1 \\ I_2 \end{bmatrix} \begin{bmatrix} R_{11} & -R_{12} & -R_{13} \\ -R_{21} & R_{22} & -R_{23} \\ -R_{31} & -R_{32} & R_{33} \end{bmatrix}$$

where $\Phi_1 = 18$ V $\qquad R_{33} = 6.8 + 9.1 = 15.9\ \Omega$

$\Phi_2 = \Phi_3 = 0 \qquad R_{12} = R_{21} = 27\ \Omega$

$R_{11} = 10 + 27 + 6.8 = 43.8\ \Omega \qquad R_{13} = R_{31} = 6.8\ \Omega$

$R_{22} = 27 + 4.7 = 31.7\ \Omega \qquad R_{23} = R_{32} = 0$

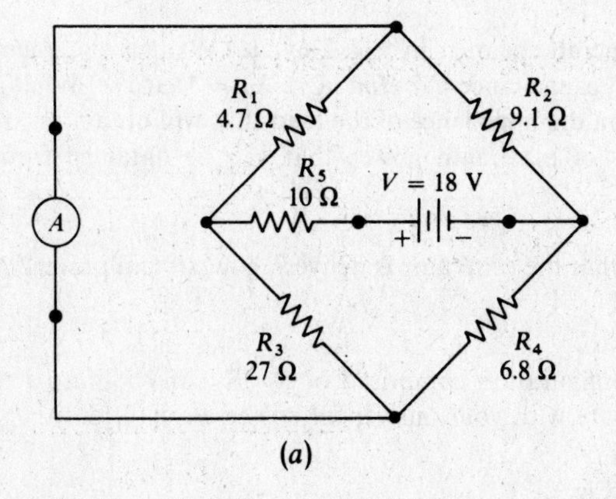

(*a*)

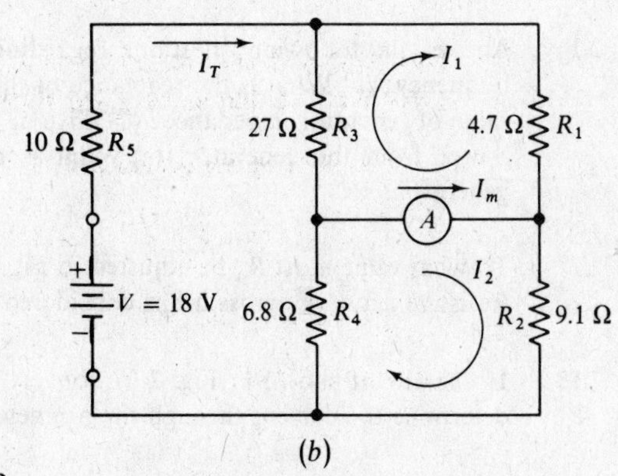

(*b*)

Fig. 2-68

Therefore

$$\begin{bmatrix} 18 \\ 0 \\ 0 \end{bmatrix} = \begin{bmatrix} I_T \\ I_1 \\ I_2 \end{bmatrix} \begin{bmatrix} 43.8 & -27 & -6.8 \\ -27 & 31.7 & 0 \\ -6.8 & 0 & 15.9 \end{bmatrix}$$

$$\Delta = (-6.8)(6.8)(31.7) + 15.9[(43.8)(31.7) - (-27)^2] = 9019.6$$

and

$$I_1 = \frac{\begin{vmatrix} 43.8 & 18 & -6.8 \\ -27 & 0 & 0 \\ -6.8 & 0 & 15.9 \end{vmatrix}}{\Delta} = \frac{-18(-27)(15.9)}{9019.6} = 0.857 \text{ A}$$

$$I_2 = \frac{\begin{vmatrix} 43.8 & -27 & 18 \\ -27 & 31.7 & 0 \\ -6.8 & 0 & 0 \end{vmatrix}}{\Delta} = \frac{18(6.8)(31.7)}{9019.6} = 0.430 \text{ A}$$

So $$I_m = I_2 - I_1 = 0.430 - 0.857 = -0.427 \text{ A}$$

We could also solve this problem by using Kirchhoff's current rule to write

$$I_1 + I_m = I_2$$

To find I_1 and I_2, we first calculate I_T:

$$I_T = \frac{V}{R_T}$$

where $$R_T = R_5 + \frac{(R_1)(R_3)}{R_1 + R_3} + \frac{(R_2)(R_4)}{R_2 + R_4} = 10 + \frac{(4.7)(27)}{4.7 + 27} + \frac{(9.1)(6.8)}{9.1 + 6.8} = 17.895 \ \Omega$$

Therefore $$I_T = \frac{18}{17.895} = 1.006 \text{ A}$$

By the current divider rule,

$$I_1 = \left(\frac{R_3}{R_1 + R_3}\right) I_T = \left(\frac{27}{4.7 + 27}\right) 1.006 = 0.857 \text{ A}$$

and $$I_2 = \left(\frac{R_4}{R_2 + R_4}\right) I_T = \left(\frac{6.8}{9.1 + 6.8}\right) 1.006 = 0.430 \text{ A}$$

So $$I_m = I_2 - I_1 = 0.430 - 0.857 = -0.427 \text{ A} \quad \text{as before}$$

Supplementary Problems

2.16 Answer the following questions regarding the generator shown in Fig. 2-69: (*a*) What is the generator frequency? (*b*) What is the reactance of the internal capacitance C_i? *Hint:* Use $X_C = 1/\omega C$. (*c*) What is the value of generator impedance? (*d*) Give in polar form the impedance of the load that will obtain maximum power from the generator. (*e*) What is the value of maximum power that can be obtained from the generator?

2.17 To what value must R_X be adjusted in Fig. 2-70 so that the generator is delivering maximum power? *Note:* Resistive network across *a-b* is considered the load.

2.18 In the circuit shown in Fig. 2-67 convert the T configuration comprised of R_1, R_3, and R_5 into a Δ and determine the current through the ammeter. Compare with your answer for solved Prob. 2.15.

2.19 How much current is there through resistor R_3 in Fig. 2-71?

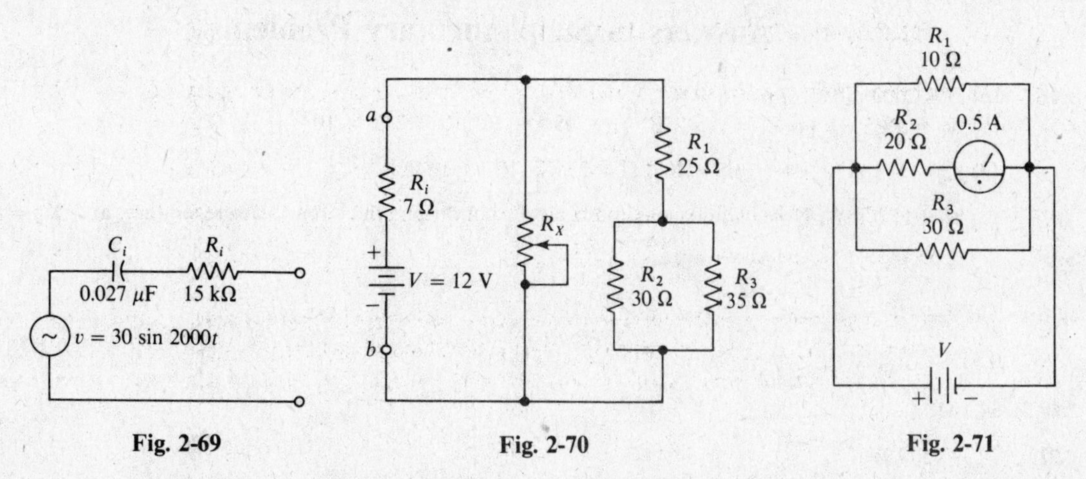

Fig. 2-69　　Fig. 2-70　　Fig. 2-71

2.20 How much resistance must be placed in series with the battery in Fig. 2-71 in order to limit the battery current to 0.25 A?

2.21 How much power is being dissipated by R_3 in the circuit shown in Fig. 2-72?

2.22 How much current is V_2 supplying to the circuit shown in Fig. 2-72?

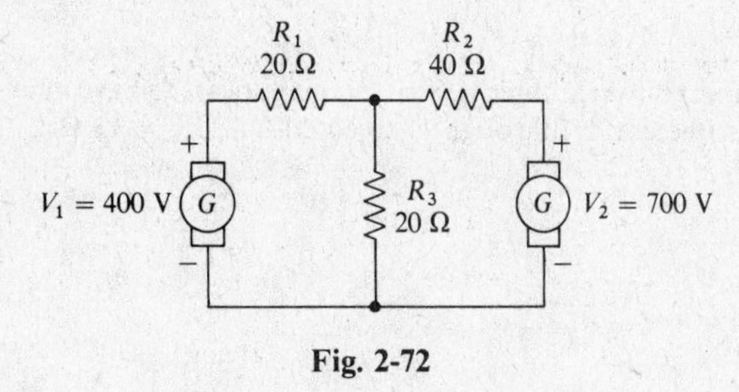

Fig. 2-72

2.23 What is the efficiency of a generator system if its load resistance is twice the value of its internal resistance?

2.24 How much load resistance is needed for a generator having an internal resistance of R_A so that the efficiency of the system is 75 percent?

2.25 Find the Norton generator for the circuit shown in Fig. 2-73.

2.26 Show that a 10-Ω resistor draws the same current from the Norton generator in the solution of Prob. 2-25 and the circuit shown in Fig. 2-73.

2.27 What is the resistance between a and b when R_X is adjusted so that no current flows through V_b in Fig. 2-74?

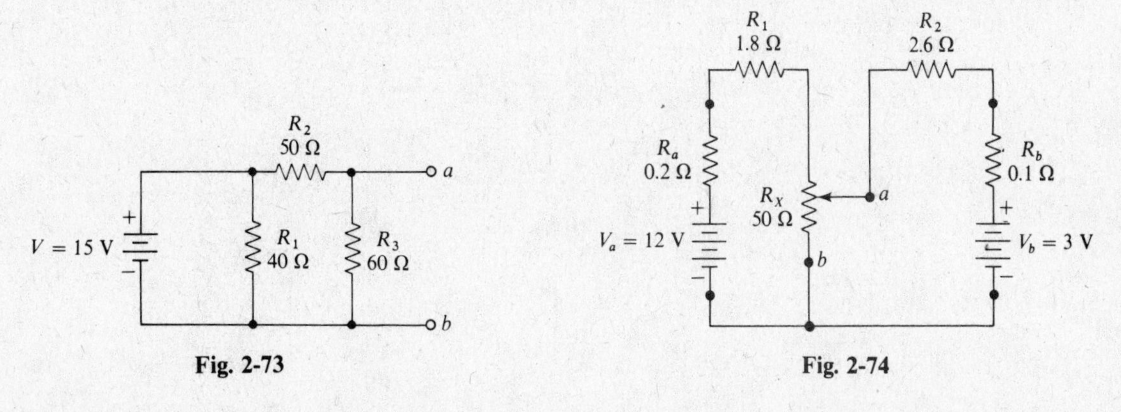

Fig. 2-73　　Fig. 2-74

Answers to Supplementary Problems

2.16 (*a*) $\omega = 2000$ rad/s, $f = 318.3$ Hz.
(*b*) $X_C = 18.5$ kΩ. (*c*) $\mathbf{Z}_L = (15 \times 10^3) - (j18.5 \times 10^3)\ \Omega = 23.8 \times 10^3 \underline{/-51^\circ}\ \Omega$.

(*d*) $\mathbf{Z}_L^* = (15 \times 10^3) + (j18.5 \times 10^3)\ \Omega = 23.8 \times 10.3 \underline{/51^\circ}\ \Omega$.

(*e*) $P = 15.0$ mW. *Note:* Under maximum power conditions the circuit is in resonance, and $X_L = X_C$.

2.17 $R_X = 8.44\ \Omega$.

2.18 $I = 0.428$ A $= 428$ mA.

2.19 0.333 A or $\frac{1}{3}$ A.

2.20 34.5 Ω.

2.21 $P_3 = 4500$ W.

2.22 14 A.

2.23 Efficiency $= 66\frac{2}{3}$ percent.

2.24 $R_L = 3R_a$.

2.25 $I_N = 0.3$ A; $R_N = 27.3\ \Omega$.

2.26 0.22 A in both cases.

2.27 Remember, if no current flows through V_b, that circuit appears to have infinite impedance. The voltage from a to b must be 3 V for the current through V_b to equal 0 A. $R_{ab} = 13\ \Omega$.

Chapter 3

Four-Terminal Networks

3.1 INTRODUCTION

The Thevenin and Norton generators described in Chap. 2 are sometimes referred to as *two-terminal networks.* This simply implies that there are two terminals by which to gain entrance to the circuit. Two-terminal networks are also referred to as *one-port networks.*

This chapter will cover *four-terminal networks*, or, as they are sometimes called, *two-port networks.* Figure 3-1 shows the symbol. One pair of the terminals serves as the voltage, current, or signal input and the other pair as the output for voltage, current, or signal. Networks like the one shown in Fig. 3-1 are sometimes referred to as *black boxes.*

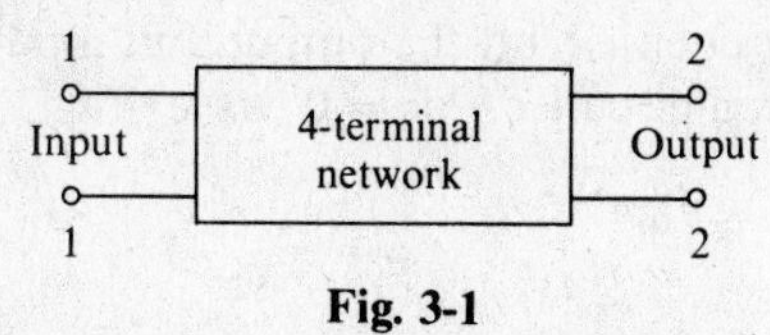

Fig. 3-1

The markings in Fig. 3-1 for the input terminals (1-1) and the output terminals (2-2) are used extensively in electronics literature.

Amplifying components, such as tubes and transistors, have an input signal and an output signal. Therefore, they are examples of two-port networks and can be analyzed by four-terminal techniques. A tube or transistor is considered to be a source of signal within the four-terminal network, so it is an example of an *active four-terminal network.*

It is possible to discuss amplifiers from the standpoint of four-terminal analysis because an amplifier is basically an active four-terminal network which follows the theory discussed in this chapter. An advantage of using the four-terminal approach is that the amplifying component becomes secondary to the circuit design. If new amplifying components are invented, the design procedures will not be changed provided the new components can be represented as active four-terminal networks.

Passive four-terminal networks have no source of voltage, current, or power. They are made up of networks having resistors, inductors, capacitors, and transformers. Transmission lines are examples of passive four-terminal networks which deliver RF energy from one point to another. Two other examples of passive four-terminal networks are

- *Filter circuits*, used for smoothing dc voltages and for eliminating some frequencies while passing others.
- *Attenuators* and *pads*, used for reducing a signal by some desired amount. The signal is fed to the attenuator or pad at the input terminals and appears at the output terminals in the same form but with reduced amplitude.

In this chapter we will discuss some common sets of four-terminal network parameters. These are the z, y, h, and $ABCD$ (sometimes called a) parameters. We will also mention some special applications of four-terminal parameters.

All the parameters are defined for ac voltages and currents and include dc voltages and currents as a special case. The parameters are also defined for active networks with bilateral or passive networks included as a special case.

3.2 IMPEDANCE PARAMETERS (z PARAMETERS)

The four-terminal network is usually set up as shown in Fig. 3-2 with v_1, v_2, i_1, and i_2 measured in the directions shown.

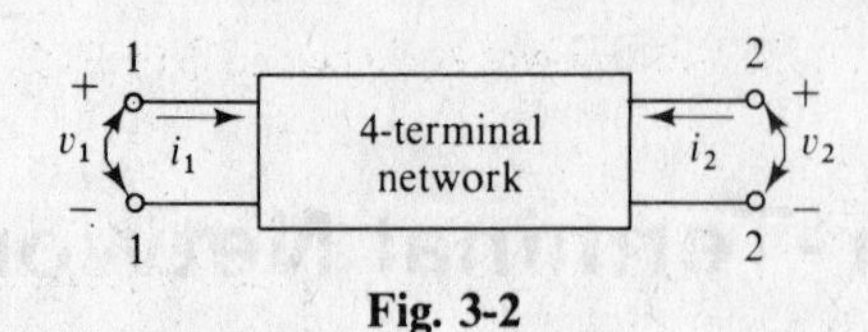

Fig. 3-2

With these directions, the general impedance parameters relate the input and output voltages and currents according to the following equations:

$$v_1 = z_{11}i_1 + z_{12}i_2 \tag{3-1}$$

$$v_2 = z_{21}i_1 + z_{22}i_2 \tag{3-2}$$

or in matrix form

$$\begin{bmatrix} v_1 \\ v_2 \end{bmatrix} = \begin{bmatrix} i_1 \\ i_2 \end{bmatrix} \begin{bmatrix} z_{11} & z_{12} \\ z_{21} & z_{22} \end{bmatrix} \tag{3-3}$$

To obtain the *z parameters*, we open-circuit the output and input terminals sequentially.

With the output terminals open-circuited, $i_2 = 0$ (see Fig. 3-3) and Eq. (*3-1*) becomes $v_1 = z_{11}i_1$. Solving for z_{11},

$$z_{11} = \left.\frac{v_1}{i_1}\right|_{i_2=0} \quad \Omega \tag{3-4}$$

Also, Eq. (*3-2*) becomes $v_2 = z_{21}i_1$. Solving for z_{21},

$$z_{21} = \left.\frac{v_2}{i_1}\right|_{i_2=0} \quad \Omega \tag{3-5}$$

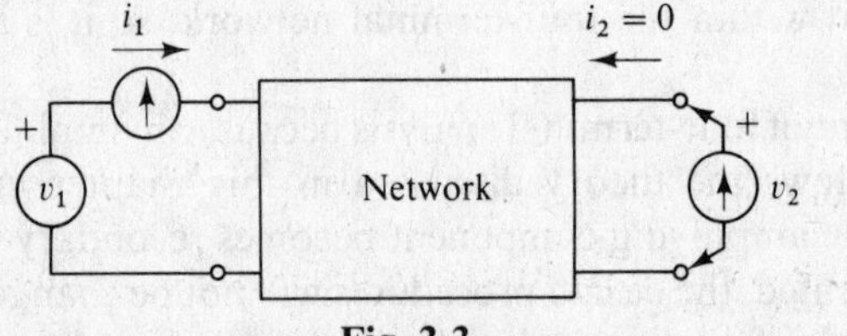

Fig. 3-3

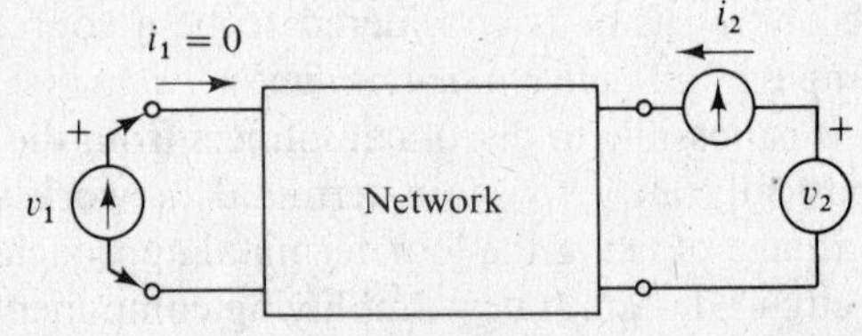

Fig. 3-4

With the input terminals open-circuited, $i_1 = 0$ (see Fig. 3-4) and Eq. (*3-1*) becomes $v_1 = z_{12}i_2$ and

$$z_{12} = \left.\frac{v_1}{i_2}\right|_{i_1=0} \quad \Omega \tag{3-6}$$

We may also write Eq. (*3-2*) as $v_2 = z_{22}i_2$ to get

$$z_{22} = \left.\frac{v_2}{i_2}\right|_{i_1=0} \quad \Omega \tag{3-7}$$

The z parameters are *open-circuit impedance parameters* and have the dimensions of (are measured in) ohms:

- z_{11} is called the *open-circuit input impedance* because it relates the input voltage to the input current.
- z_{21} is called the *open-circuit forward-transfer impedance*. "Forward" refers to the fact that we are considering an output-over-input ratio, which is the usual case. "Transfer" means that output and input quantities are considered.
- z_{12} is called the *open-circuit reverse-transfer impedance*. The term "reverse" is used to indicate that the ratio is not the normal output over input but the reverse, input over output. "Transfer" again signifies that both input and output quantities are used.
- z_{22} is called the *open-circuit output impedance* because it relates the output voltage to the output current.

Example 3.1 Find the z parameters of the network shown in Fig. 3-5.

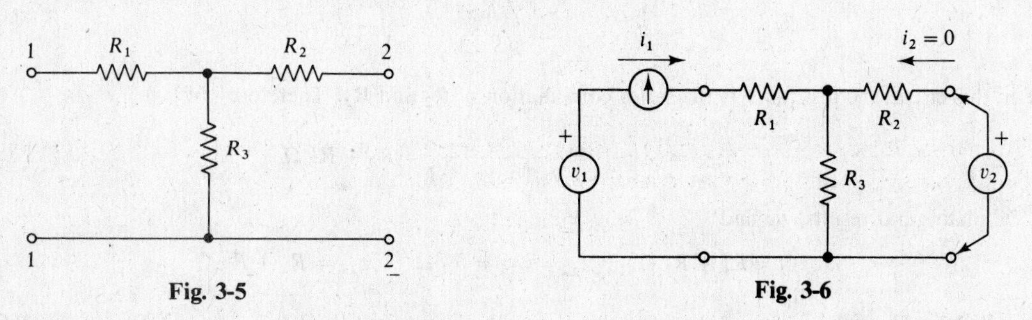

Fig. 3-5

Fig. 3-6

Using the definitions of the z parameters, we first open-circuit the output terminals and measure the voltage and current at the input terminals as shown in Fig. 3-6.

With v_1 the driving voltage, we may (by Ohm's law) write i_1 in terms of v_1 as

$$i_1 = \frac{v_1}{R_1 + R_3}$$

since the total resistance is the series combination of R_1 and R_3. Restating Eq. (3-4):

$$z_{11} = \left.\frac{v_1}{i_1}\right|_{i_2=0}$$

Substituting for i_1, we have

$$z_{11} = \frac{\cancel{v_1}}{\cancel{v_1}/(R_1 + R_3)} = R_1 + R_3 \ \Omega$$

Restating Eq. (3-5):

$$z_{21} = \left.\frac{v_2}{i_1}\right|_{i_2=0}$$

We see from Fig. 3-6 that there is no voltage drop across R_2, so v_2 may be written in terms of i_1 as $v_2 = i_1 R_3$. Substituting into Eq. (3-5), we have

$$z_{21} = \left.\frac{v_2}{i_1}\right|_{i_2=0} = \frac{\cancel{i_1} R_3}{\cancel{i_1}} = R_3 \ \Omega$$

As shown in Fig. 3-7, we now open-circuit the input terminals and write v_1 in terms of i_2 as $v_1 = i_2 R_3$ since in this case there is no voltage drop across R_1. Restating Eq. (3-6):

$$z_{12} = \left.\frac{v_1}{i_2}\right|_{i_1=0}$$

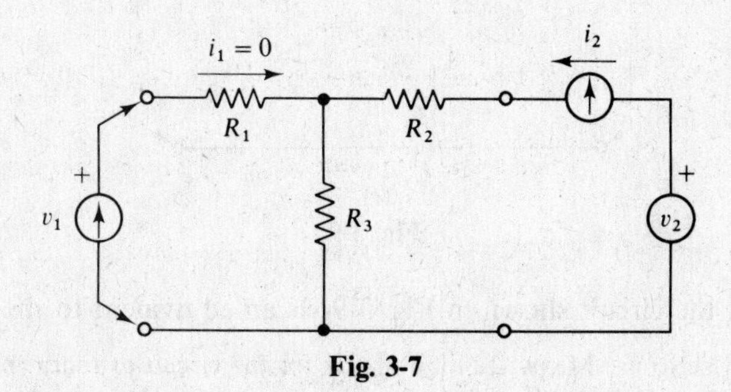

Fig. 3-7

Substituting for v_1 we have

$$z_{12} = \frac{\cancel{i_2} R_3}{\cancel{i_2}} = R_3 \ \Omega$$

Finally, by Ohm's law, writing i_2 in terms of v_2, we have

$$i_2 = \frac{v_2}{R_2 + R_3}$$

since in this circuit the resistance is the series combination of R_2 and R_3. Therefore (3-7),

$$z_{22} = \frac{v_2}{i_2}\bigg|_{i_1=0} = \frac{v_2}{v_2/(R_2 + R_3)} = R_2 + R_3\ \Omega$$

Tabulating our results, we find

$$z_{11} = R_1 + R_3\ \Omega \qquad z_{12} = z_{21} = R_3\ \Omega \qquad z_{22} = R_2 + R_3\ \Omega$$

Example 3.2 Find the z parameters for the network shown in Fig. 3-8.

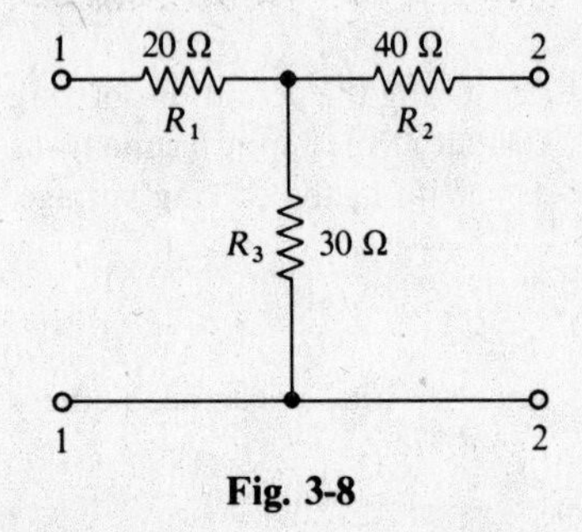

Fig. 3-8

From Fig. 3-8, we see that

$$R_1 = 20\ \Omega \qquad R_2 = 40\ \Omega \qquad R_3 = 30\ \Omega$$

Utilizing the results of Example 3.1,

$$z_{11} = R_1 + R_3 = 20 + 30 = 50\ \Omega$$

$$z_{21} = z_{12} = R_3 = 30\ \Omega$$

$$z_{22} = R_2 + R_3 = 40 + 30 = 70\ \Omega$$

Any bilateral, linear two-port, active or passive, can be defined in terms of the z parameters, and we may substitute any of the equivalent circuits shown in Fig. 3-9 for the black box. Notice that in each of the equivalent networks we are required to use a *current*-dependent voltage source.

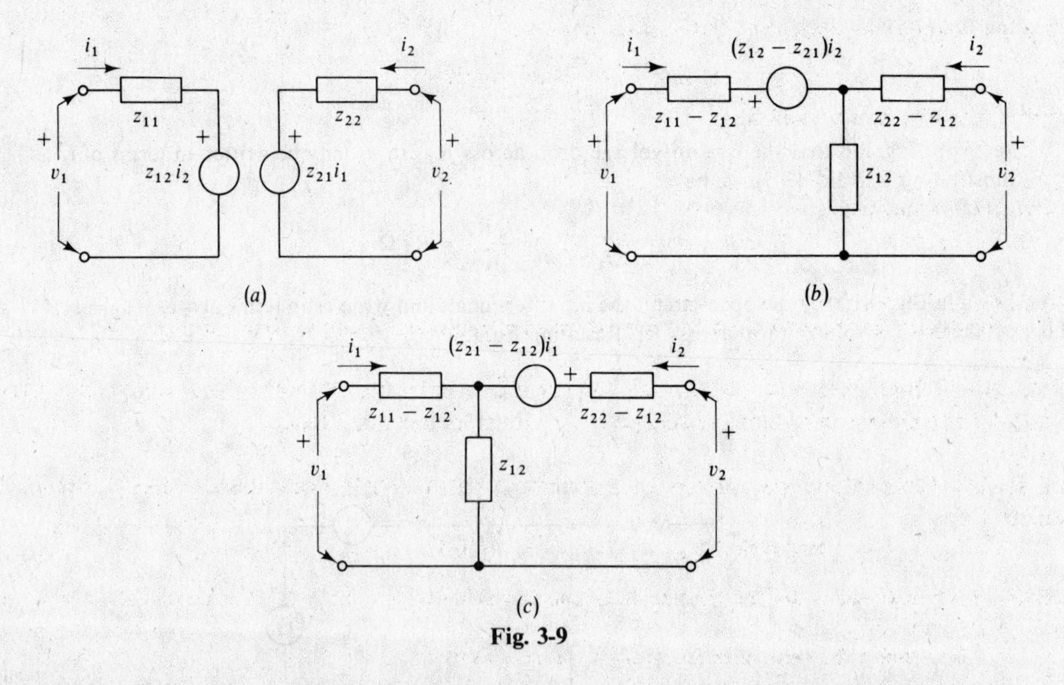

Fig. 3-9

Example 3.3 Prove that the circuit shown in Fig. 3-9c is an equivalent to the general two-port.

Our approach here will be to use Maxwell's loop matrix for the circuit as redrawn in Fig. 3-10 and see if the resulting matrix checks with Eqs. (3-1) and (3-2). When writing the matrix, take careful note that i_2 is opposite to the Maxwell conventional clockwise i_2.

$$\begin{bmatrix} \phi_1 \\ \phi_2 \end{bmatrix} = \begin{bmatrix} i'_1 \\ i'_2 \end{bmatrix} \begin{bmatrix} Z_{11} & -Z_{12} \\ -Z_{21} & Z_{22} \end{bmatrix}$$

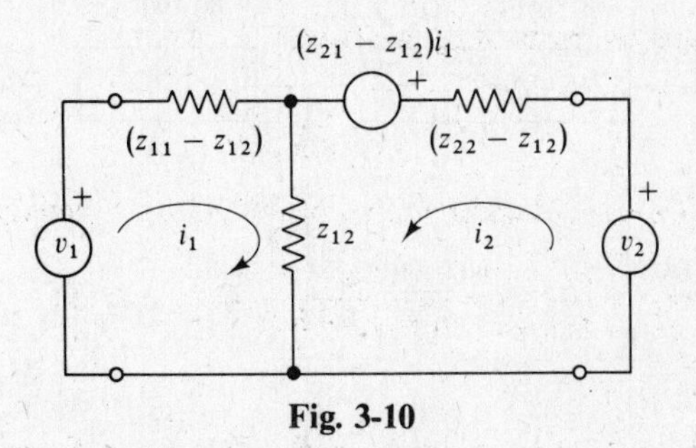

Fig. 3-10

Looking at Fig. 3-10, we see that

$$\phi_1 = v_1 \qquad \phi_2 = (z_{21} - z_{12})i_1 - v_2$$

$$i'_1 = i_1 \qquad i'_2 = -i_2$$

$$Z_{11} = (z_{11} - z_{12}) + z_{12} = z_{11}$$

$$Z_{21} = Z_{12} = z_{12}$$

$$Z_{22} = (z_{22} - z_{12}) + z_{12} = z_{22}$$

Substituting,

$$\begin{bmatrix} v_1 \\ (z_{21} - z_{12})i_1 - v_2 \end{bmatrix} = \begin{bmatrix} i_1 \\ -i_2 \end{bmatrix} \begin{bmatrix} z_{11} & -z_{12} \\ -z_{12} & z_{22} \end{bmatrix}$$

Writing the first equation:

$$v_1 = z_{11}i_1 + z_{12}i_2$$

which is the same as Eq. (*3-1*). The second equation is

$$(z_{21} - z_{12})i_1 - v_2 = -z_{12}i_1 - z_{22}i_2$$

Multiplying out and changing all signs,

$$-z_{21}i_1 + \cancel{z_{12}i_1} + v_2 = \cancel{z_{12}i_1} + z_{22}i_2$$

Canceling and solving for v_2,

$$v_2 = z_{21}i_1 + z_{22}i_2$$

which is the same as Eq. (*3-2*).

This proves that any two-port can be replaced with the equivalent T circuit shown in Fig. 3-9*c*.

We also see that when $z_{12} \neq z_{21}$ we have an active network, and when $z_{12} = z_{21}$ the current-dependent voltage source reduces to zero and we have a passive network.

Example 3.4 Given the z parameters of a four-terminal network, draw the equivalent circuit. The parameters are

$$z_{11} = 30\ \Omega \qquad z_{12} = z_{21} = -j40\ \Omega \qquad z_{22} = j50\ \Omega$$

Although we may select any of the equivalent networks shown in Fig. 3-9, let us choose the one we have analyzed in Example 3.3.

In Fig. 3-9*c*, replace the general terms with the values specified:

$$z_{11} - z_{12} = 30 - (-j40) = 30 + j40 = 50\underline{/53.2^\circ}$$

$$z_{22} - z_{12} = j50 - (-j40) = j90 = 90\underline{/90^\circ}$$

$$z_{21} - z_{12} = (-j40) - (-j40) = 0$$

$$z_{12} = -j40 = 40\underline{/-90^\circ}$$

and the equivalent circuit becomes the network shown in Fig. 3-11.

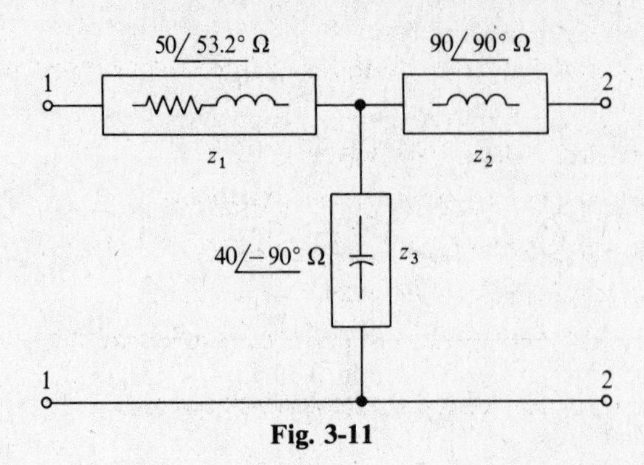

Fig. 3-11

3.3 ADMITTANCE PARAMETERS (*y* PARAMETERS)

By duality, we may also write the input and output voltage and current relationships of any four-terminal network in the form of admittances. Using the principles of duality, we may easily rewrite Eqs. (*3-1*) through (*3-3*) as follows:

$$i_1 = y_{11}v_1 + y_{12}v_2 \tag{3-8}$$

$$i_2 = y_{21}v_1 + y_{22}v_2 \tag{3-9}$$

and in matrix form,

$$\begin{bmatrix} i_1 \\ i_2 \end{bmatrix} = \begin{bmatrix} v_1 \\ v_2 \end{bmatrix} \begin{bmatrix} y_{11} & y_{12} \\ y_{21} & y_{22} \end{bmatrix} \tag{3-10}$$

As expected, to obtain the *y parameters*, we short-circuit the output and input terminals sequentially.

With the output terminals short-circuited, $v_2 = 0$ (see Fig. 3-12) and Eq. (*3-8*) can be written $i_1 = y_{11}v_1$. Solving for y_{11},

$$y_{11} = \frac{i_1}{v_1}\bigg|_{v_2=0} \quad \text{S} \tag{3-11}$$

Under the same conditions, Eq. (*3-9*) is written $i_2 = y_{21}v_1$. So

$$y_{21} = \frac{i_2}{v_1}\bigg|_{v_2=0} \quad \text{S} \tag{3-12}$$

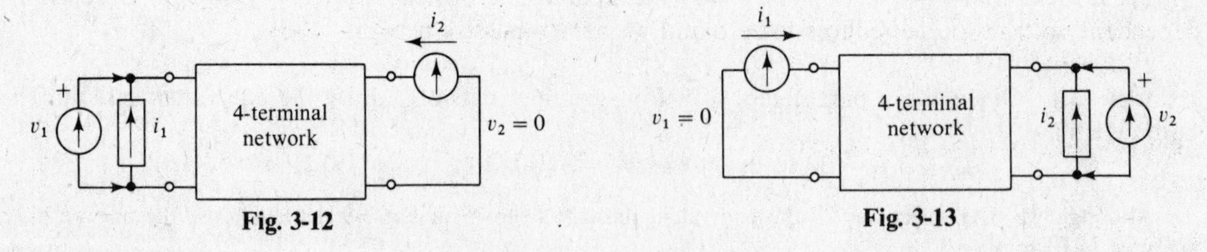

Fig. 3-12 Fig. 3-13

With the input terminals short-circuited, $v_1 = 0$ (see Fig. 3-13), and we write Eq. (*3-8*) as $i_1 = y_{12}v_2$. Therefore,

$$y_{12} = \frac{v_2}{i_1}\bigg|_{v_1=0} \quad \text{S} \tag{3-13}$$

Also, Eq. (*3-9*) becomes $i_2 = y_{22}v_2$. So

$$y_{22} = \frac{i_2}{v_2}\bigg|_{v_1=0} \quad \text{S} \tag{3-14}$$

The y parameters are known as the *short-circuit admittance parameters* and are measured in siemens (S):

- y_{11} is called the *short-circuit input admittance.*
- y_{21} is called the *short-circuit forward-transfer admittance.*
- y_{12} is called the *short-circuit reverse-transfer admittance.*
- y_{22} is called the *short-circuit output admittance.*

The terms "forward," "reverse," and "transfer" have the same meanings as with the z parameters.

Example 3.5 Find the y parameters of the network shown in Fig. 3-14.

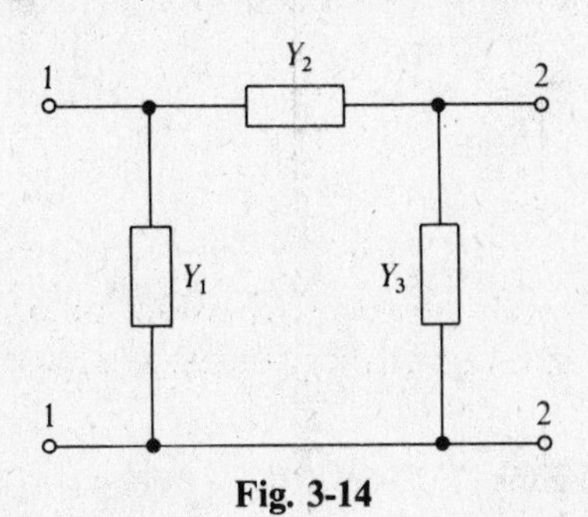

Fig. 3-14

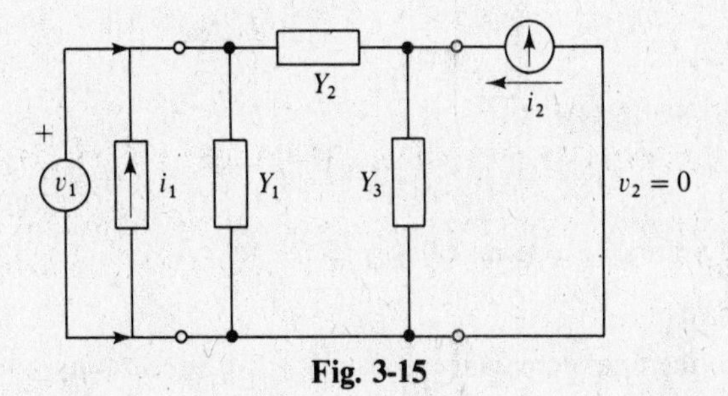

Fig. 3-15

Using the definitions for the y parameters, we first short-circuit the output terminals (see Fig. 3-15). Restating (*3-11*):

$$y_{11} = \left.\frac{i_1}{v_1}\right|_{v_2=0}$$

Since the short circuit "removes" Y_3, the total admittance is the combination of Y_1 and Y_2 in parallel, so

$$i_1 = Y_t v_1 = (Y_1 + Y_2)v_1$$

Substituting into Eq. (*3-11*) gives

$$y_{11} = \frac{(Y_1 + Y_2)\cancel{v_1}}{\cancel{v_1}} = Y_1 + Y_2 \text{ S}$$

Next, according to (*3-12*):

$$y_{21} = \left.\frac{i_2}{v_1}\right|_{v_2=0}$$

Using the current divider rule to write the current i_2 through Y_2, we have

$$i_2 = -\left(\frac{Y_2}{Y_1 + Y_2}\right)i_1$$

The negative sign results from Fig. 3-15, which shows i_2 in the opposite direction to i_1.

Writing v_1 in terms of i_1,

$$i_1 = Y_t v_1$$

But Y_t was found to be $Y_1 + Y_2$, so

$$i_1 = (Y_1 + Y_2)v_1$$

Rearranging,

$$v_1 = \frac{1}{Y_1 + Y_2}i_1$$

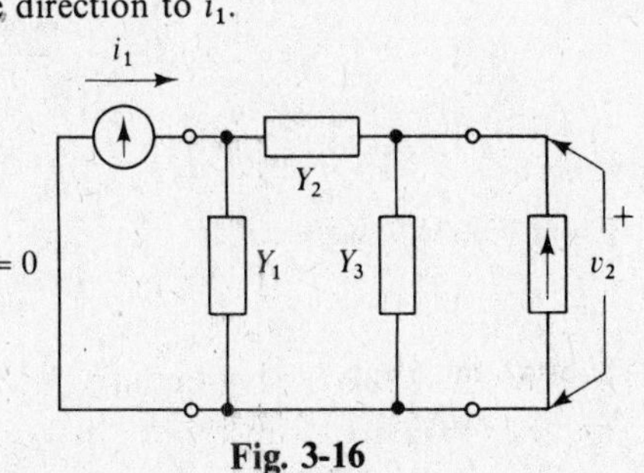

Fig. 3-16

Substituting into Eq. (*3-12*) gives

$$y_{21} = \frac{-[Y_2/(Y_1 + Y_2)]\cancel{i_1}}{[1/(Y_1 + Y_2)]\cancel{i_1}} = \frac{-Y_2}{\cancel{Y_1 + Y_2}}\,\frac{\cancel{Y_1 + Y_2}}{1} = -Y_2 \text{ S}$$

Now let us short-circuit the input terminals as shown in Fig. 3-16. According to (*3-13*):

$$y_{12} = \left.\frac{i_1}{v_2}\right|_{v_1=0}$$

We may use the current divider rule to write the current i_1 in terms of i_2:

$$i_1 = \left(\frac{-Y_2}{Y_2 + Y_3}\right) i_2$$

To write v_2 in terms of i_2, we note that the short circuit eliminates Y_1, so the total admittance is the parallel combination of Y_2 and Y_3. Rearranging,

$$i_2 = (Y_2 + Y_3) v_2$$

$$v_2 = \left(\frac{1}{Y_2 + Y_3}\right) i_2$$

Substituting into Eq. (*3-13*) gives

$$y_{12} = \frac{-[Y_2/(Y_2 + Y_3)]\cancel{i_2}}{1/(Y_2 + Y_3)\cancel{i_2}} = \frac{-Y_2}{\cancel{Y_2 + Y_3}} \frac{\cancel{Y_2 + Y_3}}{1} = -Y_2 \text{ S}$$

Finally, taking (*3-14*):

$$y_{22} = \left.\frac{i_2}{v_2}\right|_{v_1 = 0}$$

And writing i_2 in terms of v_2 gives us

$$i_2 = (Y_2 + Y_3) v_2$$

since the total admittance is just $Y_2 + Y_3$. Substituting into Eq. (*3-14*) gives

$$y_{22} = \frac{(Y_2 + Y_3)\cancel{v_2}}{\cancel{v_2}} = Y_2 + Y_3 \text{ S}$$

Tabulating our results, we find

$$y_{11} = Y_1 + Y_2 \text{ S} \qquad y_{12} = y_{21} = -Y_2 \text{ S} \qquad y_{22} = Y_2 + Y_3 \text{ S}$$

Example 3.6 Find the y parameters of the network shown in Fig. 3-17.

Making use of the results of Example 3.5,

$$y_{11} = Y_1 + Y_2 \text{ S} \qquad y_{21} = y_{12} = -Y_2 \text{ S} \qquad y_{22} = Y_2 + Y_3 \text{ S}$$

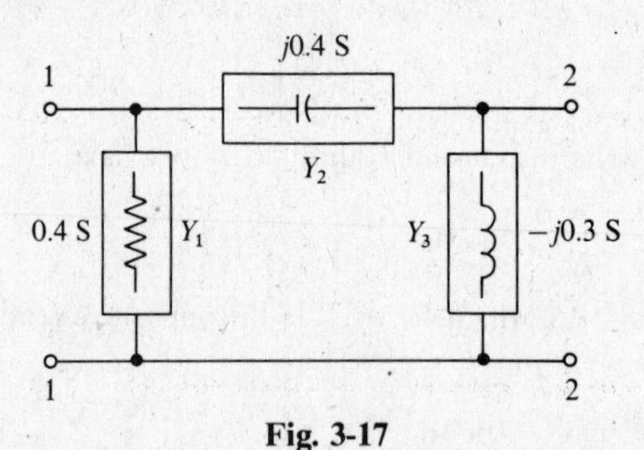

Fig. 3-17

From Fig. 3-17, we see that

$$Y_1 = 0.4 \text{ S} \qquad Y_2 = j0.4 \text{ S} \qquad Y_3 = -j0.3 \text{ S}$$

Substituting yields

$$y_{11} = 0.4 + j0.4 = 0.56\underline{/45^\circ} \text{ S}$$

$$y_{21} = y_{12} = -(j0.4) = 0.4\underline{/-90^\circ} \text{ S}$$

$$y_{22} = (j0.4) + (-j0.3) = j0.1 = 0.1\underline{/90^\circ} \text{ S}$$

As in the case of z parameters, any two port can be defined in terms of the y parameters and we may substitute any of the equivalent circuits shown in Fig. 3-18 for the black box. In each of the equivalent circuits, we are required to use a voltage-dependent current source.

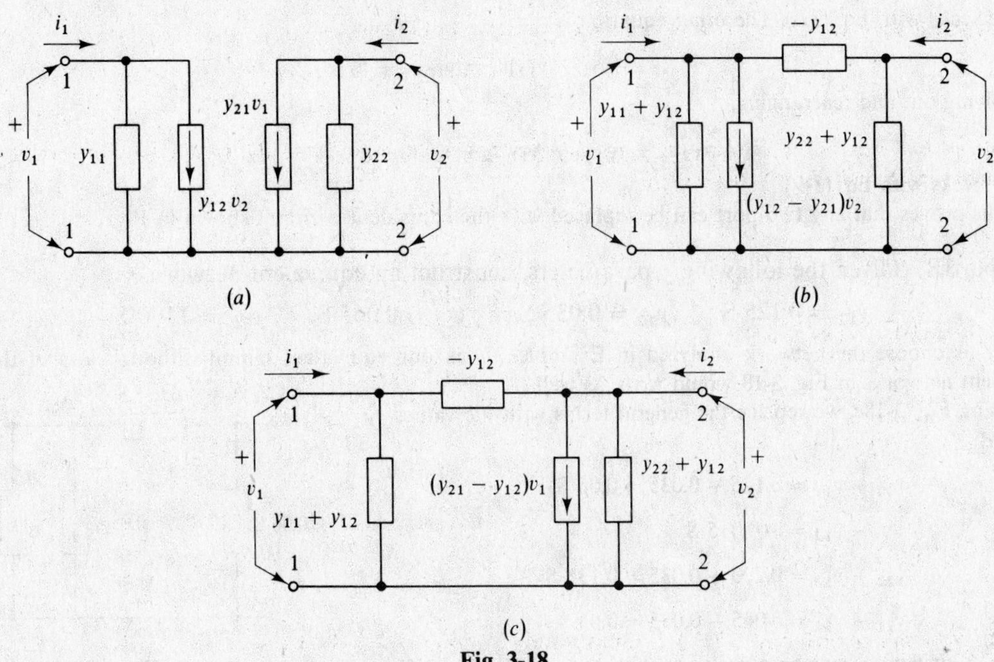

Fig. 3-18

Example 3.7 Prove that the network shown in Fig. 3-18c is an equivalent to the general two-port.

Since we are dealing with the admittance case, we will use Maxwell's nodal matrix to check the validity of the circuit as relabeled in Fig. 3-19:

$$\begin{bmatrix} i'_1 \\ i'_2 \end{bmatrix} = \begin{bmatrix} \phi_1 \\ \phi_2 \end{bmatrix} \begin{bmatrix} Y_{11} & -Y_{12} \\ -Y_{21} & Y_{22} \end{bmatrix}$$

v_1 and v_2 are the nodal voltages we want to solve for, so

$$\phi_1 = v_1 \qquad \text{and} \qquad \phi_2 = v_2$$

Fig. 3-19

Also, from Fig. 3-19,

$$Y_{11} = (y_{11} + y_{12}) + (-y_{12}) = y_{11} \qquad Y_{21} = Y_{12} = -y_{12} \qquad Y_{22} = (y_{22} + y_{12}) + (-y_{12}) = y_{22}$$

$$i'_1 = i_1 \qquad \text{and} \qquad i'_2 = i_2 - (y_{21} - y_{12})v_1$$

Substituting,

$$\begin{bmatrix} i_1 \\ i_2 - (y_{21} - y_{12})v_1 \end{bmatrix} = \begin{bmatrix} v_1 \\ v_2 \end{bmatrix} \begin{bmatrix} y_{11} & -(-y_{12}) \\ -(-y_{12}) & y_{22} \end{bmatrix}$$

For the first equation,

$$i_1 = y_{11}v_1 + y_{12}v_2$$

which checks with Eq. (*3.8*). The other equation is

$$i_2 - (y_{21} - y_{12})v_1 = y_{12}v_1 + y_{22}v_2$$

Multiplying out and rearranging,

$$i_2 = \cancel{y_{12}v_1} + y_{21}v_1 - \cancel{y_{12}v_1} + y_{22}v_2 = y_{21}v_1 + y_{22}v_2$$

which checks with Eq. (*3-9*).

This proves that any two-port can be replaced with the equivalent π circuit shown in Fig. 3-18*c*.

Example 3.8 Given the following *y* parameters, construct an equivalent network.

$$y_{11} = 0.125 \text{ S} \qquad y_{12} = 0.035 \text{ S} \qquad y_{21} = 0.065 \text{ S} \qquad y_{22} = 0.100 \text{ S}$$

Let us choose the network analyzed in Example 3.7 as our equivalent circuit although any of the other equivalent networks in Fig. 3-18 would work as well.

Using Fig. 3-18*c*, we replace the general terms with the values specified:

$$y_{11} + y_{12} = 0.125 + 0.035 = 0.06 \text{ S}$$

$$-y_{12} = -0.035 \text{ S}$$

$$y_{22} + y_{12} = 0.100 + 0.035 = 0.135 \text{ S}$$

$$y_{21} - y_{12} = 0.065 - 0.035 = 0.03 \text{ S}$$

and the equivalent circuit becomes the network shown in Fig. 3-20.

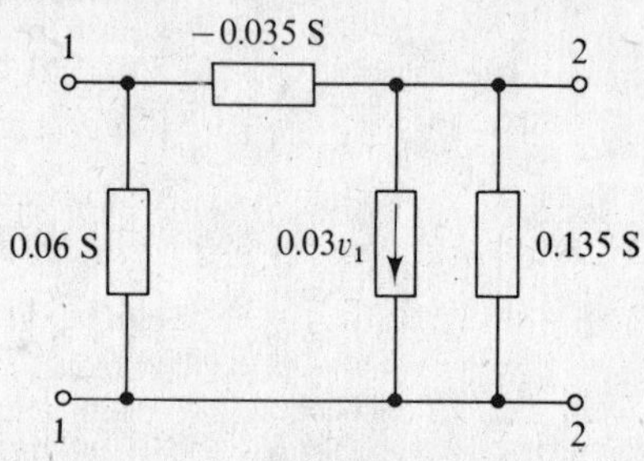

Fig. 3-20

As in the case of the impedance parameters, when $y_{12} \neq y_{21}$ we have an active network, and when $y_{12} = y_{21}$ we have a passive π. In the case of the passive networks, we could go from the impedance network to the admittance network by using the Y-Δ conversion, or from the admittance to impedance network by using the Δ-Y conversion. (Review Sec. 2.5.)

3.4 HYBRID PARAMETERS (*h* PARAMETERS)

The *z* parameters are useful if we are dealing with voltage sources, and the *y* parameters are useful when we are dealing with current sources. There are many cases where it would be expedient to intermix the types of sources, as for bipolar or FET transistors. For these cases, we derive a new set of relationships that depend on both current and voltage. The *h parameters* defined in the following equations are one possibility:

$$v_1 = h_{11}i_1 + h_{12}v_2 \tag{3-15}$$

$$i_2 = h_{21}i_1 + h_{22}v_2 \tag{3-16}$$

or in matrix form

$$\begin{bmatrix} v_1 \\ i_2 \end{bmatrix} = \begin{bmatrix} i_1 \\ v_2 \end{bmatrix} \begin{bmatrix} h_{11} & h_{12} \\ h_{21} & h_{22} \end{bmatrix} \tag{3-17}$$

Since the parameters refer to both voltage and current sources, the term "hybrid" is used.

We can arrive at the *h* parameters by sequentially short-circuiting the output terminals and open-circuiting the input terminals.

With the output shorted, $v_2 = 0$ (see Fig. 3-21) and Eq. (*3-15*) becomes $v_1 = h_{11}i_1$; so

$$h_{11} = \left.\frac{v_1}{i_1}\right|_{v_2=0} \quad \Omega \tag{3-18}$$

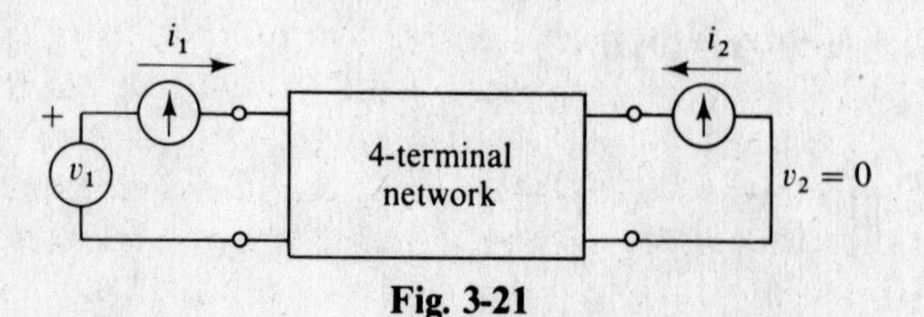

Fig. 3-21

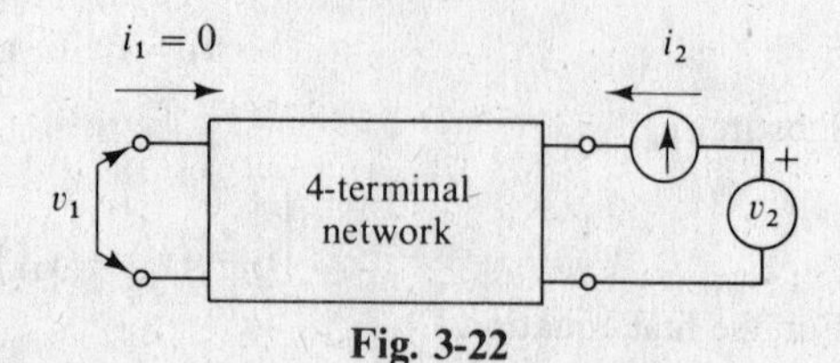

Fig. 3-22

Also, Eq. (*3-16*) can be written $i_2 = h_{21} i_1$; so

$$h_{21} = \left.\frac{i_2}{i_1}\right|_{v_2=0} \quad \text{(no units)} \tag{3-19}$$

With the input terminals open-circuited, $i_1 = 0$ (see Fig. 3-22), and we may write Eq. (*3-15*) as $v_1 = h_{12} v_2$; so

$$h_{12} = \left.\frac{v_1}{v_2}\right|_{i_1=0} \quad \text{(no units)} \tag{3-20}$$

Also, Eq. (*3-16*) now becomes $i_2 = h_{22} v_2$; so

$$h_{22} = \left.\frac{i_2}{v_2}\right|_{i_1=0} \quad \text{S} \tag{3-21}$$

In regard to h parameters,

- h_{11} is called the *short-circuit input impedance*. It has the dimension of (is measured in) ohms. h_{11} is sometimes denoted h_i.
- h_{21} is called the *short-circuit forward-transfer current ratio*. Since it is a true ratio of currents, it has no units. h_{21} is sometimes denoted h_f.
- h_{12} is called the *open-circuit reverse-transfer voltage ratio*. Again, this is a true ratio and there are no units. h_{12} is sometimes denoted h_r.
- h_{22} is called the *open-circuit output admittance*. It has the dimension of (is measured in) siemens. h_{22} is sometimes denoted h_o.

Manufacturers usually include h parameters with the specifications for a given bipolar transistor. This makes it possible to determine how the transistor will perform in any circuit.

When h parameters are given for transistors, the subscripts are modified to make them more meaningful. Tubes, transistors, and FETs are examples of three-terminal networks since one terminal is always common to the input and output and they can easily be handled with four-terminal analysis. The terminals are named differently for each type, as shown in Table 3-1.

Table 3-1 Description of terminals for amplifying components

Name of component	Name of component terminals
Vacuum tubes	Cathode, grid, plate
Bipolar transistors	Emitter, base, collector
Field-effect transistors	Source, gate, drain

The bipolar transistor can be operated in any of three configurations: common emitter, common base, and common collector. This determines whether the emitter, base, or collector is connected to the common input/output terminal of the black box. It also determines which set of parameters can be used in solving a transistor black box problem. Although a different set of parameters can be given for each amplifier configuration, the usual practice is to give either the ones for the common-emitter connection or for the common-base connection. If one set of parameters is known, the others can be derived.

By using h_i for the input parameter h_{11}, h_o for the output parameter h_{22}, h_r for the reverse parameter h_{12}, and h_f for the forward parameter h_{21}, the subscript numbers can be eliminated. Then the second letter in the subscript can determine which electrode is common in the black box. For example h_{FE} is the short-circuit forward-transfer current ratio. The letter E indicates that the emitter is connected to the common input and output terminal. In other words, the transistor is in a *common-emitter configuration*. Note that capital letters are used in this example, and this is an important designation. When capital letters are used in a subscript, it means that dc parameters are being considered, and when lowercase letters are used, it indicates that ac or signal parameters are being considered. The parameter h_{FE} tells how much output dc current change will occur when the input current is changed. If the parameter is given as h_{fe}, then it is still the short-circuit forward-transfer current ratio but it tells how much signal current will occur at the output for a given input signal current in a common-emitter configuration.

Example 3.9 Find the h parameters of the circuit shown in Fig. 3-23.

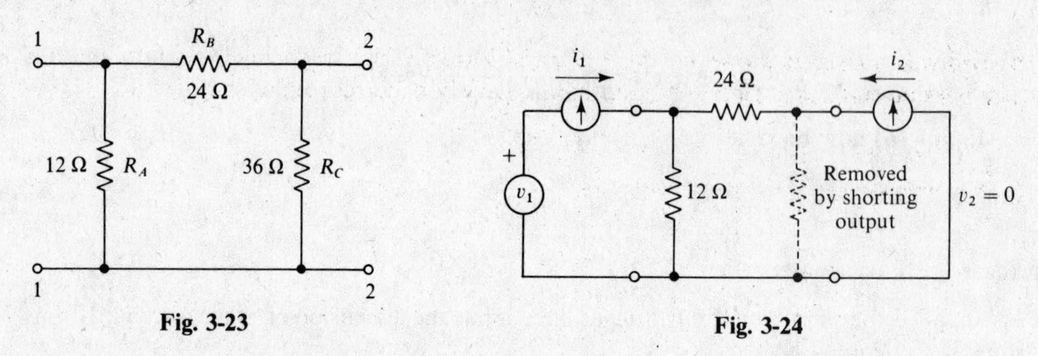

Fig. 3-23

Fig. 3-24

Using the hybrid definitions, we first short-circuit the output terminals as shown in Fig. 3-24.

Since the total resistance in this configuration comprises the 24- and 12-Ω resistors in parallel, we may write v_1 in terms of i_1:

$$v_1 = \frac{(24)(12)}{(24+12)} i_1 = 8i_1$$

Substituting into Eq. (*3-18*):

$$h_{11} = \left.\frac{v_1}{i_1}\right|_{v_2=0} = \frac{8\cancel{i_1}}{\cancel{i_1}} = 8\ \Omega$$

Using the current divider rule, we may write i_2 in terms of i_1:

$$i_2 = \frac{-12}{12+24} i_1 = -\tfrac{1}{3} i_1$$

The negative sign is used because i_1 and i_2 are in opposite directions.

Substituting into Eq. (*3-19*):

$$h_{21} = \left.\frac{i_2}{i_1}\right|_{v_2=0} = \frac{-\frac{1}{3}\cancel{i_1}}{\cancel{i_1}} = -\tfrac{1}{3} \quad \text{(no units)}$$

Now we open-circuit the input terminals as shown in Fig. 3-25.

We may write v_1 terms of v_2 by using the voltage divider rule:

$$v_1 = \frac{12}{12+24} v_2 = \tfrac{1}{3} v_2$$

Substituting into Eq. (*3-20*):

$$h_{12} = \left.\frac{v_1}{v_2}\right|_{i_1=0} = \frac{\frac{1}{3}\cancel{v_2}}{\cancel{v_2}} = \tfrac{1}{3} \quad \text{(no units)}$$

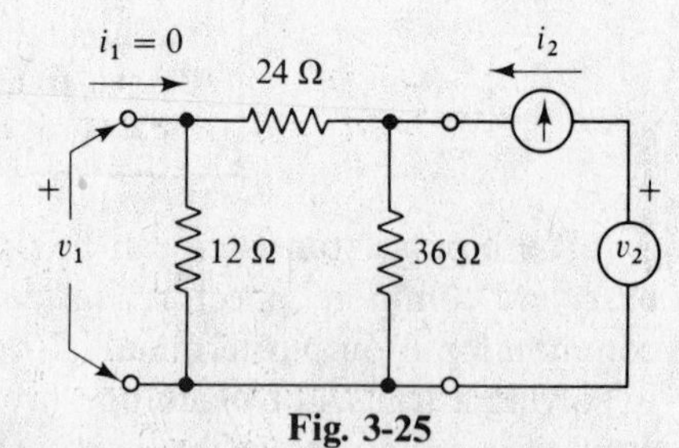

Fig. 3-25

Since this is a passive network, $h_{12} = -h_{21}$. In the z and y parameter cases, when the network was passive the mutual terms were equal to each other without a minus sign. The minus sign is characteristic for the h parameters only.

Finally, we have (*3-21*):

$$h_{22} = \left.\frac{i_2}{v_2}\right|_{i_1=0} \text{ S}$$

To write v_2 in terms of i_2, we need the total resistance R_t. From Fig. 3-25,

$$R_t = \frac{(12+24)(36)}{(12+24)+(36)} = \frac{(36)(36)}{72} = 18\ \Omega$$

So by Ohm's law,

$$v_2 = R_t i_2 = 18 i_2$$

Substituting into Eq. (*3-21*):

$$h_{22} = \frac{i_2}{18 i_2} = 0.056 \text{ S}$$

The equivalent circuits shown in Fig. 3-26*a* and *b* are useful in replacing hybrid devices such as the transistor with an equivalent circuit. Notice that we have both current and voltage sources in the circuits.

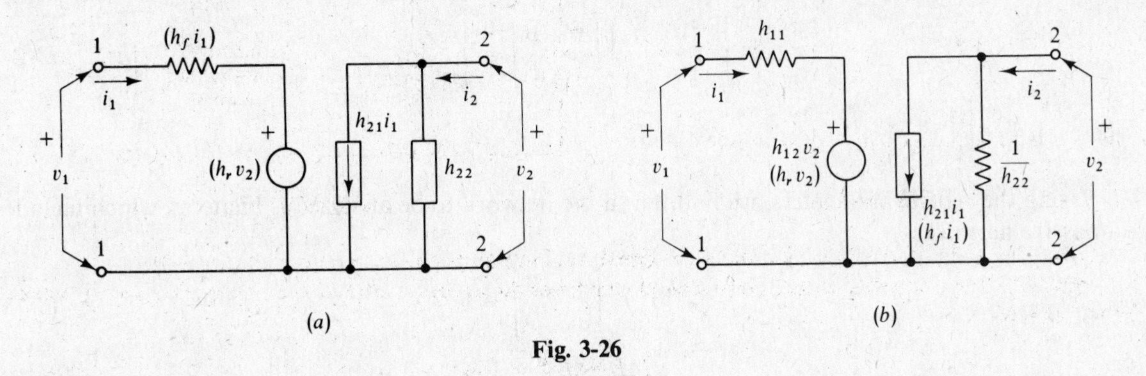

Fig. 3-26

Example 3.10 Find the equivalent circuit for a transistor with the following *h* parameters:

$$h_i = 1400\ \Omega \qquad h_f = -45.0$$

$$h_r = 0.98 \qquad h_o = 27\ \mu\text{S}$$

Since

$$h_{11} = h_i = 1400\ \Omega \qquad h_{12} = h_r = 0.98$$

$$h_{21} = h_f = -45.0 \qquad h_{22} = h_o = 27 \times 10^{-6}\ \text{S}$$

the equivalent circuit can be drawn as shown in Fig. 3-27. The polarity of the voltage source has been reversed to take the minus sign of h_{21} into account.

We will examine transistor equivalent circuits in more detail in Chap. 5.

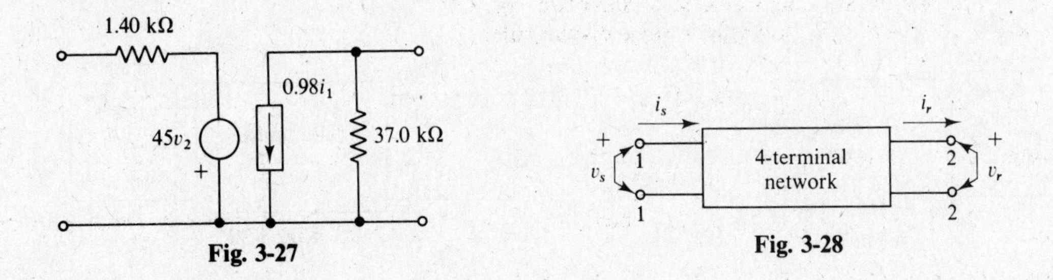

Fig. 3-27

Fig. 3-28

3.5 *ABCD* PARAMETERS (*a* PARAMETERS)

Another set of hybrid parameters for four-terminal networks was developed in the study of power transfer and transmission lines where it is useful to refer the quantities at the sending end to the quantities at the receiving end. These parameters were developed by using the arrangement shown in Fig. 3-28. This is the same system that was used in defining the *z*, *y*, and *h* parameters with the exception of the change in direction of the output current to conform more closely to a true transmission line or power distribution system. Defining these parameters:

A = inverse open-circuit voltage gain C = open-circuit forward-transfer admittance

B = short-circuit reverse-transfer impedance D = inverse short-circuit current gain

Relating the output or receiving quantities to the input or sending quantities and using the subscripts r and s to mean receive and send, we may write

$$v_s = Av_r + Bi_r \tag{3-22}$$

$$i_s = Cv_r + Di_r \tag{3-23}$$

or in matrix form

$$\begin{bmatrix} v_s \\ i_s \end{bmatrix} = \begin{bmatrix} v_r \\ i_r \end{bmatrix} \begin{bmatrix} A & B \\ C & D \end{bmatrix} = \begin{bmatrix} v_r \\ i_r \end{bmatrix} [a] \tag{3-24}$$

where $[a] = \begin{bmatrix} A & B \\ C & D \end{bmatrix}$.

Using the $ABCD$ parameters, we find that if the network to be analyzed is bilateral, which includes all passive networks,

$$[a] = \begin{bmatrix} A & B \\ C & D \end{bmatrix} = 1 \tag{3-25}$$

or

$$AD - BC = 1 \tag{3-25a}$$

If, in addition, the network is symmetrical, $A = D$, a fact that proves useful when finding image impedances.

We determine the a parameters as we found the other sets by sequentially open-circuiting and short-circuiting the output terminals.

With the output terminals open-circuited, $i_r = 0$ (see Fig. 3-29), and we write for Eq. (3-22) $v_s = Av_r$; so

$$A = \left.\frac{v_s}{v_r}\right|_{i_r=0} \quad \text{(no units)} \tag{3-26}$$

Equation (3-23) becomes $i_s = Cv_r$; so

$$C = \left.\frac{i_s}{v_r}\right|_{i_r=0} \quad \text{S} \tag{3-27}$$

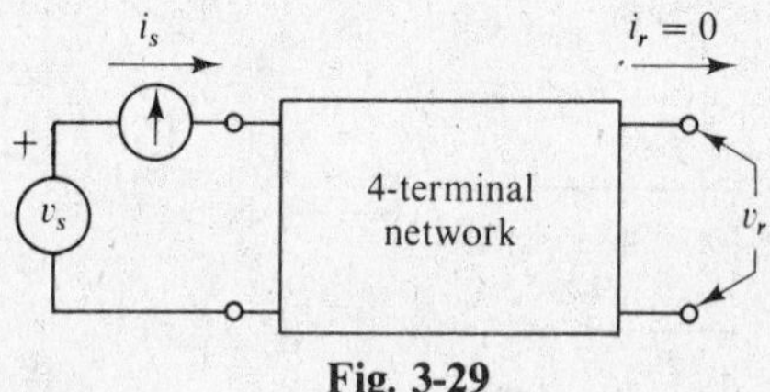

Fig. 3-29

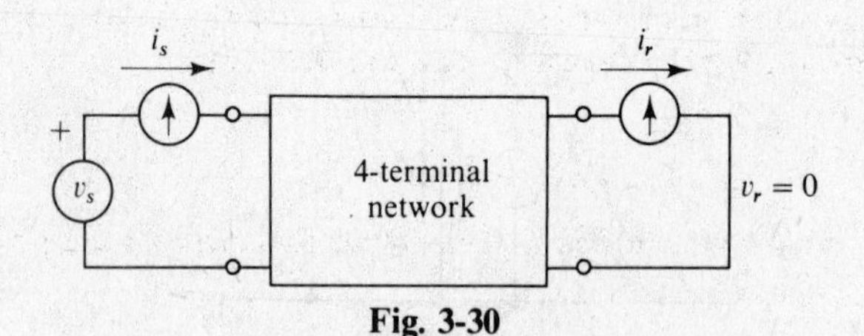

Fig. 3-30

With the output terminals short-circuited, $v_r = 0$ (see Fig. 3-30), and Eq. (3-22) becomes $v_s = Bi_r$; so

$$B = \left.\frac{v_s}{i_r}\right|_{v_r=0} \quad \Omega \tag{3-28}$$

We also write Eq. (3-23) as $i_s = Di_r$; so

$$D = \left.\frac{i_s}{i_r}\right|_{v_r=0} \quad \text{(no units)} \tag{3-29}$$

Example 3.11 Find the *ABCD* parameters of the network shown in Fig. 3-31.

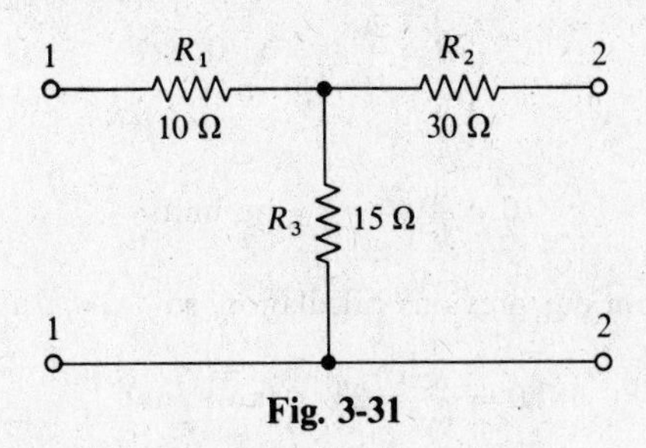

Fig. 3-31

Using the definitions, we first open-circuit the receiving terminals as shown in Fig. 3-32 so that [Eq. (3-26)]

$$A = \frac{v_s}{v_r}\bigg|_{i_r=0} \quad \text{(no units)}$$

There is no voltage drop across the 30-Ω resistor, so we can write v_r in terms of v_s by the voltage divider rule:

$$v_r = \frac{15}{10+15}v_s = \frac{15}{25}v_s = 0.6v_s$$

Substituting into Eq. (3-26):

$$A = \frac{v_s}{0.6v_s} = 1.67 \quad \text{(no units)}$$

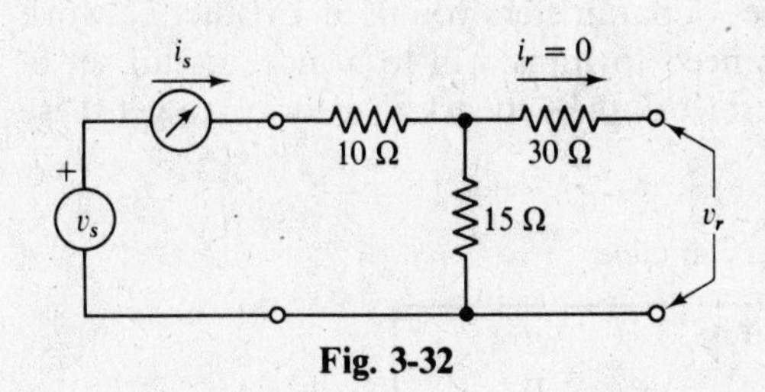

Fig. 3-32

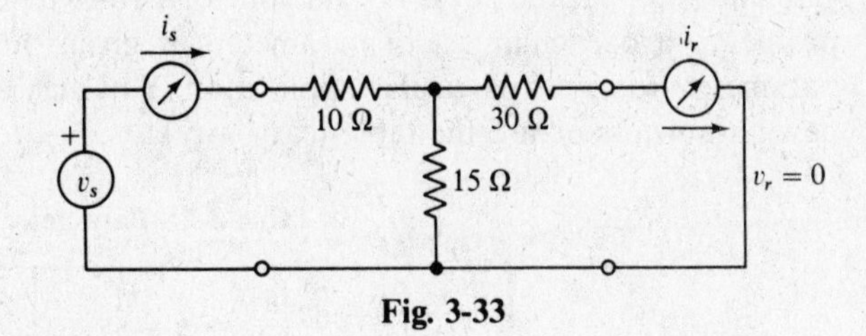

Fig. 3-33

Next [Eq. (3-27)],

$$C = \frac{i_s}{v_r}\bigg|_{i_r=0} \quad \text{S}$$

Since there is no voltage drop across the 3-Ω resistor, v_r is equal to the voltage across the 15-Ω resistor. By Ohm's law, $v_r = 15i_s$. Substituting into Eq. (3-27):

$$C = \frac{i_s}{15i_s} = 0.0667 \text{ S}$$

Now we short-circuit the output terminals as shown in Fig. 3-33 so that [Eq. (3-28)]

$$B = \frac{v_s}{i_r}\bigg|_{v_r=0} \quad \Omega$$

First we find v_s in terms of i_s by writing the total resistance R_t:

$$R_t = 10 + \frac{(15)(30)}{15+30} = 10 + 10 = 20 \ \Omega$$

and

$$v_s = R_t i_s = 20i_s$$

Now we write i_r in terms of i_s using the current divider rule:

$$i_r = \frac{15}{15+30}i_s = \frac{15}{45}i_s = \tfrac{1}{3}i_s$$

Substituting for v_s and i_r in Eq. (*3-28*):

$$B = \frac{20 i_s}{\frac{1}{3} i_s} = 60\ \Omega$$

Finally [Eq. (*3-29*)],

$$D = \left.\frac{i_s}{i_r}\right|_{v_r = 0} \qquad \text{(no units)}$$

We already know i_r in terms of i_s from our previous calculation, so $i_r = \frac{1}{3} i_s$. Substituting into Eq. (*3-29*):

$$D = \frac{i_s}{\frac{1}{3} i_s} = 3 \qquad \text{(no units)}$$

We may check our calculations by using (*3-25a*), since the network is passive:

$$AD - BC = 1$$

Substituting,

$$\tfrac{5}{3}(3) - (60)\tfrac{1}{15} = 5 - 4 = 1$$

This shows our calculations are correct.

The z, y, h, and a parameters are related to one another as shown in Table 3-2. If a network is known in terms of any one set of parameters, the other sets may be calculated by using the table.

Which set to use for a black box depends on the type of problem encountered, and with experience you will be able to select the parameters that best fit the problem. In some network problems, you are given nothing but the network and you solve for whatever set of parameters you need. In other network problems, a particular set of parameters is given but you need to transform to a more useful set of parameters for your particular application. You may either rework the network (hard way) to get these new parameters or use the table (easy way).

Table 3-2 Parameter conversion table

$z_{11} = \frac{y_{22}}{[y]} = \frac{[h]}{h_{22}} = \frac{A}{C}$	$h_{11} = \frac{[z]}{z_{22}} = \frac{1}{y_{11}} = \frac{B}{D}$
$z_{12} = \frac{-y_{12}}{[y]} = \frac{h_{12}}{h_{22}} = \frac{[a]}{C}$	$h_{12} = \frac{z_{12}}{z_{22}} = \frac{-y_{12}}{y_{11}} = \frac{[a]}{D}$
$z_{21} = \frac{-y_{21}}{[y]} = \frac{-h_{21}}{h_{22}} = \frac{1}{C}$	$h_{21} = \frac{-z_{21}}{z_{22}} = \frac{y_{21}}{y_{11}} = \frac{-1}{D}$
$z_{22} = \frac{y_{11}}{[y]} = \frac{1}{h_{22}} = \frac{D}{C}$	$h_{22} = \frac{1}{z_{22}} = \frac{[y]}{y_{11}} = \frac{C}{D}$
$y_{11} = \frac{z_{22}}{[z]} = \frac{1}{h_{11}} = \frac{D}{B}$	$A = \frac{z_{11}}{z_{21}} = \frac{-y_{22}}{y_{21}} = \frac{-[h]}{h_{21}}$
$y_{12} = \frac{-z_{12}}{[z]} = \frac{-h_{12}}{h_{11}} = \frac{-[a]}{B}$	$B = \frac{[z]}{z_{21}} = \frac{-1}{y_{21}} = \frac{-h_{11}}{h_{21}}$
$y_{21} = \frac{-z_{21}}{[z]} = \frac{h_{21}}{h_{11}} = \frac{-1}{B}$	$C = \frac{1}{z_{21}} = \frac{-[y]}{y_{21}} = \frac{-h_{22}}{h_{21}}$
$y_{22} = \frac{z_{11}}{[z]} = \frac{[h]}{h_{11}} = \frac{A}{B}$	$D = \frac{z_{22}}{z_{21}} = \frac{-y_{11}}{y_{21}} = \frac{-1}{h_{21}}$

Note: $[y]$ means $\begin{bmatrix} y_{11} & y_{12} \\ y_{21} & y_{22} \end{bmatrix}$ $[z]$ means $\begin{bmatrix} z_{11} & z_{12} \\ z_{21} & z_{22} \end{bmatrix}$

$[h]$ means $\begin{bmatrix} h_{11} & h_{12} \\ h_{21} & h_{22} \end{bmatrix}$ $[a]$ means $\begin{bmatrix} A & B \\ C & D \end{bmatrix}$

Another use of the *ABCD* parameters is to find the input impedance of a two-port in the open-circuit and short-circuit output situations.

If we open-circuit the output terminals, $i_r = 0$ (see Fig. 3-34), and we may write Eqs. (*3-22*) and (*3-23*) as $v_s = Av_r$ and $i_s = Cv_r$, respectively. Dividing, we find

$$\left.\frac{v_s}{i_s}\right|_{\text{oc}} = \frac{A\cancel{v_r}}{C\cancel{v_r}} = \frac{A}{C}$$

But

$$\left.\frac{v_s}{i_s}\right|_{\text{oc}} = z_{11}$$

Therefore

$$z_{11} = \frac{A}{C} \tag{3-30}$$

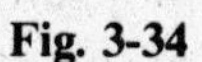

Fig. 3-34

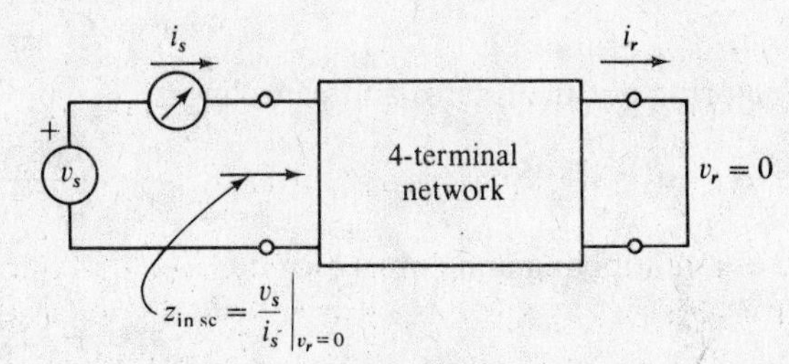

Fig. 3-35

If we short-circuit the receiving terminals, $v_r = 0$ (see Fig. 3-35), and Eqs. (*3-22*) and (*3-23*) become $v_s = Bi_r$ and $i_s = Di_r$, respectively. Dividing, we find

$$\left.\frac{v_s}{i_s}\right|_{\text{sc}} = \frac{B\cancel{i_r}}{D\cancel{i_r}} = \frac{B}{D}$$

But

$$\left.\frac{v_s}{i_s}\right|_{\text{sc}} = h_{11}$$

Therefore

$$h_{11} = \frac{B}{D} \tag{3-31}$$

which we could have determined by consulting Table 3-2.

The open-circuit input impedance z_{11} is sometimes written $z_{\text{in oc}}$, while the short-circuit input impedance h_{11} is sometimes written $z_{\text{in sc}}$.

3.6 IMAGE IMPEDANCE

For the special case of a passive symmetrical two-port network, we can find a particular value of impedance that when connected to the receiving terminals is reflected as exactly the same value at the sending terminals. This particular value of impedance is called the iterative or *image impedance* z_o. See Fig. 3-36.

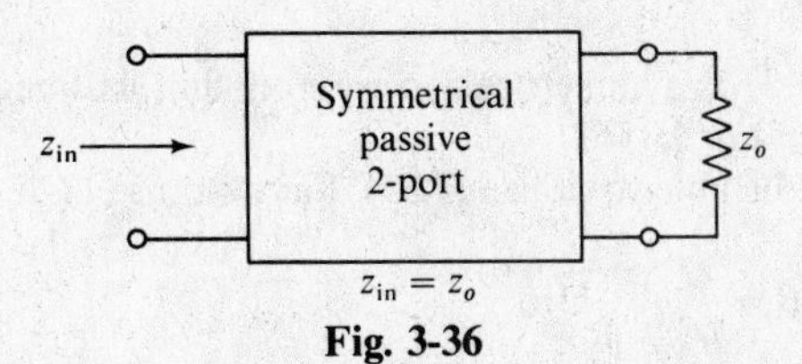

Fig. 3-36

Since our given network is symmetrical, $A = D$ and we may rewrite Eqs. (3-22) and (3-23) as

$$v_s = Av_r + Bi_r \tag{3-32}$$

$$i_s = Cv_r + Ai_r \tag{3-33}$$

We require that the output impedance $v_r/i_r = z_o$ and that the exact impedance be reflected at the input, so that $v_s/i_s = z_o$ also.

Dividing Eq. (3-32) by Eq. (3-33):

$$\frac{v_s}{i_s} = \frac{Av_r + Bi_r}{Cv_r + Ai_r} = z_o \tag{3-34}$$

Since $v_r/i_r = z_0$, we may replace v_r by $i_r z_o$ and obtain

$$\frac{v_s}{i_s} = \frac{A(i_r z_o) + Bi_r}{C(i_r z_o) + Ai_r} = z_o$$

Factoring out the i_r term and canceling:

$$\frac{v_s}{i_s} = \frac{\not{i_r}(Az_o + B)}{\not{i_r}(Cz_o + A)} = z_o$$

Cross multiplying and solving for z_o,

$$z_o^2 C + \not{A}\not{z_o} = \not{A}\not{z_o} + B$$

$$z_o^2 = \frac{B}{C}$$

$$z_o = \sqrt{\frac{B}{C}} \tag{3-35}$$

Example 3.12 Find the image impedance of a two-port in terms of its open-circuit and short-circuit input impedances.

Making use of Eqs. (3-30) and (3-31),

$$z_{11} = z_{\text{in oc}} = \frac{A}{C} \qquad \text{and} \qquad h_{11} = z_{\text{in sc}} = \frac{B}{D}$$

Since the two-port is symmetrical, $A = D$, and multiplying $z_{\text{in oc}}$ by $z_{\text{in sc}}$ gives

$$(z_{11})(h_{11}) = (z_{\text{in oc}})(z_{\text{in sc}}) = \left(\frac{A}{C}\right)\left(\frac{B}{D}\right) = \frac{\not{A}B}{C\not{A}} = \frac{B}{C}$$

But we have previously proven [Eq. (3-35)] that

$$z_o = \sqrt{\frac{B}{C}}$$

So, substituting,

$$z_o = \sqrt{(z_{\text{in oc}})(z_{\text{in sc}})} \tag{3-36}$$

$$= \sqrt{(z_{11})(h_{11})} \tag{3-36a}$$

Example 3.13 Find the image impedance of the network shown in Fig. 3-37.

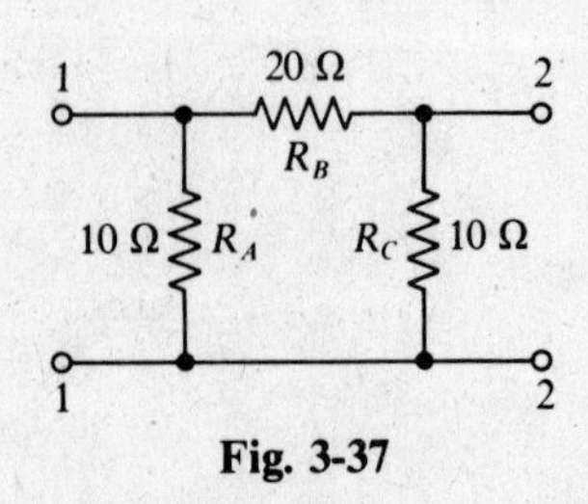

Fig. 3-37

The network shown in Fig. 3-37 is symmetrical, so we may find its image impedance by using either Eq. (3-35) or (3-36).

If we obtain B and C, we will be able to utilize (3-35). First, we use (3-28):

$$B = \left.\frac{v_s}{i_r}\right|_{v_r = 0} \quad \Omega$$

Applying the definition of B to the circuit as shown in Fig. 3-38, by the current divider rule

$$i_r = i_s\left(\frac{10}{10 + 20}\right) = \tfrac{1}{3}i_s$$

But by Ohm's law

$$i_s = \frac{v_s}{R_t}$$

From Fig. 3-38,

$$R_t = \frac{(20)(10)}{(20 + 10)} = \frac{200}{30} = 6.67\ \Omega$$

Substituting,

$$i_r = \frac{1}{3}\frac{v_s}{6.67} = \frac{v_s}{20}$$

$$B = \frac{v_s}{v_s/20} = 20\ \Omega$$

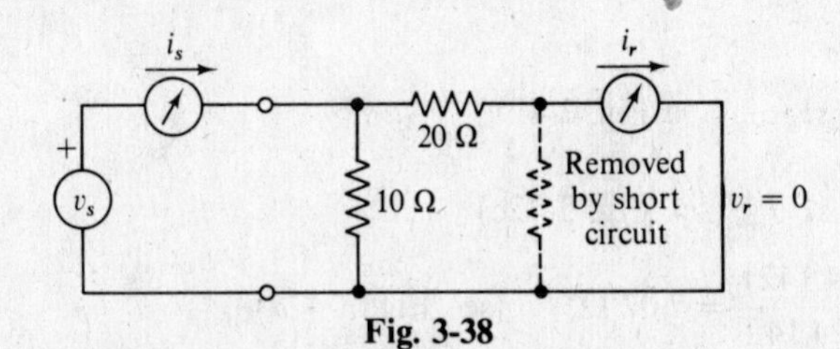

Fig. 3-38

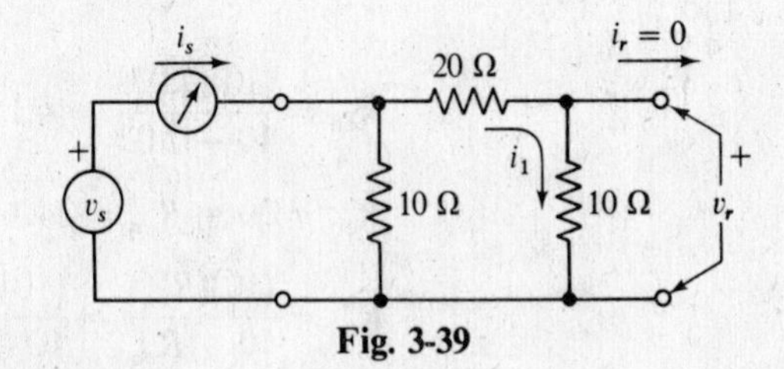

Fig. 3-39

Redrawing the circuit to find C (see Fig. 3-39), we use (3-27):

$$C = \left.\frac{i_s}{v_r}\right|_{i_r = 0}\ \text{S}$$

By the current divider rule,

$$i_1 = \frac{10}{10 + (20 + 10)}i_s = \frac{10}{40}i_s$$

From Ohm's law,

$$v_r = (i_1)(10) = \frac{100}{40}i_s$$

Substituting into Eq. (3-27),

$$C = \frac{i_s}{(100/40)i_s} = 0.4\ \text{S}$$

and (3-35)

$$z_o = \sqrt{\frac{B}{C}} = \sqrt{\frac{20}{.4}} = 7.07\ \Omega$$

Utilizing the second method, we need to find z_{11} and h_{11}.

Open-circuiting the output terminals as shown in Fig. 3-40, we may write the input impedance as

$$z_{11} = z_{\text{in oc}} = \frac{(10)(20 + 10)}{10 + (20 + 10)} = \frac{(10)(30)}{40} = 7.5\ \Omega$$

Shorting the output terminals as shown in Fig. 3-41, the input impedance becomes

$$h_{11} = z_{\text{in sc}} = \frac{(10)(20)}{10 + 20} = \frac{20}{3} = 6.67\ \Omega$$

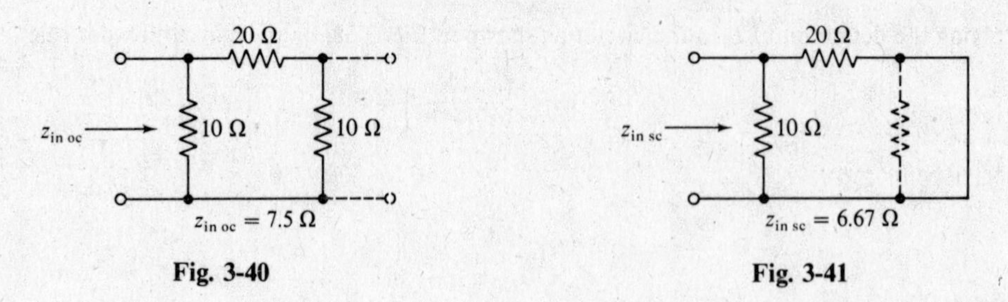

Fig. 3-40 **Fig. 3-41**

Substituting into Eq. (*3-36*),

$$z_o = \sqrt{(z_{11})(h_{11})} = \sqrt{(7.5)(6.67)} = 7.07\ \Omega$$

After one becomes adept at finding impedances of complicated networks, the second method is usually easier than the first.

Let us now prove that this value of resistance *is* reflected at the input terminals when it is connected across the output.

Looking at Fig. 3-42*a*, we can find the input impedance z_{in} by combining the given resistors in series and parallel:

$$R_{\text{eq}} = \frac{(10)(7.07)}{10 + 7.07} = 4.142\ \Omega \qquad \text{(see Fig. 3-42}b\text{)}$$

$$R'_{\text{eq}} = 20 + R_{\text{eq}} = 24.142\ \Omega \qquad \text{(see Fig. 3-42}c\text{)}$$

$$R''_{\text{eq}} = \frac{(10)(R'_{\text{eq}})}{10 + R'_{\text{eq}}} = \frac{(10)(24.142)}{10 + 24.142} = 7.07\ \Omega \qquad \text{(see Fig. 3-42}d\text{)}$$

which is the value that we connected to the output.

The concept of image impedance is important in considering types of filter and attenuator circuits where maximum power transfer is important.

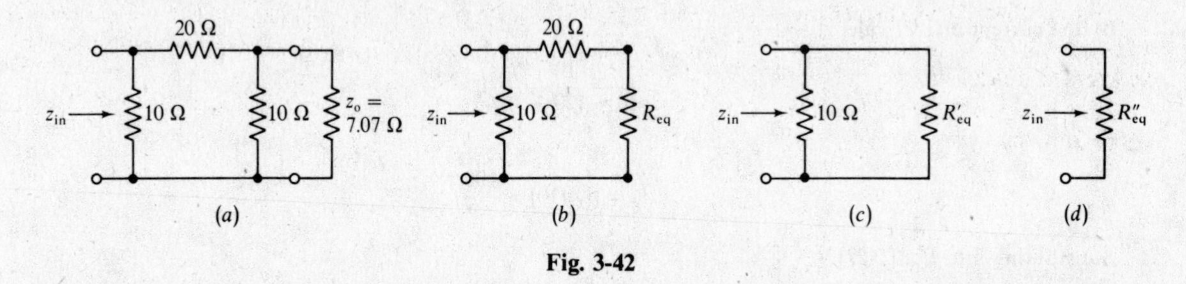

Fig. 3-42

In antenna work, the characteristic impedance for which the system is designed is usually 50, 75, or 300 Ω. In broadcast systems, the characteristic impedance is 600 Ω, while in telephone systems it is either 600 or 900 Ω.

3.7 RESISTANCE PARAMETERS (*R* PARAMETERS)

As another special case, consider a passive, resistive black box. We would like to replace the black box with an equivalent T or π network by using only an ohmmeter to measure the network characteristics.

While there are four measurements that are convenient to make with an ohmmeter, only three are needed to completely specify the network. We can then draw the equivalent T and, by using a Y-Δ transform, convert the network to a π if we desire. See Fig. 3-43.

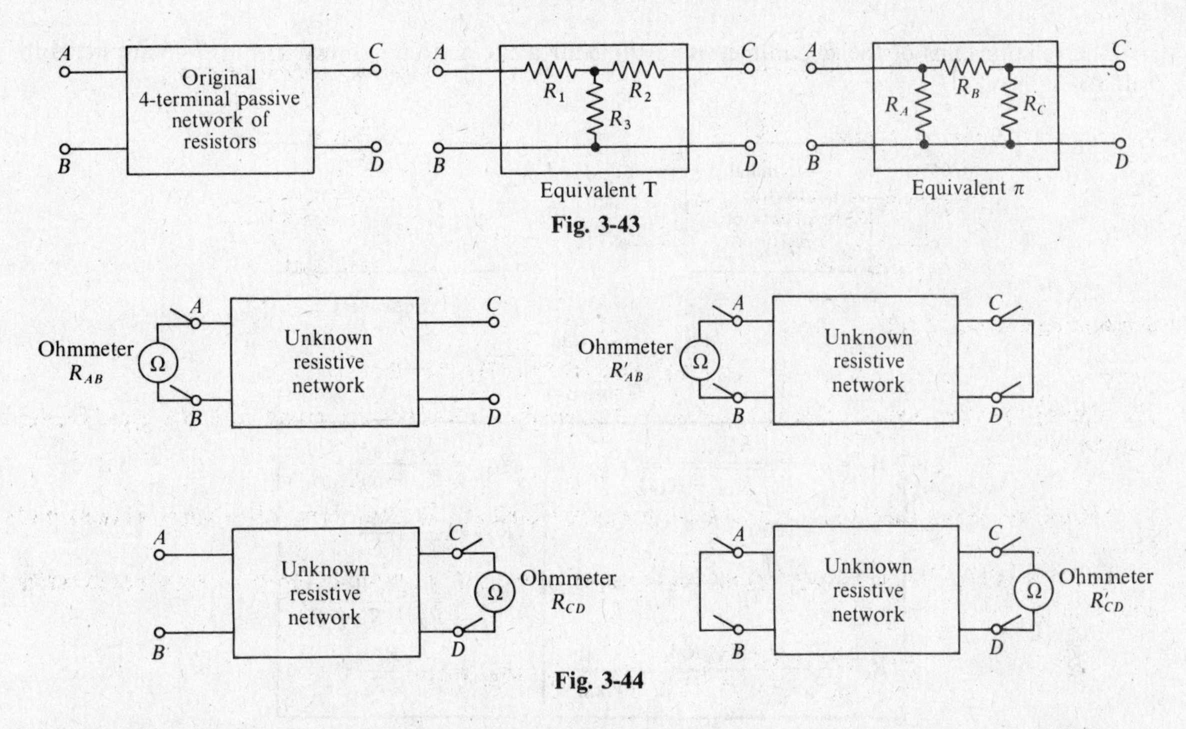

Fig. 3-43

Fig. 3-44

The particular measurements that we will choose are shown in Fig. 3-44 and given as follows:

- The resistance across terminals A and B with terminals C and D open, R_{AB}
- The resistance across terminals A and B with terminals C and D short-circuited, R'_{AB}
- The resistance across terminals C and D with terminals A and B open, R_{CD}

The fourth measurement that can be obtained is the resistance across terminals C and D with terminals A and B short-circuited, R'_{CD}. This measurement can be used as a check.

Notice that the measurements are taken at either the input or output terminals and never between input and output terminals. The ohmmeter measurements for a given black box are the *resistive parameters* of that box.

The relationships of the R parameters to the equivalent T circuit shown in Fig. 3-45 are listed in Table 3-3.

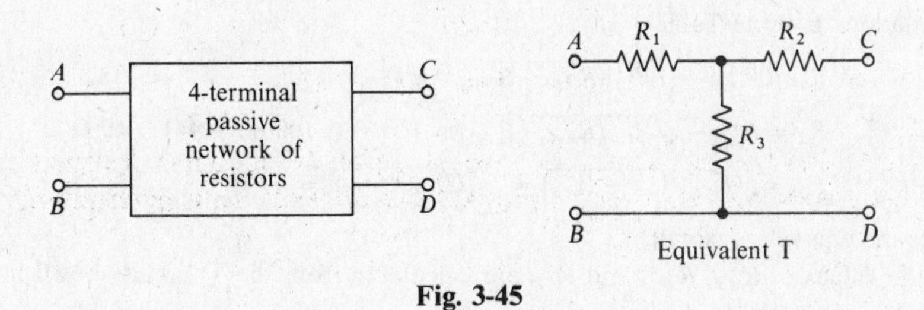

Fig. 3-45

Table 3-3

$R_1 = R_{AB} - \sqrt{R_{CD}(R_{AB} - R'_{AB})} = R_{AB} - R_3$	$R_{AB} = R_1 + R_3$
	$R'_{AB} = R_1 + R_2 R_3/(R_2 + R_3)$
$R_2 = R_{CD} - \sqrt{R_{CD}(R_{AB} - R'_{AB})} = R_{CD} - R_3$	$R_{CD} = R_2 + R_3$
$R_3 = \sqrt{R_{CD}(R_{AB} - R'_{AB})}$	$R'_{CD} = R_2 + R_1 R_3/(R_1 + R_3)$

The relationships of the R parameters to the equivalent π circuit shown in Fig. 3-46 are listed in Table 3-4.

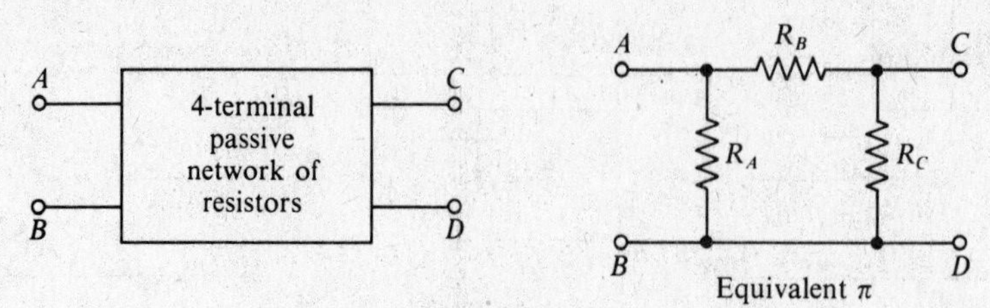

Fig. 3-46

Table 3-4

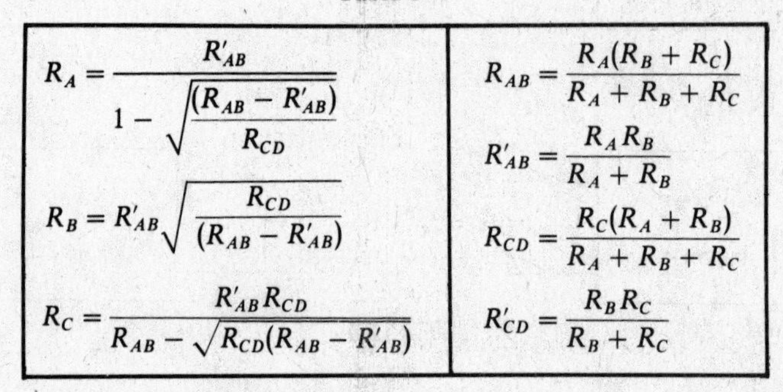

$R_A = \dfrac{R'_{AB}}{1 - \sqrt{\dfrac{(R_{AB} - R'_{AB})}{R_{CD}}}}$	$R_{AB} = \dfrac{R_A(R_B + R_C)}{R_A + R_B + R_C}$
$R_B = R'_{AB}\sqrt{\dfrac{R_{CD}}{(R_{AB} - R'_{AB})}}$	$R'_{AB} = \dfrac{R_A R_B}{R_A + R_B}$
$R_C = \dfrac{R'_{AB} R_{CD}}{R_{AB} - \sqrt{R_{CD}(R_{AB} - R'_{AB})}}$	$R_{CD} = \dfrac{R_C(R_A + R_B)}{R_A + R_B + R_C}$
	$R'_{CD} = \dfrac{R_B R_C}{R_B + R_C}$

Example 3.14 For the black box of Fig. 3-47, the following measurements are shown:

$$R_{AB} = 80\ \Omega \qquad R'_{AB} = 44\ \Omega \qquad R_{CD} = 100\ \Omega$$

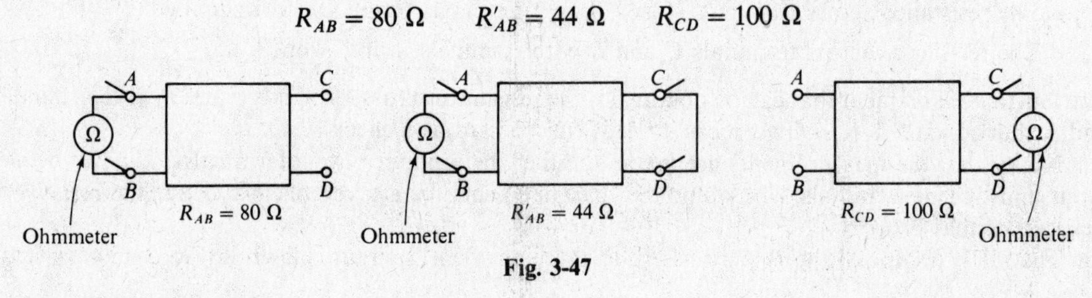

Fig. 3-47

Find the equivalent T circuit.

Using the equations given in Table 3-3,

$$R_1 = R_{AB} - \sqrt{R_{CD}(R_{AB} - R'_{AB})} = 80 - \sqrt{100(80 - 44)} = 20\ \Omega$$

$$R_2 = R_{CD} - \sqrt{R_{CD}(R_{AB} - R'_{AB})} = 100 - \sqrt{100(80 - 44)} = 40\ \Omega$$

$$R_3 = \sqrt{R_{CD}(R_{AB} - R'_{AB})} = \sqrt{100(80 - 44)} = 60\ \Omega$$

Figure 3-48 shows the equivalent circuit.

As a check, the values of R_{AB}, R'_{AB}, and R_{CD} are calculated from the T circuit shown in Fig. 3-48 by using Table 3-3:

$$R_{AB} = R_1 + R_3 = 20 + 60 = 80\ \Omega$$

$$R'_{AB} = R_1 + \frac{R_2 R_3}{R_2 + R_3} = 20 + \frac{(60)(40)}{60 + 40} = 44\ \Omega$$

$$R_{CD} = R_2 + R_3 = 40 + 60 = 100\ \Omega$$

Fig. 3-48

We can also calculate

$$R'_{CD} = R_2 + \frac{R_1 R_3}{R_1 + R_3} = 40 + \frac{(20)(60)}{20 + 60} = 55\ \Omega$$

By using an ohmmeter to measure R'_{CD} for the black box, we can check our validity.

Example 3.15 Using the measurements given in Example 3.14, find the equivalent passive four-terminal π network.

The values given are $R_{AB} = 80\ \Omega$, $R'_{AB} = 44\ \Omega$, and $R_{CD} = 100\ \Omega$. For the π circuit (see Fig. 3-46), we use Table 3-4 and find

$$R_A = \frac{R'_{AB}}{1 - \sqrt{(R_{AB} - R'_{AB})/R_{CD}}} = \frac{44}{1 - \sqrt{(80 - 44)/100}} = \frac{44}{1 - \sqrt{36/100}} = \frac{44}{0.4} = 110\ \Omega$$

$$R_B = R'_{AB}\sqrt{R_{CD}/(R_{AB} - R'_{AB})} = 44\sqrt{100/(80 - 44)} = 44\tfrac{10}{6} = 73.3\ \Omega$$

$$R_C = \frac{R'_{AB}R_{CD}}{R_{AB} - \sqrt{R_{CD}(R_{AB} - R'_{AB})}} = \frac{(44)(100)}{80 - \sqrt{100(80 - 44)}} = \frac{4400}{80 - (10)(6)} = \frac{4400}{20} = 220\ \Omega$$

Figure 3-49 shows the equivalent π circuit.

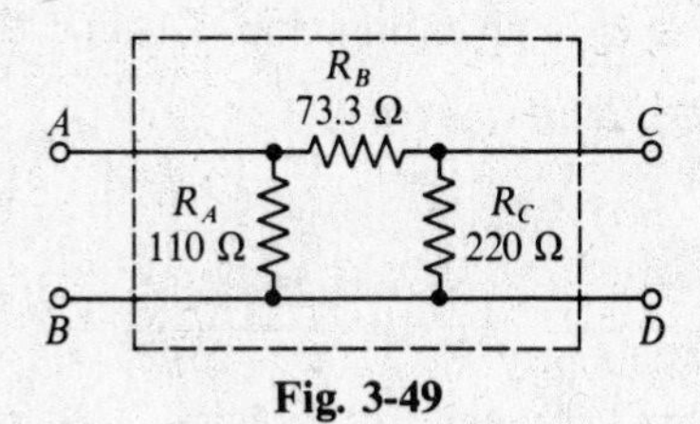

Fig. 3-49

To demonstrate the validity of the equivalent circuit, the fourth parameter is calculated from Fig. 3-49 as

$$R'_{CD} = \frac{R_B R_C}{R_B + R_C} \frac{(73.3)(220)}{73.3 + 220} = 55\ \Omega$$

This is the same value we calculated for R'_{CD} in the equivalent T circuit and again serves to check the validity of our solution.

Solved Problems

3.1 The following ohmmeter measurements are obtained from the four-terminal black box shown in Fig. 3-50: $R_{AB} = 45\ \Omega$, $R'_{AB} = 25\ \Omega$, and $R_{CD} = 35\ \Omega$. Find the power dissipated by a 10-Ω resistor connected across the output terminals when the applied voltage across the input terminals is 10 V.

The first step is to find the equivalent T circuit. Using Table 3-3,

$$R_1 = R_{AB} - \sqrt{R_{CD}(R_{AB} - R'_{AB})} = 45 - \sqrt{35(45 - 25)} = 45 - \sqrt{700} = 45 - 26.46 = 18.5\ \Omega$$

$$R_2 = R_{CD} - \sqrt{R_{CD}(R_{AB} - R'_{AB})} = 35 - \sqrt{35(45 - 25)} = 35 - 26.46 = 8.54\ \Omega$$

$$R_3 = \sqrt{R_{CD}(R_{AB} - R'_{AB})} = \sqrt{35(45 - 25)} = \sqrt{700} = 26.5\ \Omega$$

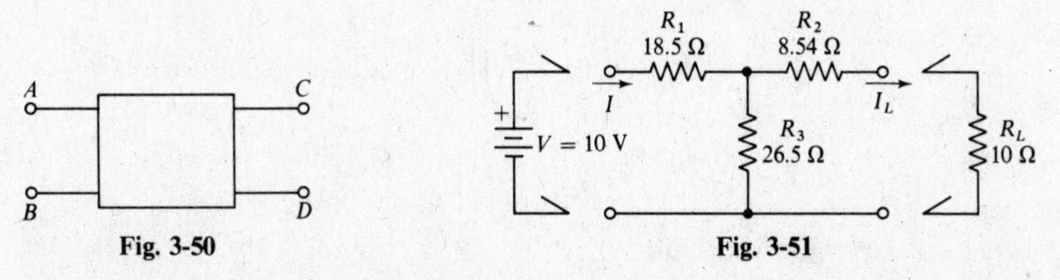

Fig. 3-50 **Fig. 3-51**

Figure 3-51 shows the equivalent T circuit with the 10-Ω resistor across the output terminals and the 10-V source across the input terminals.

The total resistance "seen" by the battery is

$$R_T = R_1 + \frac{R_3(R_2 + R_L)}{R_3 + R_2 + R_L} = 18.5 + \frac{26.5(8.54 + 10)}{26.5 + 8.54 + 10} = 29.44\ \Omega$$

The battery current can be determined by Ohm's law:

$$I = \frac{V}{R_T} = \frac{10}{29.44} = 0.34 \text{ A} = 340 \text{ mA}$$

The current I_L through R_L is found by the current divider method:

$$I_L = I\left(\frac{R_3}{R_3 + (R_2 + R_L)}\right) = 0.34\left(\frac{26.5}{26.5 + (8.54 + 10)}\right) = 0.200 \text{ A} = 200 \text{ mA}$$

Since the current I_L and the load resistance R_L are known, the power dissipated by R_L can be determined:

$$P_L = (I_L)^2 R_L = (0.2)^2(10) = 0.4 \text{ W} = 400 \text{ mW}$$

3.2 Using the ohmmeter measurements given in Prob. 3.1, find the equivalent π circuit.

Given: $R_{AB} = 45\ \Omega \qquad R'_{AB} = 25\ \Omega \qquad R_{CD} = 35\ \Omega$

From Table 3-4,

$$R_A = \frac{R'_{AB}}{1 - \sqrt{(R_{AB} - R'_{AB})/R_{CD}}} = \frac{25}{1 - \sqrt{(45 - 25)/35}} = \frac{25}{1 - \sqrt{20/35}} = \frac{25}{0.244} = 102\ \Omega$$

$$R_B = R'_{AB}\sqrt{R_{CD}/(R_{AB} - R'_{AB})} = 25\sqrt{35/(45 - 25)} = 25\sqrt{35/20} = 33.1\ \Omega$$

$$R_C = \frac{R'_{AB} R_{CD}}{R_{AB} - \sqrt{R_{CD}(R_{AB} - R'_{AB})}} = \frac{(25)(35)}{45 - \sqrt{35(45 - 25)}} = \frac{(25)(35)}{45 - \sqrt{700}} = 47.2\ \Omega$$

Figure 3-52 shows the equivalent π circuit. This solution can be checked by finding R_{AB}, R'_{AB}, and R_{CD} from the circuit shown in Fig. 3-52, then comparing the results with the values given in Prob. 3.1. Another way of checking the answer is to convert the π circuit to a T circuit by using a Δ-Y transform. The resulting T circuit should be identical to the equivalent T of Fig. 3-51.

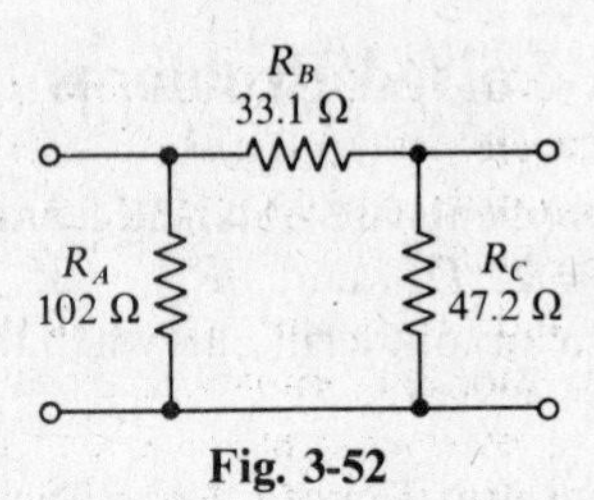

Fig. 3-52

3.3 The z parameters for a certain four-terminal network are given as follows: $z_{11} = 40\ \Omega$, $z_{12} = z_{21} = 30\ \Omega$, and $z_{22} = 50\ \Omega$. A voltage of 10 V is applied to input terminals 1-1, and 20 V is applied to output terminals 2-2. Find the currents (i_1 and i_2) at the input and output terminals.

The current values can be obtained directly by substitution into the equation with z parameters (3-3):

$$\begin{bmatrix} v_1 \\ v_2 \end{bmatrix} = \begin{bmatrix} i_1 \\ i_2 \end{bmatrix} \begin{bmatrix} z_{11} & z_{12} \\ z_{21} & z_{22} \end{bmatrix}$$

Substituting,

$$\begin{bmatrix} 10 \\ 20 \end{bmatrix} = \begin{bmatrix} i_1 \\ i_2 \end{bmatrix} \begin{bmatrix} 40 & 30 \\ 30 & 50 \end{bmatrix}$$

Solving for i_1 and i_2 by Cramer's rule,

$$i_1 = \frac{\begin{vmatrix} 10 & 30 \\ 20 & 50 \end{vmatrix}}{\Delta} \qquad i_2 = \frac{\begin{vmatrix} 40 & 10 \\ 30 & 20 \end{vmatrix}}{\Delta}$$

where $\Delta = \begin{vmatrix} 40 & 30 \\ 30 & 50 \end{vmatrix} = (40)(50) - (30)(30) = 1100$

Therefore

$$i_1 = \frac{\begin{vmatrix} 10 & 30 \\ 20 & 50 \end{vmatrix}}{1100} = \frac{(10)(50) - (20)(30)}{1100} = \frac{500 - 600}{1100} = -0.0909 \text{ A} = -90.9 \text{ mA}$$

$$i_2 = \frac{\begin{vmatrix} 40 & 10 \\ 30 & 20 \end{vmatrix}}{1100} = \frac{(40)(20) - (10)(30)}{1100} = \frac{800 - 300}{1100} = 0.455 \text{ A} = 455 \text{ mA}$$

The minus sign for i_1 indicates current opposite to the direction normally taken for i_1. In other words, with our values, current is directed out of the network into the input terminals.

3.4 Figure 3-53 shows the four-terminal model for a bipolar transistor. The current gain A_i of a transistor is defined as the output current i_2 divided by the input current i_1. Find the transistor current gain from the model.

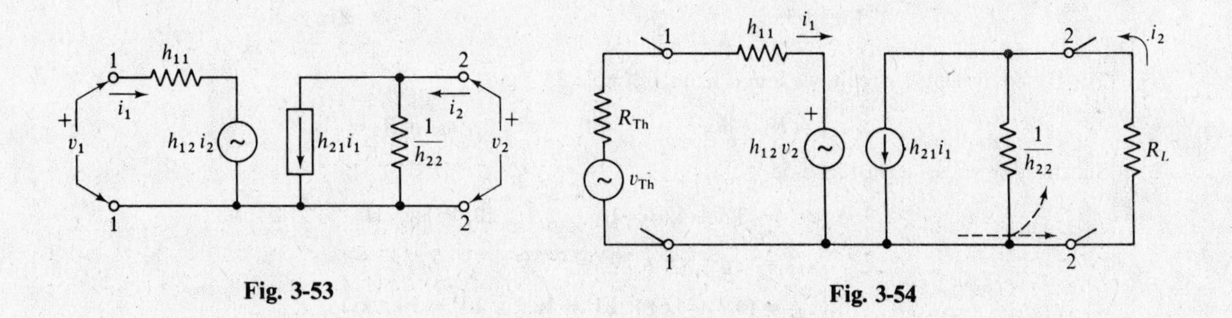

Fig. 3-53 **Fig. 3-54**

The four-terminal model for a bipolar transistor is redrawn in Fig. 3-54. A source for an input signal has been added. It is shown as a Thevenin generator (v_{Th} and R_{Th}). Also, a load resistor R_L has been added to the output terminals of the model. The complete model, including the input signal generator and the output load resistance, represents an *amplifier*.

The problem to be solved is to find the current gain A_i of the amplifier. Mathematically, $A_i = i_2/i_1$.

The input signal current i_1 is presumed to be known, so the current gain can be calculated after the value of output current i_2 is determined. The output current comes from the Norton equivalent current generator $h_{21}i_1$. The output of this generator divides, part of it flowing through the Norton generator resistor $1/h_{22}$ and the rest of the current, i_2, flowing through R_L. Using the current divider rule,

$$i_2 = \frac{1/h_{22}}{1/h_{22} + R_L} h_{21} i_1$$

Multiplying the numerator and denominator of the fraction by h_{22} gives

$$i_2 = \frac{h_{21} i_1}{1 + h_{22} R_L}$$

Now the current gain can be determined:

$$A_i = \frac{i_2}{i_1} = \frac{h_{21}\cancel{i_1}/(1 + h_{22}R_L)}{\cancel{i_1}} = \frac{h_{21}}{1 + h_{22}R_L}$$

Using the alternate symbols h_f for h_{21} and h_o for h_{22}:

$$A_i = \frac{h_f}{1 + h_o R_L}$$

This equation shows that the current gain of a transistor amplifier can be determined if the values of load resistance (or impedance), short-circuit-forward-transfer current ratio, and open-circuit output admittance are known.

3.5 Write the equations for the T circuit shown in Fig. 3-55 using z parameters. With these equations, find the current through R_1 when v_1 (10 V) and v_2 (20 V) are connected to the circuit.

The equations are written as

$$v_1 = z_{11}i_1 + z_{12}i_2 \tag{3-1}$$

$$v_2 = z_{21}i_1 + z_{22}i_2 \tag{3-2}$$

$$\begin{bmatrix} v_1 \\ v_2 \end{bmatrix} = \begin{bmatrix} i_1 \\ i_2 \end{bmatrix} \begin{bmatrix} z_{11} & z_{12} \\ z_{21} & z_{22} \end{bmatrix} \tag{3-3}$$

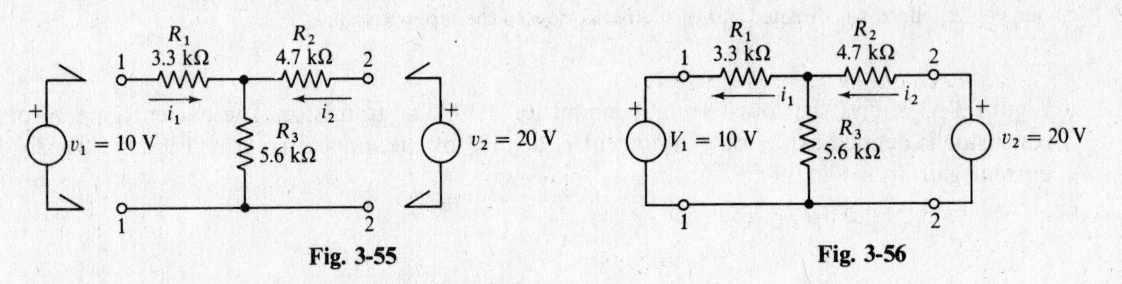

Fig. 3-55 Fig. 3-56

For the equivalent T circuit, we found from Example 3.1 that

$$z_{11} = R_1 + R_3 \qquad z_{12} = z_{21} = R_3 \qquad z_{22} = R_2 + R_3$$

Substituting values from Fig. 3-55,

$$z_{11} = (3.3 + 5.6) \times 10^3 = 8.9 \times 10^3 = 8.9\ \text{k}\Omega$$

$$z_{12} = z_{21} = 5.6 \times 10^3 = 5.6\ \text{k}\Omega$$

$$z_{22} = (4.7 + 5.6) \times 10^3 = 10.3 \times 10^3 = 10.3\ \text{k}\Omega$$

Substituting these values into Eq. (3-3),

$$\begin{bmatrix} v_1 \\ v_2 \end{bmatrix} = \begin{bmatrix} i_1 \\ i_2 \end{bmatrix} \begin{bmatrix} z_{11} & z_{12} \\ z_{21} & z_{22} \end{bmatrix} = \begin{bmatrix} i_1 \\ i_2 \end{bmatrix} \begin{bmatrix} (8.9 \times 10^3) & (5.6 \times 10^3) \\ (5.6 \times 10^3) & (10.3 \times 10^3) \end{bmatrix}$$

When $v_1 = 10$ V and $v_2 = 20$ V, we may write the matrix as

$$\begin{bmatrix} 10 \\ 20 \end{bmatrix} = \begin{bmatrix} i_1 \\ i_2 \end{bmatrix} \begin{bmatrix} (8.9 \times 10^3) & (5.6 \times 10^3) \\ (5.6 \times 10^3) & (10.3 \times 10^3) \end{bmatrix}$$

Solving for i_1 by Cramer's rule,

$$i_1 = \frac{\begin{vmatrix} 10 & 5.6 \times 10^3 \\ 20 & 10.3 \times 10^3 \end{vmatrix}}{\Delta}$$

where $$\Delta = \begin{vmatrix} (8.9 \times 10^3) & (5.6 \times 10^3) \\ (5.6 \times 10^3) & (10.3 \times 10^3) \end{vmatrix} = (8.9 \times 10^3)(10.3 \times 10^3) - (5.6 \times 10^3)^2 = 60.31 \times 10^6$$

So $$i_1 = \frac{(10)(10.3 \times 10^3) - 20(5.6 \times 10^3)}{60.31 \times 10^6} = \frac{-9 \times 10^3}{60.3 \times 10^6} = -0.149 \times 10^{-3}\ \text{A} = -0.149\ \text{mA}$$

The negative sign indicates that i_1 is in the opposite direction of the conventional input current, as shown in Fig. 3-56.

3.6 The following z parameters are given for a certain four-terminal network: $z_{11} = 8$ kΩ, $z_{12} = 4$ kΩ, $z_{22} = 8$ kΩ, and $z_{21} = 6$ kΩ. Write equations for the network using y parameters.

We will use Table 3-2 to convert the z to y parameters:

$$[z] = \begin{vmatrix} z_{11} & z_{12} \\ z_{21} & z_{22} \end{vmatrix} = \begin{vmatrix} 8 \times 10^3 & 4 \times 10^3 \\ 6 \times 10^3 & 8 \times 10^3 \end{vmatrix} = [(8)(8) - (4)(6)] \times 10^6 = 40 \times 10^6$$

Therefore,

$$y_{11} = \frac{z_{22}}{[z]} = \frac{8 \times 10^3}{40 \times 10^6} = 0.200 \times 10^{-3}\ \text{S} = 200\ \mu\text{S}$$

$$y_{12} = \frac{-z_{12}}{[z]} = \frac{-4 \times 10^3}{40 \times 10^6} = -0.100 \times 10^{-3}\ \text{S} = -100\ \mu\text{S}$$

$$y_{21} = \frac{-z_{21}}{[z]} = \frac{-6 \times 10^3}{40 \times 10^6} = -0.150 \times 10^{-3}\ \text{S} = -150\ \mu\text{S}$$

$$y_{22} = \frac{z_{11}}{[z]} = \frac{8 \times 10^3}{40 \times 10^6} = 0.200 \times 10^{-3}\ \text{S} = 200\ \mu\text{S}$$

The equations for the y parameters are

$$i_1 = y_{11}v_1 + y_{12}v_2 \qquad (3\text{-}8)$$

$$i_2 = y_{21}v_1 + y_{22}v_2 \qquad (3\text{-}9)$$

By substituting the above values:

$$i_1 = 200 \times 10^{-6}v_1 - 100 \times 10^{-6}v_2$$

$$i_2 = -150 \times 10^{-6}v_1 + 200 \times 10^{-6}v_2$$

3.7 Given the parameters of a bilateral (passive) four-terminal network as follows: $A = 20$, $C = 5$ S, $D = 1$. Find the output voltage v_r when the input voltage $v_s = 30 \sin 377t$ and the input current $i_s = 2.0 \sin 377t$.

Since both v_s and i_s are given as sine waves with the same frequency, we may use the magnitude of each in the a parameter equation (*3-24*):

$$\begin{bmatrix} v_s \\ i_s \end{bmatrix} = \begin{bmatrix} v_r \\ i_r \end{bmatrix}[a] = \begin{bmatrix} v_r \\ i_r \end{bmatrix}\begin{bmatrix} A & B \\ C & D \end{bmatrix}$$

Since this network is bilateral, from (*3-25*) and (*3-25a*):

$$[a] = AD - BC = 1$$

so

$$(20)(1) = B(5)$$

and

$$B = 3.8\ \Omega$$

Substituting the parameters into Eq. (*3-24*) with $v_s = 30$ and $i_s = 2.0$,

$$\begin{bmatrix} 30 \\ 2.0 \end{bmatrix} = \begin{bmatrix} v_r \\ i_r \end{bmatrix}\begin{bmatrix} 20 & 3.8 \\ 5 & 1 \end{bmatrix}$$

Solving for v_r by Cramer's rule and $[a] = 1$,

$$v_r = \frac{\begin{vmatrix} 30 & 3.8 \\ 2.0 & 1 \end{vmatrix}}{[a]} = (30)(1) - (2.0)(3.8) = 22.4\ \text{V}$$

3.8 A passive network is terminated in its image impedance as shown in Fig. 3-57. Find the input voltage if the input current is 25 mA.

Since the network is terminated in its image impedance z_o, the input impedance z_{in} is also equal to z_o. From Fig. 3-57, $z_o = 75\ \Omega$, so $z_{in} = z_o = 75\ \Omega$. But by Ohm's law, $z_{in} = v_{in}/i_{in}$ and we have $v_{in} = i_{in} z_{in}$. Substituting,

$$v_{in} = (25 \times 10^{-3})(75) = 1.88\ \text{V}.$$

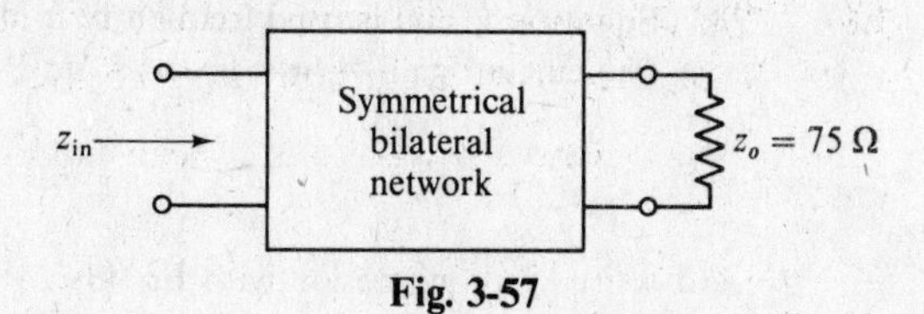

Fig. 3-57

3.9 In the hybrid equivalent circuit shown in Fig. 3-58, find the voltage gain A_v ($A_v = v_2/v_1$).

Writing Kirchhoff's voltage law for the input terminals:

$$v_1 = i_1 h_i + h_r v_2 \tag{3-37}$$

The output voltage can be written (by Ohm's law) as $v_2 = (-h_f i_1)R_t$ where the minus sign takes the direction of the $h_f i_1$ current generator into account and

$$R_t = \frac{(1/h_o)R_L}{1/h_o + R_L} = \frac{R_L}{1 + h_o R_L}$$

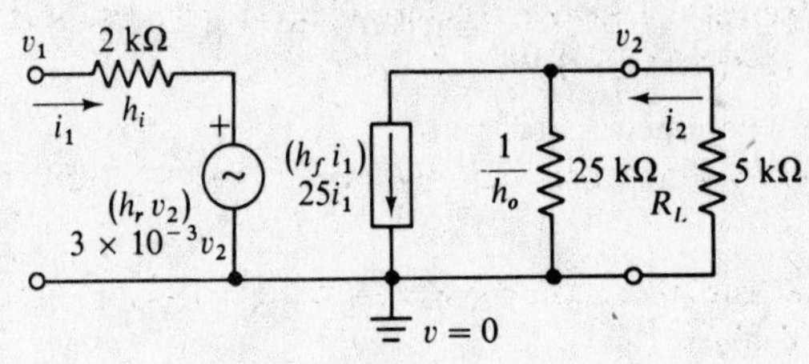

Fig. 3-58

So

$$v_2 = \frac{-h_f i_1 R_L}{1 + h_o R_L}$$

Solving for i_1,

$$i_1 = -\frac{(1 + h_o R_L)v_2}{h_f R_L}$$

Substituting this value of i_1 into Eq. (*3-37*),

$$v_1 = -\frac{(1 + h_o R_L)h_i v_2}{h_f R_L} + h_r v_2$$

Putting the right-hand side of the equation over the common denominator,

$$v_1 = \frac{-(1 + h_o R_L)h_i v_2 + h_r h_f R_L v_2}{h_f R_L}$$

Dividing through by v_2, we obtain

$$\frac{v_1}{v_2} = \frac{-(1 + h_o R_L)h_i + h_r h_f R_L}{h_f R_L}$$

Since $A_v = v_2/v_1$, we invert the fraction and get

$$A_v = \frac{v_2}{v_1} = \frac{h_f R_L}{h_r h_f R_L - h_i(1 + h_o R_L)} \tag{3-38}$$

Substituting the values from Fig. 3-58,

$$A_v = \frac{(25)(5 \times 10^3)}{(3 \times 10^{-3})(25)(5 \times 10^3) - 2 \times 10^3[1 + (5 \times 10^3/25 \times 10^3)]} = \frac{125 \times 10^3}{375 - 2400} = -61.7$$

The minus sign of the answer indicates a reversal in polarity between the input and the output and is common with amplifier circuits.

Equation (*3-38*) is used frequently in designing and analyzing transistor amplifiers.

The current gain A_i was found from Prob. 3.4 to be

$$A_i = \frac{h_f}{1 + h_o R_L} \tag{3-39}$$

and is used in conjunction with Eq. (*3-38*).

Supplementary Problems

3.10 If the parameters in Prob. 3.6 are substituted into Eqs. (3-1) and (3-2), the following equations result:

$$v_1 = 8 \times 10^3 i_1 + 4 \times 10^3 i_2 \qquad v_2 = 6 \times 10^3 i_1 + 8 \times 10^3 i_2$$

Using determinants, solve for i_1 and i_2.

3.11 Show that the relationships between the z and y parameters given in Table 3-2 are correct. *Hint*: Use Eqs. (3-1) and (3-2). Solve for i_1 and i_2 by determinants. Arrange the values of i_1 and i_2 in the same form as Eqs. (3-8) and (3-9). Set the coefficients of the i_1 and i_2 terms obtained by determinants equal to the coefficients of the i_1 and i_2 terms of Eqs. (3-8) and (3-9).

3.12 Can any four-terminal black box be represented by the y, z, and h parameters?

3.13 The following four-terminal passive measurements are taken: $R_{AB} = 20\ \Omega$, $R'_{AB} = 15\ \Omega$, and $R_{CD} = 20\ \Omega$. What is the value of R'_{CD}?

3.14 Using the same parameters given in Prob. 3.13, find the equivalent π circuit. Show that R'_{CD} for the π circuit is equal to R'_{CD} of the T circuit.

3.15 Write two different symbols for each of the following: (*a*) the open-circuit forward-transfer impedance, (*b*) the short-circuit reverse-transfer admittance, and (*c*) the short-circuit input impedance.

3.16 A certain manufacturer gives these specifications for a transistor in production:

$$h_i = 1400\ \Omega \qquad h_f = 44$$

$$h_r = 3.37 \times 10^{-4} \qquad h_o = 27 \times 10^{-6}\ \mu\text{S}$$

Write the four-terminal equations using these parameters.

3.17 The transistor in Prob. 3.16 is connected into a circuit. The input signal voltage v_1 is 1.0 V and the output signal voltage v_2 is 5.0 V. What is the value of the input signal current?

3.18 Using the transistor described in Prob. 3.16 and the circuit described in Prob. 3.17, find the value of the output current and the current gain (current gain $= A_i = i_2/i_1$).

3.19 The following measurements are made on a black box: with the *output* terminals shorted—$v_1 = 0.6$ V, $i_1 = 1$ mA, and $i_2 = 60$ mA; with the *input* terminals shorted—$i_1 = 100\ \mu$A, $i_2 = 10\ \mu$A, and $v_2 = 12$ V. Write the values of the h parameters.

3.20 Find the input impedance of a two-port network whose parameters are $A = 25$, $B = 1$ kΩ, $C = 500\ \mu$S, $D = 10$ if the output terminals are open-circuited.

3.21 Find the image impedance z_o of the network shown in Fig. 3-59.

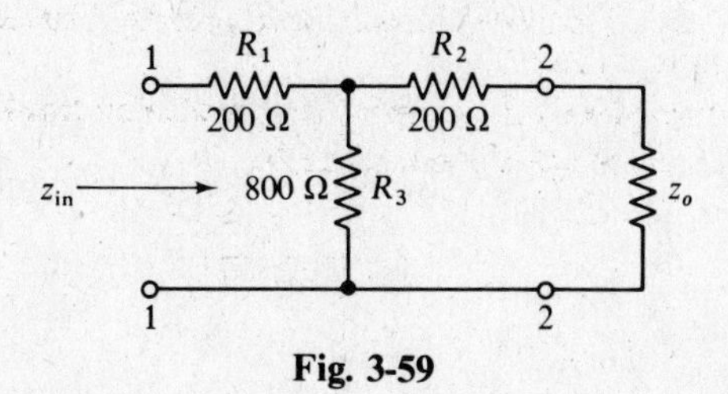

Fig. 3-59

Answers to Supplementary Problems

3.10 The equations obtained for i_1 and i_2 should be the same as the equations obtained in Prob. 3.6 by converting the z parameters to y parameters. These equations are

$$i_1 = 200 \times 10^{-6} v_1 - 100 \times 10^{-6} v_2 \qquad i_2 = -150 \times 10^{-6} v_1 + 200 \times 10^{-6} v_2$$

It may be necessary to manipulate the equations obtained by determinants to get them into this form.

3.11 All the conversion factors in Table 3-2 can be obtained in this manner.

3.12 No. The network within the four-terminal black box must contain linear, bilateral, passive components and active elements (sources). Tubes, bipolar transistors, and field-effect transistors may be considered to be ac sources, so they qualify as active elements.

3.13 $R'_{CD} = 15\ \Omega$.

3.14 In both cases $R'_{CD} = 15\ \Omega$.

3.15 (*a*) z_{21} or z_f; (*b*) y_{12} or y_r; (*c*) h_{11} or h_i or $z_{\text{in sc}}$.

3.16 $v_1 = 1400 i_1 + 3.37 \times 10^{-4} v_2$; $i_2 = 44 i_1 + 27 \times 10^{-6} v_2$.

3.17 $i_1 = 71.3$ mA. (The values assigned for v_1 and v_2 are not necessarily practical. The input signal amplitude and allowable current for the input and output depend upon factors described in Chap. 4.)

3.18 $i_2 = 0.0312\ A = 31.2$ mA; $A_i = 44.0$.

3.19 Since short-circuit measurements were given, it is logical to use the y parameters first, then convert them to h parameters by Table 3-2:

$y_{11} = 1.67 \times 10^{-3}$ S $\qquad h_{11} = 600\ \Omega$

$y_{12} = 8.33\ \mu$S $\qquad h_{12} = -5 \times 10^{-3}$ (no units)

$y_{21} = 0.10$ S $\qquad h_{21} = 60$ (no units)

$y_{22} = 0.833\ \mu$S $\qquad h_{22} = 500\ \mu$S

3.20 $z_{11} = z_{\text{in oc}} = 50\ \text{k}\Omega$.

3.21 $z_o \doteq 600\ \Omega$.

Chapter 4

Diodes and Diode Circuits

4.1 INTRODUCTION

In the earliest electronic systems, the use of diodes was restricted almost entirely to rectifier and detector circuits. (Rectifiers are used for converting ac power to dc, and detectors are used for separating RF and audio signals in receivers.) Semiconductor technology has produced a wide range of diode types, so in addition to rectification and detection, diodes are used for voltage regulation, tuning receivers, protection of sensitive instruments, and many other applications.

Diodes are nonlinear circuit elements. In other words, doubling the voltage across a diode does not necessarily double the current through it. For that reason, Ohm's law cannot be used to find the current in a diode circuit. Ohm's law is valid only in linear circuits, where a plot of voltage vs. current is a straight line.

A graphical procedure—called *plotting a load line*—can be used for solving nonlinear circuit problems. This technique not only works for diodes, but can also be used for finding current in vacuum tube and transistor amplifying devices. An explanation of the load line graphical solution is given in this chapter along with descriptions of various types of diodes and their applications. Hole flow and electron flow are described in this material to help in understanding what is happening inside a semiconductor diode or transistor.

4.2 CHARACTERISTIC CURVES

Load line solutions of problems involving diodes, transistors, and other nonlinear devices require the use of characteristic curves. These curves are not always available from manufacturers, but they can be obtained in several ways. The methods described are for PN junction diodes, but they also work for the junctions in bipolar transistors and for other types of diodes.

A single set of readings to obtain the characteristic curve for a device would be helpful only if that particular device is to be used in a circuit. There is actually a *range* of characteristics for any device manufactured, and the readings for a single curve may be out of tolerance in terms of an average set of values. In Fig. 4-1 the shaded curve shows a range of values for a particular diode which is referred to as

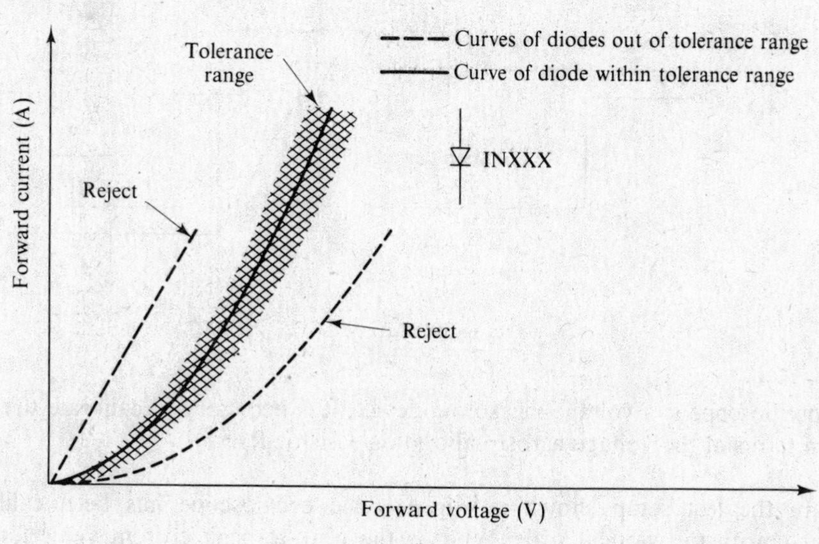

Fig. 4-1

"type INXXX." Any diode with this type number must have a characteristic curve that falls within the shaded range. If its curve does not fall within the range, it is a reject. Two examples of rejects are shown in the illustration.

When the curve is plotted for a single device, there is no way of telling if it is typical, or if it is outside the range of average values for the device. For that reason a number of different devices with the same identifying number should be tested and an average curve drawn. The greater the number of devices tested, the more accurate the curve.

The characteristic curves plot current vs. voltage and may be obtained by a point-by-point method using meters, as shown in Fig. 4-2, by an oscilloscope display as shown in Fig. 4-3, or by an X-Y plotter.

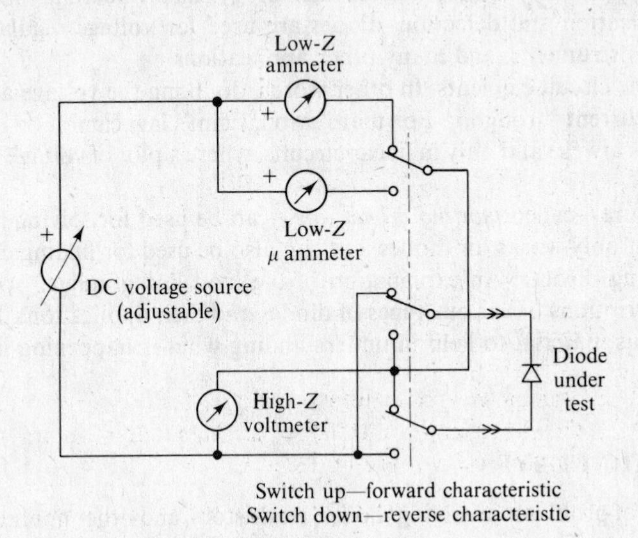

Fig. 4-2

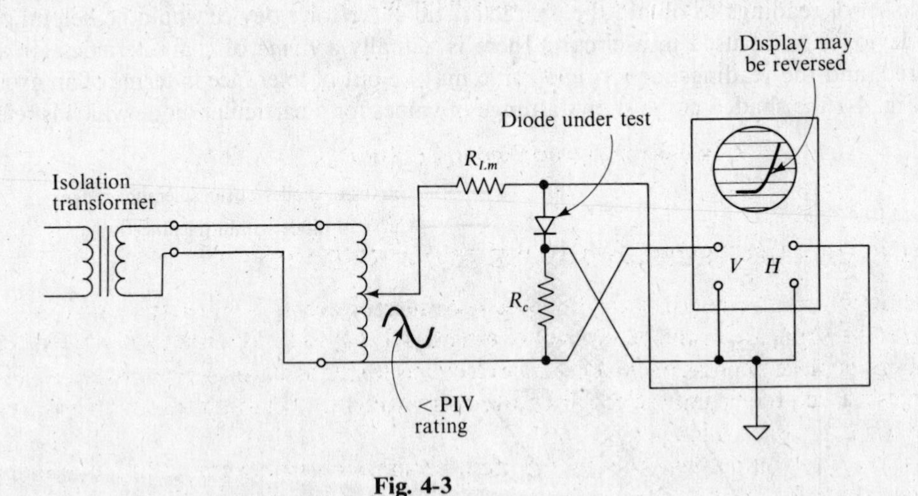

Fig. 4-3

Since the oscilloscope is a voltage-measuring device, it is necessary to calibrate the current for the vertical scale in terms of the voltage across calibration resistor R_c.

Example 4.1 In the test setup shown in Fig. 4-4, the oscilloscope has been calibrated to read 100 mV/cm per cm on the vertical scale. What is the peak-to-peak current value for the sine wave displayed?

From Fig. 4-4, we see that the vertical deflection is 3 cm and since the oscilloscope is calibrated to read 100 mV/cm, the peak-to-peak sine wave voltage value is 300 mV. The calibration resistor, R_c, is given in Fig. 4-4 as 10 Ω. Therefore, using Ohm's law,

$$I_{p-p} = \frac{V_{p-p}}{R_c} = \frac{300 \times 10^{-3}}{10} = 30 \times 10^{-3} \text{ A} = 30 \text{ mA}$$

The peak-to-peak value of the current is 30 mA.

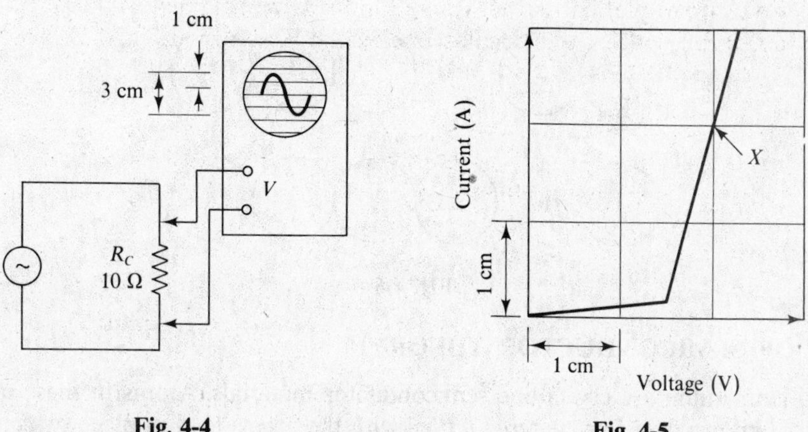

Fig. 4-4 **Fig. 4-5**

Example 4-2 The diode characteristic curve shown in Fig. 4-5 is obtained by using the oscilloscope setup shown in Fig. 4-3. The horizontal calibration is 0.1 V/cm and the vertical calibration is 100 mV/cm. Find the diode current and voltage at point X.

From Fig. 4-5 the horizontal reading at point X is 2 cm. Since the horizontal calibration is given as 0.1 V/cm,

$$V_X = (\text{deflection})(\text{calibration}) = (2)(0.1) = 0.2 \text{ V}$$

Also from Fig. 4-5, we see the vertical reading at point X is 2 cm. Since the vertical calibration is given as 100 mV/cm, we need to convert the vertical voltage calibration to a current calibration by using Ohm's law and the calibrating resistor R_c:

$$I = \frac{V}{R_c} = \frac{100 \times 10^{-3}}{10} = 10 \times 10^{-3} \text{ A/cm} = 10 \text{ mA/cm}$$

$$I_X = (\text{deflection})(\text{calibration}) = (2)(10 \times 10^{-3}) = 20 \text{ mA}$$

4.3 REVIEW OF ELECTRON AND HOLE FLOW THEORY

Electron flow occurs when a voltage source is connected across a material that has *free electrons.* Free electrons are electrons in the outer shell of a material that are not strongly bound to their atoms. When a voltage is placed across the material, the electrons leave their atoms. Since like charges repel and unlike charges attract, the negative electrons move away from the negative terminal of the voltage source and toward the positive terminal.

When a free electron leaves an atom, the atom becomes positively charged until another electron comes along to take the place of the one that left. The motion of electrons in the material is from atom to atom. A convenient way to think of this electron flow is illustrated with marbles, as shown in Fig. 4-6. The marbles, which represent electrons, are moved from left to right. One marble is moved at a time as shown by Figs. 4-6*a* to 4-6*d*. Note that when the electrons move from left to right, the hole (place left vacant by an electron) moves from right to left. In other words, as the electron moves from atom to atom, the hole also moves from atom to atom, but in the opposite direction.

The hole can be thought of as being a positive charge that moves away from the positive terminal of the source and toward the negative terminal.

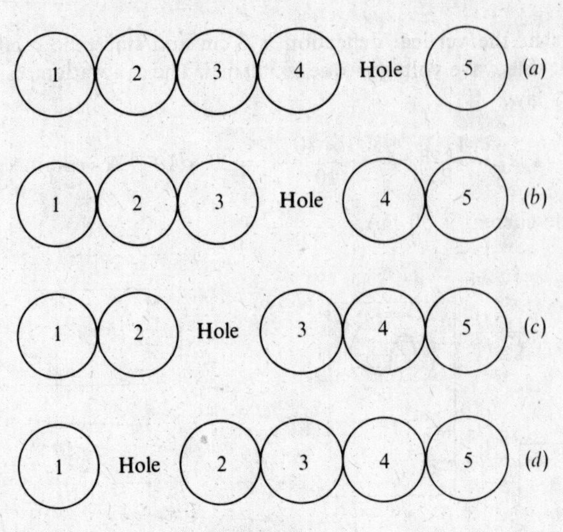

Fig. 4-6

4.4 REVIEW OF SEMICONDUCTOR THEORY

Silicon and germanium are crystalline semiconductor materials. Atoms in these materials form a crystalline pattern known as a *face-centered lattice* with the atoms held in place by *covalent bonds.*

Figure 4-7 shows a two-dimensional representation of a face-centered lattice. Each atom is illustrated in a covalent bond with four neighboring atoms. This covalent bond is a method of sharing electrons so that the outer shell of each atom appears to be filled with eight electrons. Since each electron is strongly attached to the atoms, there are apparently no free electrons.

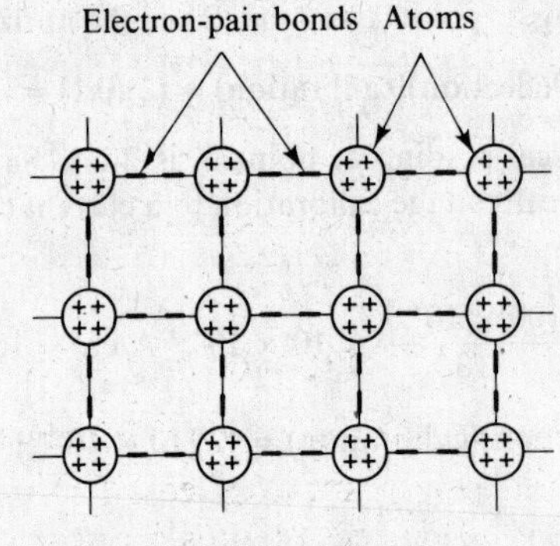

Fig. 4-7

Actually, at room temperature there are always a few electrons that break away from their bond. If this were not true, germanium and silicon would be perfect *insulators,* and absolutely no current could occur. However, since there are always a few free electrons in the material, a small current will exist when a voltage is applied. This is called *intrinsic* current.

When impurities are added to a crystal so that a few impurity atoms take the place of germanium or silicon atoms, the crystal is said to be doped. The doping produces a *P-* or *N-type* material. Figure 4-8 shows the P-type material. The impurity atoms have only three electrons that can enter into a covalent bond, so there is a hole in the crystal bonding associated with each impurity atom.

Figure 4-9 shows N-type material. In this case the impurity atom has one more electron than needed to complete the covalent bond. Because this electron is not strongly bound to its atom, it is a free electron.

If a voltage is applied across the P-type material, electrons migrate from the negative toward the positive terminal. At the same time, the holes migrate from the positive toward the negative terminal. There are no free electrons in the P-type material, so the holes are considered to be the current or *majority charge carriers.* The electrons are called the *minority* charge carriers in P-type material.

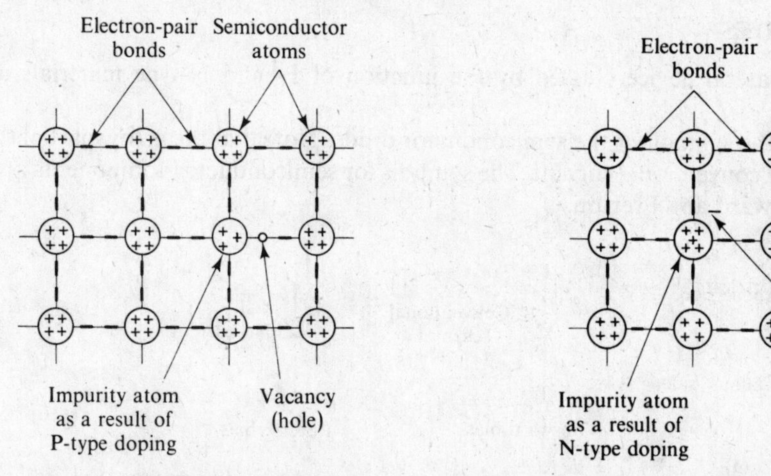

Fig. 4-8 P-type silicon or germanium

Fig. 4-9 N-type silicon or germanium

In the N-type material, the free electrons are the majority charge carriers. They move from the negative to the positive terminal. There are a small number of holes that exist in N-type material, which migrate in the opposite direction. The holes are the minority charge carriers in N-type material.

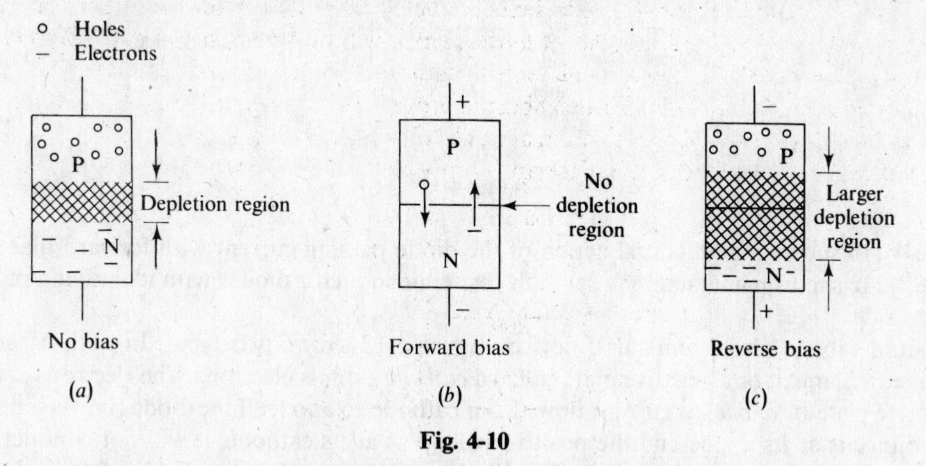

Fig. 4-10

When P- and N-type materials occur in the same crystal so that there is no break in the crystalline structure of the semiconductor material, the result is called a *PN junction*. Figure 4-10*a* shows a model for such a junction. When the junction is formed, there is a region around the interface that is called a *depletion region*. There are no charge carriers in this depletion region, so it acts like an insulator.

When a positive voltage is applied to the P material and a negative voltage to the N material, the charge carriers in the crystal move toward the junction. This is shown in Fig. 4-10*b*. The size of the depletion region decreases, and the junction is said to be *forward-biased*.

If the forward bias is increased until the depletion region disappears, the charge carriers will move across the junction and current will start. It takes approximately 0.2 V to forward-bias a PN germanium junction into conduction and approximately 0.7 V to forward-bias a silicon PN junction.

When a negative voltage is applied to the P material and a positive voltage is applied to the N material, the charge carriers move away from the junction. This is shown in Fig. 4-10*c*. The positive holes are attracted to the negative terminal and the negative electrons are attracted to the positive terminal so the size of the depletion region increases and the junction is said to be *reverse-biased*. With a nominal reverse bias, no current flows through the PN junction. However, if the reverse bias is made sufficiently high, the barrier breaks down and current flows through the junction. The point at which this occurs is called the *zener* voltage. Most germanium junctions are destroyed if they are reverse-biased to the zener point. Silicon junctions are not usually destroyed by a zener voltage, provided the reverse current is not excessive.

4.5 RECTIFIER DIODES

A two-terminal unilateral device created by the junction of P- and N-type materials is called a *semiconductor diode.*

Figure 4-11*a* shows the symbol for the semiconductor diode. Note that the arrow part of the symbol points in the direction of conventional current. The symbols for semiconductor components always have their arrows pointing toward an N region.

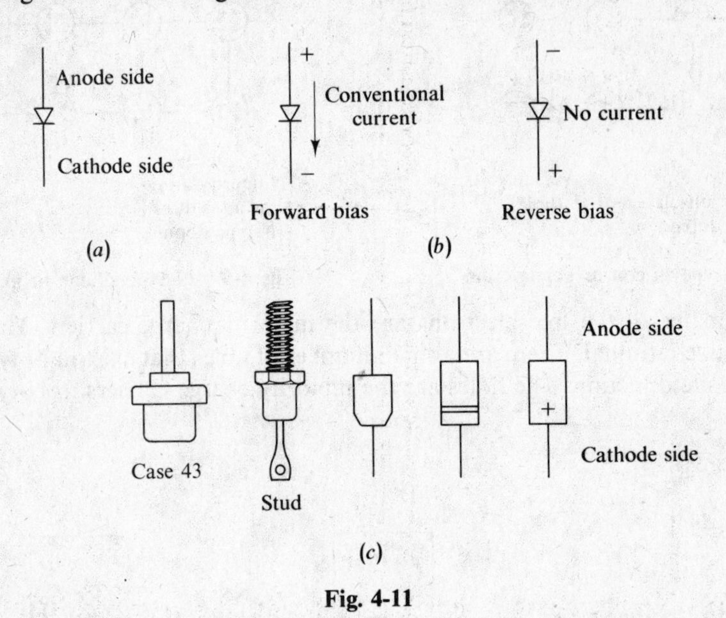

Fig. 4-11

Figure 4-11*b* shows the unilateral action of the diode passing current with forward bias only.

Figure 4-11*c* shows manufacturer's symbols for semiconductor diodes with identifying markings on the cathode side.

A vacuum tube is also a unilateral device. Figure 4-12 shows two types. In both cases a heated metal—either a filament or a negative plate called a *cathode*—emits electrons. The electrons are attracted to the positive plate or *anode*, so current flows from cathode to anode. If the diode is reversed so that the negative voltage is at its anode and the positive voltage is at its cathode, it will not conduct.

The fact that current will flow through a diode in only one direction makes it useful in rectifier circuits that convert ac to dc power. Silicon or germanium diodes are used extensively in power supply

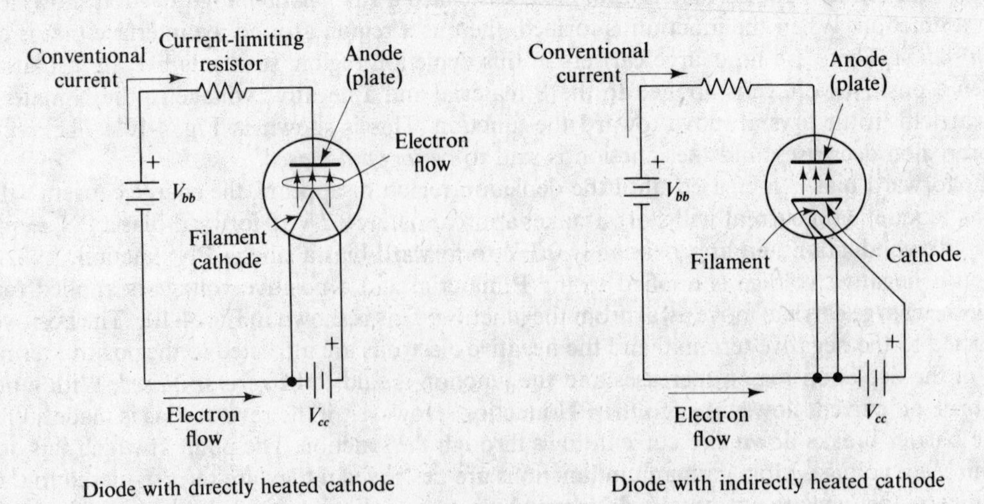

Fig. 4-12

circuits. Figure 4-13 compares the characteristic curves for these diodes. A typical tube rectifier curve is also shown.

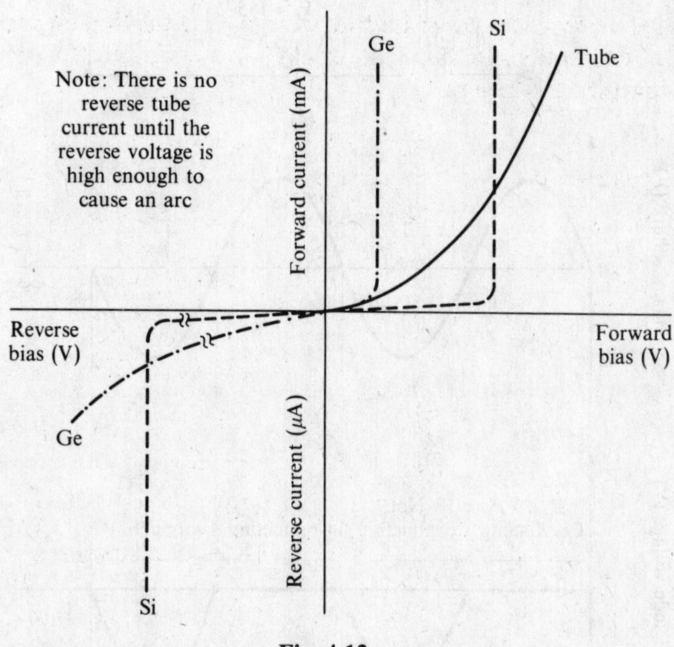

Fig. 4-13

The following parameters are usually supplied by the manufacturers of rectifier diodes:

- PRV, PIV, or VRM (see below)
- Maximum average forward current
- Surge or fault current
- Temperature range or temperature derating curve
- Forward voltage drop
- Maximum reverse current
- Power dissipation

The manufacturer may supply other information including:

- Base diagram
- Total capacitance
- Reverse recovery time
- Recommended operating ranges

PRV, PIV, or VRM stand for *peak reverse voltage*, *peak inverse voltage*, and *voltage reverse, maximum*. All mean the same thing. It is the maximum allowable reverse-bias voltage for the diode. If a diode has a PRV rating of 200 V, then it must never be used in a circuit where the cathode is more than 200 V positive with respect to the anode. If the PRV rating is exceeded, the diode will be destroyed.

Example 4.3 The transformer shown in Fig. 4-14*a* has a secondary voltage of 184 V. The PIV rating of the diode is 250 V. Can the diode be used in this circuit?

The rms value of secondary voltage given in the problem statement is 184 V. The peak or maximum value is written [from (*1-12b*)]

$$V_m = \sqrt{2}V_s = (\sqrt{2})(184) = 260.2 \text{ V}$$

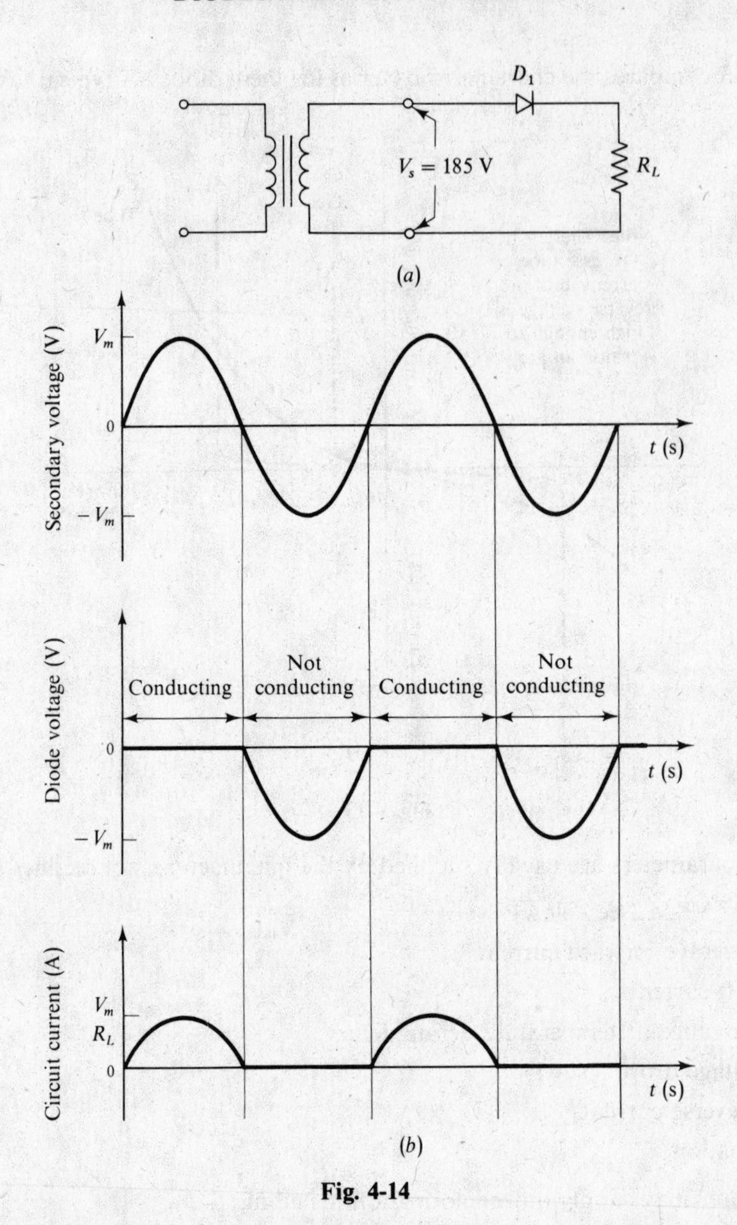

Fig. 4-14

On the half-cycle of secondary voltage that makes the anode positive with respect to the cathode, the diode conducts with very little voltage drop across it, and practically all the voltage drop appears across R_L.

On the half-cycle in which the diode is reverse-biased, there is no current. Therefore, there is no voltage drop across R_L. Since the algebraic sum of the voltages around the secondary circuit must be zero, and since there is no drop across R_L, it follows that *all the secondary voltage is across the diode during the half-cycle when it is reverse-biased*! This is shown in Fig. 4-14*b*.

The calculation shows that the peak reverse voltage is 260.2 V, which is greater than the 250 V allowed by the manufacturer. Therefore the diode cannot be used.

Maximum Average Forward Current

This is the maximum value of *full-cycle* average current I_F that the rectifier diode can safely conduct without becoming overheated.

The half-cycle average current for a sine wave is given by (*1-13a*)

$$I_{av} = \frac{2}{\pi} I_m$$

In a single diode rectifier circuit, the diode only conducts for half the cycle, so the full-cycle average I'_{av} is equal to I_{av} for the first half-cycle plus zero for the nonconducting half-cycle divided by 2:

$$I'_{av} = \frac{I_{av} + 0}{2}$$

Substituting $(2/\pi)I_m$ for I_{av},

$$I'_{av} = \frac{2I_m}{2\pi}$$

$$= \frac{I_m}{\pi} \qquad (4\text{-}1)$$

$$= 0.318 I_m \qquad (4\text{-}1a)$$

If the rms value I of circuit current is given instead of the peak value, we substitute

$$I_m = \sqrt{2}I$$

into Eq. (*4-1*) and obtain

$$I'_{av} = \frac{\sqrt{2}I}{\pi} \qquad (4\text{-}2)$$

$$= 0.450I \qquad (4\text{-}2a)$$

Generally, a safety factor of 2 to 1 is used, so a diode with an average current rating of 2 A would be used in a circuit where the actual average current will be 1 A.

Example 4.4 The maximum average forward current rating of the diode in the circuit shown in Fig. 4-15 is 1 A. Can it be used safely in that circuit?

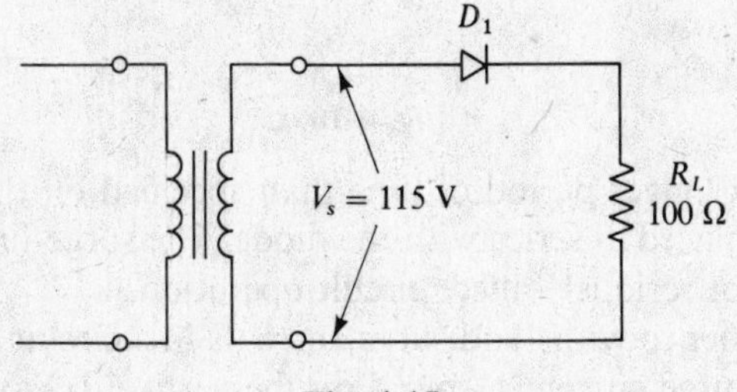

Fig. 4-15

Looking at Fig. 4-15, we see that $V_s = 115$ V, $R_L = 100\ \Omega$, and the rms maximum current that can pass through the diode D_1 in the forward condition is (by Ohm's law)

$$I = \frac{V_s}{R_L + r_{fd}}$$

where r_{fd} is the forward impedance (resistance) of the diode and $R_L + r_{fd}$ is the total impedance (resistance) of the circuit.

The maximum value of I will occur when $r_{fd} = 0$ (ideal diode). So let us use this as the "worst case" and calculate I:

$$I = \frac{V_s}{R_L} = \frac{115}{100} = 1.15 \text{ A}$$

Now from (*4-2a*) $\qquad I'_{av} = 0.450I$

Substituting, $\qquad I'_{av} = (0.450)(1.15) = 0.518 \text{ A} = 518 \text{ mA}$

The average diode current is only about one-half the value allowed by the manufacturer, so the diode can be used in the circuit shown in Fig. 4-15.

Designers usually choose a diode with a maximum average current rating equal to about twice the actual average current flow. This provides an adequate safety factor provided the temperature stays within specified bounds. Temperature dependence will be explained later in this section.

The average value of a nonsinusoidal waveform is not as easily obtained. A reasonably accurate value of average current can be obtained from an oscilloscope trace if the test setup is made carefully.

Surge Current (Also Called "Fault Current")

This is the amount of momentary overload current I_s the diode can withstand without being destroyed. Figure 4-16 shows typical surge currents for silicon diodes with ratings from 0.5 to 40 A. If the

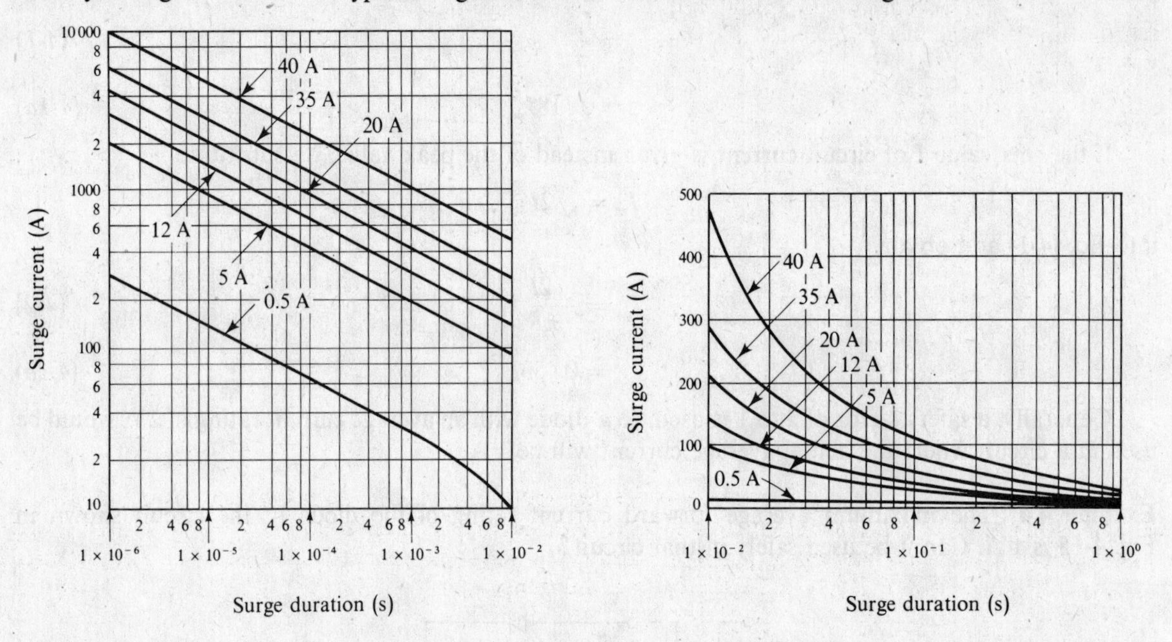

Fig. 4-16

surge current is expected for a longer period of time than specified on the chart, then a surge-limiting resistor (or a fuse) is normally placed in series with the diode. The surge-limiting resistor value is usually small (10 to 20 Ω) and does not seriously affect circuit operation.

The ability of a given rectifier to withstand surge currents in a circuit can be determined graphically from Fig. 4-16. The expected surge current is found on the vertical (y) axis and plotted as a horizontal straight line. The intersection of the surge current line with the rating curve gives the maximum allowable duration of the surge current and is read on the horizontal (x) axis. If the surge current is prolonged beyond this point without a surge-limiting resistor or fuse, the diode will be destroyed.

Example 4-5 In the circuit shown in Fig. 4-17, capacitor C will charge when the switch is closed. The charging current produces a surge of 150 A for a duration of 1×10^{-5} s. Can a 0.5-A diode be used in this circuit without protection?

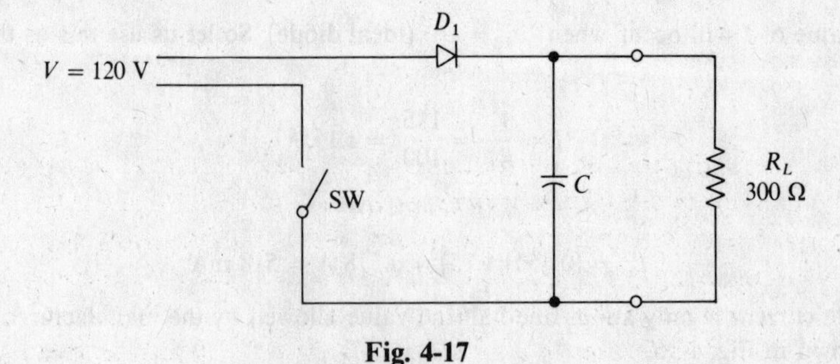

Fig. 4-17

Assuming that the diode has been properly selected for I_F and VRM, we check for surge currents by using the supplied surge current graph (Fig. 4-18).

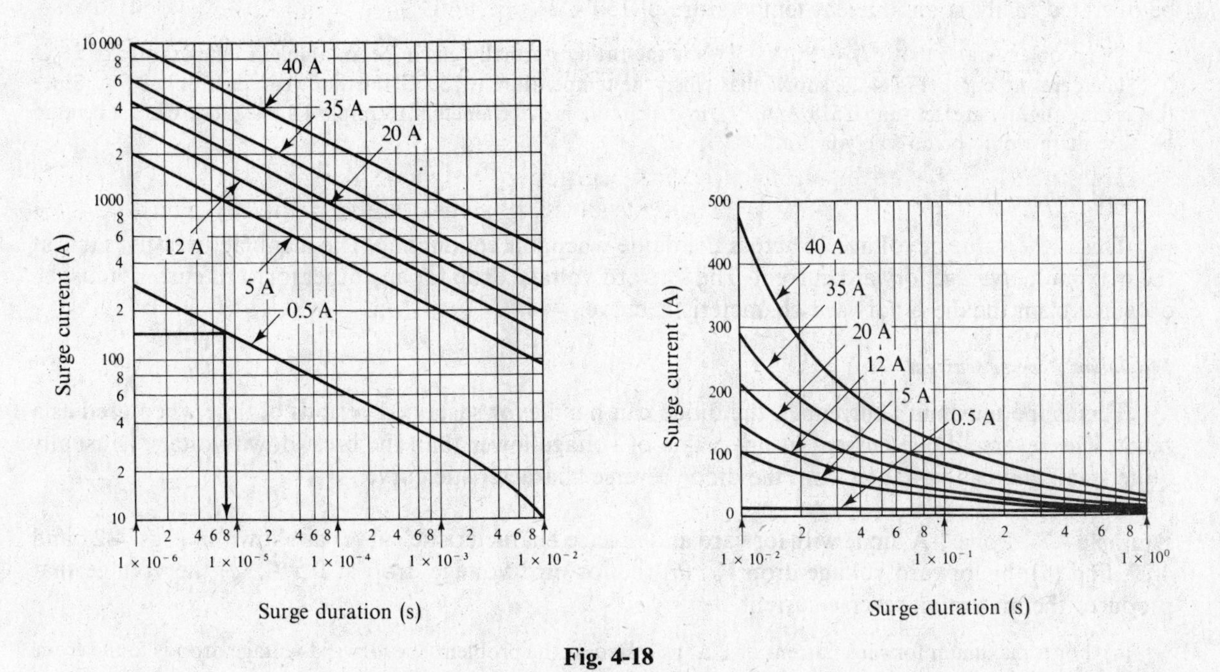

Fig. 4-18

As shown in Fig. 4-18, a line drawn horizontally at the 150-A surge current point intersects the 0.5-A curve at 8×10^{-6} s. This is a shorter time than the 1×10^{-5} s that the problem statement gives for the surge current. So, the diode cannot be used without a surge protection resistor, or it will be necessary to use a diode with a higher surge current rating. Note from Fig. 4-18 that a diode with a 5-A rating can withstand a surge current of 150 A for almost 1×10^{-2} s.

Figure 4-19 shows the circuit with a surge-limiting resistor R_S.

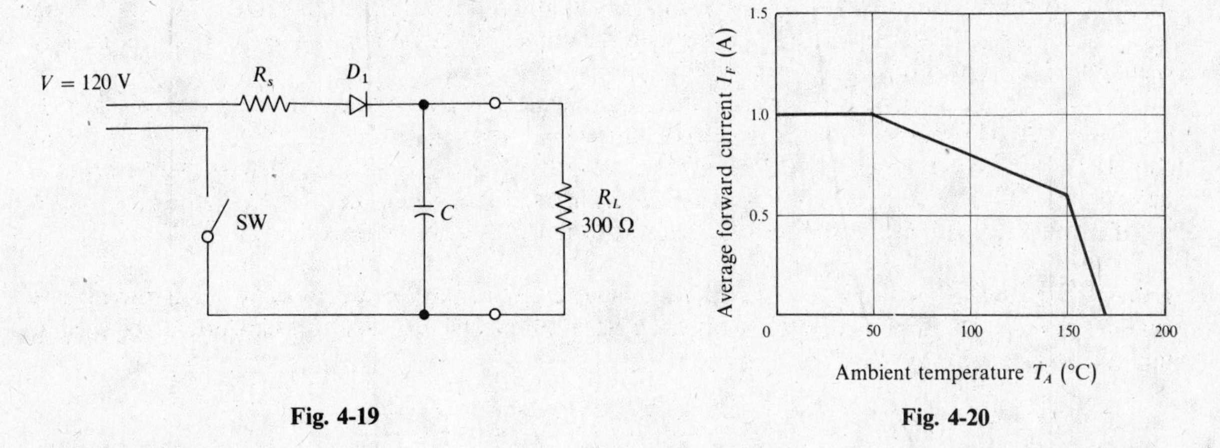

Fig. 4-19 **Fig. 4-20**

Temperature Range

This is the range of temperatures at which the diode can safely operate within the listed parameters. For most diodes, the maximum temperature range is from −25 to 85°C (−13 to 185°F). The I_F, PIV, and other parameters are usually listed for an ambient (surrounding) temperature of 25°C (77°F).

Depending on the manufacturer of the diode, a *temperature derating curve* is sometimes supplied. Since increasing the temperature of the diode reduces its ability to handle current, the curves should be used to determine the I_F at other than 25°C. Figure 4-20 shows a typical temperature derating curve.

Example 4.6 Figure 4-20 shows the temperature derating curve for the diode in the circuit shown in Fig. 4-15. The maximum average forward current rating I_F of this diode is 1.0 A. Could the diode circuit be operated safely at an ambient temperature of 150°C?

The problem states that $I_F = 1.0$ A. This is the rating normally given for an ambient temperature of 25°C. The derating curve, Fig. 4-20, shows that when the temperature is 150°C, the diode can pass only 0.6 A. Since this is only slightly greater than 0.518 A, the I'_{av} in the circuit as determined in Example 4.4, the diode cannot be used because there would be no safety factor.

Forward Voltage Drop

This is the value of voltage V_F across the diode when it is conducting. The manufacturer gives this at the maximum average forward current. The forward voltage drop for any other forward current must be obtained from the diode forward characteristic curve.

Maximum Reverse Current

This is the maximum current I_R the diode can handle for sustained periods of time when used as a zener. The reverse diode current at any value of voltage lower than the breakdown voltage is usually quite small and can be taken from the diode reverse characteristic curve.

Example 4.7 For a 1-A diode with forward and reverse characteristic curves as shown in Figs. 4-21 and 4-22, find (*a*) the forward voltage drop V_F, (*b*) the forward voltage drop at 0.5 A, (*c*) the voltage that produces the maximum reverse current.

(*a*) For a maximum forward current of 1 A, as stated in the problem, the forward voltage drop is found from Fig. 4-21.

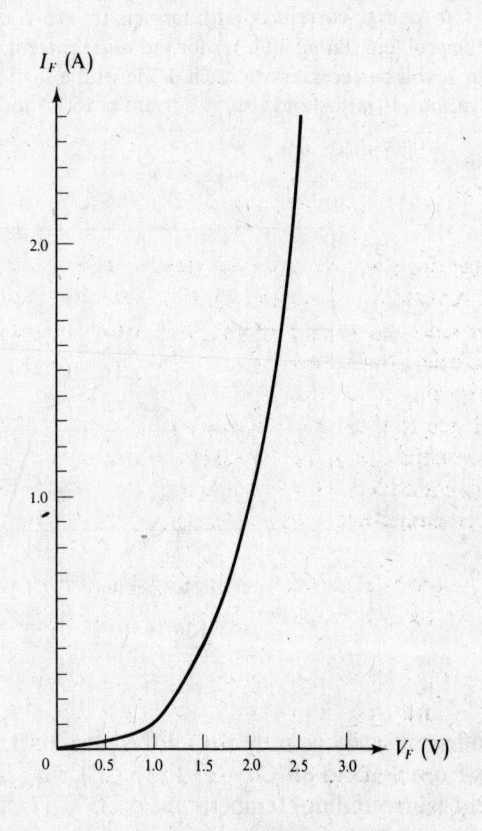

Fig. 4-21 Forward characteristic curve, 1.0 A diode

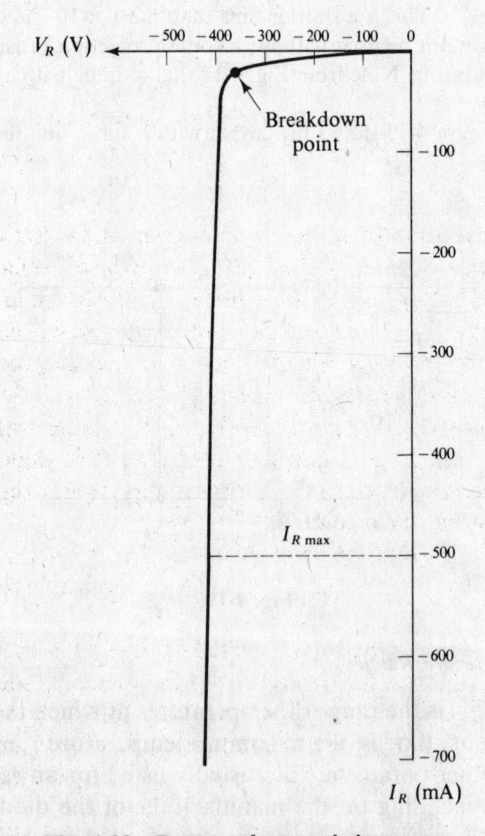

Fig. 4-22 Reverse characteristic curve, 1.0 A diode

We draw a horizontal line at 1 A on the vertical current scale. At the point where the horizontal line intersects the characteristic curve, we drop a vertical line to the horizontal voltage scale and read the value from Fig. 4-21. This value for 1 A is 2.0 V.

(*b*) Using Fig. 4-21 again, we see that a horizontal line at 0.5 A intersects the curve at a forward voltage drop of 1.6 V.

(*c*) The maximum reverse current as shown in Fig. 4-22 is 500 mA. Using the same method as in part (*a*), we see that the reverse voltage for 500 mA is −410 V.

4.6 RECTIFIER DIODES IN SERIES AND PARALLEL

Figure 4-23*a* and *b* shows how rectifier diodes are connected in series and parallel. The series connection is used to increase the peak inverse voltage. The PIV ratings are additive, so if a diode rated at 500 PIV is connected in series with a diode rated at 1000 PIV, the PIV rating of the series combination will be 1500 V. Normally, however, diodes used in series are chosen to have the same PIV rating.

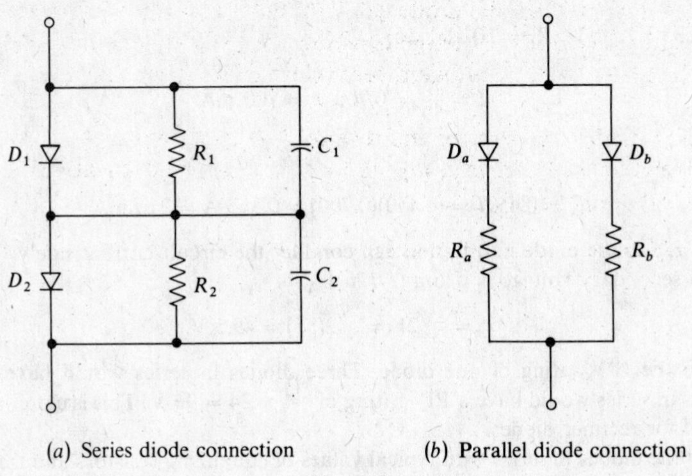

(*a*) Series diode connection (*b*) Parallel diode connection

Fig. 4-23

Resistors R_1 and R_2 are used to equalize the voltage drop across the series diodes. Even though the diodes have the same PIV rating, a slight difference in reverse leakage can result in a greater-than-rated reverse voltage across one diode. The resistors are chosen to have high values of resistance so they will not draw much current and are placed across each diode as shown in Fig. 4-23*a*. This assures equal voltages across the diodes—hence, they are called *equalizing resistors*.

Transient voltages on the ac line can exceed the diode PIV rating. The capacitors protect the series diodes by bypassing the transients. When selecting these capacitors, it is important to remember that the larger voltage drop occurs across the lower value of capacitance. This is true regardless of whether the voltage is ac or dc. It is important, therefore, to connect capacitors with equal capacitance values across diodes that have equal PIV ratings.

The current rating of series-connected diodes is equal to the lowest current rating among the diodes. For example, when a 1-A diode is connected in series with a 2-A diode, the rating of the combination is 1 A.

When diodes are connected in parallel as shown in Fig. 4-23*b*, the current rating is the sum of each diode rating. The resistors in series with the diodes assure that each carries its share of the current. Without the resistors, one diode may start to conduct first, and the voltage drop across it will be so low that the other diode(s) cannot start. Keep in mind the fact that the approximate forward drop is 0.2 V across a germanium diode and 0.7 V across a silicon diode.

Example 4.8 A certain diode has the following specifications: VRM = 24 V; $I_F = 0.6$ A. Show how this type of diode can be used in the circuit shown in Fig. 4-24.

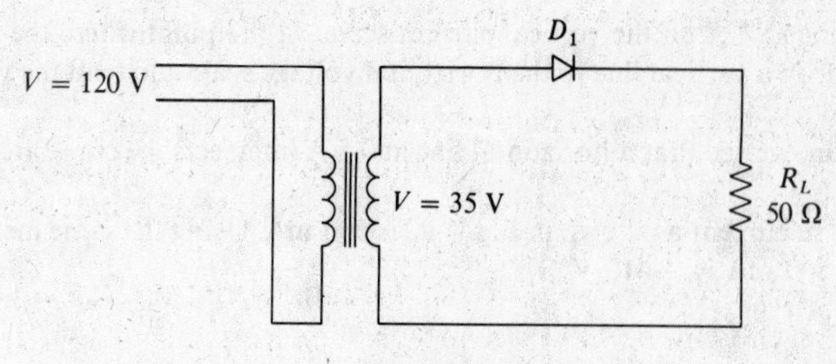

Fig. 4-24

The rms current in the circuit during the first half-cycle, and assuming a perfect diode, is (by Ohm's law)

$$I = \frac{V}{R_L}$$

From Fig. 4-24, $V = 35$ V and $R_L = 50\ \Omega$. So,

$$I = \frac{35}{50} = 0.700 \text{ A} = 700 \text{ mA}$$

Then from (*4-2a*),

$$I'_{av} = 0.450I = (0.450)(0.700) = 0.315 \text{ A} = 315 \text{ mA}$$

This is about half I_F, so the diode in question can conduct the circuit current safely.

The peak value of secondary voltage is [from (*1-12b*)]

$$V_m = \sqrt{2}V = \sqrt{2}(35) = 49.5 \text{ V}$$

This is over twice the PIV rating of one diode. Three diodes in series would have a PIV rating of $3 \times 24 = 72$ V. Four diodes in series would have a PIV rating of $4 \times 24 = 96$ V. This is closer to the two-to-one safety factor usually preferred for rectifier diodes.

Figure 4-25 shows the diodes in series with typical values of equalizing resistors and transient filter capacitors. The total reverse resistance is $4 \times 470\ \text{k}\Omega = 1.88\ \text{M}\Omega$. The total reverse capacitance is

$$\frac{0.001 \times 10^{-6}}{4} = 0.00025\ \mu\text{F}$$

The high resistance and low capacitance form a large reverse impedance. The small reverse ac current can be neglected.

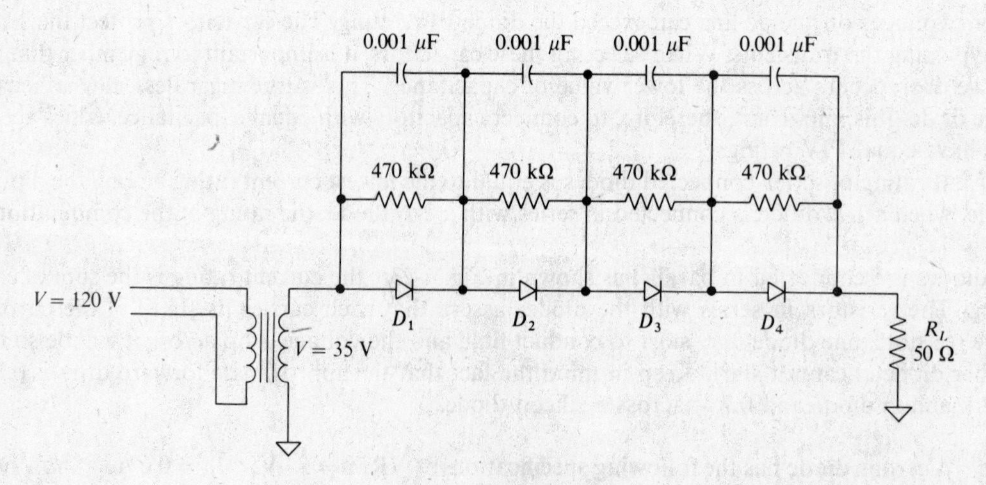

Fig. 4-25

4.7 LOAD LINE SOLUTIONS FOR DIODES AND NONLINEAR LOAD IMPEDANCES

Diodes are nonlinear components. Doubling the voltage across a diode does not necessarily double the current through it. Since they are nonlinear, it is not possible to use Ohm's law. However, a graphical solution can usually be obtained. The procedure is better understood by first reviewing the procedure of graphing voltages and currents in linear circuits.

Consider the simple circuit shown in Fig. 4-26*a*. Assuming that the resistor R_x is linear, a graph of the voltage and current for the circuit will be a straight line as in Fig. 4-26*b*. This line can be called a *characteristic curve* for R_x.

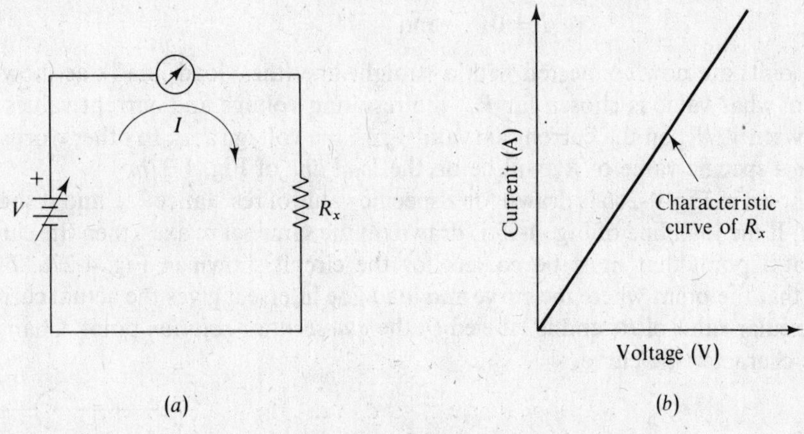

Fig. 4-26

If R_x is connected into a circuit with a 10-V battery and a 100-Ω resistor, as shown in Fig. 4-27*a*, the current I in the circuit will depend upon the actual resistance of R_x. Even without knowing the actual resistance value of R_x, it is certain that it must be some value between *zero* and *infinity*. These two conditions will be used to plot the *load line* as follows.

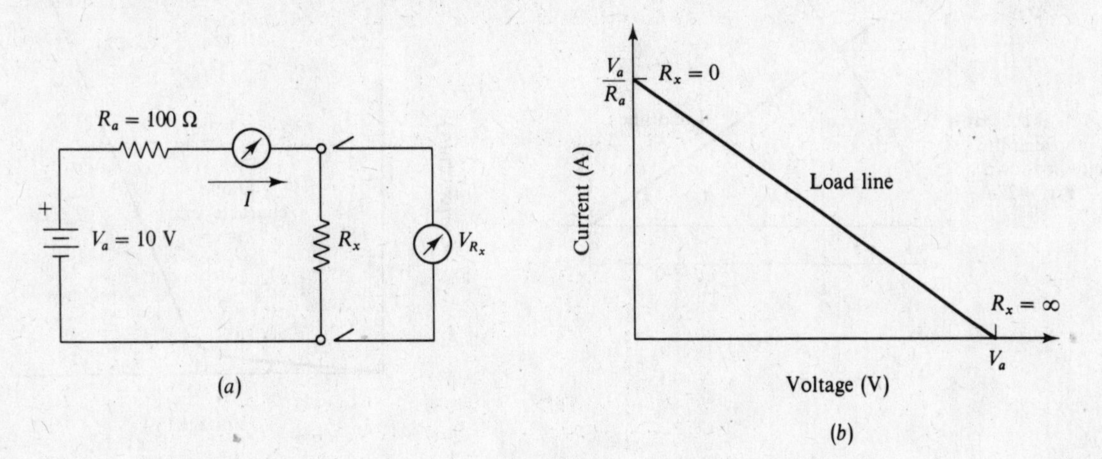

Fig. 4-27

First Condition

If the resistance of R_x is 0 Ω, then it is a short circuit. The short-circuit current is

$$I = \frac{V_a}{R_a}$$

The voltage across R_x must be zero for this short-circuit condition.

Second Condition

If the resistance of R_x is infinity, then it is an open circuit and the voltage across it will be V_a. The current through R_a must be zero for this open-circuit condition.

Two points are now known:

- When $R_x = 0$ (first condition),

$$I = \frac{V_a}{R_a} \quad \text{and} \quad V_x = 0$$

- When $R_x = \infty$ (second condition),

$$I = 0 \quad \text{and} \quad V_x = V_a$$

These two points are now connected with a straight line (the "load line"), as shown in Fig. 4-27*b*.

Regardless of what value is chosen for R_x, the resulting voltage and current values must lie on the straight line between V_a/R_a on the current axis and V_a on the voltage axis. In other words, a plot of the V and I points for a specific value of R_x will be on the *load line* of Fig. 4-27*b*.

The curve shown in Fig. 4-26*b* is drawn for a specific value of resistance R_x, and it shows the possible values of V vs. I. If the load line of Fig. 4-27 is drawn on the same set of axes, then the curve and the load line will cross at a point that must be correct for the circuit shown in Fig. 4-27*a*. This is shown in Fig. 4-28. Note that the point where the curve and load line intersect gives the actual current and voltage values for a particular value of R_x and is labeled Q, the *quiescent* or *solution* point. Changing R_x changes the slope of the characteristic curve.

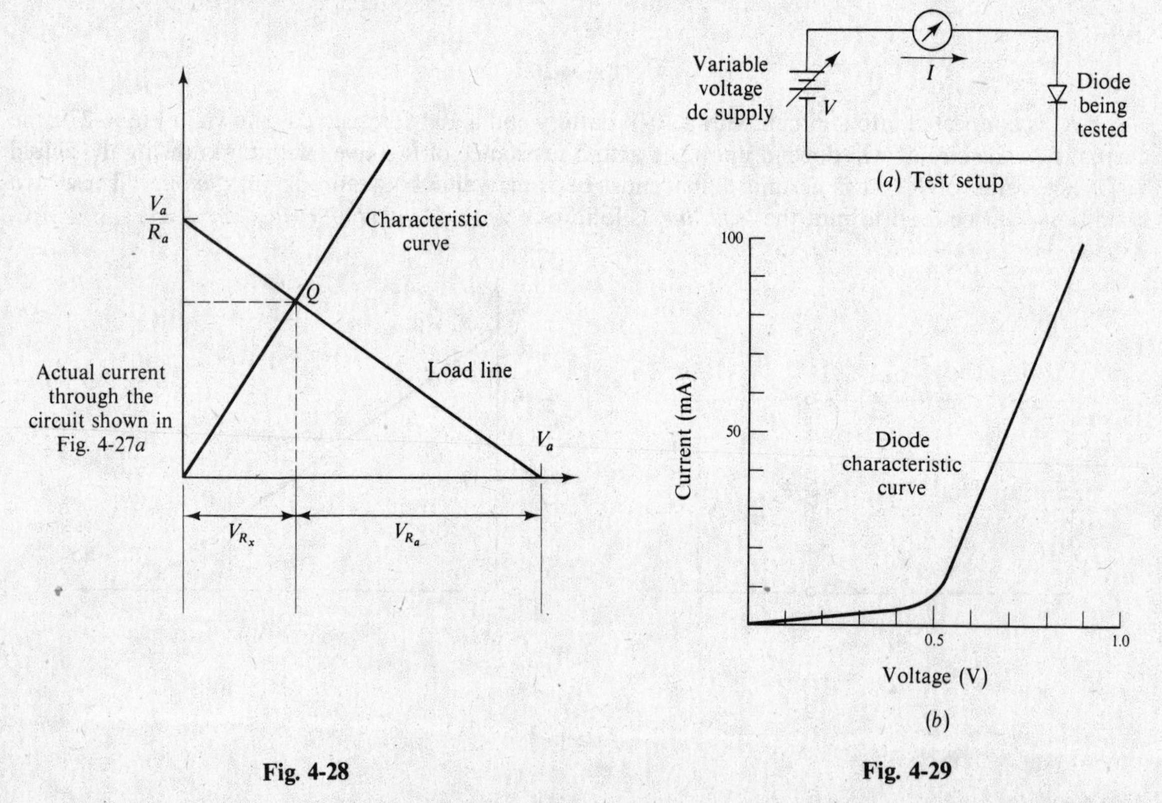

Fig. 4-28

Fig. 4-29

The graphical solution just described for linear resistors can also be used for nonlinear resistors and other electronic components.

For the test setup shown in Fig. 4-29*a*, the current I is measured for various values of voltage V, and the data plotted. From the data, a characteristic curve is obtained, as shown in Fig. 4-29*b*. Then, for any particular value of voltage, the current can be determined graphically, and vice versa. The term *static characteristic curve* is used for such curves because they are obtained with dc voltages and currents.

Example 4.9 Using the diode characteristic curve shown in Fig. 4-29*b*, determine the current when the voltage across the diode is 0.6 V.

The graphical solution is shown with dotted lines in Fig. 4-30. With a forward voltage of 0.6 V, the diode current is 27.5 mA.

Example 4.10 The curve shown in Fig. 4-29*b* is for a silicon rectifier that has $I_F = 52.5$ mA. What is V_F?

Figure 4-31 shows the solution with dotted lines. The graphical solution shows that $V_F = 0.7$ V.

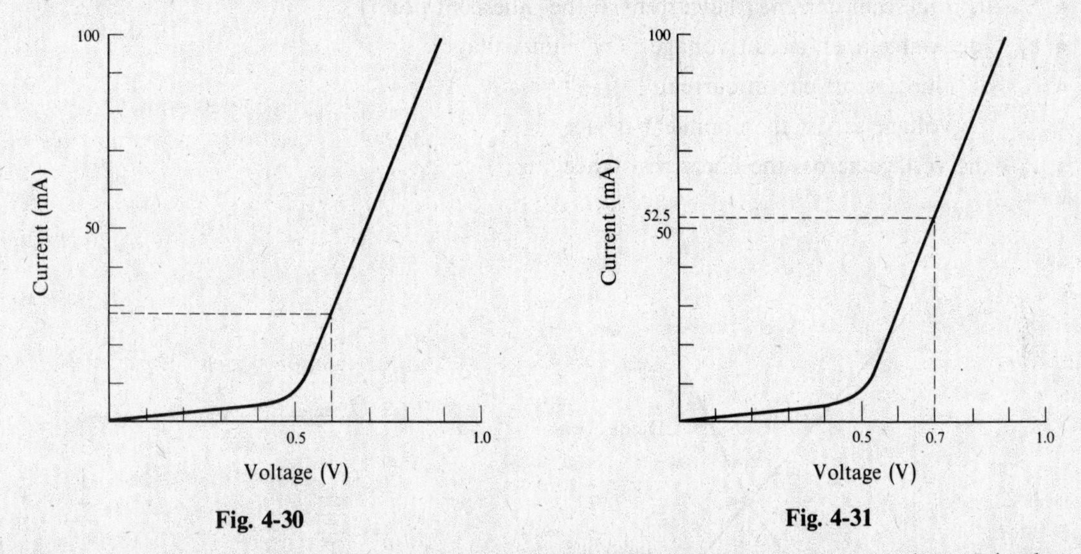

Fig. 4-30

Fig. 4-31

Figure 4-32*a* shows a somewhat more involved problem. In this case, a linear resistor R is placed in series with the nonlinear diode. Now there will be two voltage drops around the circuit: V_R and V_b. It would be difficult to find the current in this circuit analytically, and so a graphical solution, using a load line, is used to find the diode voltage and current.

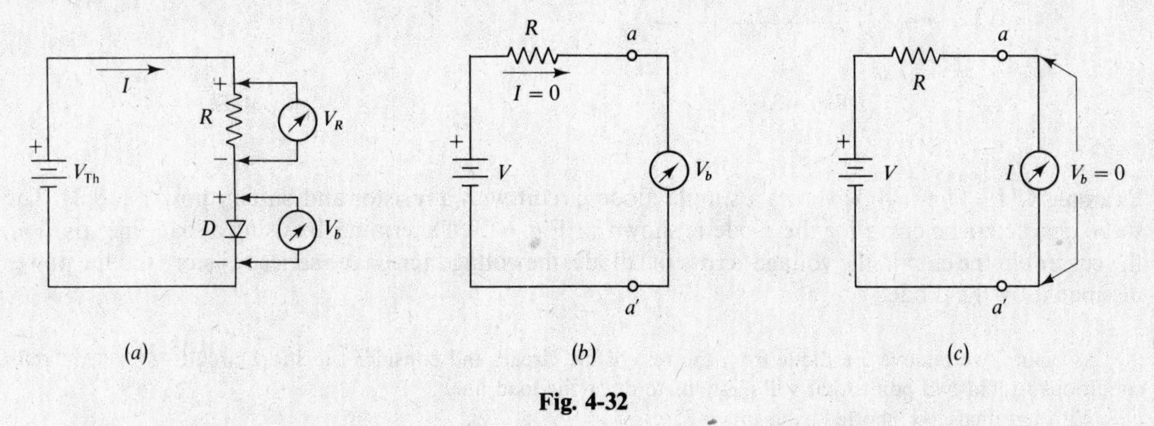

Fig. 4-32

Let us consider the diode as a nonlinear load and draw the load line for the remaining circuit. Two conditions are of interest:

- First condition: Terminals *a*-*a*′ open-circuited, Fig. 4-32*b*,

$$I = 0 \quad \text{and} \quad V_b = V$$

- Second condition: Terminals *a*-*a*′ short-circuited, Fig. 4-32*c*,

$$V_b = 0 \quad \text{and} \quad I = \frac{V}{R}$$

When the two points just obtained are connected by a straight line, we have the load for the circuit with the diode removed.

If we now plot the diode's static characteristic curve on the same set of axes, the intersection of the line and curve will represent the specific value of current with the diode connected and will also show the voltage across the diode and the resistor. Figure 4-33 shows the line and curve superimposed with various circuit values identified.

Figure 4-33 shows that a load line, when superimposed on the characteristic curve of a nonlinear device, will give the following information:

- I_Q—the quiescent current (the current at the quiescent point)
- V_{Th}—dc source open-circuit voltage (Thevinin voltage)
- I_N—dc source short-circuit current
- V_b—the voltage across the nonlinear device
- V_R—the voltage across the linear resistance

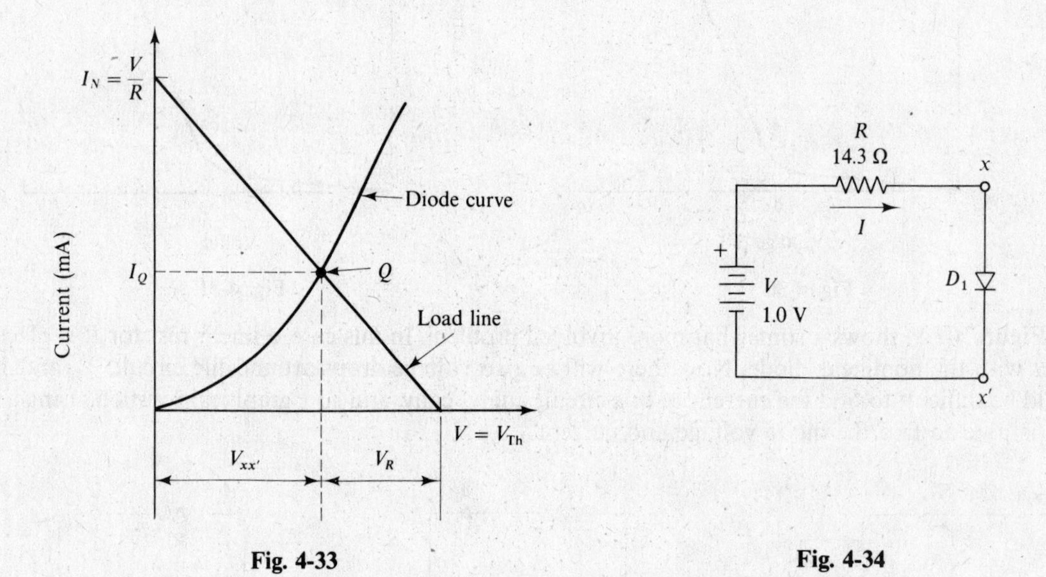

Fig. 4-33

Fig. 4-34

Example 4.11 Figure 4-34 shows a simple diode circuit with a resistor and battery power supply. The static characteristic curve for the diode is shown in Fig. 4-29. Determine, by using a load line solution, the current in the circuit, the voltage across the diode, the voltage across the series resistor, and the power dissipated by the diode.

As before, we remove the diode from the rest of the circuit and consider the short-circuit and open-circuit conditions to find two points that will allow us to draw the load line.

With terminals x-x' shorted as in Fig. 4-35a:

$$V_{x\text{-}x'} = 0 \quad \text{and} \quad I = \frac{V}{R}$$

Substituting values,

$$I = \frac{1.00}{14.3} = 69.9 \times 10^{-3}\ \text{A} = 69.9\ \text{mA}$$

When terminals x-x' are open-circuited as in Fig. 4-35b:

$$I = 0 \quad \text{and} \quad V_{x\text{-}x'} = V = 1.00\ \text{V}$$

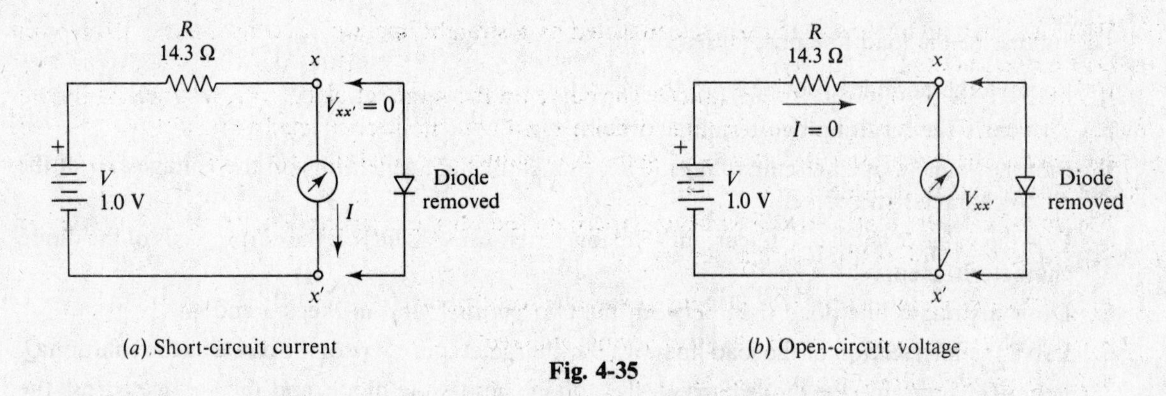

(a) Short-circuit current (b) Open-circuit voltage

Fig. 4-35

These two points are plotted and the load line drawn between them. The static characteristic curve for the diode is then drawn as shown in Fig. 4-36 using the same set of axes. The point where the load line and characteristic curve intersect is labeled Q.

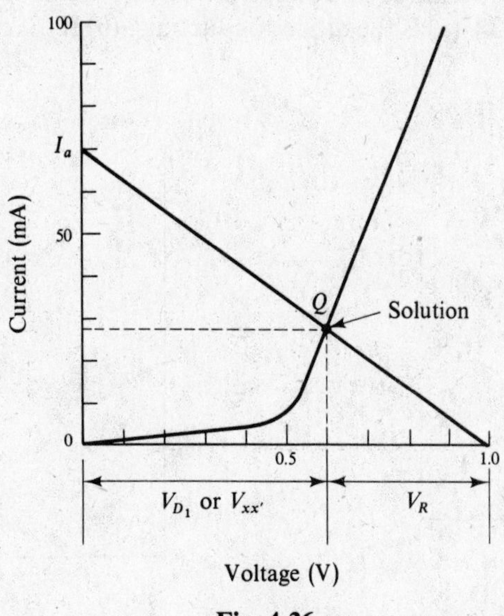

Fig. 4-36

The voltage $V_{xx'}$ across the diode is the horizontal distance from the origin to the solution. Figure 4-36 shows this to be 0.6 V.

The voltage V_R across the resistor is the horizontal distance from the quiescent point to the applied voltage V. Figure 4-36 shows this to be 0.4 V.

Note that the voltage across the diode and the voltage across the resistor add to equal the applied voltage in accordance with Kirchhoff's voltage law.

The current through the circuit with the diode connected is the vertical distance from the origin to the solution. Figure 4-36 shows this to be 27.5 mA.

The power dissipated by the diode is equal to the voltage across it times the current through it. Using Eq. (*1-29a*),

$$P = VI = (0.6)(27.5 \times 10^{-3}) = 16.5 \times 10^{-3}\ \text{W} = 16.5\ \text{mW}$$

Although a diode has been used as an example of a nonlinear device, the procedure would be the same if any nonlinear device (such as a thermistor) were used in place of the diode in the circuit shown in Fig. 4-34.

The diode in Fig. 4-34 is actually connected to a Thevenin generator. This means that the current for the nonlinear diode can be determined in a circuit by removing that diode and Thevenizing the remaining circuit. This assumes, of course, that the remainder of the circuit is linear so that it can be Thevenized.

To summarize the load line procedures:

1. Remove the nonlinear device from the circuit.
2. Thevenize the resulting two-terminal circuit.
3. Plot the value of open-circuit voltage (Thevenin equivalent generator) on the voltage axis of the diode characteristic curve.
4. Plot the value of short circuit current (Norton generator current) on the current axis of the diode characteristic curve.
5. Draw a straight line (load line) between the two points found in Steps 3 and 4.
6. Label the intersection of the load line with the characteristic curve of the diode as the solution Q.
7. From Q, determine the diode current, the voltage across the diode, and the voltage across the Thevenin equivalent resistance that will occur when the diode is reinserted in the original circuit, by reading the required values from the graph.

Example 4.12 Figure 4-37 shows a diode in a circuit with two voltage sources. The characteristic curve for the diode is shown in Fig. 4-38. (*a*) Is the diode conducting? (*b*) If the diode is conducting, what is the value of diode current?

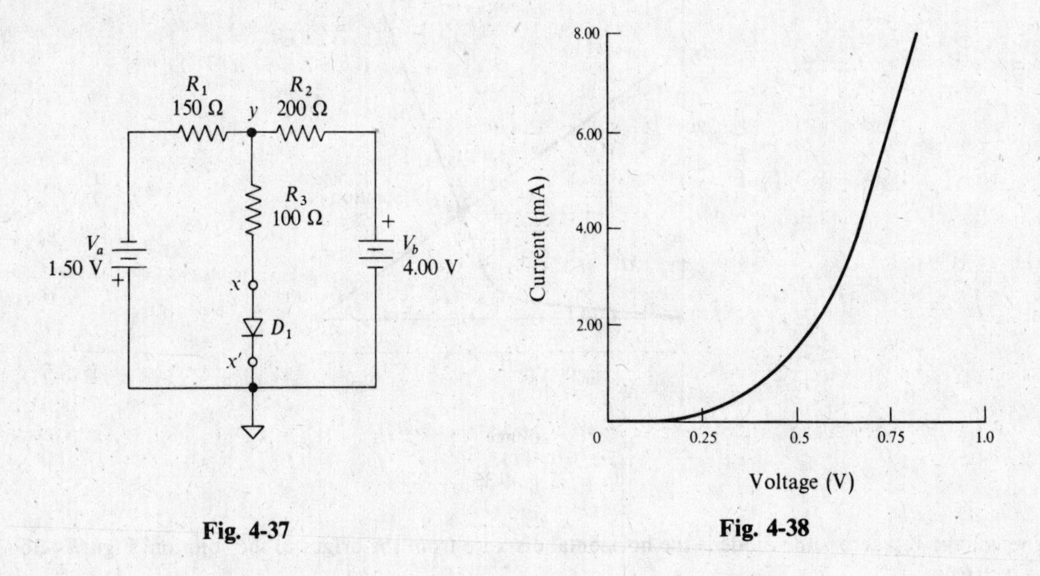

Fig. 4-37

Fig. 4-38

We will use the step by step summary to solve this problem:

1. Remove the diode from the circuit.
2. Thevenize the remaining circuit, which we will redraw (see Fig. 4-39) to make the solution easier.

The Thevenin resistance R_{Th} is the resistance looking back into the circuit with all voltage sources shorted and replaced with their internal resistances, if any. This is shown in Fig. 4-40*a*. From Fig. 4-40*a*, we can determine that

$$R_{\text{Th}} = R_3 + \frac{R_1 R_2}{R_1 + R_2} = 100 + \frac{(150)(200)}{150 + 200} = 100 + 85.7 = 186\ \Omega$$

The Thevenin voltage is the open-circuit voltage $V_{xx'}$.

While there are many ways to obtain the Thevenin voltage, let us use the Maxwell nodal solution.

Assign the unknown Thevenin voltage V_{Th} to point y in Fig. 4-39. x' is the reference point and since terminals x-x' are open-circuited, $I_3 = 0$, there is no voltage drop across R_3, and the voltage at point y is equal to V_{Th}.

The currents leaving node y are

$$I_1 = \frac{V_{\text{Th}} - (-V_a)}{R_1} = \frac{V_{\text{Th}} + V_a}{R_1} \qquad I_2 = \frac{V_{\text{Th}} - (V_b)}{R_2} \qquad I_3 = 0$$

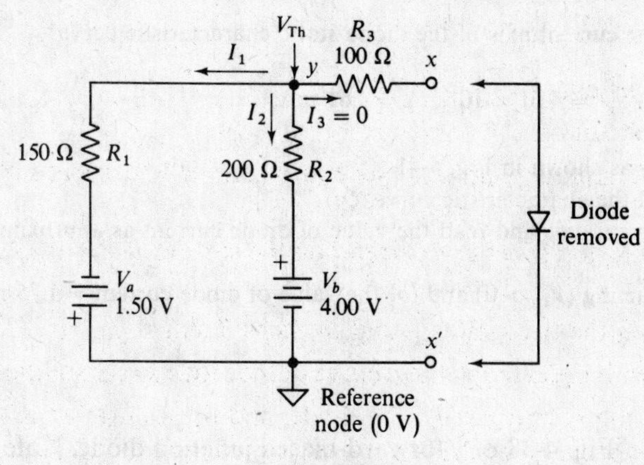

Fig. 4-39

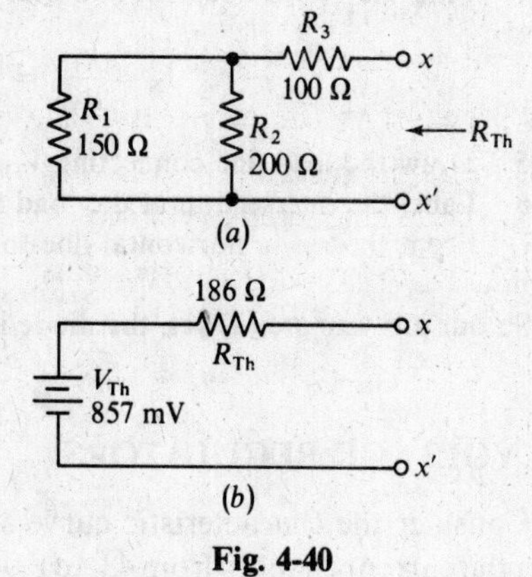

Fig. 4-40

Writing the nodal equation at point y,

$$I_1 + I_2 + I_3 = 0$$

$$\frac{V_{Th} + V_a}{R_1} + \frac{V_{Th} - V_b}{R_2} = 0$$

Grouping terms and factoring,

$$V_{Th}\left(\frac{1}{R_1} + \frac{1}{R_2}\right) = \frac{V_b}{R_2} - \frac{V_a}{R_1}$$

Solving for V_{Th},

$$V_{Th} = \frac{V_b/R_2 - V_a/R_1}{1/R_1 + 1/R_2}$$

Substituting values,

$$V_{Th} = \frac{4.00/200 - 1.50/150}{1/150 + 1/200} = \frac{(2 \times 10^{-2}) - (1 \times 10^{-2})}{1.167 \times 10^{-2}} = 8.57 \times 10^{-1} = 857 \times 10^{-3}\ \text{V} = 857\ \text{mV}$$

3. With the Thevenized circuit as shown in Fig. 4-40b, plot V_{Th} on the voltage axis of the diode static characteristic curve.

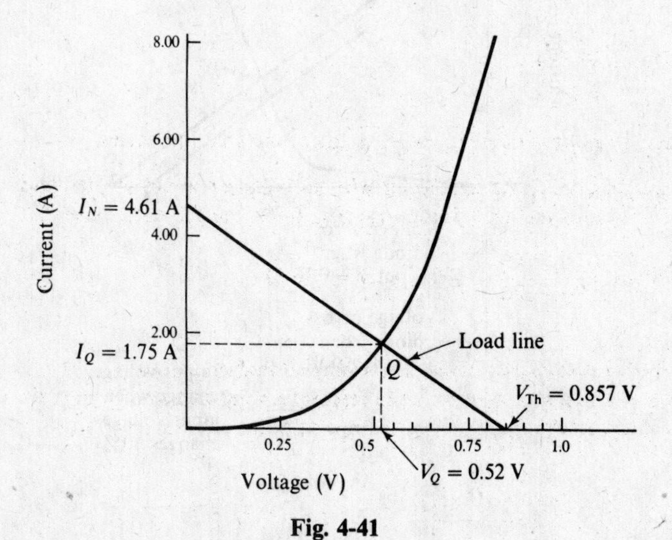

Fig. 4-41

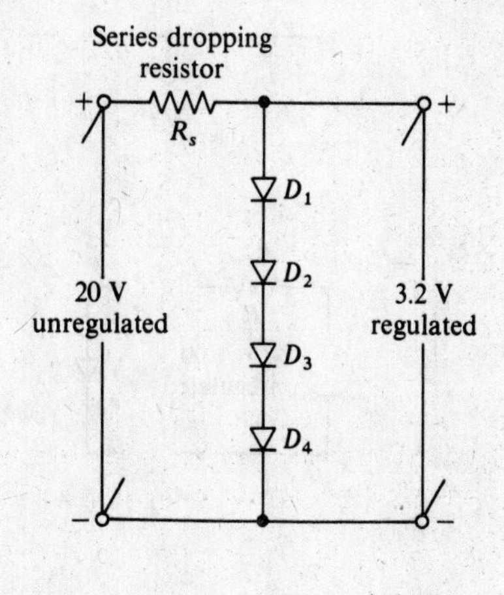

Fig. 4-42

4. Find the Norton current and plot it on the current axis of the diode static characteristic curve:

$$I_N = \frac{V_{Th}}{R_{Th}} = \frac{857 \times 10^{-3}}{186} = 4.61 \times 10^{-3}\ \text{A} = 4.61\ \text{mA}$$

5. Draw the load line connecting V_{Th} and I_N as shown in Fig. 4-41.
6. Label the intersection of the load line and the characteristic curve, Q.
7. From Q, draw a horizontal line to the current axis and read the value of diode current as approximately 1.75 mA.

So our answers are (*a*) yes, the diode is conducting ($V_{Th} > 0$) and (*b*) the value of diode current = 1.75 mA.

4.8 VOLTAGE REGULATORS

Consider the characteristic curve shown in Fig. 4-38 of a forward-biased junction diode. Note that when the current changes from 4.0 to 8.0 mA, the voltage across the diode changes only about 0.18 V. Thus, doubling the current produces only a small change in voltage. This makes the diode useful as a voltage regulator for low-voltage values.

Figure 4-42 shows four diodes D_1 to D_4 each with a forward voltage drop V_F equal to 0.8 V connected in series to obtain a regulated value of 3.2 V. Reasonably large changes in the unregulated applied voltage in this circuit will not produce a significant change in the 3.2-V regulated output. Currents and voltages in the regulated supply of Fig. 4-42 can be determined by the load line method. Stacking four diodes in series increases the peak inverse voltage rating to four times the rating of one diode—assuming the diodes are identical. The method of regulating voltage as shown in Fig. 4-42 is used extensively in integrated circuit technology.

Example 4.13 Figure 4-43*a* shows a simple regulator for a 0.6-V output voltage. This circuit uses a 1-V unregulated input, a series resistor, and a diode with the characteristic curve shown in Fig. 4-29*b*. Find the following: (*a*) the required value of *R*; (*b*) the change in output voltage if the unregulated input voltage drops to 0.9 V.

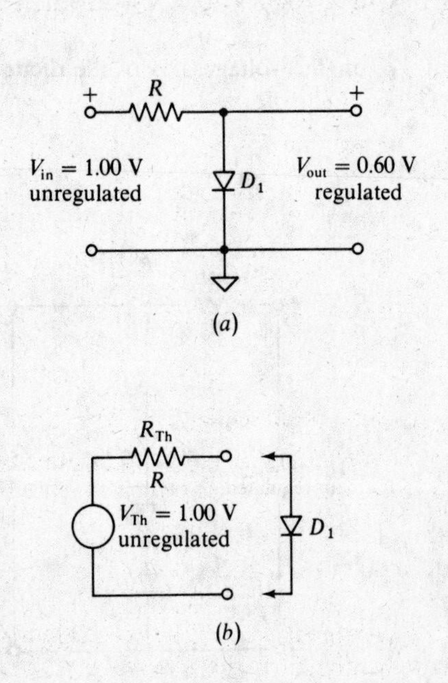

Fig. 4-43

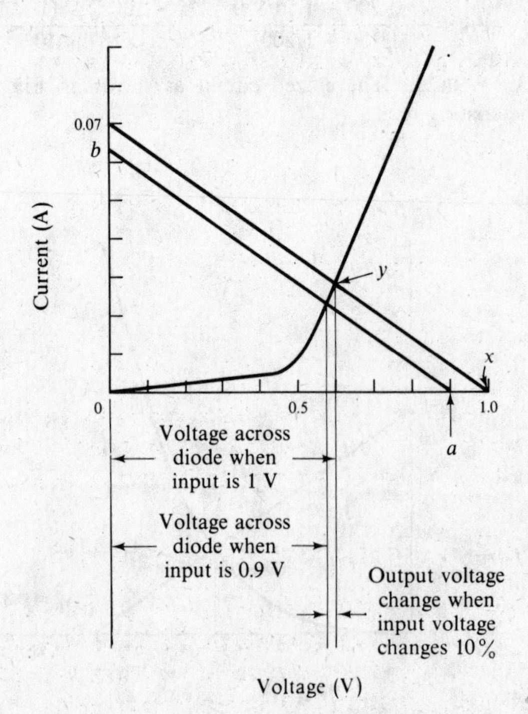

Fig. 4-44

(*a*) Two points are known for drawing the load line, and they are shown on Fig. 4-44. The open-circuit voltage V_{Th} is the unregulated input of 1.0 V. This point is plotted on the voltage axis and marked *x*. Also, the load line must cross the diode curve at a point where the voltage across the diode is 0.6 V. This point is marked *y*.

The load line is drawn through points *x* and *y* and intersects the current axis where $V = 0$ at approximately 0.07 A.

This point on the current axis is the Norton current I_N given by V_{Th}/R_{Th}. Redrawing Fig. 4-43*a* in Thevinin form as Fig. 4-43*b*, we see that V_{Th} is 1.00 V and R_{Th} is *R*. So from Ohm's law,

$$I_N = \frac{V_{Th}}{R_{Th}} = \frac{V_{Th}}{R}$$

Solving for *R* and substituting values,

$$R = \frac{V_{Th}}{I_N} = \frac{1.00}{7 \times 10^{-2}} = 14.3\ \Omega$$

(*b*) Before proceeding with the solution of the problem, it should be explained that 1 V is assumed to be a Thevenin voltage in the solution just worked. Obviously, it is not a true Thevenin voltage because it is marked *unregulated* in Fig. 4-43*a*. However, at any time when the input is 1 V, it can be considered to be a fixed value. It will now be shown that, if the input voltage changes to 0.9 V but the value of *R* remains the same, the load line will shift down, but parallel to the load line previously drawn.

When $V = 0.9$ V, a new point is found along the *x* axis. It is marked point *a*. The short-circuit current I_N is

$$I_N = \frac{V}{R} = \frac{0.9}{14.3} = 0.063 \text{ A} = 63 \text{ mA}$$

This is marked point *b* on the graph. The load line through points *a* and *b* is parallel to the original one through points *x* and *y*. Note that the voltage across the diode, which is the regulated output, drops only about 0.02 V or 3.3% when the unregulated input drops 10 per cent or 0.1 V. Thus, the output is *regulated* against voltage changes. If the diode curve were steeper, the regulation would improve.

The solution for the circuit shown in Fig. 4-43 assumes there is no load resistance connected across the output terminals. This type of circuit may be used as a voltage calibrator for an oscilloscope. The output terminals of the calibrator (Fig. 4-43) are connected to the dc input terminals of the scope, as shown in Fig. 4-45. The vertical attenuator is adjusted for a convenient scale and the scope is then usable for measuring voltages. Most modern oscilloscopes have some form of built-in calibrator, but in some cases an external calibrator may extend their use.

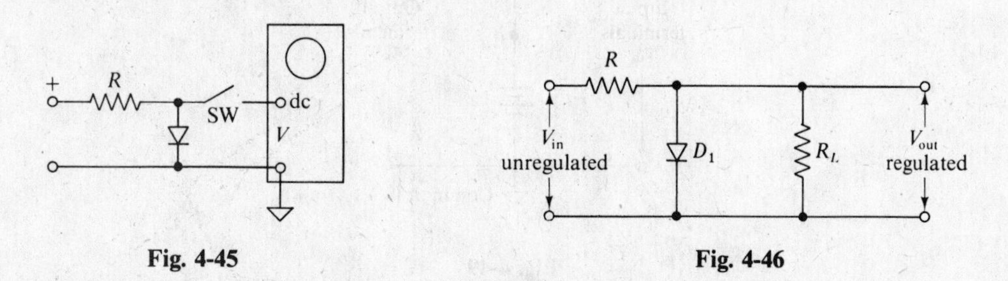

Fig. 4-45 **Fig. 4-46**

When there is a load resistance connected to the output terminals of the circuit as in Fig. 4-46, then the load current plus the diode current flows through resistor *R* and the value of *R* would have to be reduced. Problems of this type are discussed in Prob. 4.9 and also later in Chap. 8.

4.9 ZENER DIODE REGULATORS

As shown on the curve in Fig. 4-47, the reverse breakdown or zener voltage of a silicon diode is very sharp. Once this point is reached, the curve is nearly vertical with reference to the voltage axis. If the diode curve is perfectly vertical at the zener voltage point, there would be *no change in output voltage* when the unregulated input increases past the zener voltage point. This condition is very nearly reached if the diode is operated in the zener region.

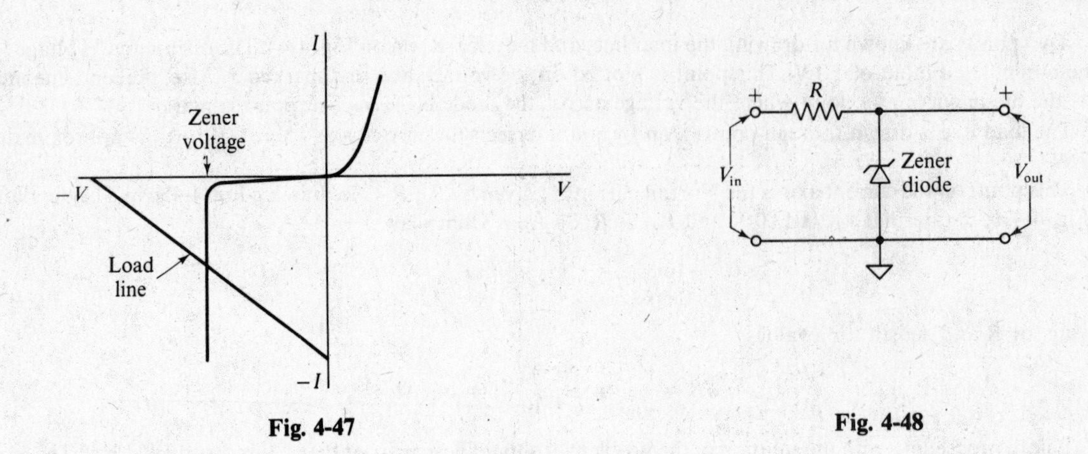

Fig. 4-47

Fig. 4-48

Figure 4-48 shows a zener diode connected as a voltage regulator. Note that the zener diode is connected into the circuit so that it is reverse-biased for operation on the reverse characteristic curve. A load line is used to determine the operating point in the same manner as when the diode is forward-biased. It is assumed that there is no load resistance across the output terminals. Regulated supplies with zeners and load resistances are discussed in Chap. 8.

4.10 LIMITERS (CLIPPERS)

Figure 4-49 shows a diode connected as a positive *limiter* or *clipper*. Neglecting the diode forward voltage drop, about 0.2 V for germanium and 0.7 V for silicon diodes, we can operate the diode as a switch.

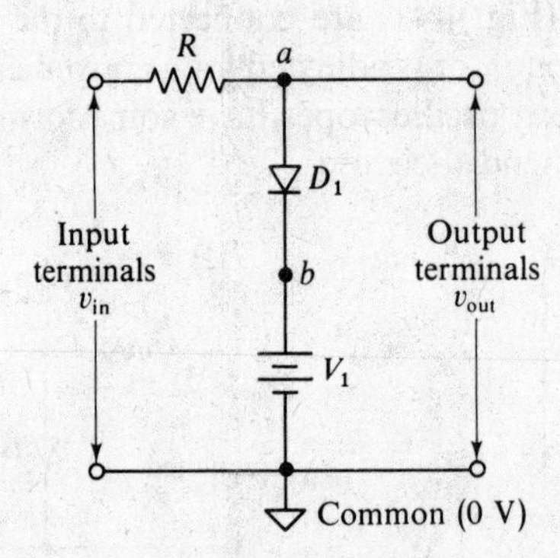

Fig. 4-49

When the anode (point *a*) is more positive than the cathode (point *b*), the diode conducts and is essentially a short circuit from *a* to *b*. When this occurs, the battery voltage *V* appears at the output terminals, regardless of the magnitude of the input voltage. This condition exists whenever the input voltage is sufficiently high to make the voltage at point *a* greater than the battery voltage *V*.

When the anode is less positive than the cathode, the diode is cut off and is essentially an open circuit. Since the output terminals are open also, no current flows through resistor *R* and the input voltage appears at the output terminals. Figure 4-50 shows how the limiter circuit (Fig. 4-49) clips the positive peaks of a sine wave applied to the input terminals.

By using two limiters or clippers as shown in Fig. 4-51, the input sine wave is converted to an approximate square wave. The vertical parts of this waveform are actual segments of sine waves, as

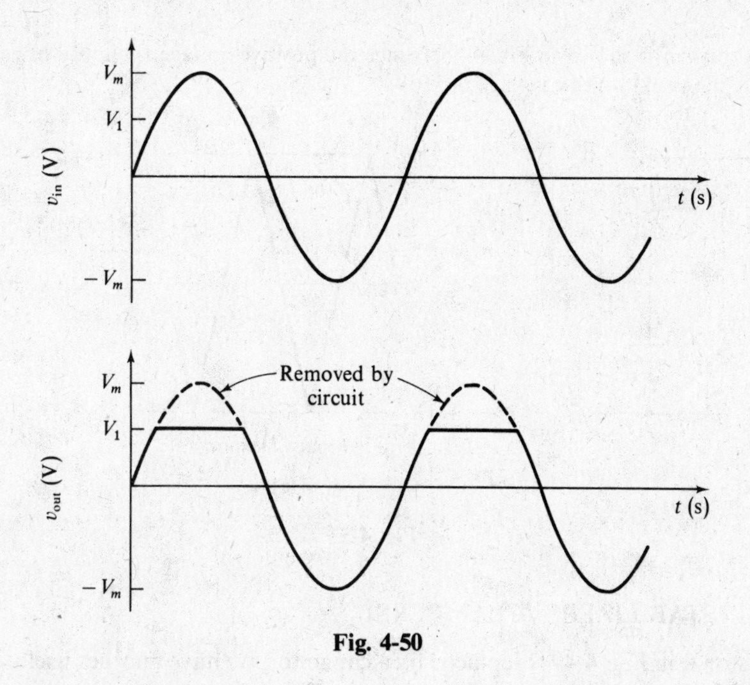

Fig. 4-50

shown in Fig. 4-52, and this prevents the wave from being a true square wave. Note that the anode of D_2 is at negative potential V_2. Therefore the voltage at point a must be slightly more negative than V_2 in order for the diode to conduct.

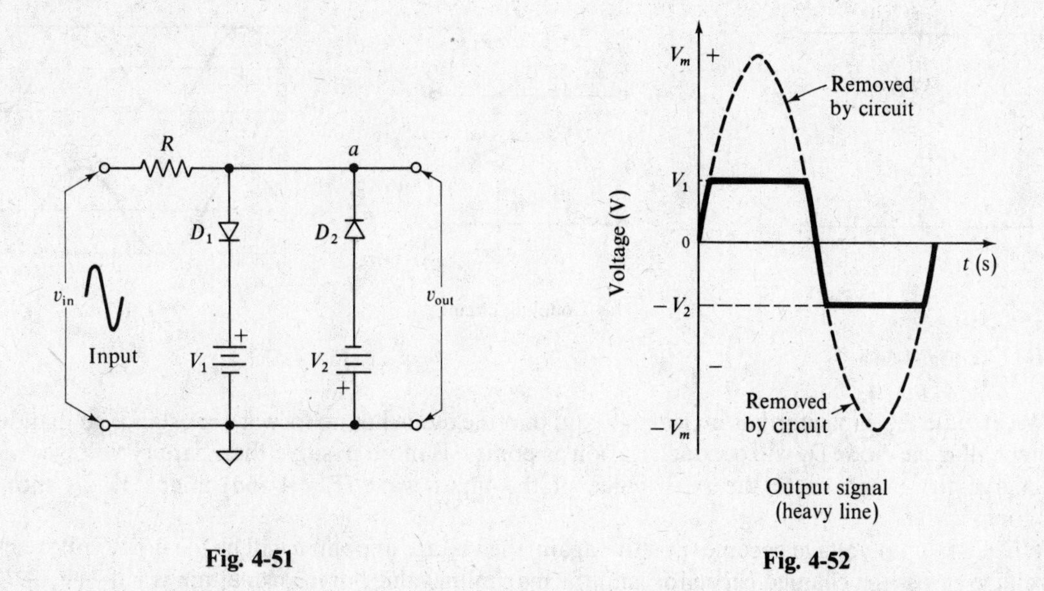

Fig. 4-51 Fig. 4-52

Example 4.14 Neglecting the forward voltage drop of the diodes and the internal resistance of the battery, determine the waveshape and find the peak-to-peak voltage for the circuit shown in Fig. 4-53a with an input voltage waveform as shown in Fig. 4-53b.

Any value of *positive voltage* will cause diode D_1 to conduct and short-circuit the output to the 0-V *line*.

Any value of negative input voltage that exceeds -6 V will cause D_2 to conduct, shorting the output terminal to the negative battery terminal and limiting the negative peaks to the -6-V battery voltage.

Since the given input voltage has a peak value of -5 V, this condition is not reached, diode D_2 does not conduct, and the negative values at the output are the same as the input wave.

The resulting waveform is shown in Fig. 4-53*c*. Since the positive voltage is clipped at zero and the negative voltage is unchanged, the peak-to-peak voltage is 5.0 V.

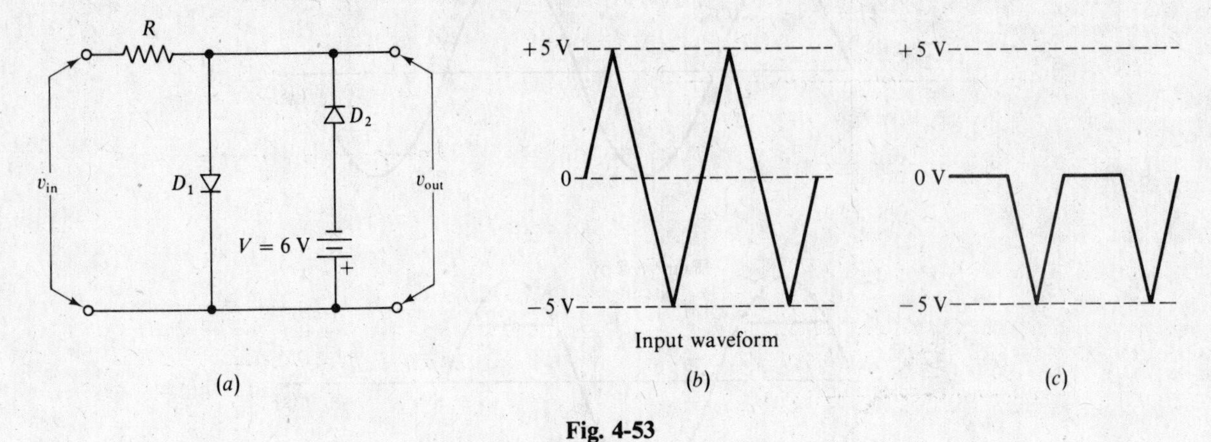

(*a*) (*b*) (*c*)

Fig. 4-53

4.11 BASE LINE STABILIZERS (CLAMPERS)

When the resistor R in Fig. 4-49 is replaced by a capacitor, we have another useful circuit known as a *base line stabilizer*, *dc restorer*, or *clamper*, as shown in Fig. 4-54.

An RC coupling circuit (Fig. 4-55*a*) blocks direct current but lets the alternating current pass unchanged, as shown in Fig. 4-55*b*. To restore the dc component, we need a clamper such as shown in Fig. 4-54.

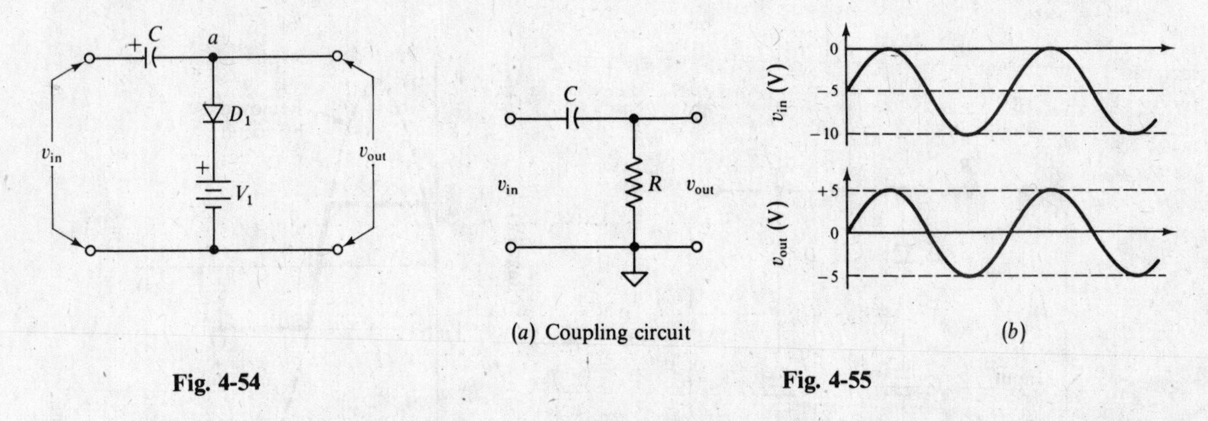

(*a*) Coupling circuit (*b*)

Fig. 4-54 **Fig. 4-55**

We assume the capacitor has no leakage, and that the diode has no forward resistance so that during the first half-cycle diode D_1 will conduct as soon as point a is more positive than battery voltage V_1. This will charge the capacitor to the peak value of the input wave (Fig. 4-56*a*) minus V_1, as shown in Fig. 4-56*b*.

When the input voltage becomes positive again, the voltage at point a will be the input voltage minus the voltage across the charged capacitor, and the maximum value point a can attain is $V_m - (V_m - V_1)$ or V_1. This means the diode will not conduct after the capacitor is fully charged, and we may effectively remove D_1 and the battery from the circuit and redraw it as shown in Fig. 4-57*a*.

The charged capacitor has the same effect as a battery of voltage $V_m - V_1$ in series with the input, as shown in Fig. 4-57*b*.

Using the principle of superposition, we can say that the output voltage is the ac input voltage minus a dc value of $V_m - V_1$, as shown in Fig. 4-58*a*. The maximum positive output voltage is then $V_m - (V_m - V_1)$ or V_1. The maximum negative output voltage is $-V_m - (V_m - V_1) = V_1 - 2V_m$, and the output wave is drawn in Fig. 4-58*b*. Note that the circuit has lowered the ac voltage at all points by an amount equal to the voltage across the capacitor $(V_m - V_1)$.

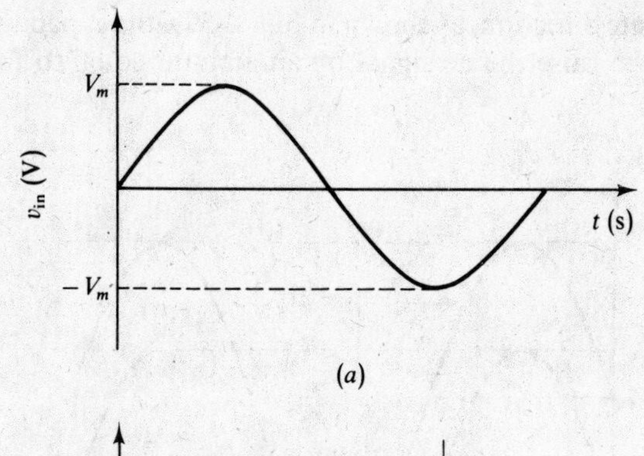

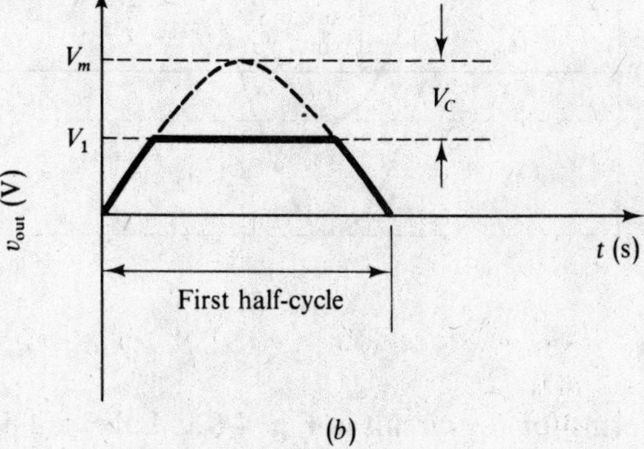

Fig. 4-56

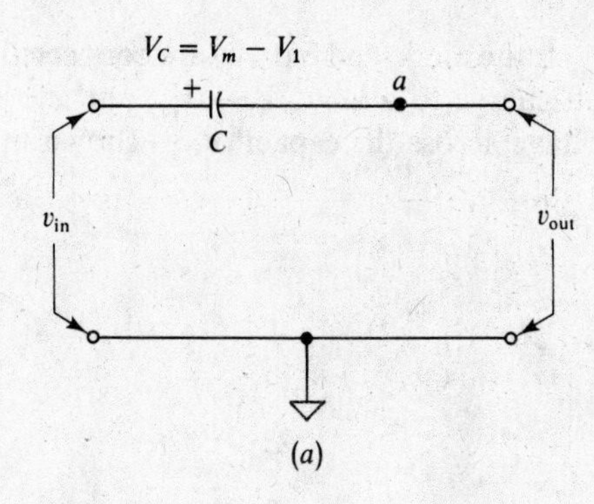

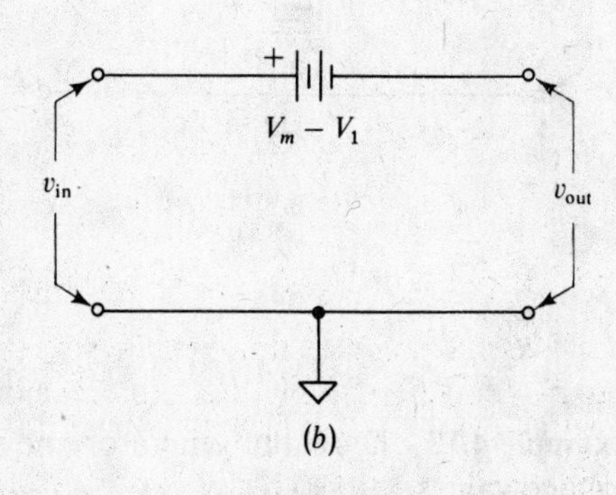

Fig. 4-57

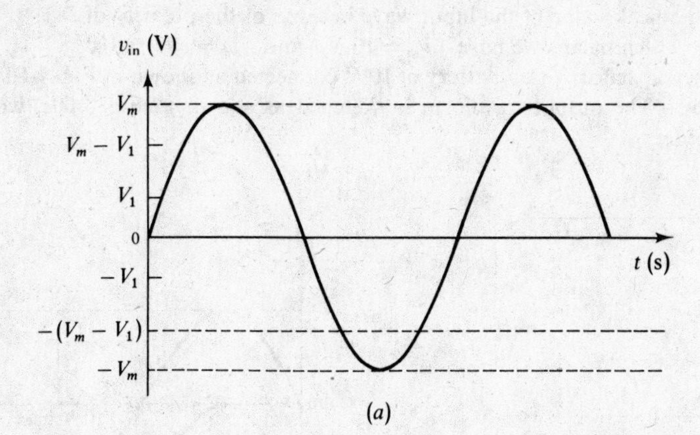

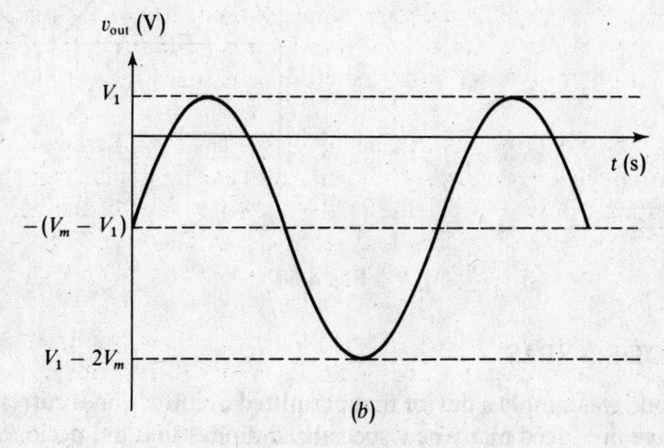

Fig. 4-58

If the diode and battery are connected in the opposite direction, as shown in Fig. 4-59*a*, the capacitor will charge to a value $V_1 - V_m$, and the effect will be to raise the ac signal by an amount equal to the voltage across the capacitor, as shown in Fig. 4-59*b*.

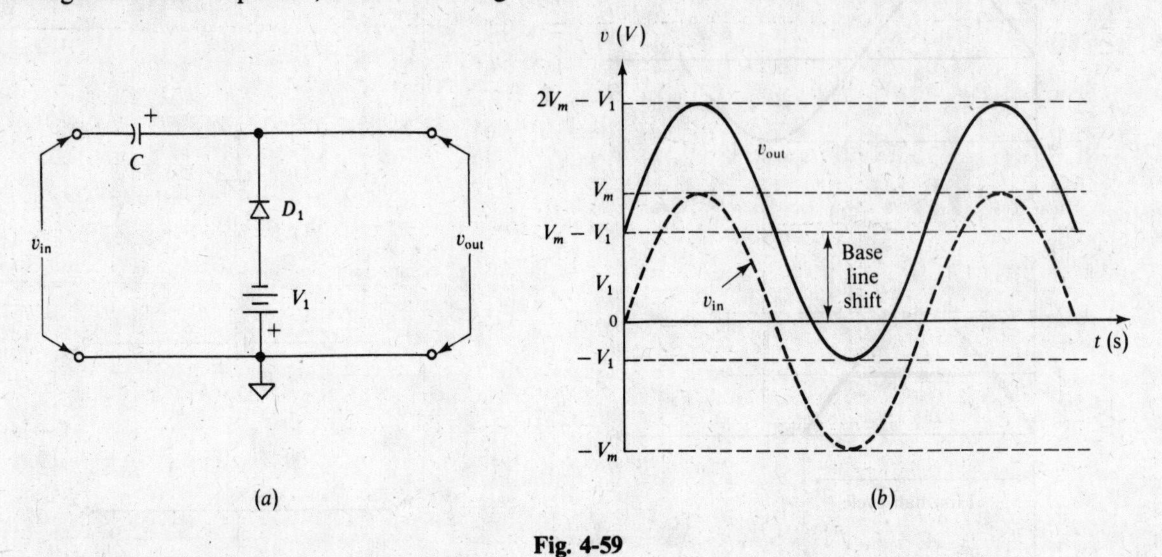

Fig. 4-59

Example 4.15 Draw the generator and output waveforms for the circuit in Fig. 4-60*a*. Label the peak voltage values.

Since there is no battery in the clamper circuit shown in Fig. 4-60*a*, the capacitor will charge on the first negative half-cycle to the peak value of the input wave because of the polarity of D_1.

$v_{in} = 10 \sin 377t$, which means we have $V_m = 10$ V and $\omega = 377$ rad/s.

We can replace the capacitor with a battery of 10 V connected as shown in Fig. 4-60*b*, and the output voltage will add to this dc level. The output waveform is sketched as shown in Fig. 4-60*c*, with the generator (input) waveform shown dashed.

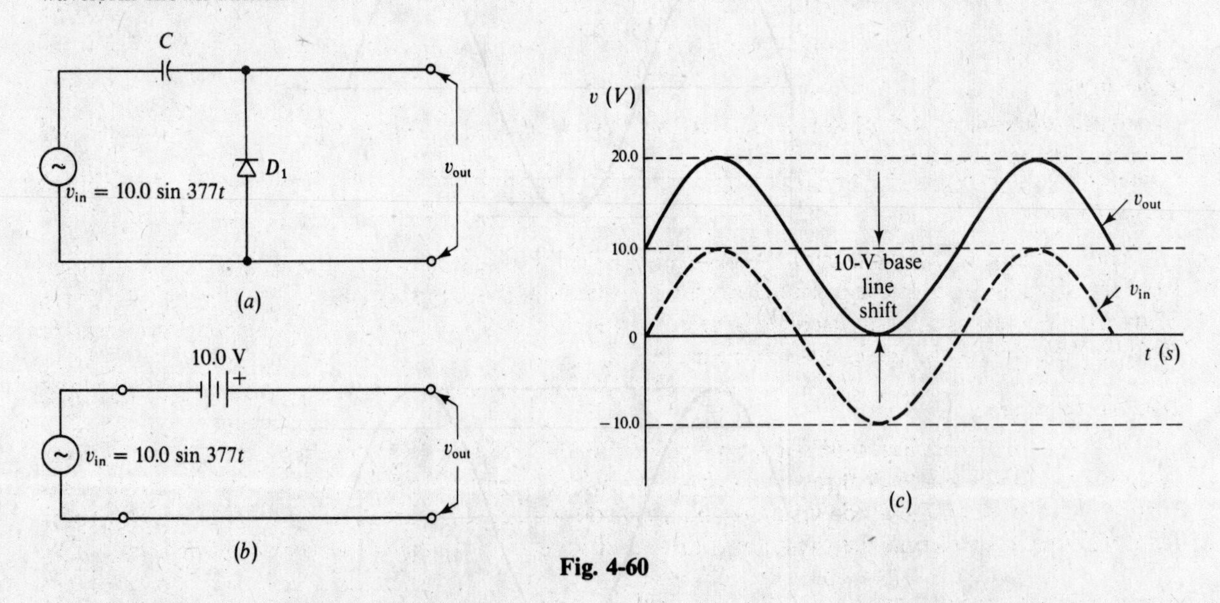

Fig. 4-60

4.12 SPECIAL DIODE TYPES

At one time a diode was simply a device that permitted unidirectional current. Advances in semiconductor technology have produced many new specialized diodes that are no longer limited in application to detection and rectification.

Figure 4-61 shows the symbols for some of the more important new types, along with the characteristic curves for all except the light-activating and light-emitting diodes (LAD and LED). A short description of the applications for each diode follows.

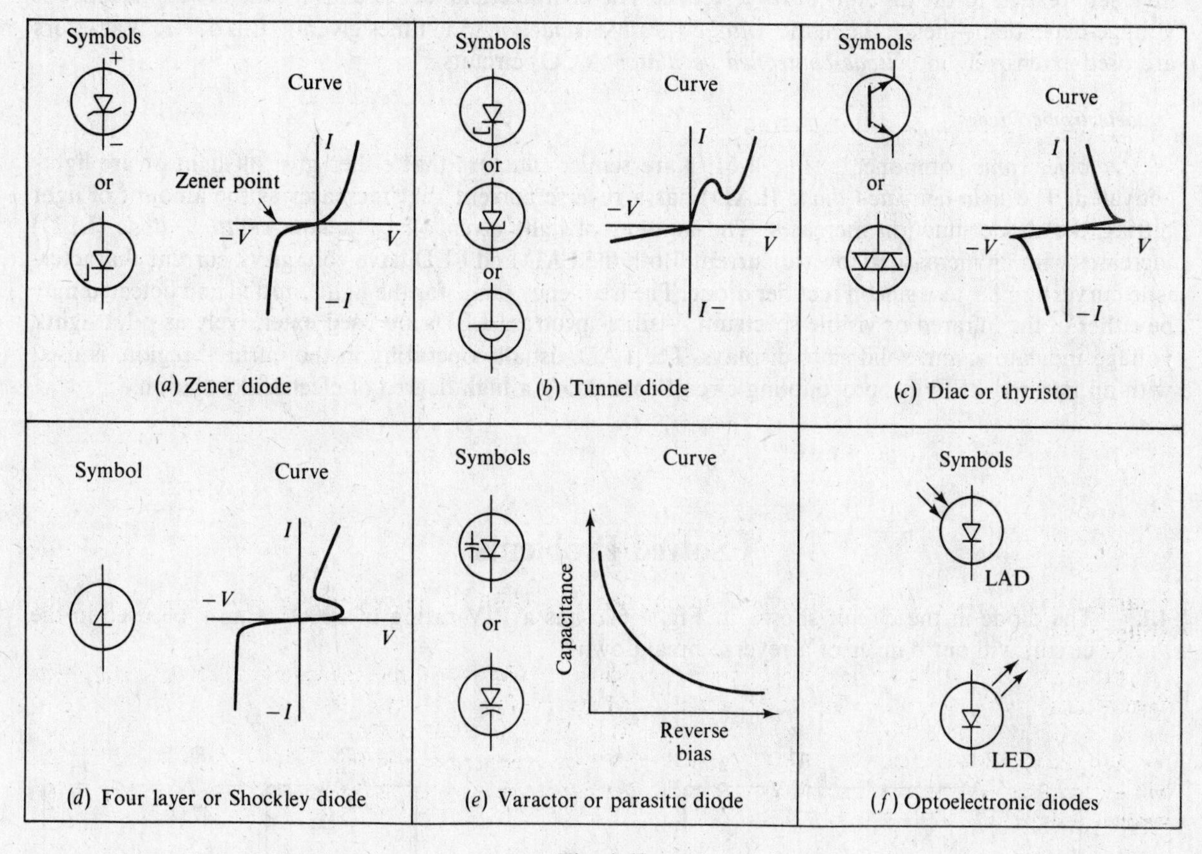

Fig. 4-61

Zener Diodes

Zener diodes (Fig. 4-61*a*) are used as voltage regulators and as voltage references. Zener diodes are discussed again in Chap. 8.

Tunnel Diodes

The dip in the *tunnel diode* curve (Fig. 4-61*b*) is called the *negative resistance region.* The name comes from the fact that an *increase* in voltage across the diode in that region produces a *decrease* in current. Tunnel diodes are capable of oscillating at VHF frequencies. Also, they are extremely fast switches. In fact, they are one of the fastest switches known.

Diacs or Thyristors

A *diac* or *thyristor* (Fig. 4-61*c*) is a type of semiconductor switching device. By definition, a thyristor is a semiconductor made of three or more layers and has two stable states: ON or OFF. Essentially, no current flows through the diode until the voltage across it reaches the breakover point. Once this point is reached, current flows and the thyristor is saturated. As shown by the curve, a diac is bilateral.

Four-Layer or Shockley Diode

A *four-layer* or *Shockley diode* (Fig. 4-61*d*) is a breakover type similar to the diac. The most important difference is that it is unilateral in its action. In other words, it will break over in the forward direction. In the reverse direction, however, it remains in a nonconducting state until the zener point is reached. It is used as a one-way switch or trigger.

Varactor or Parasitic Diode

A *varactor diode* (Fig. 4-61*e*) acts like a capacitor when it is reverse-biased. The capacitance is inversely related to the amount of reverse bias. The characteristic curve shows that the capacitance is voltage-dependent—hence the name *voltage-variable capacitor* sometimes given to this device. Varactors are used extensively in *voltage-controlled oscillator* (VCO) circuits.

Optoelectronic Diodes

Optoelectronic components (Fig. 4-61*f*) are semiconductors that either give off light or are light-activated. The *light-activated diode* (LAD) has a reverse current that increases as the amount of light hitting the diode junction increases. The amount of light given off by a *light-emitting diode* (LED) increases with an increase in foward current. Both the LAD and LED have voltage vs. current characteristic curves similar to a silicon rectifier diode. The frequency range for the light emitted and detected may be either in the infrared or visible spectrum. Visible-spectrum LEDs are used extensively as pilot lights, voltage indicators, and solid state displays. The LAD, usually operating in the infrared region, is used with an infrared LED in optocoupling circuits providing a high degree of electrical isolation.

Solved Problems

4.1 The diode in the circuit shown in Fig. 4-62*a* has a PIV rating of 100 V. Can it be used in the circuit without danger of a reverse breakdown?

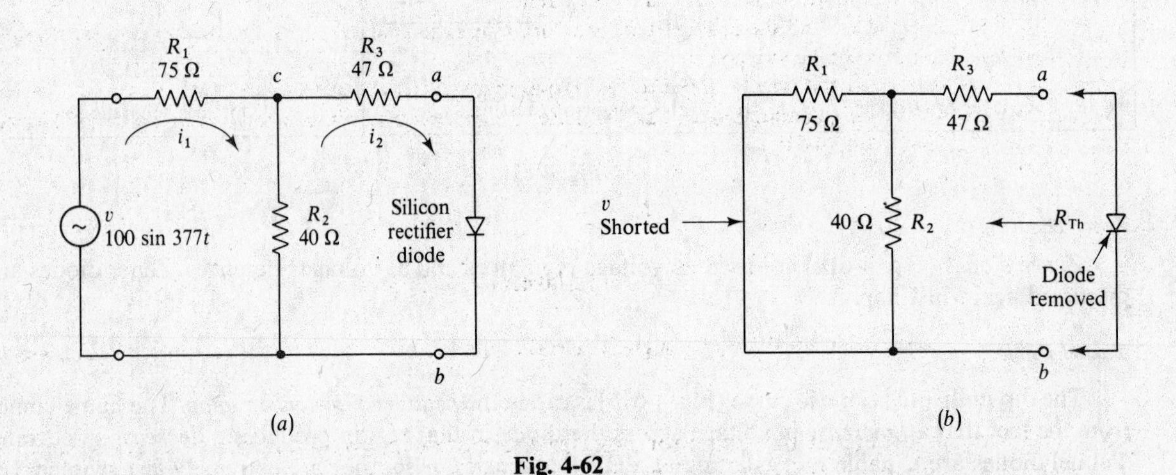

Fig. 4-62

When a reverse voltage is placed across a diode, it does not conduct and can be treated as an open circuit. Therefore, the voltage across the diode will be equal to the voltage V_{ab} across terminals *a* and *b* during the negative half-cycle of the applied voltage. Also, during this time $i_2 = 0$, so $V_{ab} = V_{cb}$.

From Fig. 4-62*a*, the applied voltage is in the form

$$v = V_m \sin \omega t = 100 \sin 377t$$

Therefore

$$V_m = 100 \text{ V}.$$

By the voltage divider rule:

$$V_{ab} = V_{cb} = V_m \frac{R_2}{R_1 + R_2} = 100\left(\frac{40}{75 + 40}\right) = 34.8 \text{ V}$$

This voltage is well within the allowable peak inverse voltage rating of 100 V, and the diode can be used.

4.2 The I_F rating of the diode in the circuit shown in Fig. 4-62*a* is 0.5 A. Can the diode be used?

It must be assumed that the forward resistance or voltage drop of the diode is negligible since the values are not given in the problem.

When the diode is conducting, terminals *a* and *b* are short-circuited, and we may solve for the maximum value of i_2. From Fig. 4-62*a* we see that the applied voltage is in the form

$$v = V_m \sin \omega t = 100 \sin 377t$$

Therefore $V_m = 100$ V.

In Prob. 4.1 we found that for a V_m of 100 V, the open-circuit (Thevenin) voltage was equal to 34.8 V. If we now solve for the Thevenin resistance of the circuit, the short-circuit (Norton) current is obtained from

$$I_N = \frac{V_{Th}}{R_{Th}}$$

Looking in at terminals *a-b* with v shorted gives us the Thevenin resistance R_{Th}. From Fig. 4-62*b*

$$R_{Th} = R_3 + \frac{R_1 R_2}{R_1 + R_2}$$

Substituting values,

$$R_{Th} = 47 + \frac{(75)(40)}{75 + 40} = 47 + 26.1 = 73.1\ \Omega$$

Using this in the expression for I_N, with $V_{Th} = 34.8$ V:

$$I_N = \frac{34.8}{73.1} = 0.476 \text{ A}$$

This is the peak value of short-circuit (diode) current.

If at this point we find the peak current equal to or less than half the I_F rating, the diode would be safe, and calculations could be stopped.

Since this is not the case, we need to find the average circuit current and compare it to I_F.

The diode is only conducting for half the cycle. This means the average circuit current is the half-cycle average given by Eq. (4-1):

$$I_{av} = \frac{1}{\pi} I_m$$

$$= \left(\frac{1}{\pi}\right)(0.476) = 0.152 \text{ A}$$

This is well within the 0.5-A rating given, and the diode can be used in the circuit shown in Fig. 4-62*a*.

4.3 A transient voltage at the generator input of the circuit in Fig. 4-62*a* causes a 20-A surge of current through the 0.5-A diode. The surge current has a duration of 0.2 ms. Will the diode be destroyed?

The surge current graph shown in Fig. 4-16 is used to determine the value of surge current for the given duration of 0.2 ms.

Since the time axis in Fig. 4-16 is given in seconds rather than milliseconds, we convert:

$$0.2 \text{ ms} = 0.2 \times 10^{-3} \text{ s} = 2 \times 10^{-4} \text{ s}$$

The intersection of a vertical line drawn at 2×10^{-4} s and the curve for a 0.5-A diode gives a surge current value of slightly greater than 50 A. The 20-A surge given in the problem statement is well within this limit, so the diode will not be destroyed.

4.4 The characteristic curve for a certain vacuum-tube diode is shown in Fig. 4-63*a*. The tube is connected into the circuit shown in Fig. 4-63*b*. What is the peak value of current in this circuit?

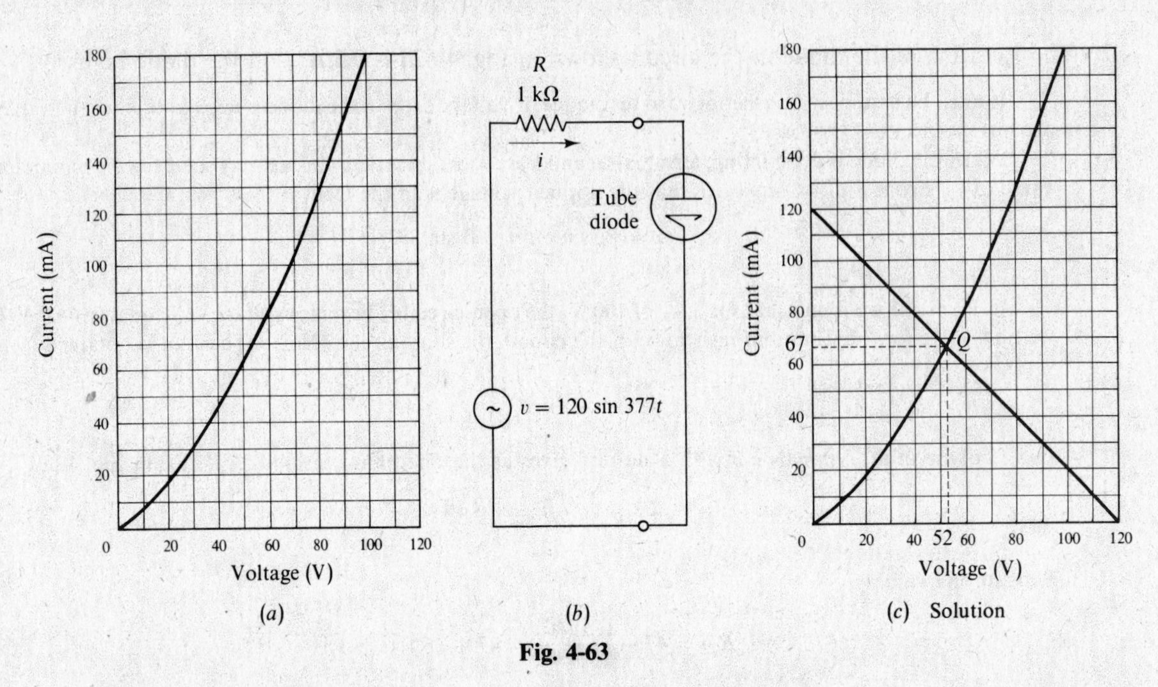

(a) (b) (c) Solution

Fig. 4-63

Since the tube diode characteristic is nonlinear, as shown in Fig. 4-63*a*, a load line solution is called for. Removing the diode from the circuit, the peak open-circuit voltage is V_m. Figure 4-63*b* gives v in the form

$$v = V_m \sin \omega t = 120 \sin 377t$$

Therefore $$V_m = 120 \text{ V.}$$

The peak short-circuit current is therefore [from Ohm's law]

$$I_m = \frac{V_m}{R} = \frac{120}{1000} = 0.12 \text{ A} = 120 \text{ mA}$$

Plotting these points on the x and y axis of the tube diode characteristic and drawing the load line as shown in Fig. 4-63*c*, we find that the intersection Q of the load line and the tube diode characteristic curve gives a peak voltage across the diode of 52 V and a peak current of 67 mA.

Therefore the peak circuit current equals 67 mA.

4.5 An oscilloscope is connected as shown in Fig. 4-64*a* to obtain the characteristic curve of a certain diode. Figure 4-64*b* shows the resulting curve. Answer the following questions regarding

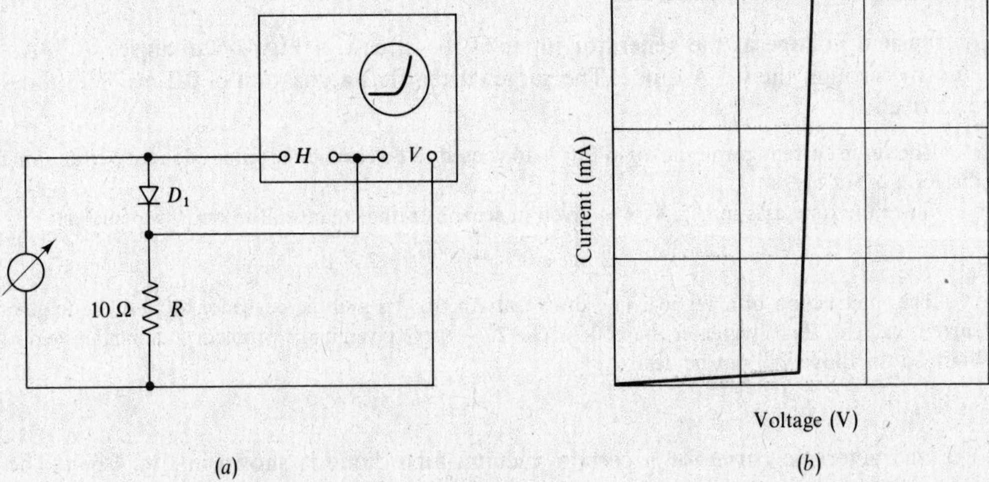

(a) (b)

Fig. 4-64

the diode: (*a*) If the horizontal amplifier of the oscilloscope is calibrated for 0.4 V per division, what is the breakdown voltage of the diode? (*b*) If the vertical amplifier of the oscilloscope is calibrated for 1 V per division, what is the maximum amount of current being displayed?

(*a*) The breakdown point is half-way between the first and second division mark, or 1.5 divisions. Since each division is equal to 0.4 V according to the problem statement, the breakdown voltage value is

$$(1.5 \text{ divisions})(0.4 \text{ V/division}) = 0.6 \text{ V}$$

(*b*) The maximum current displayed is three divisions. According to the problem statement, each division represents 1.0 V.

Therefore the maximum voltage across the calibration resistor R is

$$(3 \text{ divisions})(1.0 \text{ V/division}) = 3.0 \text{ V}$$

Ohm's law is then used to find the current through the resistor, which is the current being displayed:

$$I = \frac{V}{R} = \frac{3.0}{10} = 0.300 \text{ A} = 300 \text{ mA}$$

4.6 The diode in Prob. 4.5 is used in the circuit shown in Fig. 4-65*a*. Find the current through the diode.

The diode characteristic in Prob. 4.5 is redrawn as Fig. 4-65*b* since a load line solution is required.

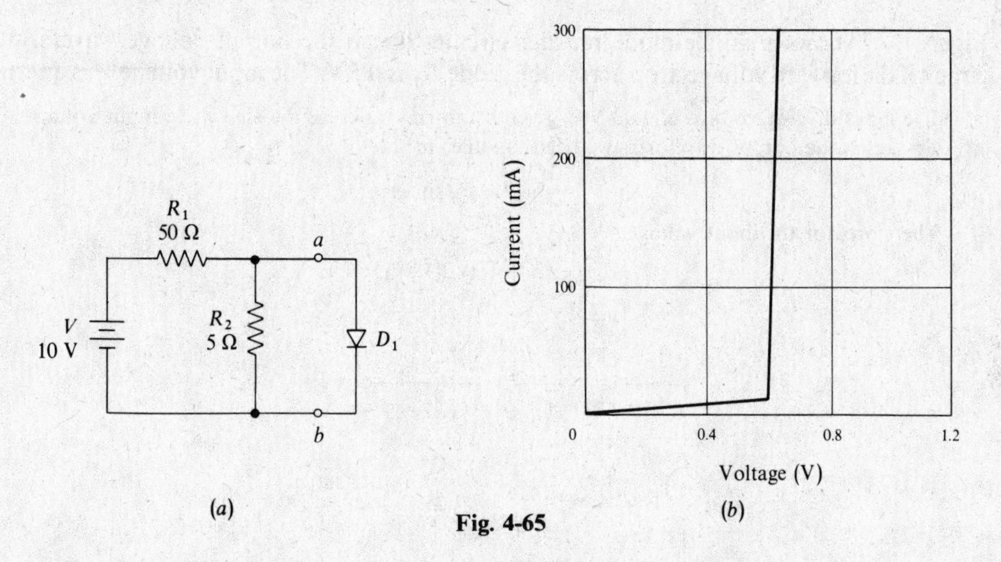

Fig. 4-65

The first step is to Thevenize the circuit with the diode removed. The open-circuit voltage across terminals a and b, with the diode removed, is obtained by the voltage divider method:

$$V_{\text{Th}} = V_{ab} = V\frac{R_2}{R_1 + R_2} = 10\left(\frac{5}{50 + 5}\right) = 0.909 \text{ V}$$

The Thevenin resistance is calculated with the battery replaced by a short circuit. As seen from the open terminals:

$$R_{\text{Th}} = \frac{R_1 R_2}{R_1 + R_2} = \frac{(50)(5)}{50 + 5} = 4.55 \ \Omega$$

Figure 4-66*a* shows the Thevenin equivalent circuit. The open-circuit voltage $V_{\text{Th}} = 0.909$ V, and the short-circuit current is

$$I_{\text{N}} = \frac{V_{\text{Th}}}{R_{\text{Th}}} = \frac{0.909}{4.55} = 0.200 \text{ A} = 200 \text{ mA}$$

The load line is drawn between V_{Th} and I_{N} as shown in Fig. 4-66*b*. The dotted line shows that the forward current is 70 mA.

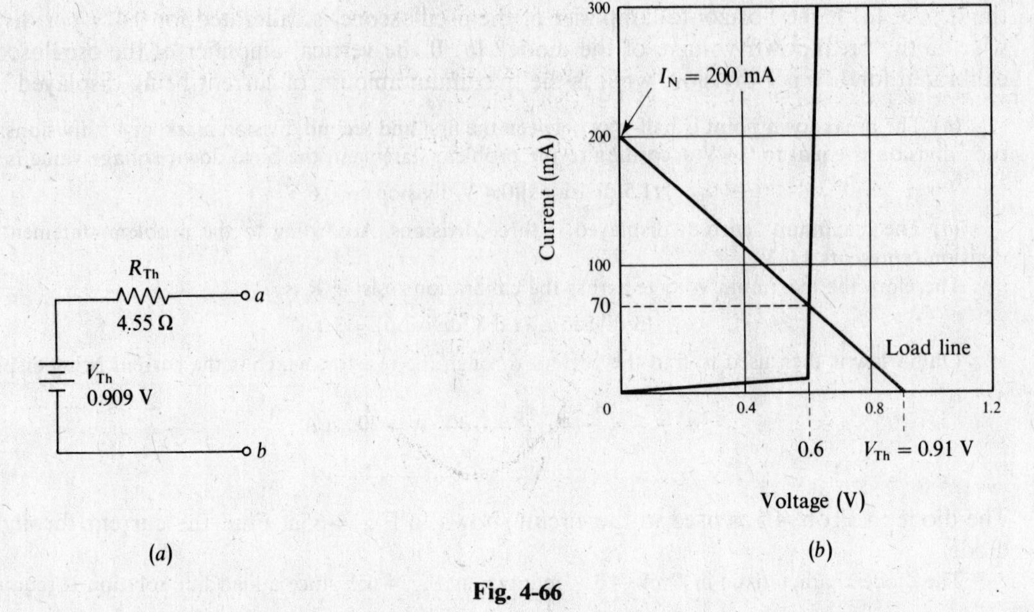

Fig. 4-66

4.7 Figure 4-67 shows a simple diode rectifier circuit. Sketch the output voltage waveform for the circuit if the forward voltage drop across the diode D_1 is 0.5 V. The input voltage V is an rms value.

The input driving voltage of 1.00 V is given as an rms value of the sine wave input voltage. To sketch the output waveform, we need voltages in the form

$$v_{in} = V_m \sin \omega t$$

Therefore, for the input voltage,

$$V_m = \sqrt{2}\ V = \sqrt{2}(1.00) = 1.41\ \text{V}$$

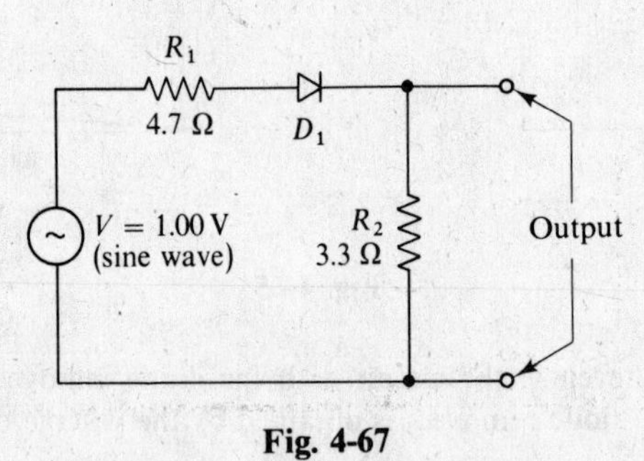

Fig. 4-67

If the diode were not present, the output voltage would have the same form and phase as the input voltage but proportionally lower values because of the voltage divider formed by R_1 and R_2:

$$v_{out} = \frac{R_2}{R_1 + R_2} v_{in} \tag{4-3}$$

$$= \frac{3.3}{4.7 + 3.3} v_{in} = 0.4125 v_{in} \tag{4-4}$$

Since $v_{in} = V_m \sin \omega t = 1.41 \sin \omega t$, substituting into Eq. (4-4):

$$v_{out} = 0.4125\,(1.41 \sin \omega t) = 0.582 \sin \omega t$$

This is shown in Fig. 4-68*a*.

The effect of a diode with no forward voltage drop would be to pass only the positive half-cycle of the input wave, so the output would appear as shown in Fig. 4-68*b*.

However, diode D_1 is given as having a forward voltage drop of 0.5 V, so the diode only conducts when the input voltage v_{in} is greater than 0.5 V.

Translating this to the output, the output voltage starts at zero when the input is 0.5 V and is maximum at $v_{in} = V_m - 0.5 = 1.41 - 0.5 = 0.91$ V. Using Eq. (*4-4*):

$$v_{out} = (0.4125)(v_{in}) = (0.4125)(0.91) = 0.375 \text{ V}$$

This is shown in Fig. 4-68*c*.

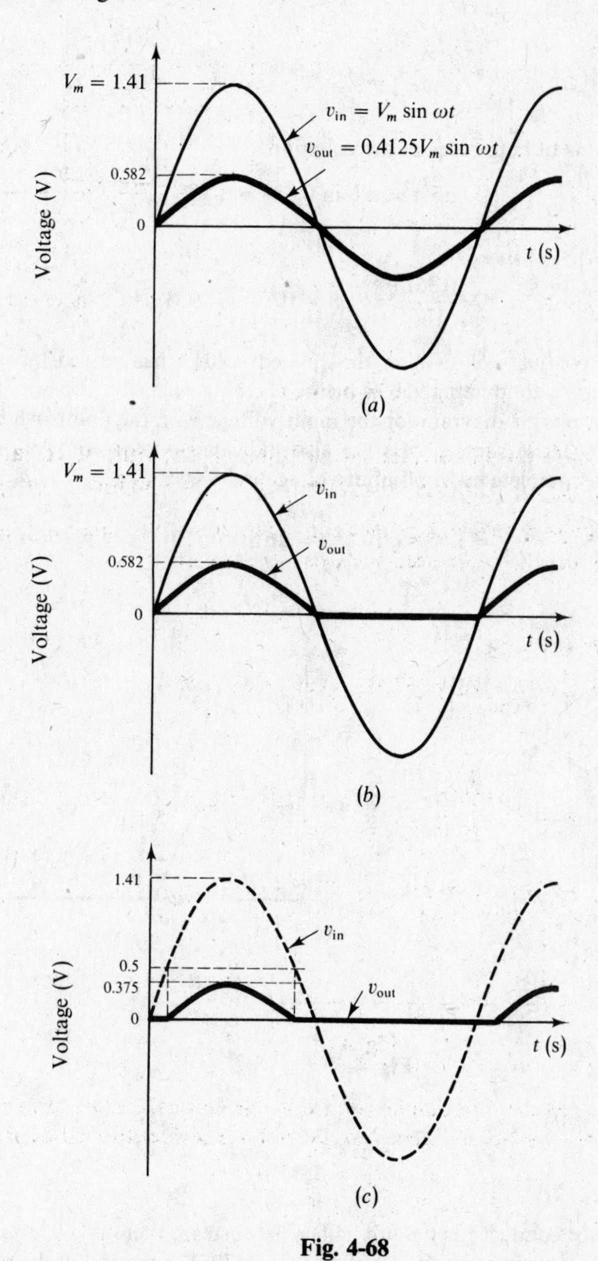

Fig. 4-68

4.8 Assume that the ac generator frequency is 60 Hz in the same circuit shown in Fig. 4-67, and all other values are the same. Determine the amount of time that the diode is actually conducting during each cycle.

The equation for the sinusoidal voltage input is (from Prob. 4.7)

$$v = V_m \sin \omega t$$

where $V_m = 1.41$ and $\omega = 2\pi f = (2\pi)(60) = 377$ rad/s. Therefore

$$v = 1.41 \sin 377t$$

The diode begins to conduct and stops conducting when $v = 0.5$ V. Substituting this value into the equation:

$$0.5 = 1.41 \sin 377t$$

Solving for sin $377t$,

$$\sin 377t = \frac{0.5}{1.41} = 0.354$$

Taking the arcsin in radians of both sides of the equation:

$$377t = \arcsin 0.354 = 0.361$$

Solving for t,

$$t = \frac{0.361}{377} = 9.58 \times 10^{-4} \text{ s} = 958 \ \mu\text{s}$$

The diode begins to conduct 958 μs after the applied voltage has passed through zero going in the positive direction, and it stops conducting 958 μs before reaching zero after it passes the first positive peak.

Figure 4-69*a* shows a phasor diagram for the input voltage with the points where the diode starts and stops conducting marked. These same points are also indicated on the conventional diagram shown in Fig. 4-69*b*. The time for a complete cycle of input voltage is

$$T = \frac{1}{f} = \frac{1}{60} = 1.667 \times 10^{-2} \text{ s} = 16670 \times 10^{-6} \text{ s} = 16670 \ \mu\text{s}$$

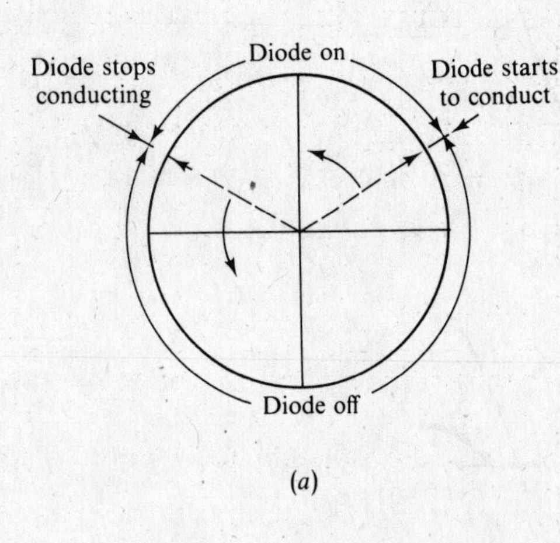

(*a*)

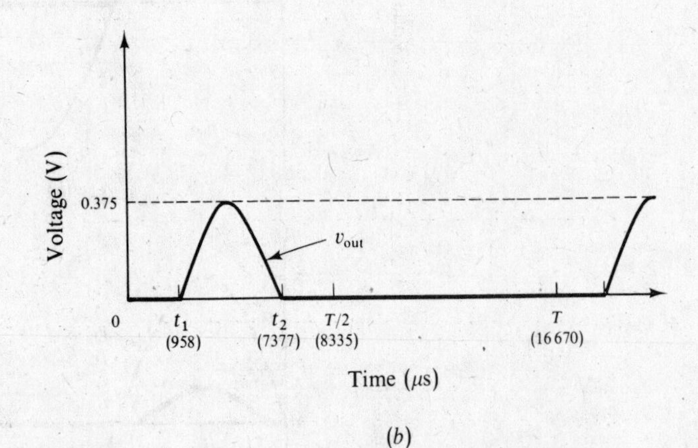

(*b*)

Fig. 4-69

The time for one half-cycle is

$$\frac{T}{2} = 8335 \ \mu\text{s}$$

But the diode does not conduct for the full half-cycle. Instead, there is a 958-μs period (0 to t_1) at the beginning of the positive half-cycle and a 958-μs period (t_2 to $T/2$) at the end of the positive half-cycle when the diode is not conducting. The total time during the first half-cycle that the diode is conducting is $t_2 - t_1$. Since $t_1 - 0 = 958$ μs, then $t_1 = 958$ μs and

$$\frac{T}{2} - t_2 = 958 \ \mu\text{s}$$

$$t_2 = \frac{T}{2} - 958 = 8335 - 958 = 7377 \ \mu\text{s}$$

Therefore the time that the diode is conducting is

$$t_2 - t_1 = 7377 - 958 = 6419\ \mu s$$

4.9 Figure 4-70a shows a 3-V regulating circuit with five identical diodes D_1 through D_5, connected in series across a load resistor R_L. When $V_L = 3.0$ V, the current through the diodes I_D is one-fifth the value of load current I_L. The characteristic curve of one diode is shown in Fig. 4-70b. (a) Determine the value of R_L. (b) Determine the value of R_1 when $V_{in} = 7.00$ V. (c) Make a graph showing the voltage across R_L as V_{in} changes from 6 to 8 V.

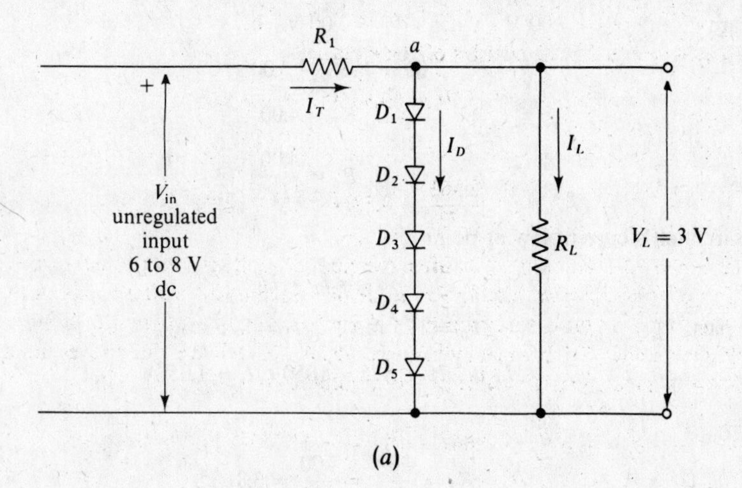

(a)

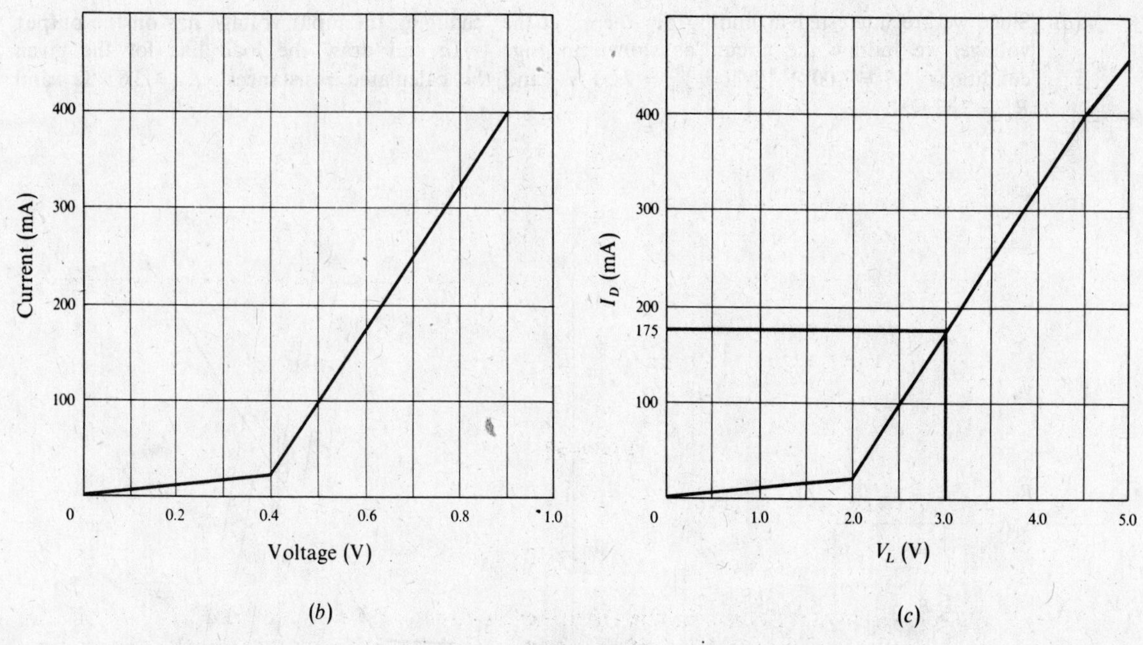

Fig. 4-70

(a) The characteristic curve shown in Fig. 4-70b is for one diode. For five diodes in series, it requires five times as much forward voltage drop to get the same amount of current. Therefore, the characteristic curve can be redrawn with each value along the x axis multiplied by five to get a curve that applies to five diodes in series. Figure 4-70c shows the redrawn curve.

Figure 4-70c shows that $I_D = 175$ mA when the voltage V_L across the diodes and across R_L is 3 V.

From the problem statement, $I_D = I_L/5$. So

$$I_L = 5I_D = 5(175 \times 10^{-3}) = 875 \times 10^{-3} \text{ A} = 875 \text{ mA}$$

Now, by Ohm's law,

$$R_L = \frac{V_L}{I_L} = \frac{3}{875 \times 10^{-3}} = 3.43 \ \Omega$$

(*b*) Writing Kirchhoff's voltage law for the circuit,

$$V_{in} - I_T R_1 - V_L = 0 \tag{4-5}$$

Since we are given $V_{in} = 7.00$ V when $V_L = 3.00$ V,

$$7.00 - I_T R_1 - 3.00 = 0$$

and

$$I_T R_1 = 4.00$$

$$R_1 = \frac{4.00}{I_T}$$

Writing Kirchhoff's current law at point *a*,

$$I_T = I_L + I_D \tag{4-6}$$

and from part (*a*) $\quad I_D = 175$ mA $\quad I_L = 875$ mA

So

$$I_T = 175 + 875 = 1050 \text{ mA} = 1.05 \text{ A}$$

Therefore

$$R_1 = \frac{4.00}{I_T} = \frac{4.00}{1.05} = 3.81 \ \Omega$$

(*c*) Since we are interested in finding out the effect that changing the input voltage has on the output voltage, we redraw the circuit as shown in Fig. 4-71*a* and draw the load line for the given conditions $V_L = 3.00$ V when $V_{in} = 7.00$ V and the calculated resistances $R_1 = 3.81 \ \Omega$ and $R_L = 3.43 \ \Omega$.

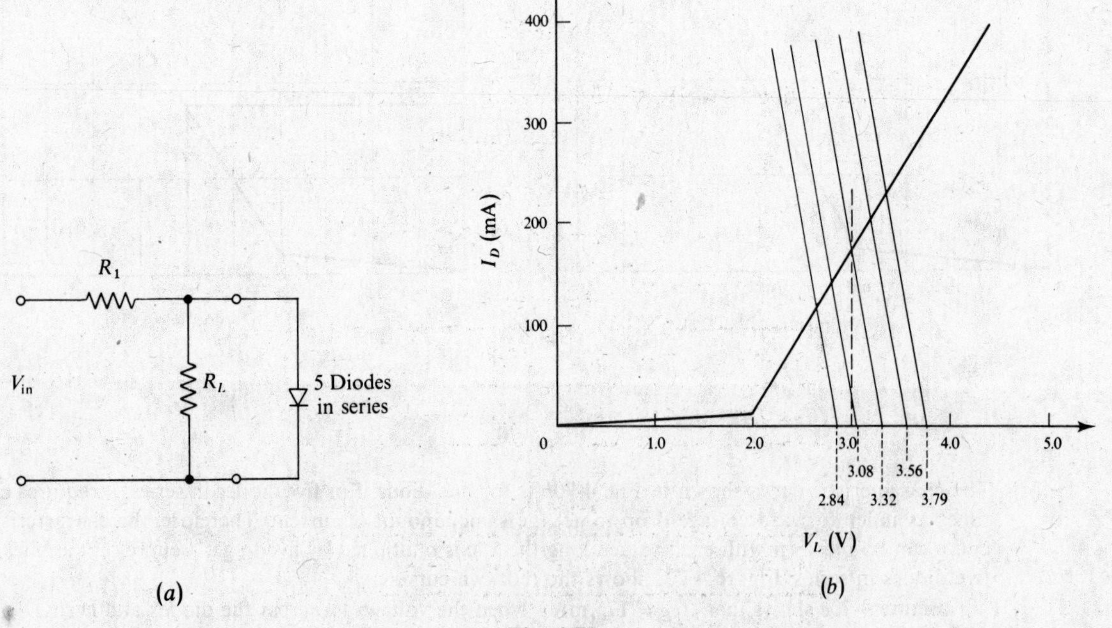

Fig. 4-71

From Fig. 4-71*a*, we find the open-circuit voltage V_{Th} with the diode removed from the voltage divider rule:

$$V_{Th} = \left(\frac{R_L}{R_1 + R_L}\right)V_{in} \tag{4-7}$$

$$= \left(\frac{3.43}{3.81 + 3.43}\right)V_{in} = 0.474V_{in} \tag{4-7a}$$

With $V_{in} = 7.00$ V, we know that $V_L = 3.00$ V, so we know two points that determine the load line. The first point is V_{Th}. With $V_{in} = 7.00$ V, using (*4-7a*):

$$V_{Th} = (0.474)(7.00) = 3.32 \text{ V}$$

The other point is the intersection of the diode curve with a vertical line at 3.00 V.

A change in input voltage will result in different load lines, all parallel to the one we have just drawn. So if we find the V_{Th} points for the values of input voltage we want (Table 4-1), we can use those points along the V_L axis and draw load lines parallel to the original one. This will allow us to pick off the new V_L points from the intersection of the load line and the five-diode characteristic curve. This is shown in Fig. 4-71*b*, and the values of V_{in} vs. V_L are tabularized in Table 4-2 and graphed in Fig. 4-72.

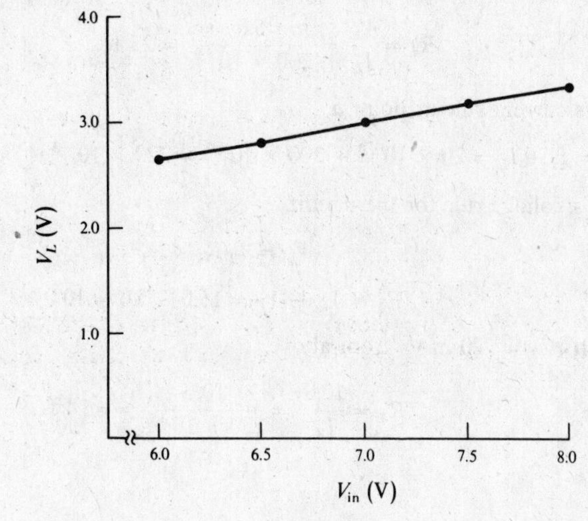

Fig. 4-72

Sample calculation for Table 4-1 [from (*4-7a*)]:

$$V_{Th} = 0.474V_{in}$$

with

$$V_{in} = 6.50 \text{ V}$$

$$V_{Th} = (0.474)(6.50) = 3.08 \text{ V}$$

Table 4-1

V_{in} (V)	6.00	6.50	7.00	7.50	8.00
V_{Th} (V)	2.84	3.08	3.32	3.56	3.79

Table 4-2

V_{in} (V)	6.00	6.50	7.00	7.50	8.00
V_L (V)	2.65	2.80	3.00	3.20	3.35

Note that a change in input voltage of 2 V causes a change in output voltage of about 0.7 V.

Actually, the diodes used are not good for the purpose of regulating the voltage. An ideal diode would have a vertical line at the breakdown point so that the voltage across the diode would not change as the current through it is changed. Such characteristics can be closely approached in practice.

Forward-biased diodes are used in integrated circuits as voltage regulators and are also used in transistor bias networks.

4.10 Figure 4-73*a* shows the characteristic curve for a particular zener diode type. Find the values of R_L and R_1 for the circuit shown in Fig. 4-73*b* that uses this diode to regulate the voltage across a 200 mA load when V_{in} is an unregulated dc voltage of 15 V. As a rule of thumb, the zener diode must pass a minimum current I_z that is roughly 0.1 of the load.

In power supply terminology, the word load refers to the current I_L through the load resistance. For the problem given,

$$I_z = 0.1 I_L = (0.1)(200 \times 10^{-3}) = 20 \times 10^{-3}\ \text{A} = 20\ \text{mA}$$

The breakdown point on the characteristic curve is 5.0 V, so we want the voltage V_L across the load to be 5.0 V. The value of R_L can then be determined by Ohm's law:

$$R_L = \frac{V_L}{I_L} = \frac{5.0}{200 \times 10^{-3}} = 25\ \Omega$$

Writing Kirchhoff's current rule at point *a*,

$$I_T = I_z + I_L = 20 \times 10^{-3} + 200 \times 10^{-3} = 220 \times 10^{-3}\ A = 220\ \text{mA}$$

Writing Kirchhoff's voltage rule for the circuit,

$$V_{in} = I_T R_1 + V_L$$

So

$$I_T R_1 = V_{in} - V_L = 15.0 - 5.0 = 10.0\ \text{V}$$

Solving for R_1 with $I_T = 220$ mA from above,

$$R_1 = \frac{10.0}{I_T} = \frac{10.0}{220 \times 10^{-3}} = 45.4\ \Omega$$

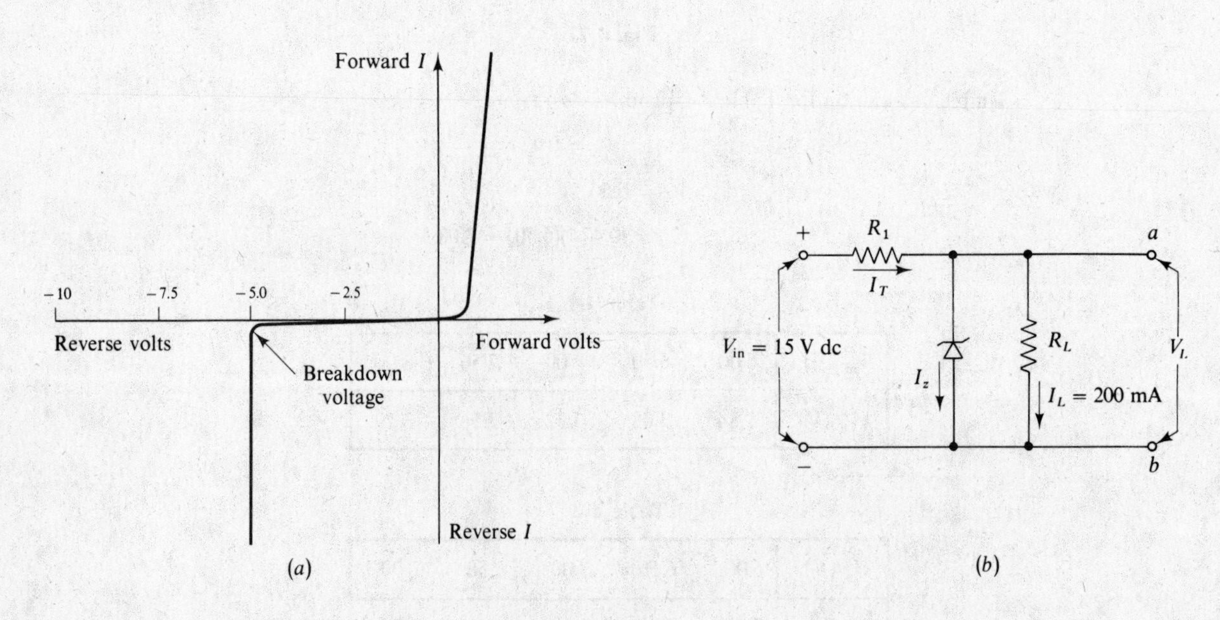

Fig. 4-73

Note that the zener diode in the circuit shown in Fig. 4-73*b* is reverse-biased, as required for proper operation.

4.11 Could a 5.0-V zener diode, rated at 0.5 W, be used in the circuit shown in Fig. 4-73*b*?

The current rating of a 5.0-V, 0.5-W zener diode can be obtained from the equation $P = VI$. Solving for I,

$$I = \frac{P}{V} = \frac{0.5}{5.0} = 0.1 \text{ A} = 100 \text{ mA}$$

At 15-V input, the zener current is 20 mA (from Prob. 4.10). Using a safety factor of 2, the allowable rating would be 40.0 mA. This is well under 100 mA, so the diode can be used as far as its ability to pass current is concerned. However, it may be more expensive than a diode with a lower rating, so it may not be an economical choice.

4.12 Use a load line on the characteristic curve shown in Fig. 4-73*a* to determine if the circuit shown in Fig. 4-73*b* has the correct value of R_1 solved for in Prob. 4-10.

This load line must be drawn with negative voltage values on the curve indicating reverse bias, and negative currents indicate cathode-to-anode current.

The Thevenin circuit is obtained to get the values of V_{Th} and I_N for the load line. The circuit is redrawn, as shown in Fig. 4-74*a*, to make it easier to Thevenize.

With terminals *a-b* open-circuited, V_{Th} is obtained from the voltage divider rule:

$$V_{Th} = \left(\frac{R_L}{R_1 + R_L}\right)V_{in} = \left(\frac{25}{45.5 + 25}\right)15 = 5.32 \text{ V}$$

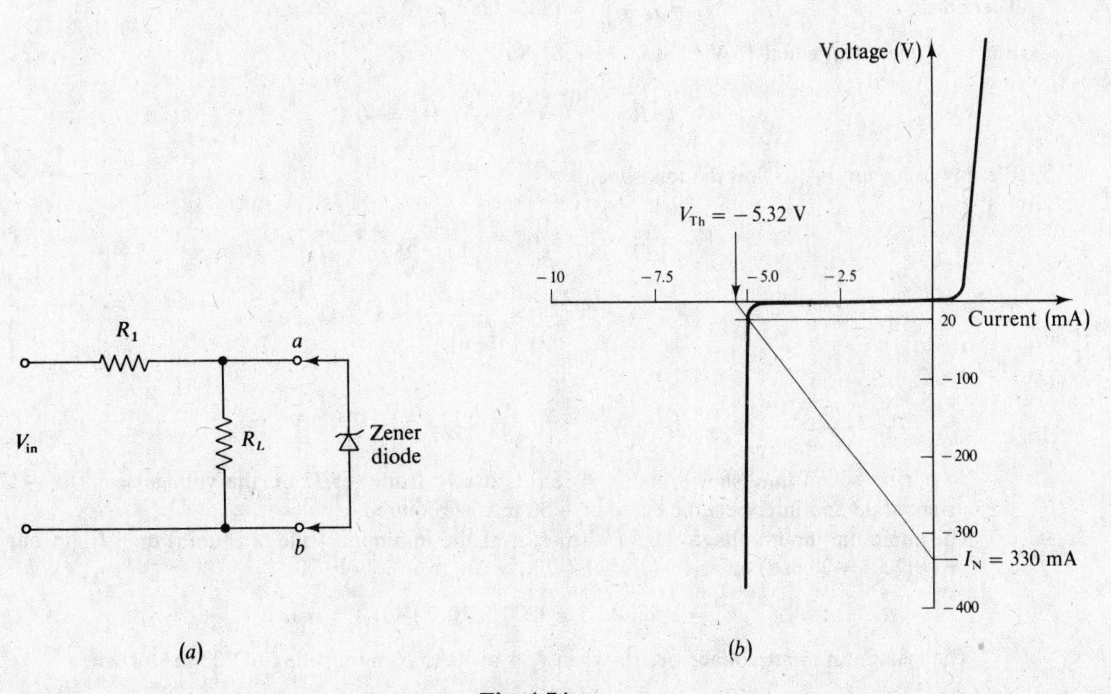

Fig. 4-74

As seen from terminals a and b, the Thevenin resistance consists of R_1 and R_L in parallel, with V_{in} replaced with a short circuit:

$$R_{\text{Th}} = \frac{R_1 R_L}{R_1 + R_L} = \frac{(45.5)(25)}{45.5 + 25} = 16.1\ \Omega$$

The value of I_{N} can now be determined:

$$I_{\text{N}} = \frac{V_{\text{Th}}}{R_{\text{Th}}} = \frac{5.32}{16.1} = 0.330\text{ A} = 330\text{ mA}$$

The load line is shown in Fig. 4-74b. Although the values for diode voltage and current are correct for this design, and it would appear R_1 is correct, operation at or near the knee of the curve is *not* desirable. Remember that the input voltage is given as 15 V unregulated. If that voltage should drop to 10.0 V, the Thevenin voltage would drop to

$$V_{\text{Th}} = \left(\frac{R_L}{R_1 + R_L}\right)V_{\text{in}} = \left(\frac{25}{45.5 + 25}\right)10 = 3.55\text{ V}$$

This would put the voltage across the diode at a point below (to the right of) the 5-V zener voltage.

This indicates an advantage of the graphical solution. Even though the circuit values may be correct, as in this case, the graphical solution by load lines may show that a circuit should not be used.

4.13 Use a load line solution to obtain a more desirable value for R_1 in the circuit shown in Fig. 4-73b. Assume that 15 V is the nominal value of input voltage. Calculate the minimum and maximum allowable input voltage with the new design.

The following circuit values must not be changed in the new design:

$$V_{\text{in}} = 15\text{ V}\quad \text{nominal or average value} \qquad I_L = 200\text{ mA}$$

$$V_L = 5\text{ V} \qquad R_L = 25\ \Omega$$

Since the maximum current rating for our diode is 100 mA, we would like the load line to intersect the zener curve at 50 mA, 5 V on the reverse characteristic to allow for a safety margin of 2 at the nominal input voltage.

Therefore
$$I_T = I_z + I_L = (50 + 200) \times 10^{-3} = 250\text{ mA}$$

But $I_T R_1$ must still equal 10 V (that is, 15 – 5). So

$$R_1 = \frac{10}{I_T} = \frac{10}{250 \times 10^{-3}} = 40\ \Omega$$

We now solve for V_{Th} to find the load line:

$$V_{\text{Th}} = \left(\frac{R_L}{R_1 + R_L}\right)V_{\text{in}} = \left(\frac{25}{40 + 25}\right)15 = 5.77\text{ V}$$

Then
$$R_{\text{Th}} = \frac{R_L R_1}{R_L + R_1} = \frac{(25)(40)}{25 + 40} = 15.4\ \Omega$$

and

$$I_{\text{N}} = \frac{V_{\text{Th}}}{R_{\text{Th}}} = \frac{5.77}{15.4} = 0.375\text{ A} = 375\text{ mA}$$

So our new load line (shown in Fig. 4-75a) is drawn from -5.77 on the voltage scale to -375 on the current scale and intersects the curve at -50 mA as required.

The minimum input voltage occurs when I_z is at the minimum (rule of thumb) or $\frac{1}{10}I_L$ (in our case, $I_{z\,\text{min}} = \frac{1}{10}(200) = 20$ mA) or when $I_T = 20 + 200 = 220$ mA. So

$$V_{\text{in}} = I_T R_1 + V_L = (220 \times 10^{-3})(40) + 5 = 13.8\text{ V}$$

The maximum input voltage occurs when I_z is at its maximum rating of 100 mA or when

$$I_T = I_z + I_L = 100 + 200 = 300\text{ mA}$$

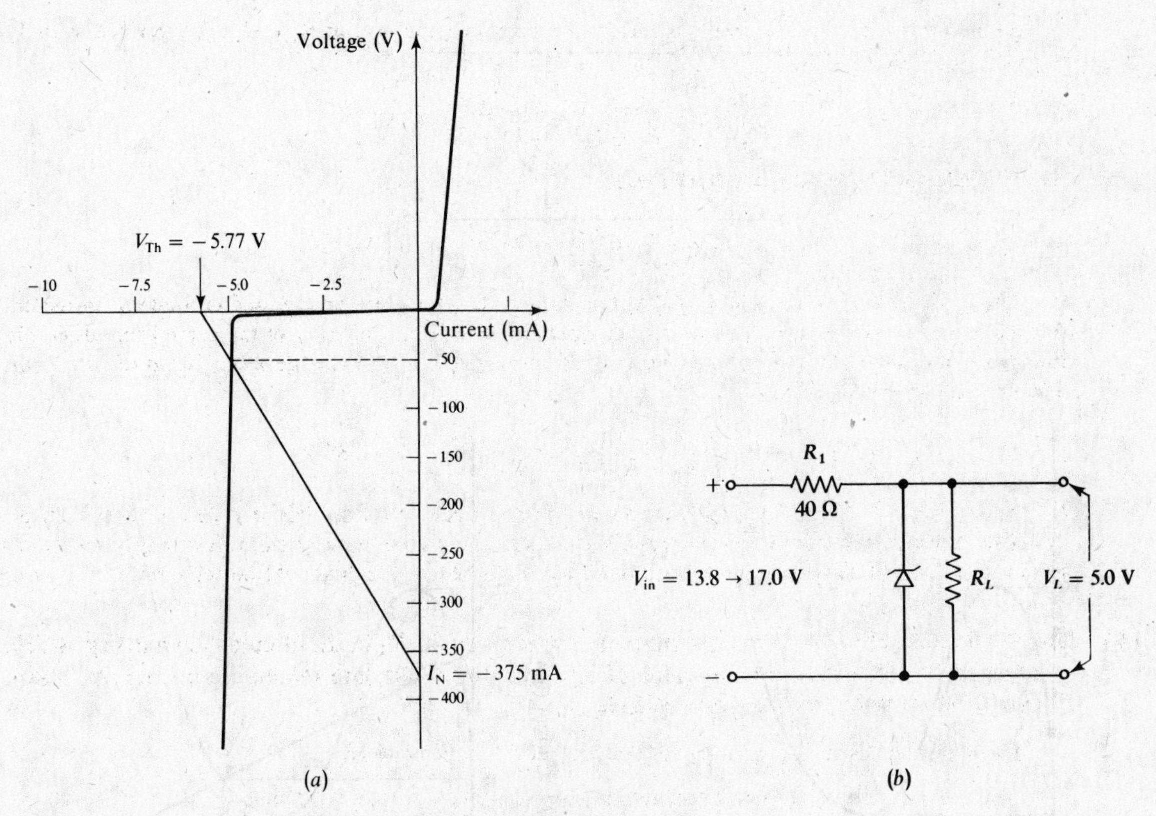

Fig. 4-75

So
$$V_{in} = (300 \times 10^{-3})(40) + 5 = 17.0 \text{ V}$$

The new circuit is shown in Fig. 4-75*b*.

4.14 Draw a load line for the tunnel diode circuit and characteristic curve shown in Fig. 4-76*a* and *b*. (*a*) How much peak current is required to switch the tunnel diode from V_1 to V_2? (*b*) How much peak current is required to switch the tunnel diode from V_2 to V_1?

The first step is to draw the load line. The V_{Th} point on the load line is the open-circuit voltage V_{in} for this circuit. The I_N point on the line is equal to the open-circuit voltage divided by the Thevinin resistance, which is R_1 in this circuit. From Fig. 4-76*a*, $V_{in} = 1.00$ V and $R_1 = 1333\ \Omega$.

$$I_N = \frac{V_{Th}}{R_{Th}} = \frac{1.00}{1333} = 75 \times 10^{-5} \text{ A} = 0.75 \text{ mA}$$

The two points (1 V and 0.75 mA) are marked on the characteristic curve and the load line is drawn. See Fig. 4-76*c*.

(*a*) From the graph of Fig. 4-76*c*, the circuit current I_1 is about 0.72 mA when the tunnel diode is at V_1. To switch to V_2, the current must exceed I_p, so the amount of current pulse is

$$+I_{(pulse)} = I_p - I_1 = 1.0 - 0.72 = 0.28 \text{ mA}$$

(*b*) The diode current I_2 is about 0.25 mA when the diode is in condition V_2. The current must be switched from 0.25 mA to below I_{min} (0.12 mA) in order to get it back to voltage V_1. Therefore, the negative current pulse must be

$$-I_{(pulse)} = I_2 - I_{min} = 0.25 - 0.12 = 0.13 \text{ mA}$$

Actually, a positive pulse greater than 0.28 mA and a negative pulse greater than 0.13 mA would be used for switching. The reason is that those values only bring the currents to the switching points, but positive switching requires greater values.

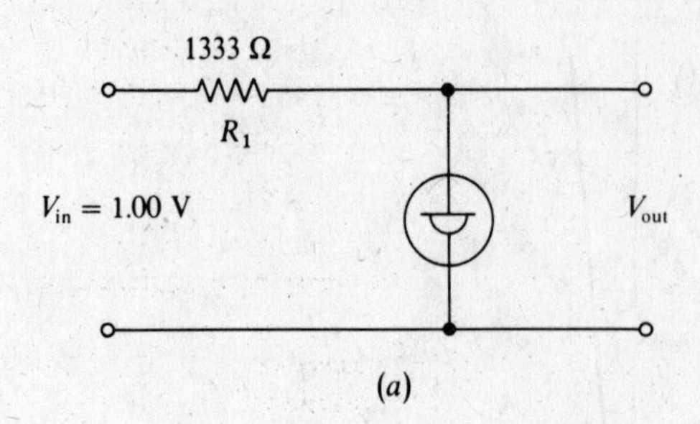

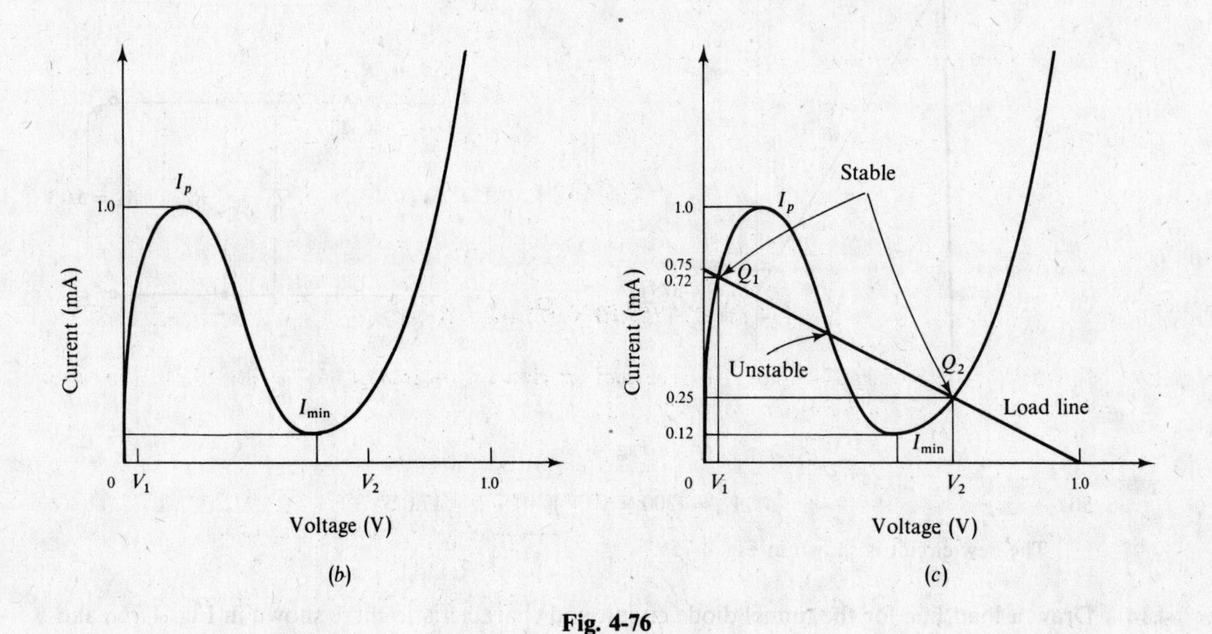

Fig. 4-76

4.15 The circuit shown in Fig. 4.2 is used to obtain the listed values. Plot the forward and reverse diode characteristic curves. Label the breakdown points.

Switch up

Voltage (V)	0.0	0.5	1.5	2.5	3.0	4.0	5.0
Current (mA)	0.0	30.0	100	200	300	800	1500

Switch down

Voltage (V)	0	−20	−40	−50	−51	−52	−53
Current (mA)	0	10	25	40	400	1000	Off scale

The required curves are shown in Fig. 4-77*a* and *b*.

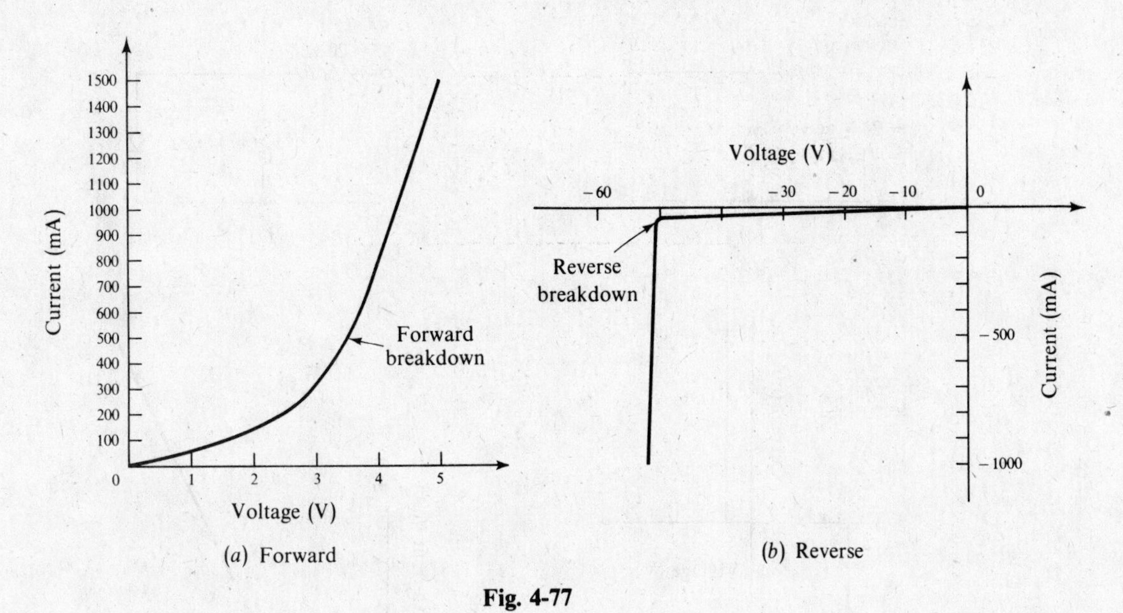

Fig. 4-77

Supplementary Problems

4.16 Find the maximum output voltage v_{out} of the clamper circuit shown in Fig. 4-78*a* when the input peak-to-peak voltage is 5.0 V as shown in Fig. 4-78*b*.

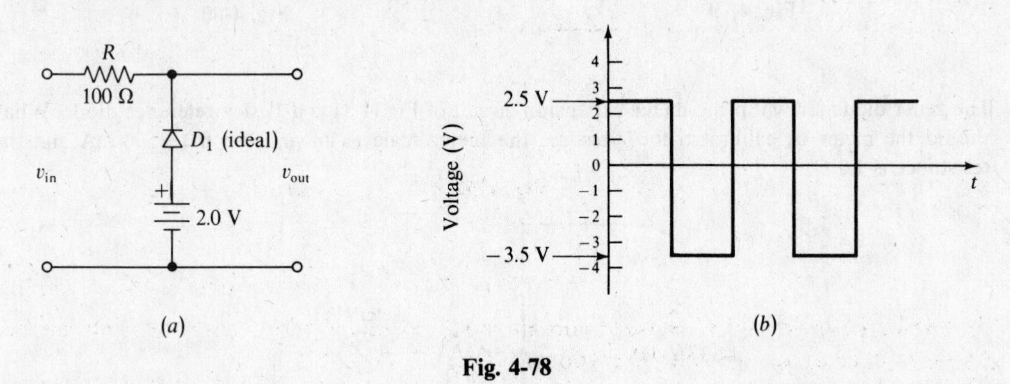

Fig. 4-78

4.17 Figure 4-79*a* shows an ac circuit with a zener diode across the output. The characteristic curve of the diode is shown in Fig. 4-79*b*. (*a*) Sketch the output waveform. (*b*) Determine the amount of time the diode conducts during each cycle.

4.18 The circuit shown in Fig. 4-80*a* uses a diode with the characteristic curve shown in Fig. 4-80*b*. What is the rms voltage of the generator in order for the diode forward current to be 25 mA?

4.19 Assume that the curve shown in Fig. 4-80*b* is obtained from an oscilloscope with a 15-Ω calibrating resistor for the vertical scale. What is the V/division setting for the vertical scale of the oscilloscope?

4.20 A diode rated at 0.5 A is subject to a surge current of 40 A for 0.2 ms. See Fig. 4-16. (*a*) Can the diode pass this surge current without being destroyed? (*b*) For what period of time could this diode pass a 100-A current without being destroyed?

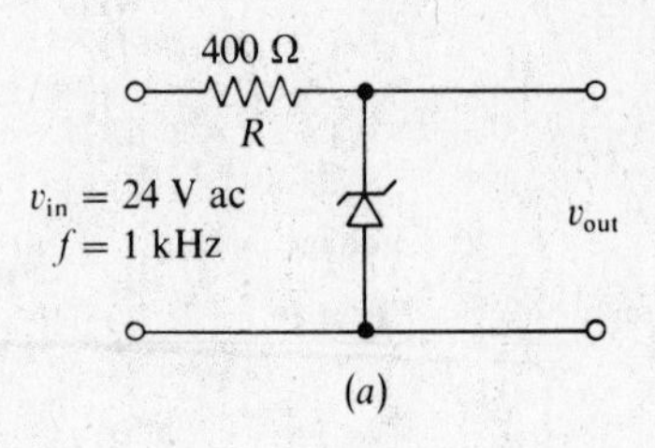

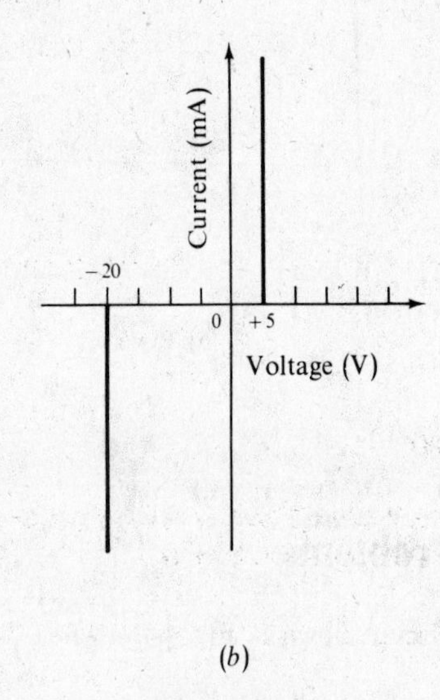

Fig. 4-79

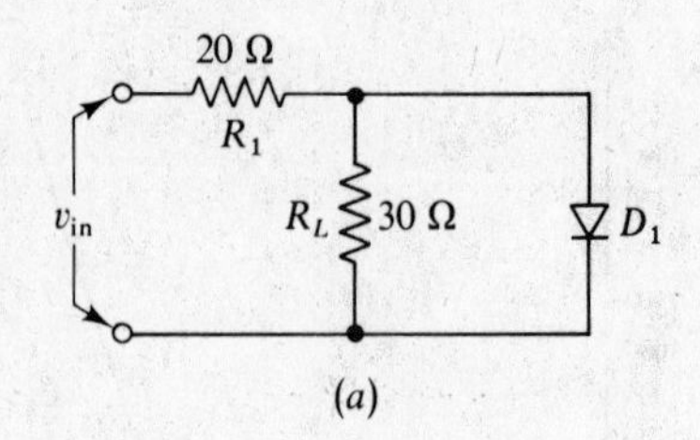

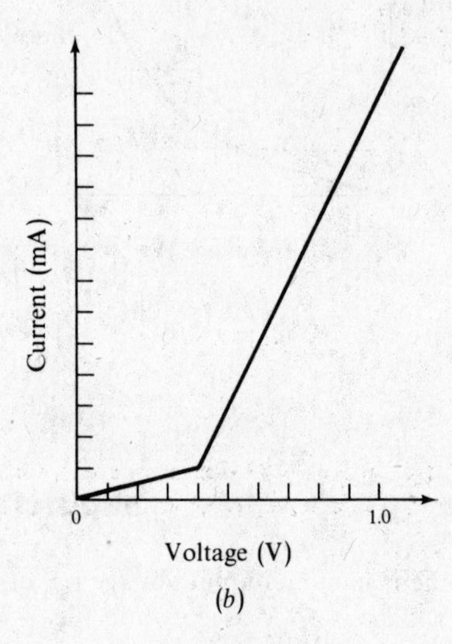

Fig. 4-80

4.21 The zener diode shown in the meter expansion circuit of Fig. 4-81 is a 12.0-V reference diode. What values should the meter be calibrated for? Assume the meter scale is linear from 0 to 500 mA and the meter resistance is 1.0 Ω.

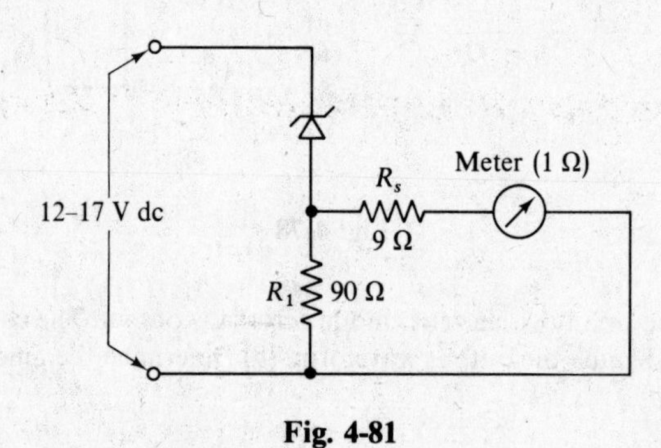

Fig. 4-81

4.22 A certain diode is found to have a drop of 0.8 V in the forward direction at its rated current of 100 mA. What should the power dissipation rating be?

4.23 A diode with a PIV rating of 200 V has a reverse recovery time of 1.0 ms. Can this diode be used to rectify an input wave of the form $v_{in} = 125 \sin 2\pi ft$ where $f = 10$ MHz?

4.24 Figure 4-82*a* shows a tunnel diode circuit and Fig. 4-82*b* shows the diode characteristic. Find the operating points of the diode in this circuit and the values of the positive and negative current pulses to switch operating points.

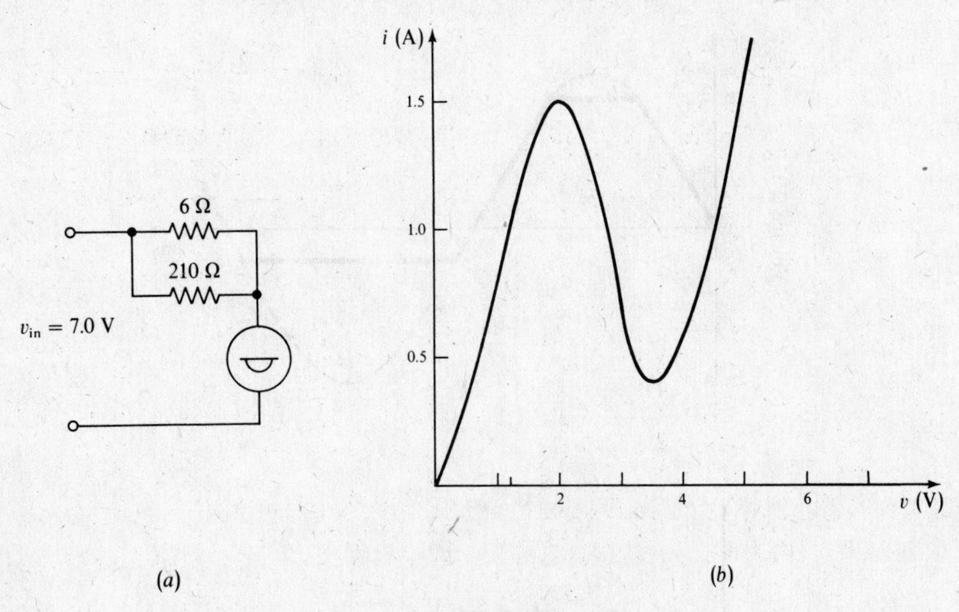

Fig. 4-82

4.25 The LED in Fig. 4-83 operates with a forward drop of 1.6 V when 20 mA is being conducted. Find the value of the dropping resistor R so this circuit can act as the pilot light for 24 V dc.

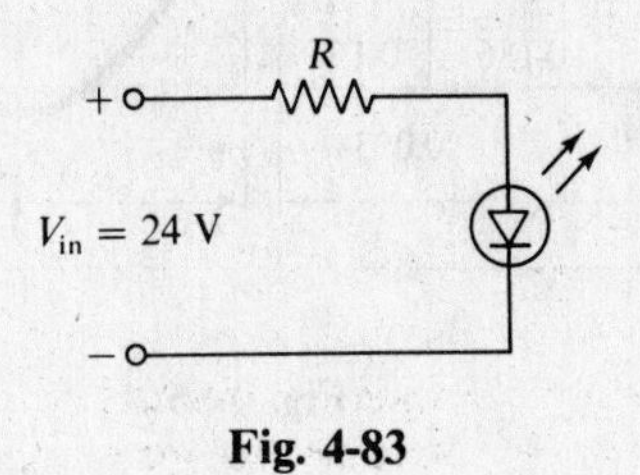

Fig. 4-83

Answers to Supplementary Problems

4.16 Refer to Fig. 4-84. $V_m = 4.5$ V.

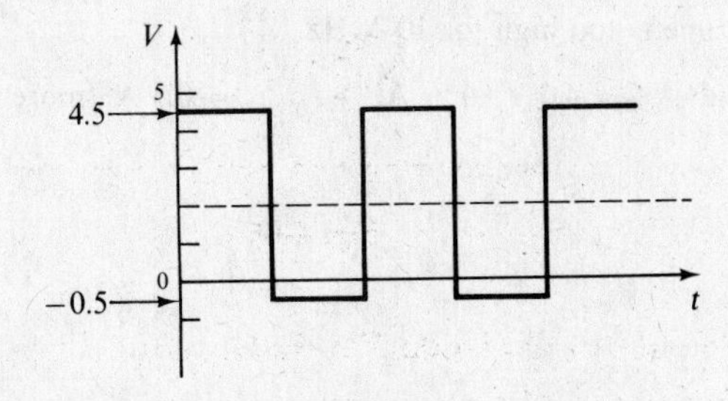

Fig. 4-84

4.17 (*a*) Refer to Fig. 4-85. (*b*) Conducting time = 0.62 ms.

4.18 1.06 V.

4.19 75 mV/division.

4.20 (*a*) Yes. (*b*) 0.20×10^{-4} s = 0.02 ms.

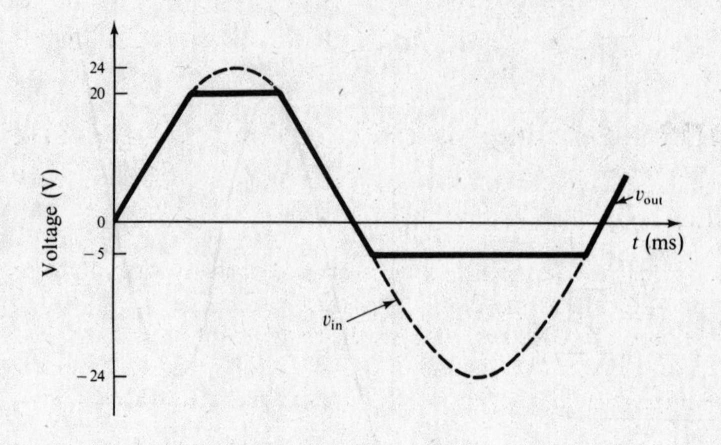

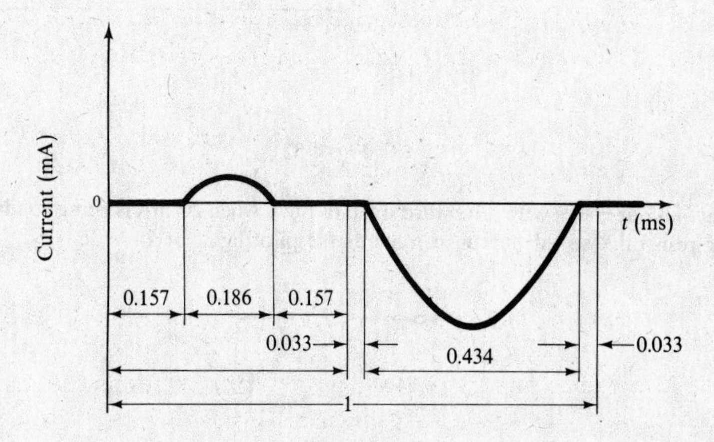

Fig. 4-85

4.21

Meter (mA)	0	100	200	300	400	500
Calibrated voltage	12.0	13.0	14.0	15.0	16.0	17.0

4.22 Minimum of 80 mW.

4.23 No. Reverse recovery time is too high for 10 MHz.

4.24 $Q_1 = 1.2$ V, 1.0 A and $Q_2 = 4.0$ V, 0.57 A. $+I_{(\text{pulse})} = 0.5$ A (more positive than); $-I_{(\text{pulse})} = 0.17$ A (more negative than).

4.25 1120 Ω.

Chapter 5

Transistors, Tubes, and Amplifiers

5.1 INTRODUCTION

Vacuum tube triodes ushered in the age of electronics. For many years they were the only way to achieve amplification and are still in use in many modern systems.

Tubes are relatively simple to construct, can handle large amounts of power easily, and are forgiving of short bursts of overrating current or voltage.

They have several important disadvantages when compared with transistors:

1. They generate a great deal of heat, even in low-power circuits.
2. The physical construction of tubes makes them susceptible to shock and vibration damage.
3. They are inefficient. Much of the input power is used for cathode heating.
4. They are physically large.
5. Their maximum useful life is limited by the life of the filament, but other factors (such as a loose grid wire or an open plate wire) often reduce their life expectancy as well.
6. Their characteristics deteriorate with age.

Because of this combination of disadvantages, the tube has not been able to prevail against the transistor.

There are two major types of transistors: *bipolar* and *field effect*. Bipolar types use both holes and electrons as charge carriers.

An FET, or *F*ield *E*ffect *T*ransistor, uses either holes or electrons for its operation, but not both.

We generally use a transistor for amplification but another important application is for switching. In fact, there are specially designed transistors used only for that purpose.

In this chapter we will examine the basic operation of tubes, bipolar transistors, and FETs, present their equivalent circuits and common circuit configurations, and extend the graphical technique of the load line to include the solution of amplifier problems.

5.2 TUBES

The first major breakthrough in electronics after Edison's 1883 discovery of the rectifying ability of a heated filament and a positive plate was Lee De Forest's invention of the Audion in 1906. He added a control grid to the Fleming diode, and although he originally intended the Audion to be an electronic relay to switch large amounts of power with a small signal voltage, the general amplifying features of the Audion led to the use of this *triode* (three elements) in many other applications. Further refinements brought about the tetrode (four elements), pentode (five elements), and other multielement tubes that we know today.

The ability of a small voltage change in the grid circuit to cause a larger change in the cathode to anode (or plate) circuit can be explained by noting the fact that the grid is closer than the anode to the source of electrons (called the cathode). Consequently, a 1-V change in the grid potential has a greater effect on electron flow than a 1-V change in the plate potential.

A graph of the plate current i_b vs. plate voltage e_b for a given grid voltage e_c as shown in Fig. 5-1 is called a *plate characteristic curve.*

The plate characteristic curves for several different grid voltages may then be drawn on a single set of axes. This type of graph is usually included in tube data and is sometimes called the plate family of curves. The plate family for a triode, tetrode, and pentode are shown as Fig. 5-2*a*, *b*, and *c*, respectively.

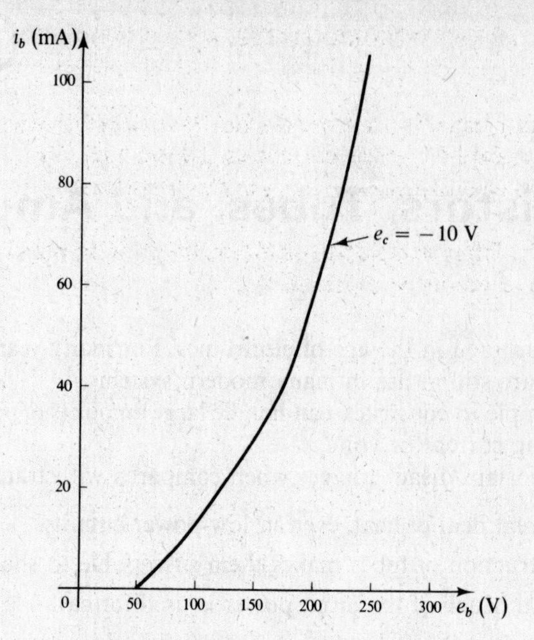

Fig. 5-1 Triode plate characteristic

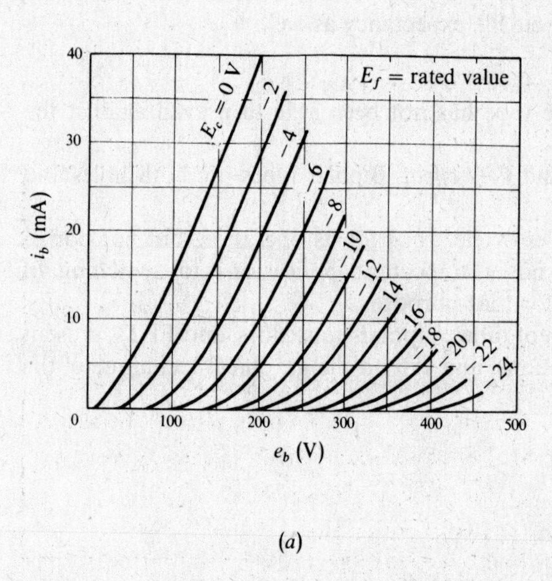

(*a*)

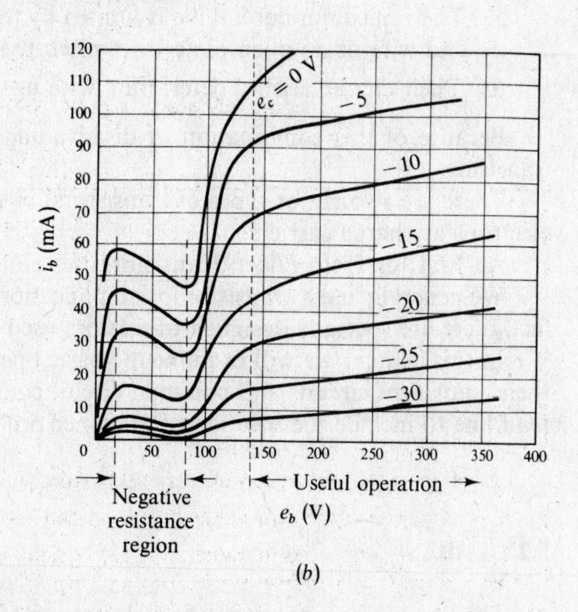

(*b*)

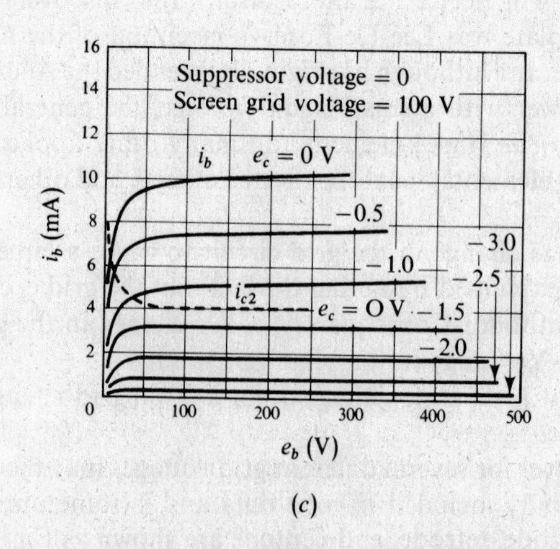

(*c*)

Fig. 5-2

Since we are usually interested in how much the plate *current* varies with a given change in grid voltage, we usually include graphs of plate current vs. grid voltage. The curves shown on these graphs are called the *average transfer characteristic curves.* This is shown for three different plate voltages in Fig. 5-3.

In tube terminology, the letter e is used to denote voltage rather than the customary v, and the terms most commonly encountered are listed in Table 5-1.

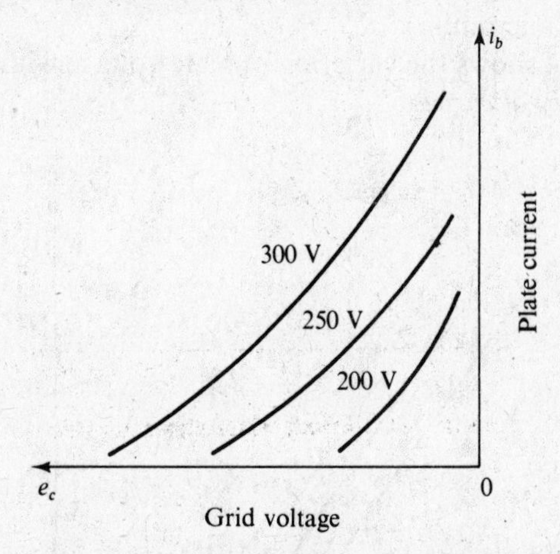

Fig. 5-3 Average transfer characteristics for 3 values of plate voltage

Table 5-1

e_b = total plate voltage	E_g = rms value of ac signal voltage
e_c = total grid-cathode voltage	E_p = rms value of ac anode voltage
e_g = ac grid-cathode voltage	E_s = rms value of ac grid circuit input voltage
e_s = ac input voltage to the grid circuit	E_o = rms value of ac component of load voltage
E_{BB} = dc plate supply voltage	i_b = total anode current
E_{CC} = dc grid supply voltage	i_p = ac component of anode current
E_{AA} = dc heater supply voltage	I_p = rms value of ac component of anode current
E_b = average or quiescent value of anode voltage	g_m = mutual conductance (transconductance)
E_c = average or quiescent value of grid-cathode voltage	r_p = dynamic plate resistance
I_b = average or quiescent value of anode current	μ = amplification factor

The slope of the average plate characteristic curve varies with the amount of plate voltage, as shown in Fig. 5-1, and provides us with a parameter that is frequently used, the *dynamic plate resistance* r_p:

$$r_p = \frac{1}{\text{slope of plate characteristic}}$$

And since the slope is given by

$$\left.\frac{\Delta i_b}{\Delta e_b}\right|_{e_c=\text{constant}}$$

$$r_p = \left.\frac{1}{\Delta i_b/\Delta e_b}\right|_{e_c=\text{constant}} = \left.\frac{\Delta e_b}{\Delta i_b}\right|_{e_c=\text{constant}} \tag{5-1}$$

r_p is given in ohms, and typical values are in the vicinity of 4 to 100 kΩ.

The slope of the average transfer characteristic curve is given by

$$\text{Slope} = \left.\frac{\Delta i_b}{\Delta e_c}\right|_{e_b=\text{constant}}$$

Since the dimensions of this slope are in conductance units, S (siemens), this tube parameter is named the *grid-plate transconductance* or *mutual conductance* g_m:

$$g_m = \left.\frac{\Delta i_b}{\Delta e_c}\right|_{e_b = \text{constant}} \tag{5-2}$$

The range of typical values for g_m are from 200 to 40 000 μS although most tube data may still use the older unit of conductance, μ℧ (micromho).

The graph shown in Fig. 5-4 shows the variation of plate voltage with grid voltage and is sometimes called the *constant-current family of curves.*

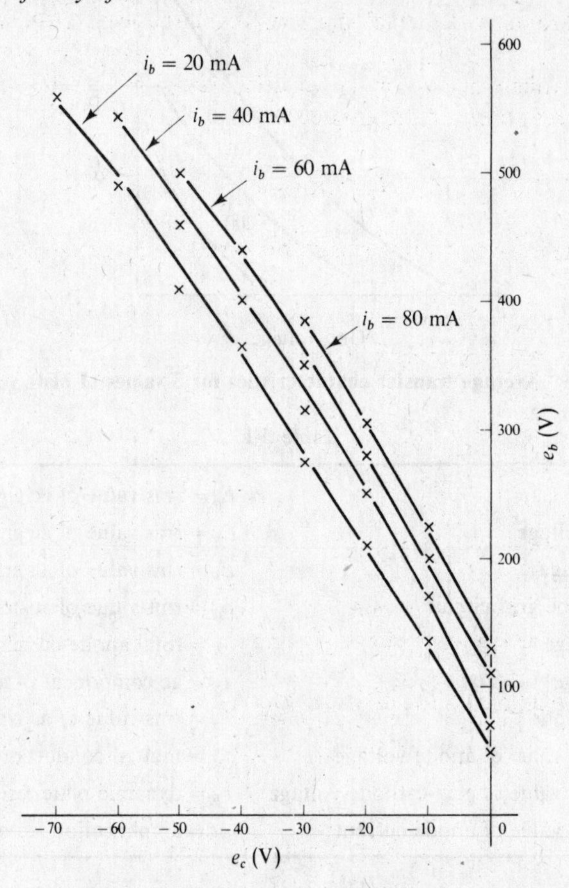

Fig. 5-4 Constant-current characteristics

The slope of this curve gives us the third tube parameter, the *amplification factor* μ.

$$\mu = \text{slope} = -\left.\frac{\Delta e_b}{\Delta e_c}\right|_{i_b = \text{constant}} \tag{5-3}$$

where the negative sign indicates that the change in plate voltage is opposite in polarity to the change in grid voltage that caused it.

Typical values of μ, which is dimensionless, range from 10 to 200.

The interdependence of the three tube coefficients or parameters can be seen by the relationship

$$\mu = g_m r_p \tag{5-4}$$

Example 5.1 Prove that Eq. (*5-4*) is correct.

From Eqs. (*5-1*) and (*5-2*), we can substitute the definitions for g_m and r_p:

$$r_p = \frac{\Delta e_b}{\Delta i_b} \qquad \text{and} \qquad g_m = \frac{\Delta i_b}{\Delta e_c}$$

Multiplying, we obtain

$$g_m r_p = \left(\frac{\Delta i_b}{\Delta e_c}\right)\left(\frac{\Delta e_b}{\Delta i_b}\right) = \frac{\Delta e_b}{\Delta e_c}$$

But if we neglect the minus sign, from (5-3)

$$\frac{\Delta e_b}{\Delta e_c} = \mu$$

So

$$\mu = g_m r_p$$

We can use the tube parameters to develop a low-frequency ac equivalent circuit to represent the tube.

The current source equivalent circuit for the vacuum tube is shown in Fig. 5-5.

Writing Kirchhoff's current rule for the circuit, we have

$$I_p = g_m E_g + \frac{E_o}{r_p} \tag{5-5}$$

Note that this equivalent circuit only exists for variable quantities and does not exist for dc quantities.

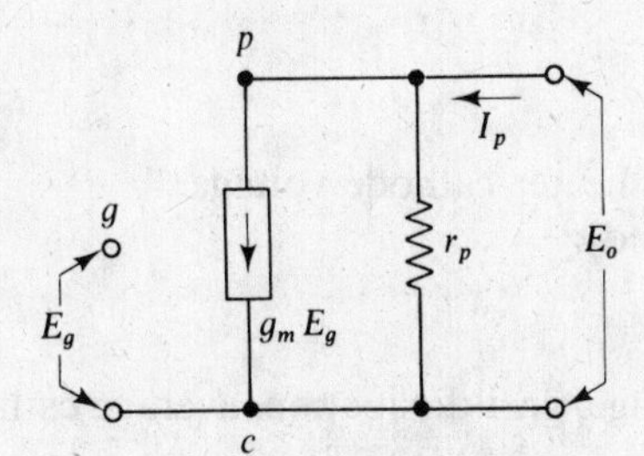

Fig. 5-5 Tube: current source ac equivalent circuit

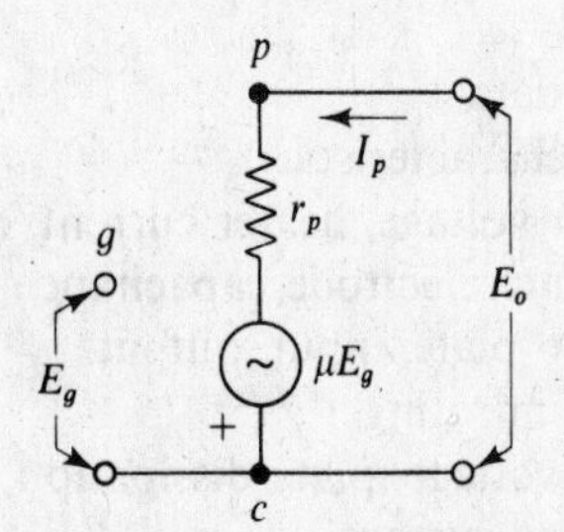

Fig. 5-6 Tube: voltage source ac equivalent circuit

We can also rearrange Eq. (5-5) and solve for E_o:

$$E_o = r_p I_p - g_m r_p E_g$$

but from (5-4)

$$g_m r_p = \mu$$

So

$$E_o = -\mu E_g + r_p I_p \tag{5-6}$$

and this suggests the voltage source equivalent circuit shown in Fig. 5-6. The voltage source equivalent is used most commonly for tube amplifier circuits.

In both equivalent circuits, the 180° phase shift is shown by the inward direction of I_p, and the input and output circuits are isolated.

The ac equivalent circuits are dependent upon the assumption that g_m, r_p, and μ are constant, a good assumption for small signals, and that the frequencies are such that the interelectrode capacitances can be ignored.

Example 5.2 Prove that the voltage source equivalent circuit shown in Fig. 5-6 satisfies Eq. (5-6).

Writing Kirchhoff's voltage rule for the equivalent circuit shown in Fig. 5-6, we obtain

$$E_o = I_p r_p - \mu E_g$$

which is identical to Eq. (5-6).

Manufacturer's tube data may include

- Mechanical data (see Fig. 5-7) including bulb type, base diagram, mounting position.

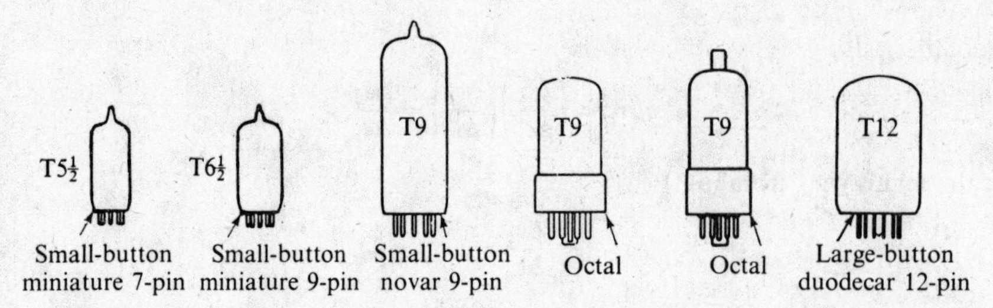

(*a*) Tube outlines

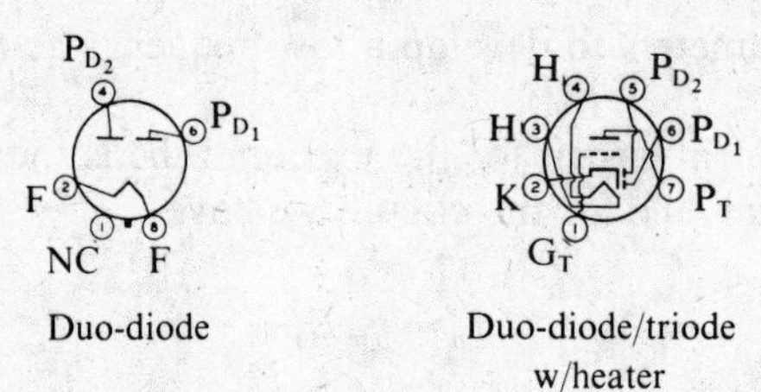

(*b*) Two types of tube base diagrams

Fig. 5-7

- Electrical data

  Heater characteristics:

  heater voltage, heater current, maximum heater-cathode voltage.

  Direct interelectrode capacitance (unshielded):

  grid to plate, input, output.

  Maximum ratings:

  plate voltage, plate dissipation, grid voltage, grid dissipation, average cathode current, peak cathode current.

  Characteristics and typical operation:

  plate voltage, grid voltage, plate current, plate resistance (r_p), transconductance (g_m), amplification factor (μ).

- Graphs

  Average plate characteristics (see Fig. 5-2*a*, *b*, and *c*)

  Average transfer characteristics (see Fig. 5-3)

Most of the tube characteristics are self-explanatory.

The interelectrode capacitances, even though they are usually 5 pF or less, become important when tubes are used at high frequencies.

The construction of the tube, Fig. 5-8, is a major factor in controlling the interelectrode capacitances and the g_m as well.

5.3 BIPOLAR TRANSISTORS

Bipolar transistors are named for the fact that two types of majority charge carriers (holes and electrons) are used for their operation. As shown in Fig. 5-9, the PNP and NPN types are constructed with layers of P material and N material. The models are used for explaining their operation but, as with most models of electronic devices, they are not replicas of their actual construction.

The majority charge carriers in P-type semiconductor material are holes. In the PNP transistor, the holes are attracted across the forward-biased emitter-base junction. A few of these go to the negative base supply, but the majority fall under the influence of the greater negative collector supply. The arrows in the model show how the hole flow is divided.

The majority charge carriers in NPN transistors are electrons. The arrows in the NPN transistor, then, represent electron flow. In both the NPN and PNP transistors, most of the majority charge carriers

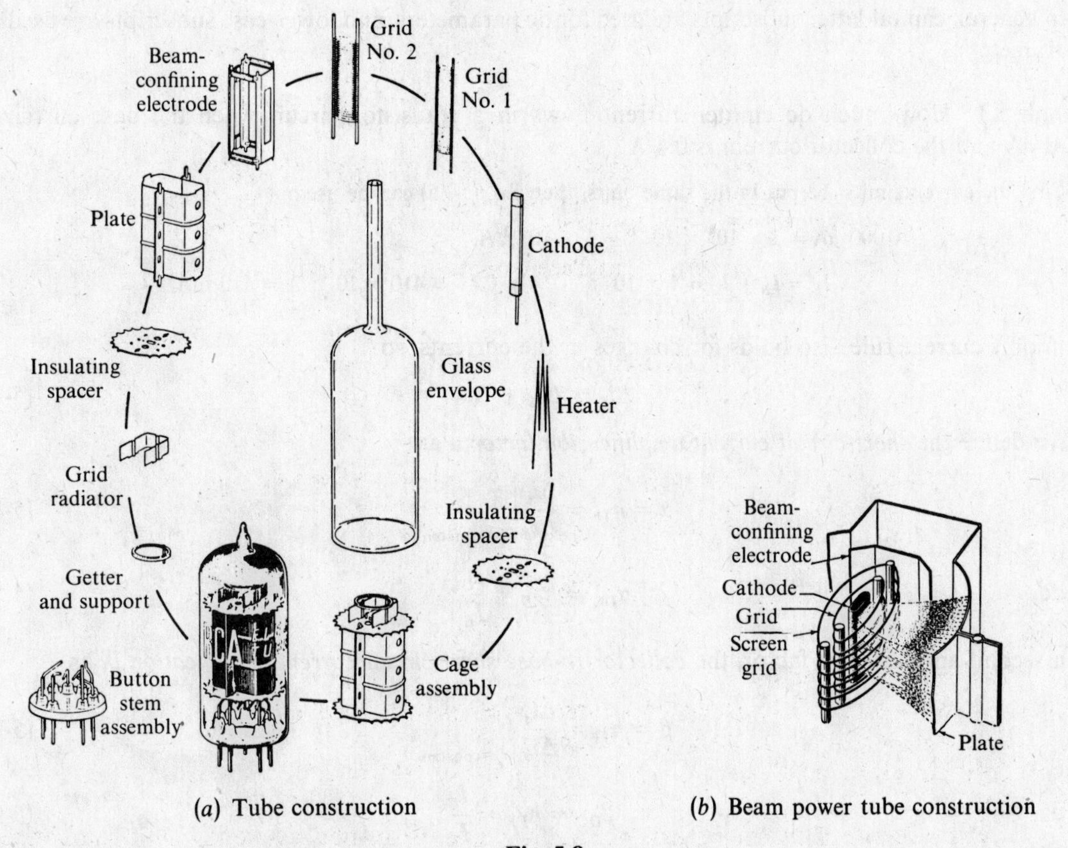

(*a*) Tube construction (*b*) Beam power tube construction

Fig. 5-8

move across the base region into the collector. Transistors amplify by virtue of the fact that *a small change in base current produces a large change in collector current.*

As pointed out in Chap. 4, *the arrow in the symbol points toward an N region* in both the NPN and PNP transistors.

The relationships in transistors between the emitter, base, and collector currents is established by Kirchhoff's current law:

$$I_E = I_B + I_C \tag{5-7a}$$

where I_E = dc emitter current
I_B = dc base current
I_C = dc collector current

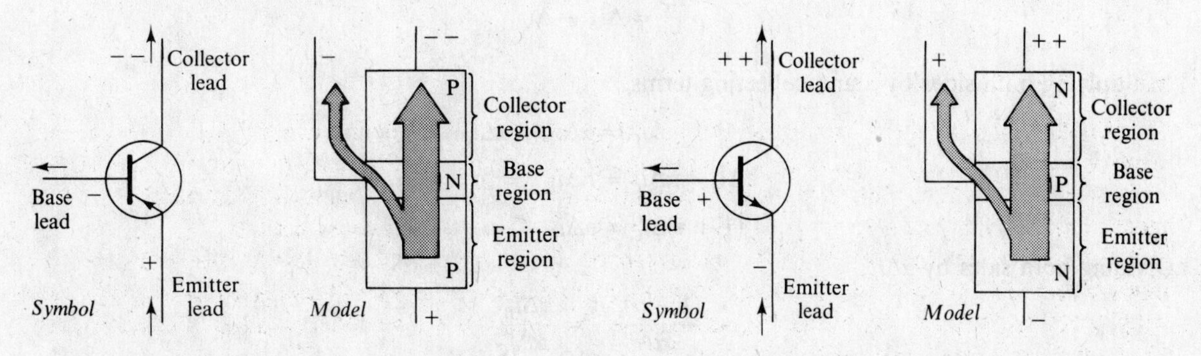

(*a*) Conventional (hole) currents in a PNP transistor (*b*) Electron currents in an NPN transistor

Fig. 5-9

In general, capital letter subscripts are used for dc parameters, and lower-case subscripts are used for ac parameters.

Example 5.3 How much dc emitter current flows in a transistor circuit when the base current is 10 000 μA and the collector current is 0.4 A?

First the currents must be put in the same units, then Eq. (5-7*a*) can be used.

$$10\,000\ \mu\text{A} = 1 \times 10^4 \times 10^{-6} = 1 \times 10^{-2}\ \text{A}$$

$$I_E = I_B + I_C = 1 \times 10^{-2} + 0.4 = 0.41 = 410 \times 10^{-3}\ \text{A} = 410\ \text{mA}$$

Kirchhoff's current rule also holds for changes in the currents, so

$$\Delta i_E = \Delta i_B + \Delta i_C \tag{5-7b}$$

We define the *short-circuit current amplification factor* α as

$$\alpha = h_{fb} = \left.\frac{\Delta i_c}{\Delta i_e}\right|_{v_c=\text{constant}} \tag{5-8a}$$

For dc,

$$\alpha_{DC} = h_{FB} = \frac{I_C}{I_E} \tag{5-8b}$$

and a second amplification factor, the *collector-to-base short circuit current amplification* β, as

$$\beta = h_{fe} = \left.\frac{\Delta i_c}{\Delta i_b}\right|_{v_c=\text{constant}} \tag{5-9a}$$

For dc,

$$\beta_{DC} = h_{FE} = \frac{I_C}{I_B} \tag{5-9b}$$

The relationship between α and β can be seen by substitution into Eq. (5-7*b*):

$$\Delta i_E = \Delta i_B + \Delta i_C$$

From (5-8*a*),

$$\alpha = \frac{\Delta i_C}{\Delta i_E}$$

Solving for Δi_E yields

$$\Delta i_E = \frac{\Delta i_C}{\alpha}$$

Substituting,

$$\frac{\Delta i_C}{\alpha} = \Delta i_B + \Delta i_C$$

Multiplying both sides by α and collecting terms,

$$\Delta i_C = \alpha \Delta i_B + \alpha \Delta i_C$$

$$\Delta i_c - \alpha \Delta i_C = \alpha \Delta i_B$$

$$(1 - \alpha)\Delta i_C = \alpha \Delta i_B$$

Dividing both sides by $\alpha \Delta i_c$,

$$\frac{(1-\alpha)\cancel{\Delta i_C}}{\alpha \cancel{\Delta i_C}} = \frac{\cancel{\alpha}\Delta i_B}{\cancel{\alpha}\Delta i_C}$$

$$\frac{1-\alpha}{\alpha} = \frac{\Delta i_B}{\Delta i_C}$$

Inverting both sides, $$\frac{\alpha}{1-\alpha} = \frac{\Delta i_C}{\Delta i_B}$$

But from (*5-9a*) $$\frac{\Delta i_C}{\Delta i_B} = \beta$$

So $$\beta = \frac{\alpha}{1-\alpha} \tag{5-10}$$

Example 5.4 β_{DC} of a certain transistor is 100. If the collector current is 100 mA, what is the emitter current value?

The first step is to find the base current from (*5-9b*):

$$\beta_{DC} = \frac{I_C}{I_B}$$

Solving for I_B and substituting values,

$$I_B = \frac{I_C}{\beta_{DC}} = \frac{100 \times 10^{-3}}{100} = 1.00 \times 10^{-3} \text{ A} = 1.00 \text{ mA}$$

The emitter current can now be found using (*5-7a*):

$$I_E = I_B + I_C = 1.00 \times 10^{-3} + 100 \times 10^{-3} = 101 \times 10^{-3} \text{ A} = 101 \text{ mA}$$

Equation (*5-10*) shows that α and β are mathematically related. A common range of α values is between 0.95 and 0.999, while β is often between 50 and 150. The graph shown in Fig. 5-10 is drawn from the information given in Table 5-2. It shows how a small increase in α can produce a large change in β.

Table 5-2 α **and corresponding** β **values**

α_{DC}	β_{DC}
0	0
0.95	19
0.96	24
0.97	32.33
0.98	49
0.99	99
0.999	999

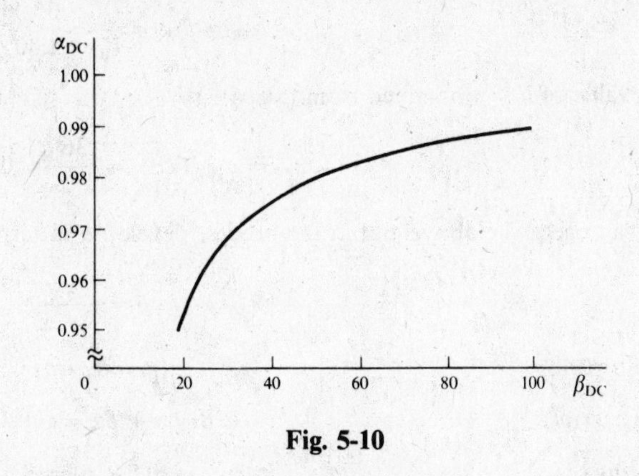

Fig. 5-10

From Chap. 4 on four-terminal networks, it is obvious that both α and β are forward hybrid parameters (h_{21}). This parameter is commonly designated h_F for dc circuits and h_f for ac. Since α and β are both forward current ratios, a distinction must be made between the one obtained with the common-emitter configuration (h_{FE}) and the one obtained with the common-base configuration (h_{FB}). The symbolism is summarized here:

$$\beta_{DC} = h_{FE} \qquad \beta = h_{fe}$$

$$\alpha_{DC} = h_{FB} \qquad \alpha = h_{fb}$$

Clearly, a complete dc analysis could be made if all of the hybrid parameters are known. The definitions of the h parameters as related to dc currents and voltages of a bipolar transistor are given in Table 5-3. In general, the parameters related to ac signals are the same, except that lowercase letters are used. Of course, resistances become impedances and conductances become admittances in ac considerations.

Table 5-3 DC parameters for transistors

h parameter	Name	Common-emitter symbol	Common-base symbol	Common-collector symbol
h_{11}	Input resistance, output shorted	h_{IE}	h_{IB}	h_{IC}
h_{12}	Reverse voltage gain, input open	h_{RE}	h_{RB}	h_{RC}
h_{21}	Forward current gain, input shorted	h_{FE}	h_{FB}	h_{FC}
h_{22}	Output conductance, input open	h_{OE}	h_{OB}	h_{OC}

Example 5.5 A certain transistor has a dc collector current of 36.0 mA and a dc emitter current of 37.8 mA. Find the following: the base current, the value of h_{FE}, and the value of h_{FB}.

The base current can be obtained from (*5-7a*),

$$I_E = I_B + I_C$$

$$37.8 = I_B + 36.0$$

$$I_B = 37.8 - 36.0 = 1.80 \text{ mA}$$

The value of h_{FE} is obtained from (*5-9b*):

$$h_{FE} = \beta_{DC} = \frac{I_C}{I_B} = \frac{36.0}{1.80} = 20.0$$

The value of h_{FB} is obtained from (*5-8b*):

$$h_{FB} = \alpha_{DC} = \frac{I_C}{I_E} = \frac{36}{37.8} = 0.952\,38$$

We can check the above and in the process develop α in terms of β by solving Eq. (*5-10*) for α:

$$\beta = \frac{\alpha}{1-\alpha}$$

Cross multiplying,
$$\beta - \beta\alpha = \alpha$$

Collecting terms,
$$\beta = \alpha + \beta\alpha = \alpha(1+\beta)$$

Rearranging,
$$\alpha(1+\beta) = \beta$$

Dividing both sides by $1 + \beta$,

$$\alpha = \frac{\beta}{1+\beta} \qquad (5\text{-}10a)$$

Substituting values as a check,

$$\alpha = \frac{20.0}{1 + 20.0} = 0.952\,38$$

There are no units for $\alpha(h_{FB})$ and $\beta(h_{FE})$. Normally, the value of α_{DC} is not carried out to five places; however, in this problem the accuracy is better checked by doing so.

Example 5.6 A transistor with $h_{FB} = 0.97$ has an emmitter current of 100 mA. What is the value of collector current, and what is the value of h_{FE}?

The collector current can be obtained from (*5-8b*):

$$h_{FB} = \alpha_{DC} = \frac{I_C}{I_E}$$

Solving for I_C and substituting values,

$$I_C = \alpha_{DC} I_E = h_{FB} I_E = (0.97)(100 \times 10^{-3}) = 97.0 \times 10^{-3} = 97.0 \text{ mA}$$

The value of β can be obtained from (*5-10*):

$$\beta_{DC} = h_{FE} = \frac{\alpha_{DC}}{1 - \alpha_{DC}} = \frac{0.97}{1 - 0.97} = 32.3$$

When transistors are analyzed for circuit design, the three-terminal transistor can be thought of as a four-terminal network with one of the elements common to the input and output terminals. Figure 5-11 shows the three possibilities. The common-base configuration was analyzed first in terms of ac equivalent circuits and the other two ac equivalent circuits derived from it.

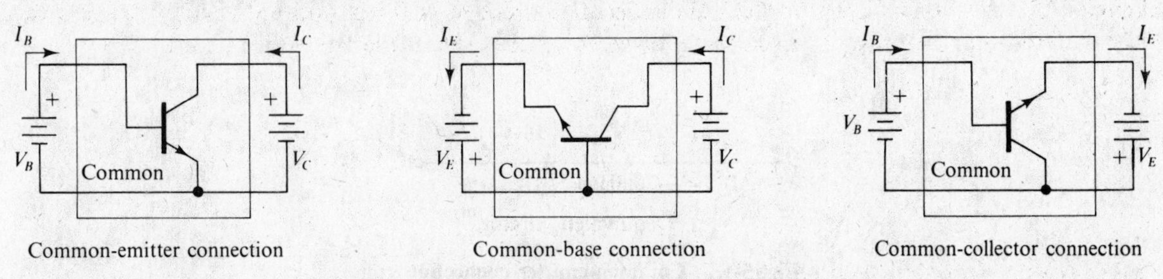

Fig. 5-11 Connections for an NPN transistor

The common-base ac equivalent circuits are shown in Fig. 5-12*a*, *b*, and *c*. The common-emitter ac equivalent circuits are shown in Fig. 5-13*a*, *b*, and *c*, and the common-collector ac equivalent circuits in Fig. 5-14*a*, *b*, and *c*.

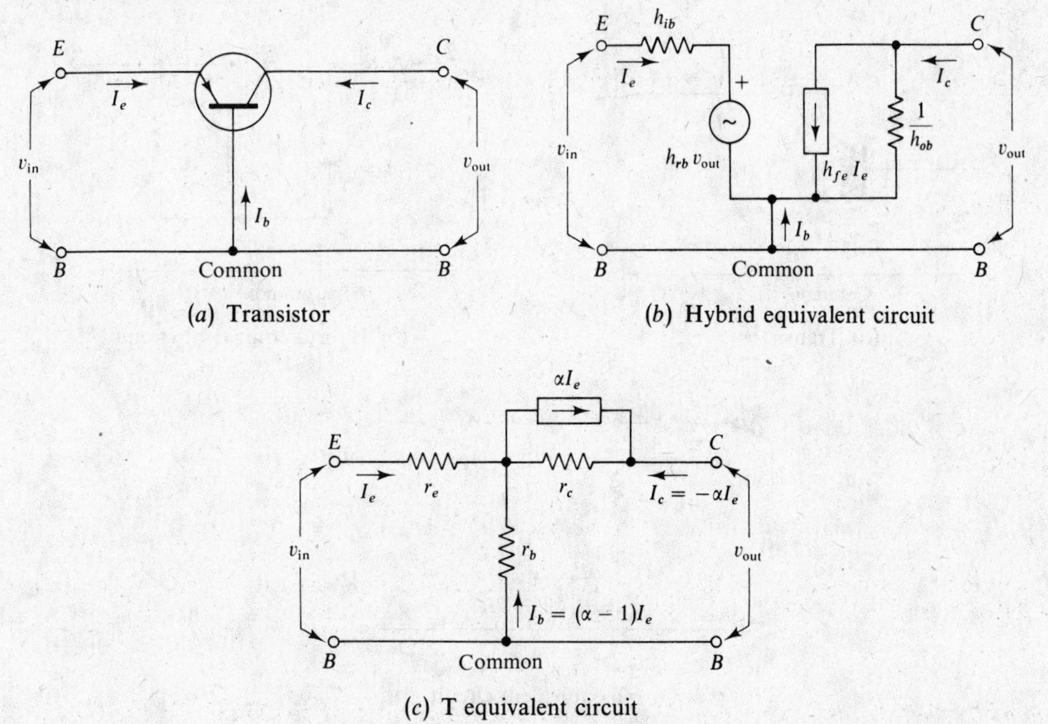

Fig. 5-12 Common-base connection

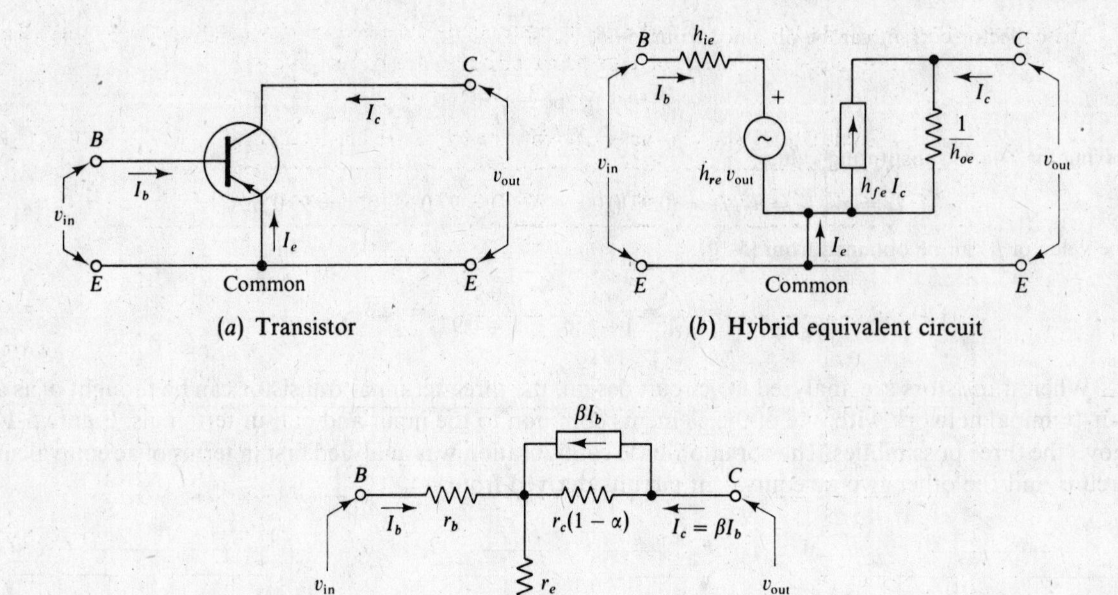

(*a*) Transistor (*b*) Hybrid equivalent circuit

(*c*) T equivalent circuit

Fig. 5-13 Common-emitter connection

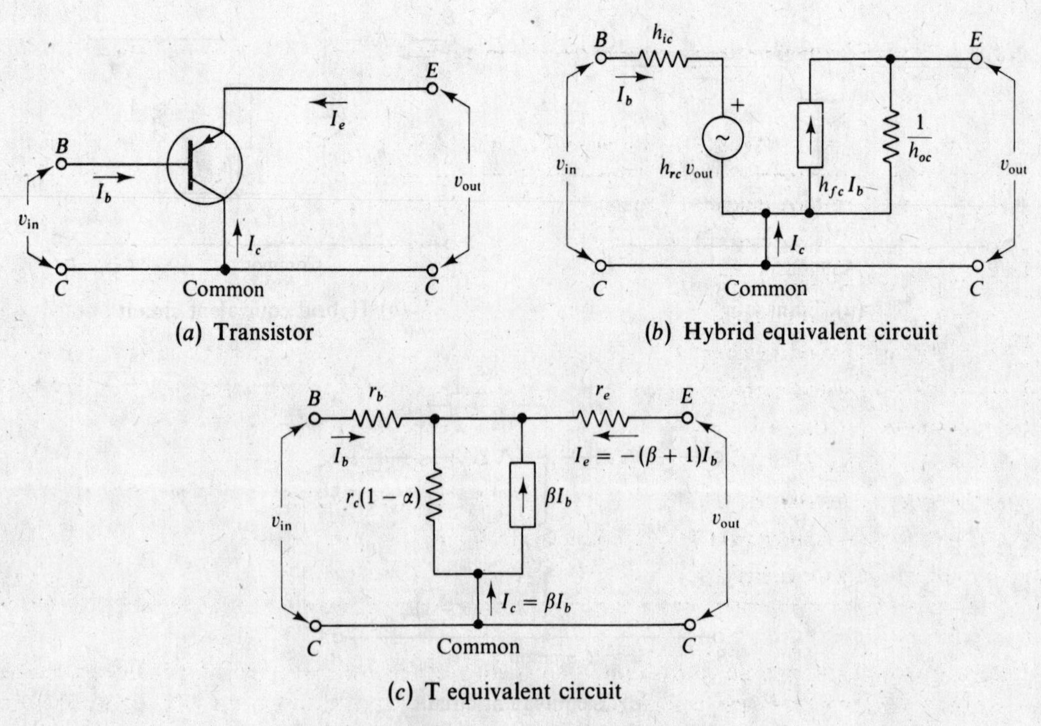

(*a*) Transistor (*b*) Hybrid equivalent circuit

(*c*) T equivalent circuit

Fig. 5-14 Common-collector configuration

The approximate values for the hybrid and T equivalent circuits for low frequencies are listed in Table 5-4, while the nomenclature used for transistor circuits is listed in Table 5-5.

Table 5-4 Values for ac equivalent circuits—three configurations

Parameter	Common base	Common emitter	Common collector
r_b	$\frac{h_{rb}}{h_{ob}}$	$h_{ie} - \frac{h_{re}}{h_{oe}}(1 + h_{fe})$	$h_{ic} + \frac{h_{fc}}{h_{oc}}(1 - h_{rc})$
r_e	$h_{ib} - \frac{h_{rb}}{h_{ob}}(1 + h_{fb})$	$\frac{h_{re}}{h_{oe}}$	$\frac{1 - h_{rc}}{h_{oc}}$
r_c	$\frac{1 - h_{rb}}{h_{ob}}$	$\frac{1 + h_{fe}}{h_{oe}}$	$-\frac{h_{fc}}{h_{oe}}$
r_m	$-\frac{h_{fb}}{h_{ob}}(1 - h_{rb})$	$\frac{h_{re} + h_{fe}}{h_{oe}}$	$-\frac{(1 + h_{fe})}{h_{oc}}$
h_i	$h_{ib} \cong r_e + (1 - \alpha)r_b$	$h_{ie} \cong r_b + \frac{r_e}{1 - \alpha}$	$h_{ic} \cong r_b + \frac{r_e}{1 - \alpha}$
h_o	$h_{ob} \cong \frac{1}{r_c}$	$h_{oe} \cong \frac{1}{(1 - \alpha)r_c}$	$h_{oc} \cong \frac{1}{(1 - \alpha)r_c}$
h_r	$h_{rb} \cong \frac{r_b}{r_c}$	$h_{re} \cong \frac{r_e}{(1 - \alpha)r_c}$	$h_{rc} \cong 1$
h_f	$h_{fb} \equiv -\alpha$	$h_{fe} \equiv \beta$	$h_{fc} \cong -(\beta + 1)$
α	$-h_{fb} = -\frac{r_m}{r_c}$	$\frac{h_{fe}}{1 + h_{fe}} = \frac{r_m}{r_c}$	$\frac{1 + h_{fc}}{h_{fc}} = \frac{r_m}{r_c}$

Table 5-5 Commonly used transistor nomenclature

V_{CB} = collector-to-base quiescent voltage	i_b = total base current
V_{EB} = emitter-to-base quiescent voltage	i_e = total emitter current
I_C = collector quiescent current	I_c = rms of ac component of collector current
I_B = base quiescent current	I_b = rms of ac component of base current
I_E = emitter quiescent current	I_e = rms of ac component of emitter current
I_{CO} = collector cutoff current ($i_e = 0$)	v_{in} = ac input voltage (signal)
r_e = emitter resistance	v_{out} = ac output voltage
r_b = base resistance	V_{CC} = dc collector voltage
r_c = collector resistance	V_{BB} = dc base voltage
r_m = equivalent emitter-to-collector resistance	V_{EE} = dc emitter voltage
i_c = total collector current	

There are voltage source ac equivalent circuits for the three-transistor configurations as shown in Fig. 5-15*a*, *b*, and *c*, but these are not used as often as the current source ac equivalent circuits since the bipolar transistor is basically a *current* device.

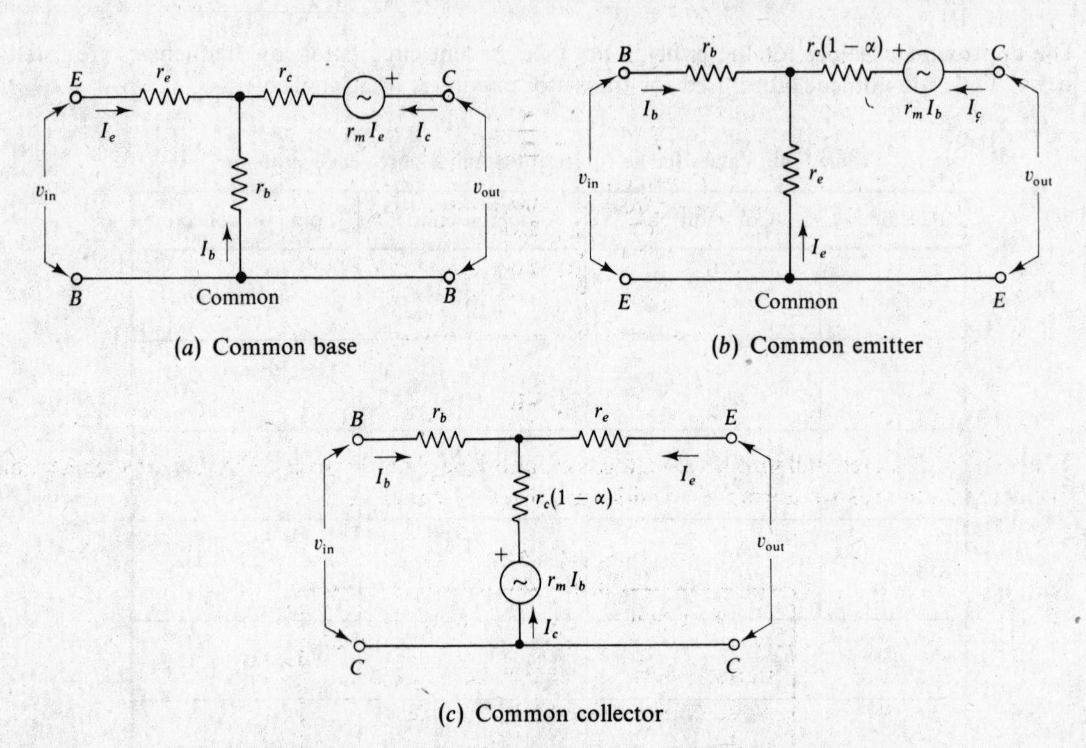

Fig. 5-15 Transistor: voltage source ac equivalent circuits

Example 5.7 The hybrid characteristics of a transistor are measured and found to be

$$h_{ie} = 1500\ \Omega \qquad h_{re} = 3.50 \times 10^{-4} \qquad h_{fe} = 45.0 \qquad h_{oe} = 30.0\ \mu\text{S}$$

Draw the T equivalent circuit shown in Fig. 5-13*c* and find α and β.

From Table 5-4, we may calculate r_b, r_e, r_c, and α as follows, noting that the hybrid parameters are for the common-emitter configuration as noted by the *e* used for the second subscript:

$$r_b = h_{ie} - \frac{h_{re}}{h_{oe}}(1 + h_{fe})$$

Substituting,

$$r_b = 1500 - \frac{3.5 \times 10^{-4}}{30.0 \times 10^{-6}}(1 + 45.0) = 2037\ \Omega = 2.04\ \text{k}\Omega$$

$$r_e = \frac{h_{re}}{h_{oe}} = \frac{3.50 \times 10^{-4}}{30.0 \times 10^{-6}} = 11.7\ \Omega$$

$$r_c = \frac{1 + h_{fe}}{h_{oe}} = \frac{1 + 45.0}{30 \times 10^{-6}} = 1.53 \times 10^{6}\ \Omega = 1.53\ \text{M}\Omega$$

$$\alpha = \frac{h_{fe}}{1 + h_{fe}} = \frac{45.0}{1 + 45.0} = 0.978\,26$$

Note: The final equation for α is identical to (*5-10a*).

To use the equivalent circuit shown in Fig. 5-13*c*, we also need to calculate

$$r_c(1 - \alpha) = (1.53 \times 10^{6})(1 - 0.978\,26) = 33.3 \times 10^{3}\ \Omega = 33.3\ \text{k}\Omega$$

Since β is identical to h_{fe},

$$\beta = 45.0$$

Or calculate β from (*5-10*):

$$\beta = \frac{\alpha}{1 - \alpha} = \frac{0.978\,26}{1 - 0.978\,26} = 45.0$$

The required equivalent circuit is shown in Fig. 5-16.

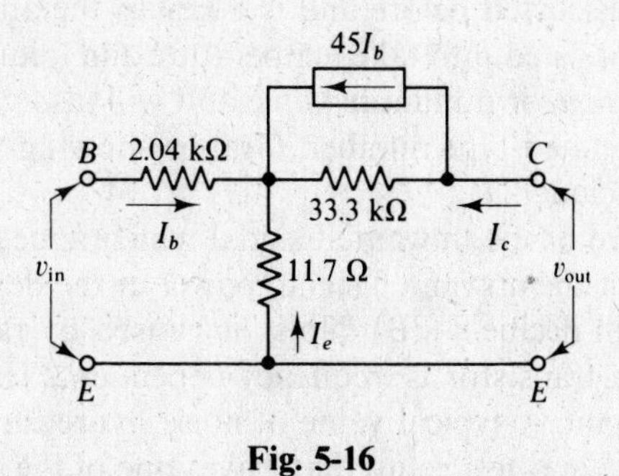

Fig. 5-16

Transistor parameters fall into seven categories, although there are over 150 standard specifications. The seven standards we normally consider are listed in Table 5-6.

Table 5-6

Function	Symbols
Current gain	h_{FE}, h_{fe}, β
Noise figure	NF
Leakage current	I_{CBO}(min), I_{EBO}, I_{CEO}
Breakdown voltage	V_{CBO}(min), V_{EBO}(min)
Conductivity	V_{CE}(sat), V_{BE}
Power dissipation	P_T, I_C(max)
Operating speed	t_{on}, t_{off}, f_T C_{EB}; C_{EC}; C_{BC}

Bipolar transistor specifications include

- *Mechanical specifications* as shown in Fig. 5-17. These include base configuration, TO type, and outline drawing.
- *Electrical specifications* as listed in Table 5-6.
- *Graphs*, including transfer characteristics, collector characteristics, and variations of parameters with temperature, frequency, and amount of load current.

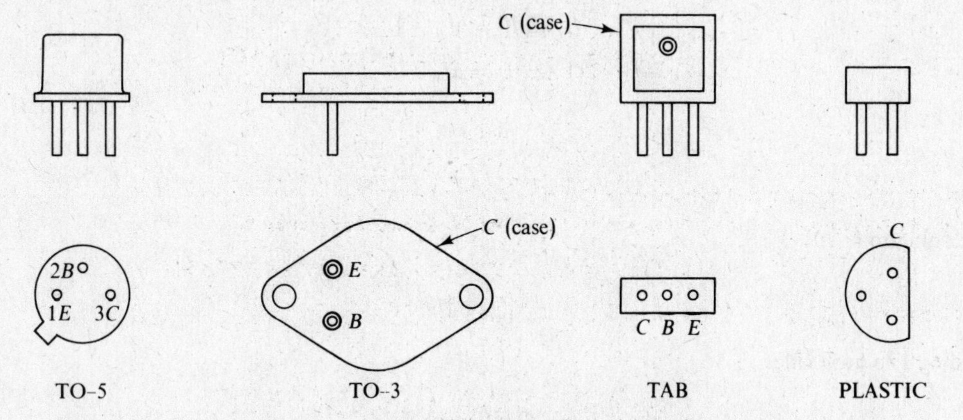

Fig. 5-17 Transistor outlines and base diagram

Let us examine the seven basic electrical characteristics in more detail.

The *current gain* has been discussed before and is a key to the amplifying ability of the transistor. Care must be taken when using a listed β to take temperature and frequency variations into effect. The β usually specified is either an average or minimum value and will also vary with the individual transistor even though it is the same designated type number. Graphs showing the β variation with temperature and frequency are sometimes included.

The *noise figure* is a measure of the unwanted signal which appears in the output of an electronic device and is usually the ratio of the unwanted signal power at the device input to the unwanted signal power at the output measured in decibels (dB). Noise is caused by random fluctuations in the device currents and voltages and for a transistor is frequency-dependent. The lower the noise figure, the less noise the device adds to the circuit. A typical value of noise figure for a junction transistor is 6 dB.

Leakage current is measured in a test setup that leaves one of the elements open and the remaining junction reverse-biased. Leakage current depends on construction, operating voltage, junction temperature, and purity concentration. These same factors also influence the α and β of a transistor.

The presence of leakage currents means we must modify the α and β equations as follows: From (*5-8b*),

$$\alpha = \frac{I_C}{I_E}$$

Cross multiplying,

$$I_C = \alpha I_E$$

For the common-base configuration, we must add the collector leakage current I_{CBO} from the base,

$$I_C = \alpha I_E + I_{CBO} \tag{5-11}$$

For the common-emitter configuration,

$$I_C = \beta I_B + (\beta + 1) I_{CBO} \tag{5-12}$$

Example 5.8 Derive Eq. (*5-12*) from Eq. (*5-11*).

From (*5-11*):

$$I_C = \alpha I_E + I_{CBO}$$

Since by (*5-7a*)

$$I_E = I_B + I_C$$

substitution yields

$$I_C = \alpha(I_B + I_C) + I_{CBO} = \alpha I_B + \alpha I_C + I_{CBO}$$

Collecting terms,

$$I_C - \alpha I_C = \alpha I_B + I_{CBO}$$

$$(1 - \alpha) I_C = \alpha I_B + I_{CBO}$$

Dividing by $1 - \alpha$,

$$I_C = \frac{\alpha}{1-\alpha} I_B + \frac{1}{1-\alpha} I_{CBO}$$

But according to (*5-10*)

$$\frac{\alpha}{1-\alpha} = \beta$$

And adding 1 to both sides,

$$1 + \frac{\alpha}{1-\alpha} = \beta + 1$$

$$\frac{1 - \not{\alpha} + \not{\alpha}}{1 - \alpha} = \beta + 1$$

So
$$\frac{1}{1 - \alpha} = \beta + 1$$

Substituting,

$$I_C = \beta I_B + (\beta + 1)I_{CBO}$$

which is Eq. (*5-12*).

Equation (*5-12*) can be thought of as a modification of Eq. (*5-9b*):

$$\beta = \frac{I_C}{I_B}$$

Cross multiplying,

$$I_C = \beta I_B$$

and for the common-emitter configuration we must add the amount of collector current I_{CEO} that leaks from collector to emitter:

$$I_C = \beta I_B + I_{CEO} \tag{5-13}$$

Comparing this result to Eq. (*5-12*), we see that

$$I_{CEO} = (\beta + 1)I_{CBO} \tag{5-14}$$

A summary of leakage current abbreviations is given in Table 5-7.

Table 5-7

I_{CBO} or I_{CO}	DC collector current when collector junction is reverse-biased and emitter is open-circuited*
I_{CEO}	DC collector current with collector junction reverse-biased and base open-circuited*
I_{CER}	DC collector current with collector junction reverse-biased and a resistor of value R between base and emitter*
I_{CES}	DC collector current with collector junction reverse-biased and base shorted to emitter*
I_{CEV}	DC collector current with collector junction reverse-biased and with a specified base-emitter voltage*
I_{CEX}	DC collector current with collector junction reverse-biased and with a specified base-emitter circuit connection*

* Test conditions must be specified.

Example 5.9 A certain transistor has the following ratings: $h_{FE} = 30$ and $I_{CBO} = 14\ \mu A$. When the base current is 225 μA, find the collector leakage current I_{CEO} and the true collector current when the transistor is connected into the common-emitter configuration.

Given:

$$h_{FE} = \beta_{DC} = 30$$

$$I_{CBO} = 14 \times 10^{-6}\ \text{A}$$

Collector leakage current [from (*5-14*)]:

$$I_{CEO} = (1 + \beta_{DC})I_{CBO} = (1 + 30)(14 \times 10^{-6}) = 434 \times 10^{-6}\ \text{A}$$

True collector current [from (5-13)]:

$$I_C = \beta_{DC} I_B + I_{CEO}$$
$$= (30)(225 \times 10^{-6}) + (434 \times 10^{-6}) = (6750 \times 10^{-6}) + (434 \times 10^{-6}) = 7184 \times 10^{-6} \text{ A} = 7184 \ \mu\text{A}$$

By comparison, the ideal collector current $I_{C(\text{ideal})}$ can be determined from (5-9b):

$$I_{C(\text{ideal})} = (\beta_{DC})(I_B) = (30)(225 \times 10^{-6}) = 6750 \times 10^{-6} \text{ A} = 6750 \ \mu\text{A}$$

This illustrates that the leakage current can produce a significant increase in collector current in the common-emitter circuit. The amount of increase is directly related to the value of β_{DC} since it equals $(1 + \beta_{DC})I_{CBO}$.

The PN junctions in a bipolar transistor are susceptible to voltage breakdown when the reverse voltage exceeds the manufacturer's ratings. If this occurs in a germanium-type transistor, the junction may be destroyed, but silicon types can safely conduct zener currents provided the circuit resistance is sufficiently high to prevent a destructive current flow. In fact, the base-collector junctions of some silicon transistors make an acceptable zener diode for emergency replacement.

Breakdown voltage terminology is listed in Table 5-8 and is standard throughout the semiconductor industry.

Table 5-8

BV_{CBO}	DC breakdown voltage collector-to-base junction reverse-biased, emitter open-circuited.* Specify I_C.
BV_{CEO}	DC breakdown voltage, collector to emitter, with base open-circuited.* Specify I_C.
BV_{CER}	DC breakdown voltage, similar to BV_{CEO} except a resistor value R between base and emitter.*
BV_{CES}	DC breakdown voltage, similar to BV_{CEO} but base shorted to emitter.*
BV_{CEV}	DC breakdown voltage, similar to BV_{CEO} but emitter-to-base junction reverse-biased.*
BV_{CEX}	DC breakdown voltage, similar to BV_{CEO} but emitter-to-base junction reverse-biased through a specified circuit.*
BV_{EBO}	DC breakdown voltage, emitter-to-base junction reverse-biased, collector open-circuited.* Specify I_E.
BV_R	DC breakdown voltage, reverse-biased diode.*

* Test conditions must be specified.

Conductivity is indicated by V_{BE}, the base saturation voltage of the diode formed in the control circuit. V_{CES} or V_{CE}(sat) is the collector saturation voltage or the voltage necessary for the full current in the working (collector-to-emitter) circuit.

The *power dissipation rating* is very important since transistors are temperature-sensitive, and the collector current I_C is an indication of how much current the transistor may safely pass.

The *allowable power dissipation* P_T is the maximum safe dissipation within the transistor and is dependent on the ambient temperature. In general, the higher the ambient (surrounding) temperature, the lower the allowable dissipation.

Example 5.10 Using a typical power-temperature derating curve (Fig. 5-18), find the maximum allowable power dissipation P_T at 25, 30, and 75°C.

From Fig. 5-18, the values are

25°C	$P_T \cong 175$ mW
30°C	$P_T \cong 160$ mW
75°C	$P_T \cong 30$ mW

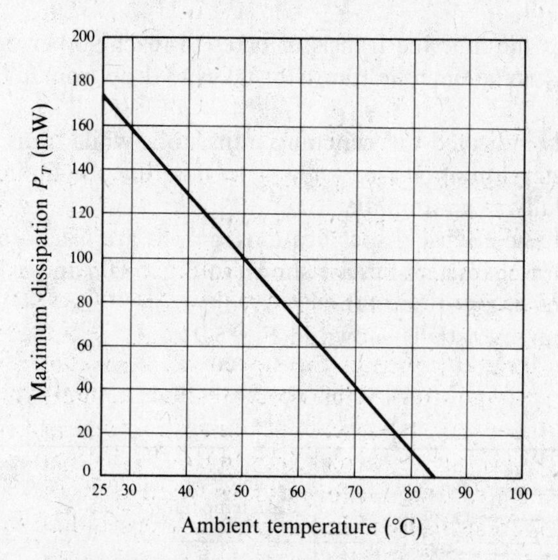

Fig. 5-18 Power-temperature derating curve

Both P_T and I_C are usually specified at an ambient temperature of 25°C (77°F) and must be derated for higher temperatures. Temperature derating curves are often included in transistor specification.

A useful procedure is to find the derated value of power at the temperature at which the transistor is to be operated and add the *maximum power dissipation curve* to the transistor characteristic curves.

This set of curves (for various temperatures) is derived from the power dissipation formula

$$P_T = V_{CE} I_C \tag{5-15}$$

and is usually added to the plate characteristic curves with dashed lines.

The construction of a transistor affects its leakage and gain as well as its operating speed. The interelement capacitances limit high frequency response, and this is usually indicated by describing the rise and decay times for a square wave of full operating voltage as shown in Fig. 5-19.

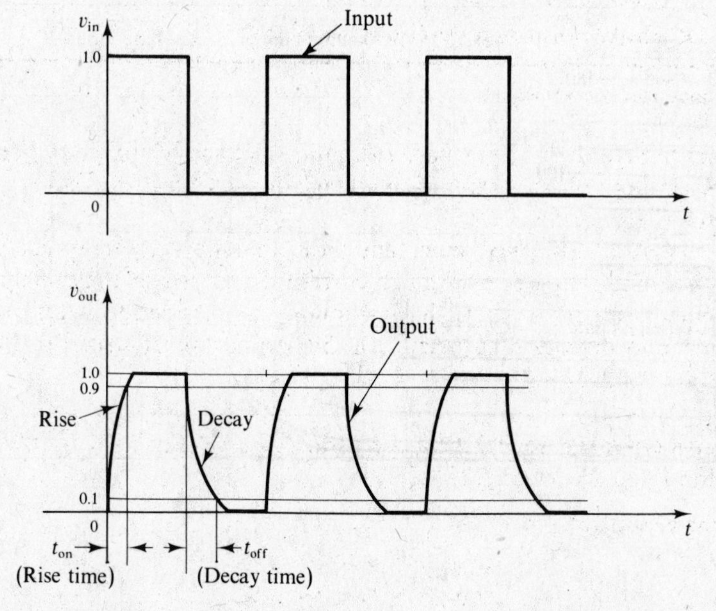

Fig. 5-19

The rise time t_r or t_{on} is the time the transistor takes to reach 90 percent of the full voltage.

The decay time t_d or t_{off} is the time the transistor takes to drop from full voltage to 10 percent of full voltage.

These times are usually indicated for switching transistors, while transistors designed primarily as amplifiers use the parameter *transition frequency* f_T to describe the limiting frequency at which the transistor can operate usefully as an amplifier.

Along with the seven basic electrical specifications, graphs are usually included.

Typical average collector characteristics are shown in Fig. 5-20*a* and *b*. They are very similar to the average plate characteristics of a pentode tube.

An average transfer characteristic is shown in Fig. 5-21.

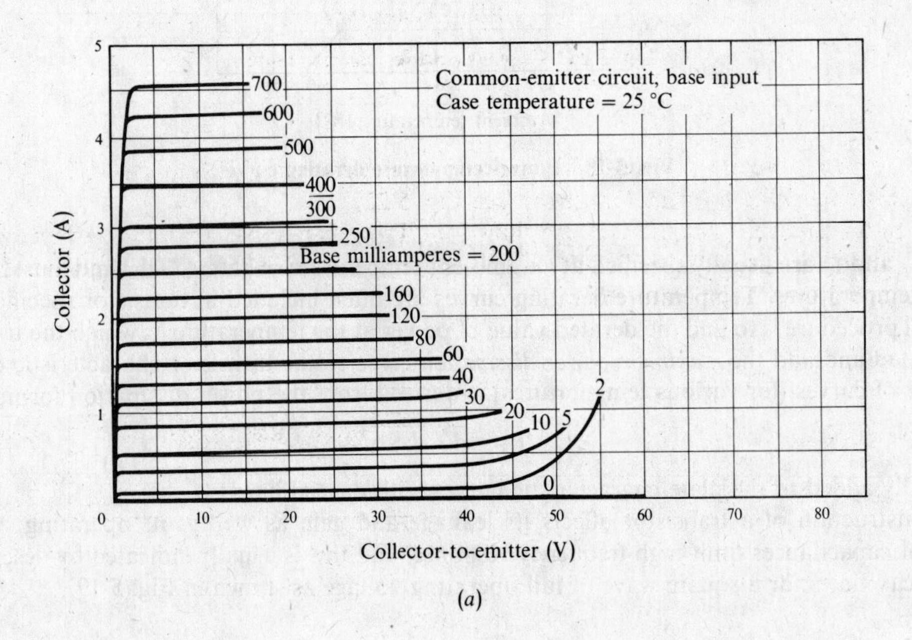

(*a*)

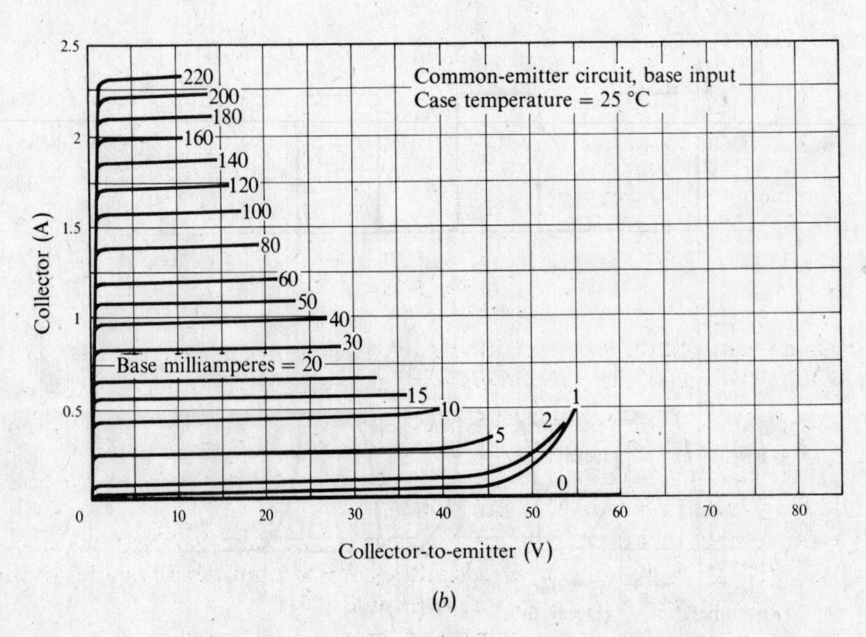

(*b*)

Fig. 5-20 Typical collector characteristics

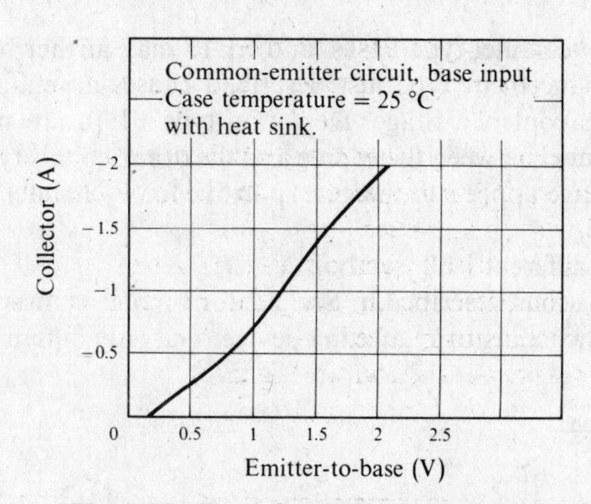

Fig. 5-21 Typical transfer characteristic

5.4 FETs

There are two basic types of *field-effect transistors*: the discrete FET and the MOSFET or *M*etal-*O*xide-*S*emiconductor *F*ield-*E*ffect *T*ransistor (see Fig. 5-22).

A MOSFET takes up much less room than a PNP bipolar transistor since only one diffusion is needed instead of two (see Fig. 5-23) and this has caused the MOSFET to be of great value in high-density integrated circuit (IC) construction.

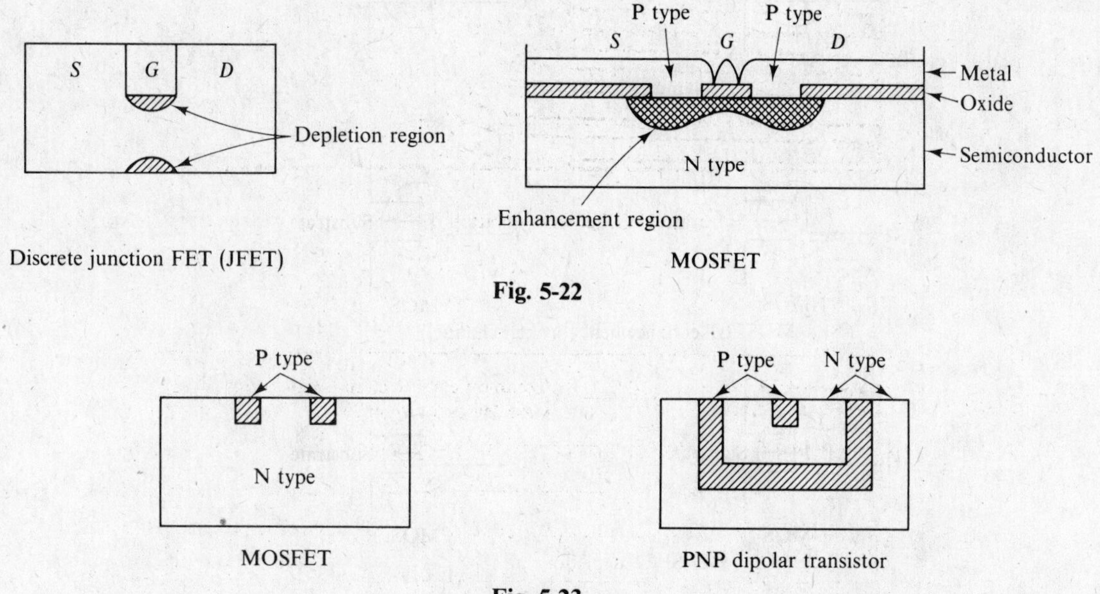

Fig. 5-22

Fig. 5-23

The two P regions, called the *source* and the *drain*, perform the same functions as the emitter and collector of a PNP transistor, while the gate performs the same function as the base.

With no voltage on the gate, no working current will exist. With a negative voltage applied to the gate, a spurt of current produces a field which spans the source and drain areas and permits holes to form a P-type channel through which the working current can pass. The more negative the gate, the thicker the channel and the more current the FET can handle.

It is important to note that a basic difference between FETs and bipolar transistors is that FETs control working current by responding to *voltage* at the control terminal and bipolar transistors control working current by responding to *current* at the control terminal.

The fact that FETs have working current passing through only one type of semiconductor material gives them the designation *unipolar*.

There are both P- and N-channel-type FETs, and FETs may further be classified as enhancement and depletion type. The enhancement type just described causes an enhancement or increase in the working current by the gate control voltage. Depletion-type FETs are manufactured by growing a continuous N- or O-type channel between the source and the drain, so this type of FET is always on. It is turned off by the proper voltage applied to the gate: positive for P-channel depletion type and negative for N-channel depletion type.

Figure 5-24 shows eight different FET symbols.

The seven major specifications described in Sec. 5.2 for bipolar transistors are also valid for FETs, and in fact are valid for *any* switching or regulating device. The only difference may be in the nomenclature used.

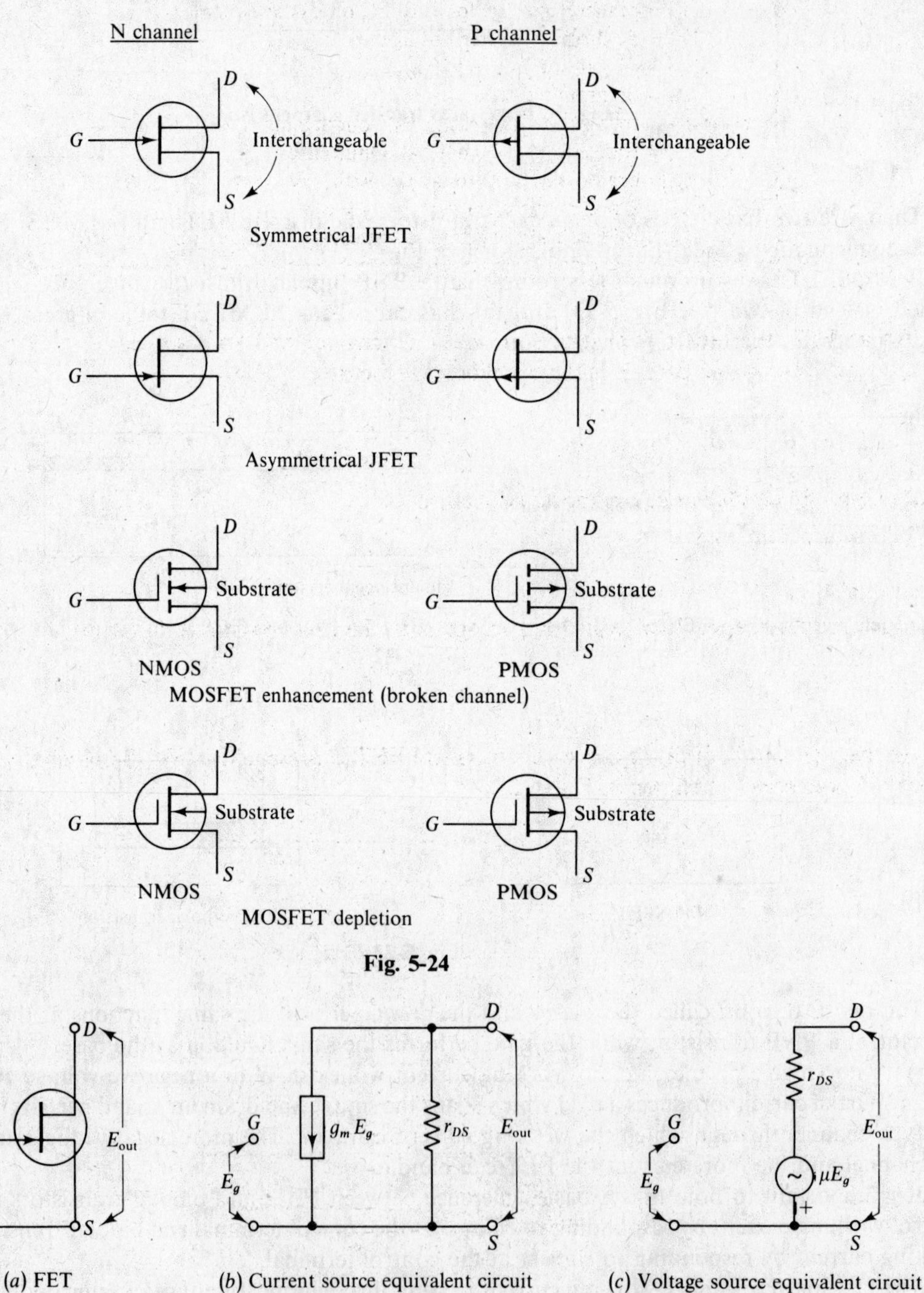

Fig. 5-24

(*a*) FET (*b*) Current source equivalent circuit (*c*) Voltage source equivalent circuit

Fig. 5-25

The ac equivalent circuits for an FET are shown in Fig. 5-25*a*, *b*, and *c*, and are the same as the tube ac equivalent circuits except for the notation.

Table 5-9 lists commonly encountered FET nomenclature.

Table 5-9 Commonly used FET nomenclature

I_{DSS}	= drain current for $V_{GS} = 0$
I_{GSS}	= gate reverse leakage current (reverse bias must be specified)
$I_D(\text{off})$	= drain cutoff current
BV_{GSS}	= gate-to-source breakdown voltage
V_P	= pinchoff voltage for I_{DSS} 1% or less of $I_{DSS}(\text{max})$
V_{DS}	= dc drain-to-source voltage
V_{GS}	= dc gate-to-source voltage
I_D	= dc drain current
g_m or g_{fs}	= transconductance, mutual conductance
y_{fs}	= transadmittance, mutual admittance
r_{DS}	= drain-source ON resistance with $V_{GS} = 0$
C_{iss}	= input capacitance with $V_{DS} = 0$
C_{rss}	= reverse transfer (drain-to-gate) capacitance with $V_{DS} = 0$

For an FET, the subscript *D* stands for *drain*, *G* for *gate*, and *S* for *source*.

As in the case of the tube, μ is the amplification factor,

$$\mu = \frac{\Delta V_{DS}}{\Delta V_{GS}} \tag{5-16}$$

g_m is the mutual conductance or transconductance,

$$g_m = \frac{\Delta I_D}{\Delta V_{GS}} \tag{5-17}$$

and r_{DS} is the drain resistance, which is analogous to the plate resistance of a tube,

$$r_{DS} = \frac{\Delta V_{DS}}{\Delta I_D} \tag{5-18}$$

An average family of drain curves for a typical FET is presented in Fig. 5-26, while a typical FET transfer characteristic is shown in Fig. 5-27*a*.

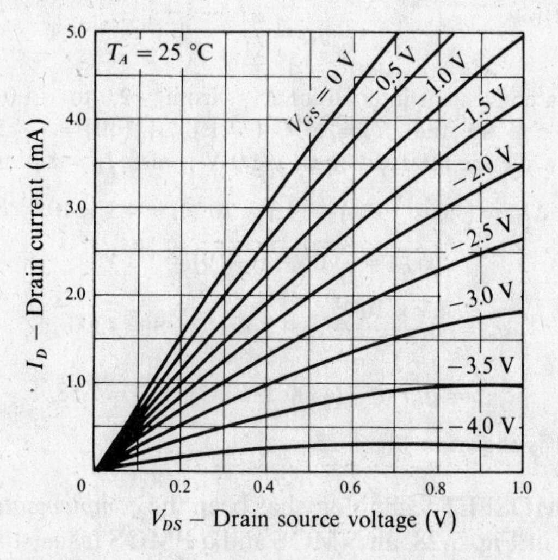

Fig. 5-26 Common drain-source characteristics

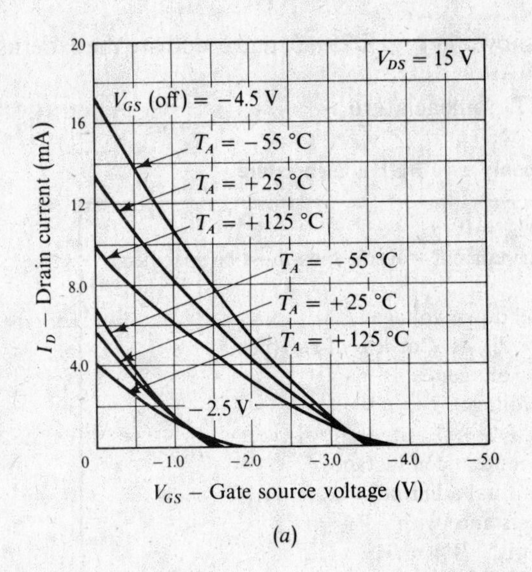

(a)

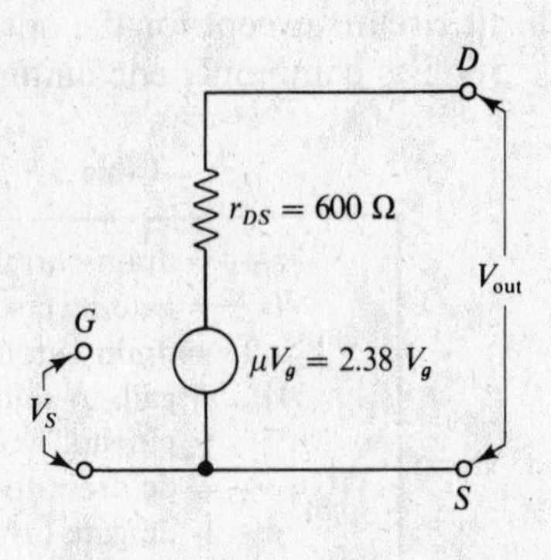

(b)

Fig. 5-27

Example 5.11 An FET has a common drain source characteristic curve, as shown in Fig. 5-26, and a transfer characteristic, as shown in Fig. 5-27*a*. Using Fig. 5-26 with $V_{GS} = -3.0$ V, find r_{DS}. Using Fig. 5-27*a* with $T_A = 25°C$ (77°F) and V_{GS} (off) = −4.5 V, find g_m. Draw the FET equivalent circuit with the calculated values. Use $\mu = g_m r_{DS}$ to calculate the value of μ.

Since by (5-18)

$$r_{DS} = \frac{\Delta V_{DS}}{\Delta I_D}$$

and the problem statement says to use $V_{GS} = -3.0$ V, we select a convenient interval for ΔI_D from 1.0 to 1.5 mA on Fig. 5-26.

The intersection of the $V_{GS} = -3.0$ V curve and $I_D = 1.0$ mA gives a V_{DS} of 0.4 V. The intersection of the $V_{GS} = -3.0$ V curve and $I_D = 1.5$ mA gives a V_{DS} of 0.7 V. So

$$\Delta V_{DS} = 0.7 - 0.4 = 0.3 \text{ V}$$

and

$$\Delta I_D = (1.5 \times 10^{-3}) - (1.0 \times 10^{-3}) = 0.5 \times 10^{-3} \text{ A}$$

Therefore

$$r_{DS} = \frac{0.3}{0.5 \times 10^{-3}} = 0.6 \times 10^3 = 600 \ \Omega$$

Since by (5-17)

$$g_m = \frac{\Delta I_D}{\Delta V_{GS}}$$

we go to Fig. 5-27*a* and select a convenient interval for ΔV_{GS} from −2.0 to −1.0 V.

The problem statement says to use the $T_A = 25°C$ (77°F) V_{GS}(off) = −4.5 V curve. The intersection at −2.0 V gives $I_D = 4.3$ mA, while the intersection at −1.0 V gives $I_D = 8.6$ mA. So

$$\Delta I_D = (8.6 \times 10^{-3}) - (4.3 \times 10^{-3}) = 4.3 \times 10^{-3} \text{ A}$$

and

$$\Delta V_{GS} = -1.0 - (-2.0) = 1.0 \text{ V}$$

Therefore

$$g_m = \frac{4.3 \times 10^{-3}}{1.0} = 4.3 \times 10^{-3} \text{ S} = 4300 \ \mu\text{S}$$

Finally, from (5-4)

$$\mu = g_m r_{DS} = (4300 \times 10^{-6})(600) = 2.58$$

The required equivalent circuit is shown in Fig. 5-27*b*.

A natural outgrowth of MOSFET technology has been the *complementary metal-oxide semiconductor* or CMOS device. As shown in Fig. 5-28, an NMOS and a PMOS transistor can be combined to form an inverter.

The waveforms for this configuration are shown in Fig. 5-29 and, as can be seen, the drain current I_D exists only as a series of pulses at the transition points of the input signal—i.e., when the signal changes level.

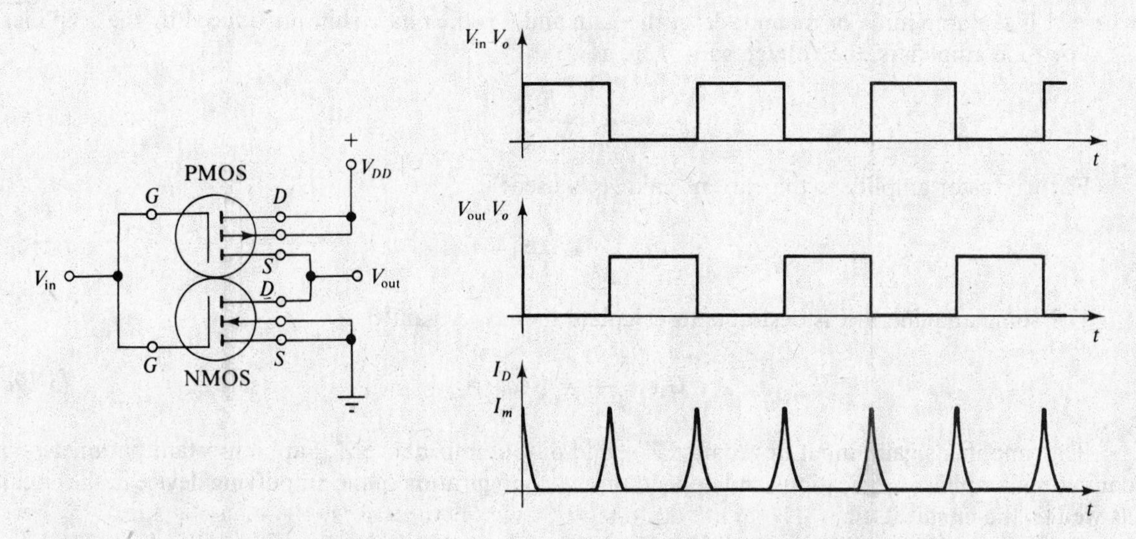

Fig. 5-28 Complementary MOSFET

Fig. 5-29 CMOS waveforms

As a result, the power consumption of CMOS devices is very low, typically on the order of 5 nW per circuit.

Another advantage of CMOS configurations is their high noise immunity.

Construction of a CMOS circuit is more complex than regular PMOS or NMOS devices since a P-type "tub" must be created to form the N-channel part of the device, as shown in Fig. 5-30. Note the channel steps that have been added to prevent undesirable interaction or *parasitic* channeling.

CMOS devices are also more sensitive to overvoltage and can even be destroyed by the static charge present in clothing. Care must be taken to ensure that CMOS devices are unpacked and placed into circuits properly.

Because the price of CMOS devices is comparable with other ICs, a great number of logic, multifunction, and linear circuits are now constructed using CMOS technology.

We will deal with CMOS and other types of logic circuits in Chap. 7.

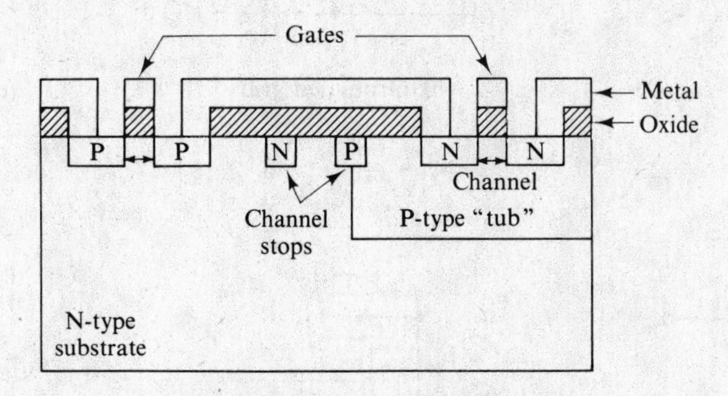

Fig. 5-30 CMOS construction

5.5 AMPLIFIER CONFIGURATIONS

We have seen that the tube, bipolar transistor, and FET have the ability to control a working current with a small control current or voltage. Used in this manner, the tube, bipolar transistor, or FET becomes an *amplifier*.

An ideal amplifier does nothing to a signal but increase its magnitude. The ratio of the output signal voltage, current, or power to the input signal voltage, current, or power is called the *gain*.

Since the input and output signals are time-varying or phasor quantities, the gain is written as $A\underline{/\theta}$, where A is the amplitude or magnitude of the gain and θ is the phase shift introduced by the amplifier.

For tube amplifiers, the voltage gain A_v is used:

$$A_v = \frac{E_{\text{out}}}{E_{\text{in}}} \tag{5-19a}$$

For transistor amplifiers, the current gain A_i is used:

$$A_i = \frac{I_{\text{out}}}{I_{\text{in}}} \tag{5-19b}$$

For some amplifiers, it is desirable to calculate the power gain G:

$$G = \frac{P_{\text{out}}}{P_{\text{in}}} = |A_v|\,|A_i| \tag{5-19c}$$

The amplifier's gain, input impedance Z_{in}, and output impedance Z_{out} are important parameters in comparing amplifier applications and depend on the configuration of the amplifying device in the circuit as well as the circuit itself.

For tube amplifiers, the six possible configurations in terms of signal are shown in Fig. 5-31.

Only the grounded or common cathode (Fig. 5-31*a*), grounded grid (Fig. 5-31*b*), and common anode or cathode follower (Fig. 5-31*c*) have had any major use.

The characteristics of these three circuits are listed in Table 5-10.

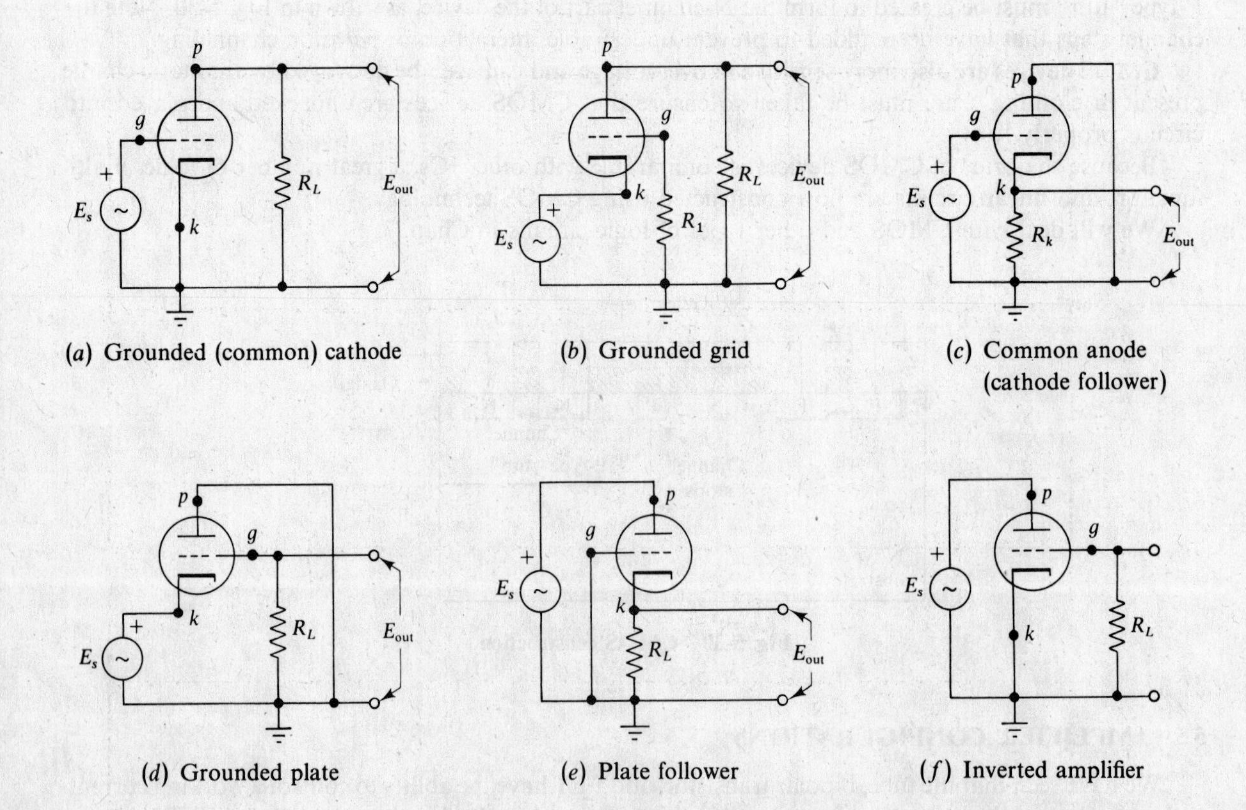

Fig. 5-31 Tube amplifier configurations

Table 5-10

Type		Z_{in}	Z_{out}	Gain	Output	Remarks
Tube circuits	Common cathode	High capacitance $\cong C_{gk} + C_{gp}(1 + g_m R_L)$	$R_L \parallel r_p$ $\cong r_p$	$A_v = \frac{-\mu R_L}{r_p + R_L} = \frac{-g_m}{1/r_p + 1/R_L}$ $A_i = -g_m R_L$	180° phase shift	Highest voltage gain; used for voltage amplification; Fig. 5-31*a*
	Grounded grid	$\cong \frac{r_p + R_L}{\mu'}$ Low capacitance $\cong C_{gk} \parallel z_{in}$	$\cong r_p + \mu' R_1$ $= r'_p$	$A_v = \frac{\mu' R_L}{r'_p + R_L}$ $A_i = \frac{r_p + R_L}{r'_p + R_L} \cong A_v$	No phase shift	$\mu' = \mu + 1$; $r'_p = r_p + \mu' R_1$; used for high-frequency amplifiers and low-*Z* to high-*Z* impedance transformation; Fig. 5-31*b*
	Cathode follower (grounded plate)	Medium capacitance $\cong C_{gp} + C_{gk}\left(\frac{1}{1 + g_m R_k}\right)$	$r''_p \parallel R_k$	$A_v = \frac{\mu'' R_k}{r''_p + R_k} \cong 1$	Same as grid signal input	$\mu'' = \frac{\mu}{\mu + 1}$; $r''_p = \frac{r_p}{\mu + 1} = \frac{r_p \mu''}{\mu}$; used for current amplification and high-*Z* to low-*Z* impedance transformation; Fig. 5-31*c*
Transistor circuits	Common emitter	$= r_b + \frac{r_e(r_c + R_L)}{R_L + (1 - \alpha) r_c}$ $\cong r_b + \frac{r_e}{1 - \alpha}$ 200 Ω to 2 kΩ	$= r_c(1 - \alpha)$ $+ \frac{\alpha r_e r_c}{r_b + r_e + R_s}$ 5 to 100 kΩ	$A_i = \frac{\alpha r_c - r_e}{r_e + r_c(1 - \alpha) + R_L} \cong \beta$ $A_v = \frac{-R_L}{R_{in} + R_s} A_i = \frac{-\alpha R_L}{r_e + (1 - \alpha)(r_b + R_s)}$	180° phase shift	Analogous to common cathode; easily cascaded without loss of gain; highest $G = \lvert A_i \rvert \lvert A_v \rvert \cong 40$ to 50 dB; Fig. 5-34*b*
	Common base	$= r_e + r_b(1 - \alpha)$ 30 to 1000 Ω	$= r_c - \frac{\alpha r_b r_c}{r_e + r_b + R_L}$ 100 kΩ to 1 MΩ	$A_i = \frac{-(r_b + \alpha r_c)}{r_c + r_b + R_L} \cong -\alpha$ $A_v = \frac{\alpha R_L}{r_e + R_s + r_b(1 - \alpha)}$	No phase shift	Analogous to grounded grid; highest Z_{out}, lowest Z_{in}; moderate $G \cong 15$ to 30 dB; Fig. 5-34*a*
	Common collector (emitter follower)	$= \frac{r_c R_L}{r_c(1 - \alpha) + R_L}$ $\cong \frac{R_L}{1 - \alpha}$ 100 kΩ to 1 MΩ	$= r_e + (1 - \alpha)(r_b + R_s)$ 10 to 500 Ω	$A_i = \frac{-r_c}{r_e + r_c(1 - \alpha) + R_L}$ $\cong -(\beta + 1)$ $A_v = \frac{R_L}{R_L + (1 - \alpha)(r_b + R_s)} \cong 1$	Same as base signal input	Analogous to cathode follower; used for high-*Z* to low-*Z* impedance transformation; lowest $G \cong 10$ to 20 dB; lowest Z_{out}, highest Z_{in}; Fig. 5-34*c*

The actual circuit and the ac equivalent circuits for a common cathode amplifier are shown in Fig. 5-32*a*, *b*, and *c*.

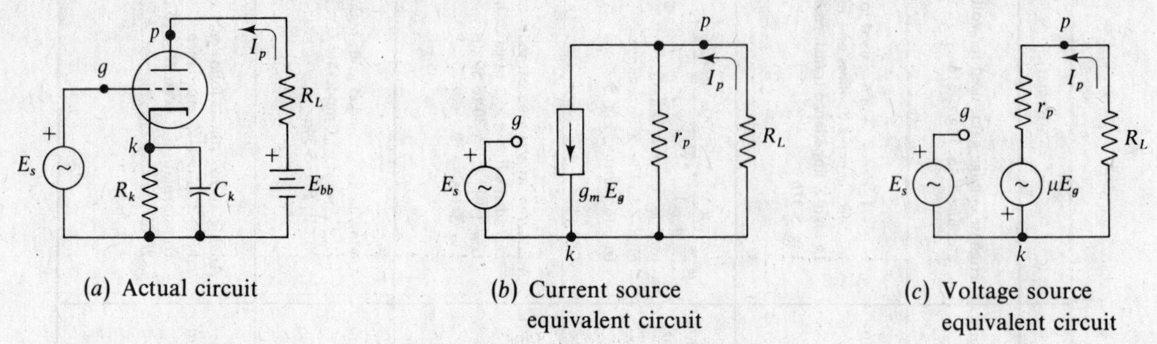

(*a*) Actual circuit (*b*) Current source equivalent circuit (*c*) Voltage source equivalent circuit

Fig. 5-32 Common cathode amplifier

Example 5.12 Find the gain and output impedance of the common cathode amplifier.

Since the common cathode amplifier is essentially a voltage device, we use the voltage source equivalent circuit shown in Fig. 5-32*c*.

The grid voltage E_g is connected directly to the signal voltage E_s so

$$E_g = E_s = E_{\text{in}}$$

The output voltage E_{out} for the ac equivalent circuit may be found by the voltage divider rule:

$$E_{\text{out}} = (\mu E_g)\frac{R_L}{r_p + R_L}$$

But from (*5-19a*)

$$A_v = \frac{E_{\text{out}}}{E_{\text{in}}} = \frac{E_{\text{out}}}{E_g}$$

Substituting for E_{out},

$$A_v E_g = \frac{\mu E_g R_L}{r_p + R_L}$$

$$A_v = \frac{\mu E_g R_L/(r_p + R_L)}{E_g} = \frac{\mu R_L}{r_p + R_L} \qquad (5\text{-}20)$$

r_p Z_{in} R_L μE_g

Fig. 5-33

The output impedance can be found by Thevenin's theorem. Removing the load resistor R_L, shorting the μE_g generator, and looking back into the output as shown in Fig. 5-33, we find

$$Z_{\text{out}} = r_p$$

For bipolar transistors, the three common amplifier connections and their ac equivalents are shown in Fig. 5-34. They consist of the three possible transistor connections with signal input V_s, its associated internal resistance R_s, and a load resistor R_L.

A comparison of the three configurations is shown in Table 5-10.

For the FET amplifier shown in Fig. 5-35, the voltage gain A_v is given by

$$A_v = \frac{-\mu R_L}{r_{DS} + R_L} \qquad (5\text{-}21)$$

Equation (*5-21*) is derived from the equivalent circuit shown in Fig. 5-36.

Notice that R_g, C_g, R_s, and C_s do not appear in the equivalent circuit. This is because for ac, C_g and C_s are essentially short circuits. So C_s shorts the ac signal to ground around R_s, and C_g acts to place V_s directly on the gate.

Keep in mind that the ac equivalent circuits presented in this chapter are low-frequency models. For high frequencies, more complex models that include the interelement capacitances must be used, and they will be discussed in Chap. 6.

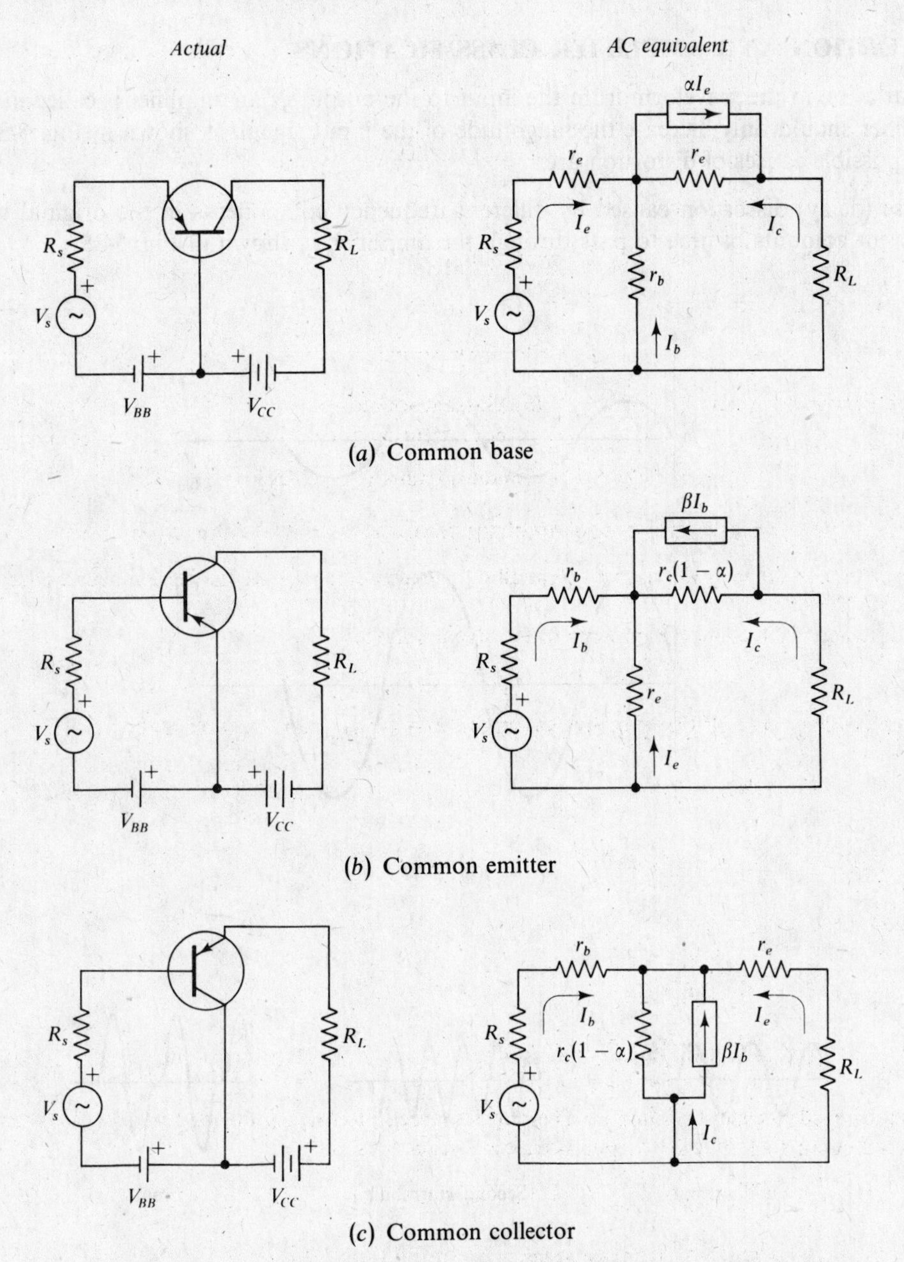

(*a*) Common base

(*b*) Common emitter

(*c*) Common collector

Fig. 5-34 Actual and equivalent circuits for transistor amplifiers

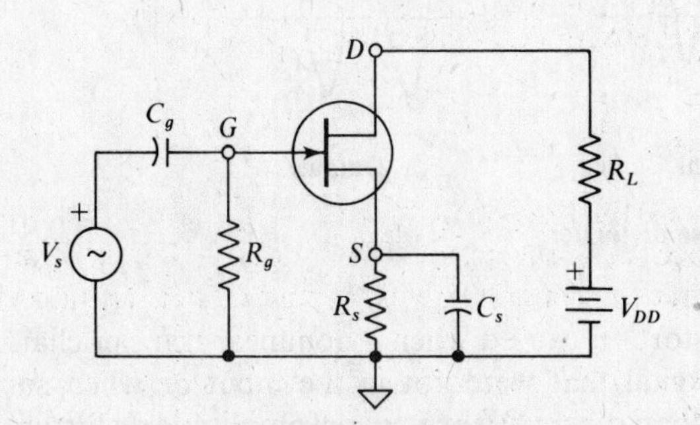

Fig. 5-35 FET amplifier

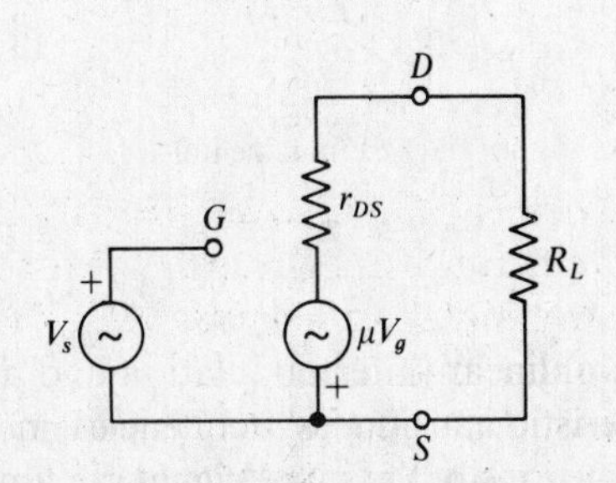

Fig. 5-36 FET amplifier equivalent circuit

5.6 DISTORTION AND AMPLIFIER CLASSIFICATIONS

Any variation in the waveform from the input to the output of an amplifier is called *distortion*. An ideal amplifier should only increase the magnitude of the input signal, as shown in Fig. 5-37.

Three possible causes of distortion are

- Phase (delay) distortion caused by different frequency components in the original wave taking different amounts of time to pass through the amplifier as shown in Fig. 5-38.

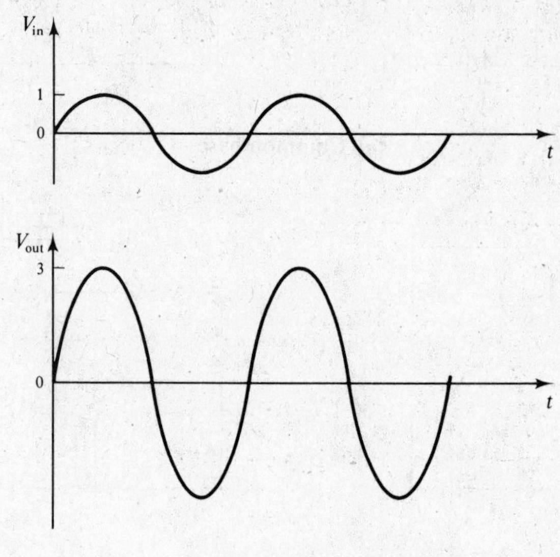

Fig. 5-37

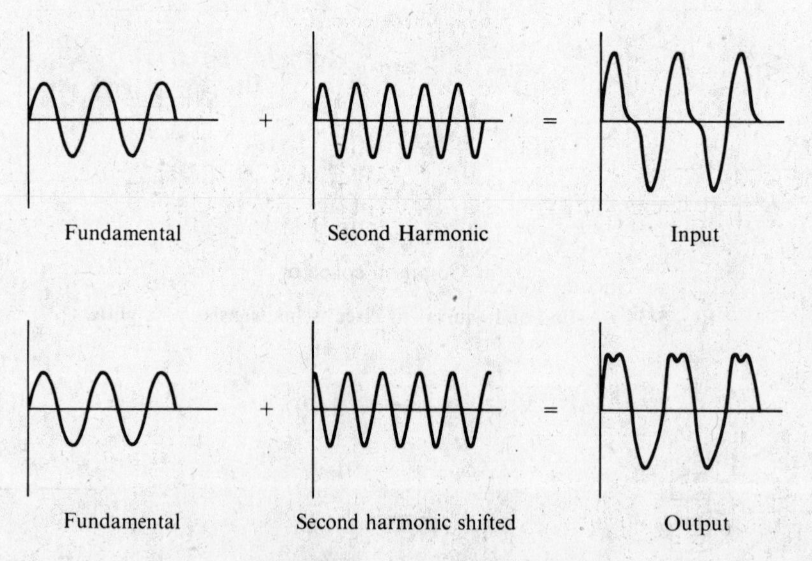

Fig. 5-38 Phase distortion

- Nonlinear (intermodulation and amplitude) distortion caused when a nonlinear transfer characteristic introduces frequencies in the output signal that were not in the input or when some portions of the input signal are amplified more than others. An example of amplitude distortion is shown in Fig. 5-39*a*.

It is easy to detect intermodulation distortion by using a *panoramic* or *spectrum* analyzer which displays the waveform on a frequency axis rather than on the time axis used for most oscilloscope displays. The spectrum analyzer displays shown in Fig. 5-39*b* illustrate intermodulation distortion.

- Frequency (harmonic) distortion caused when the different frequency components of the input wave are unequally amplified, as shown in Fig. 5-40.

Amplifier frequency distortion is usually shown by presenting the gain-frequency curve of the output when a constant-amplitude sine wave is applied to the input. The frequency axis is drawn on a logarithmic scale to include the wide range of frequencies usually covered. A typical wide-band amplifier gain-frequency curve is shown in Fig. 5-41 along with an ideal amplifier gain-frequency curve.

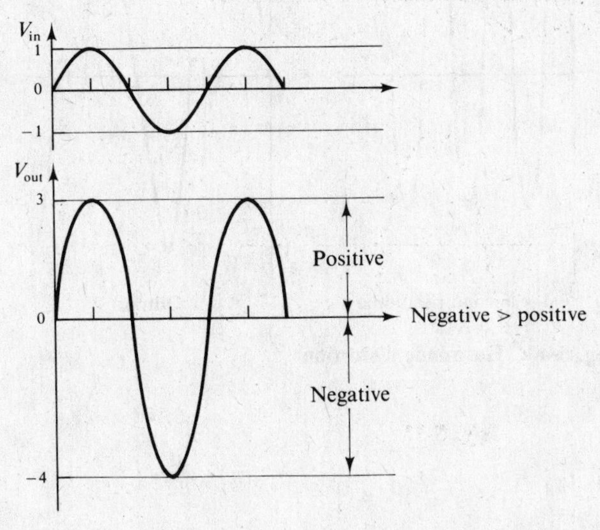

(*a*) Amplitude distortion

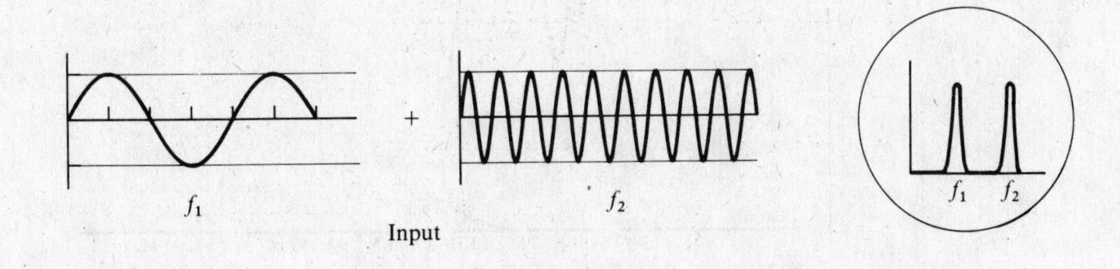

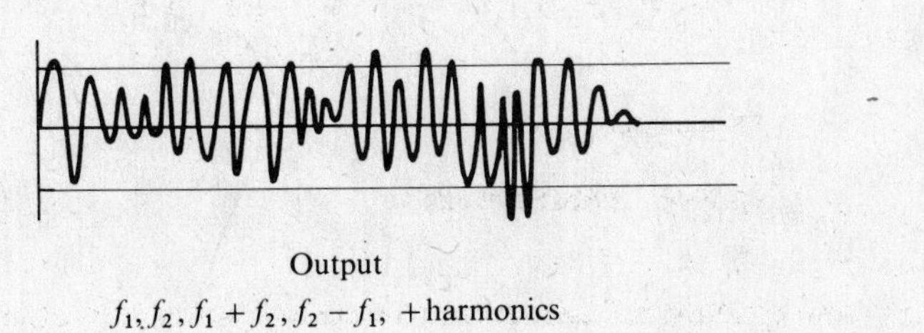

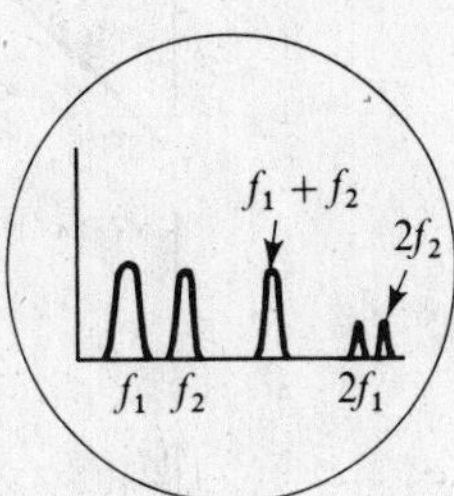

(*b*) Intermodulation (IM) distortion

Fig. 5-39

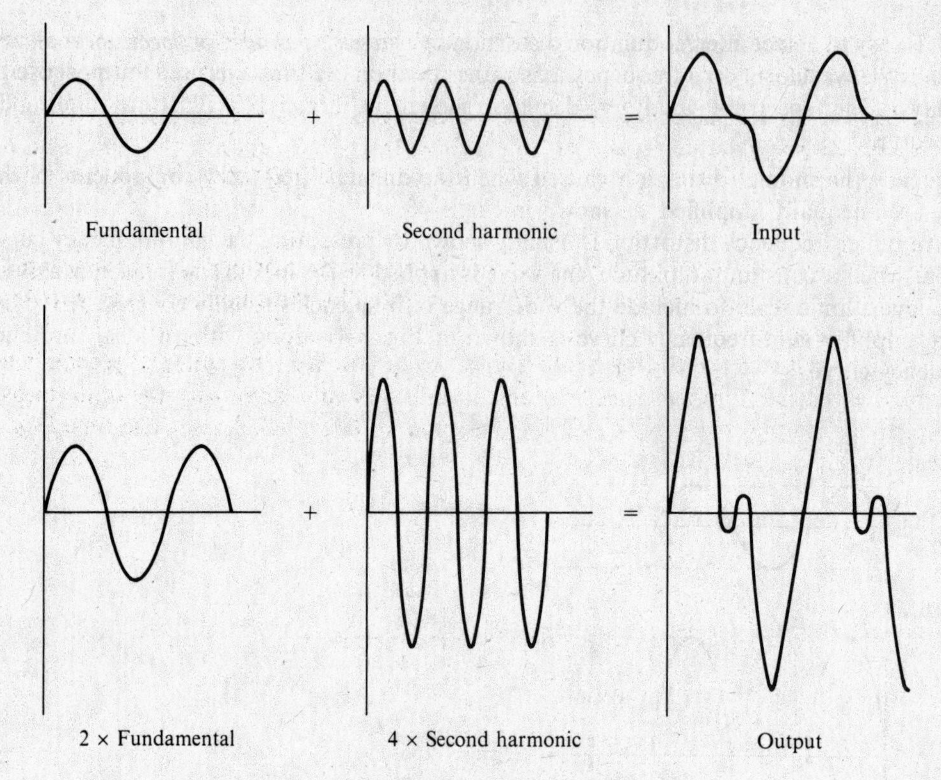

Fig. 5-40 Harmonic distortion

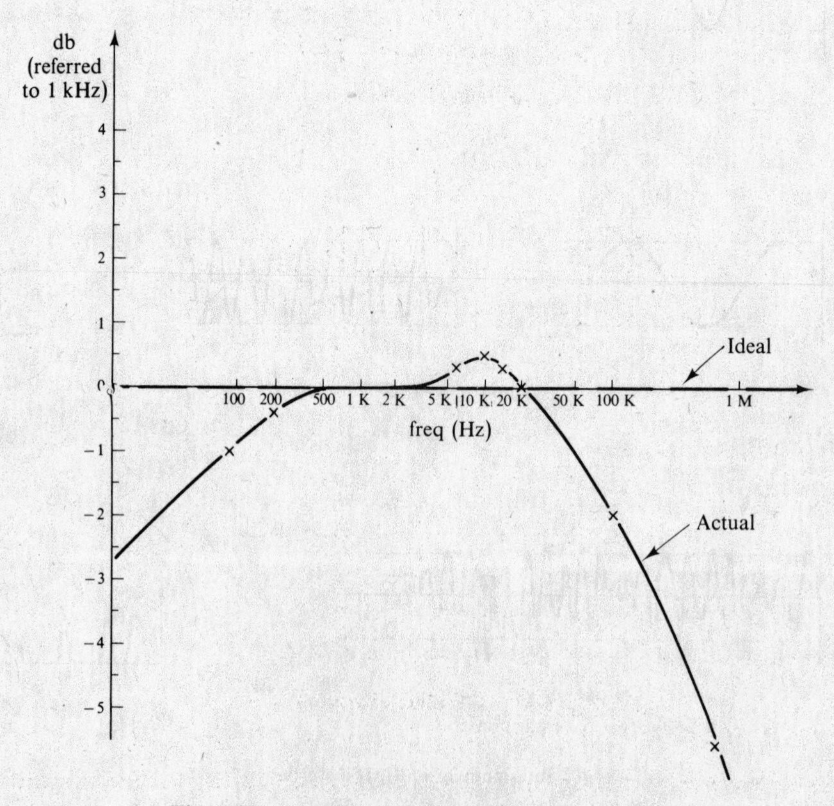

Fig. 5-41 Amplifier gain-frequency characteristic

Example 5.13 An input signal having frequencies of 100, 1000, and 10 000 Hz is used with an amplifier whose gain-frequency curve is shown in Fig. 5-41. Does harmonic distortion occur in the output?

Looking at Fig. 5-41, we see that at 100 Hz there is a drop of about 1 dB from the output reference level. At 10 kHz, there is a 0.5-dB gain, indicating that a uniform input containing the frequencies 100, 1000, and 10 000 Hz *will* exhibit harmonic distortion since the gain at 100 is less and the gain at 10 000 is slightly greater than the gain at 1000.

An input that includes signals occurring for a short period of time or that change very rapidly may give rise to transient distortion. This type of distortion is tested by the amplifier's ability to accurately reproduce a square-wave.

The *slew-induced distortion* or SID is another measure of how well the amplifier responds to quick changes in signal and some amplifier specifications include slew rate along with the other measures of distortion. Examples of poor transient response include *ringing* (Fig. 5-42*a*), poor rise time (Fig. 5-42*b*), and poor decay time (Fig. 5-42*c*).

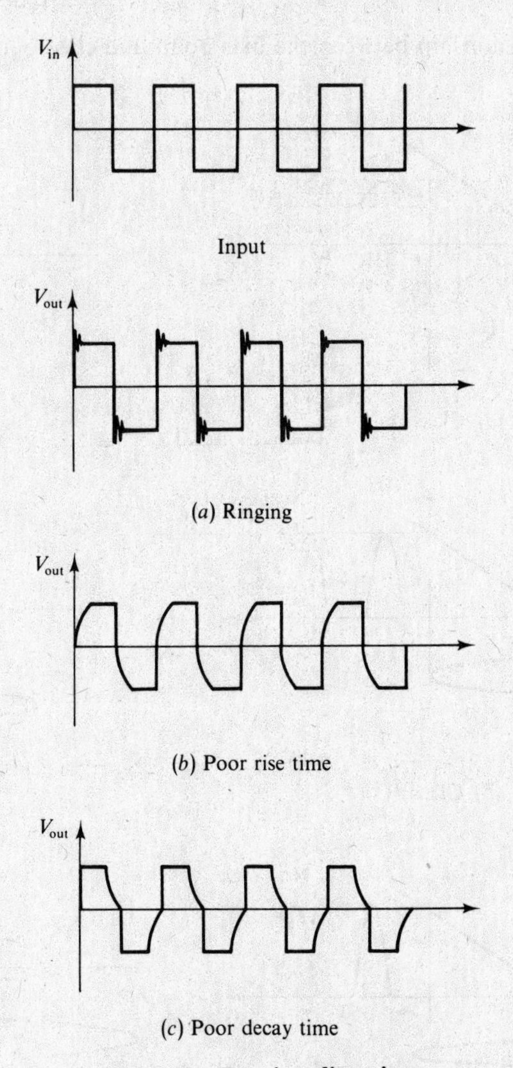

Fig. 5-42 Transient distortion

Amplifiers may be classified on the basis of operating conditions for the bias circuit. Some commonly seen classifications and their explanations are listed in Table 5-11.

Table 5-11 Amplifier classifications

Class	Bias point	Output time i_b or i_c	Distortion	Input	DC-AC power efficiency Theoretical	Practical	Diagram
A	Below cutoff	360°	Low	Small	50%	2–25% tube; 48–49% transistor	Fig. 5-43*a* Fig. 5-44*a*
AB AB_1 AB_2	Near cutoff	$360° < t > 180°$ No grid current Grid current	Higher than class A, lower than class B	Between small and medium	60–70%	60%	
B	Cutoff	180°	High	Medium	78%	75%	Fig. 5-43*b* Fig. 5-44*b*
C	Two times cutoff	$t < 180°$	Very high	Large	100%	98%	Fig. 5-43*c* Fig. 5-44*c*

Figure 5-43 shows the relationship between the bias point and class of operation for a typical tube transfer characteristic.

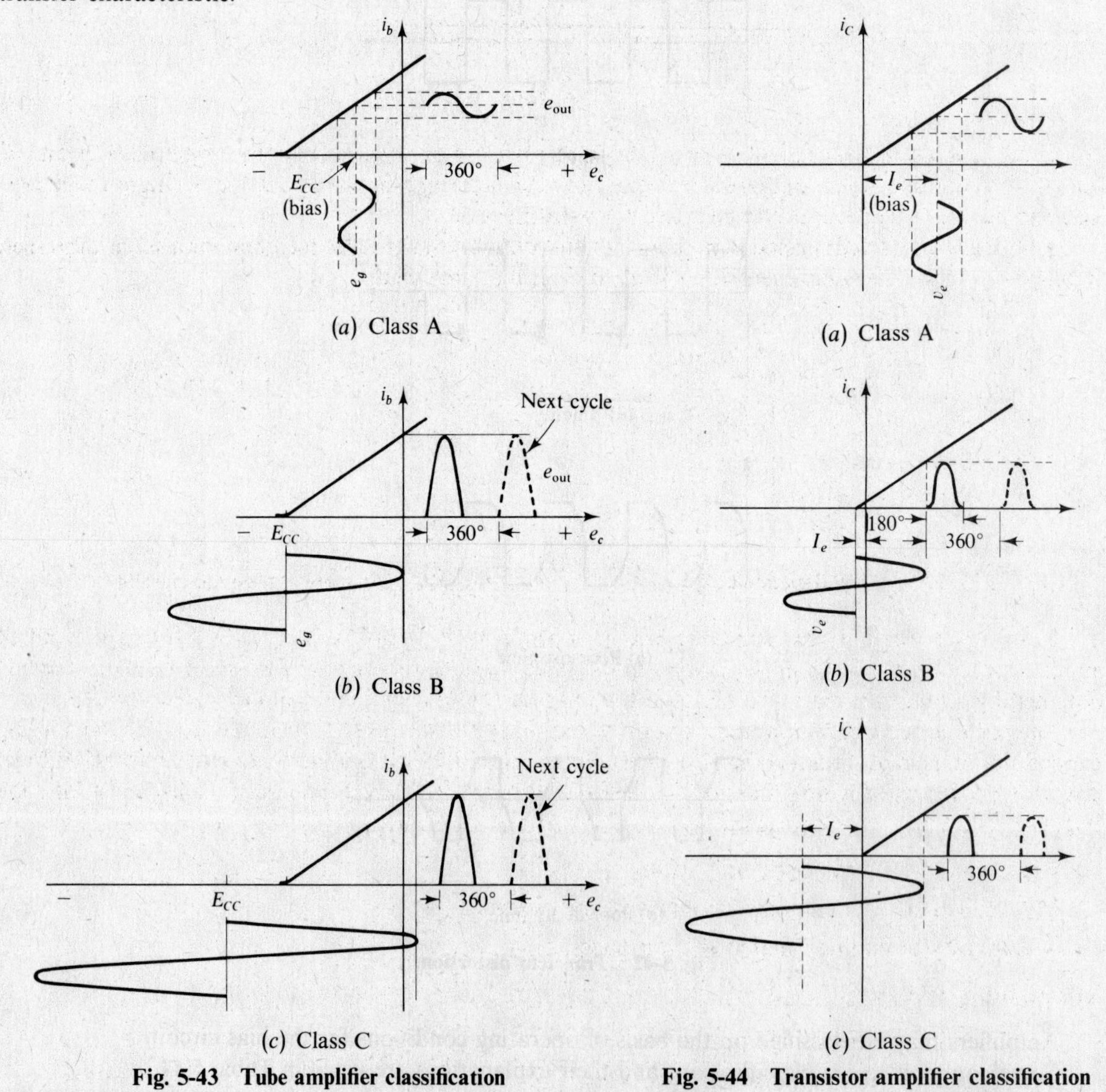

Fig. 5-43 Tube amplifier classification

Fig. 5-44 Transistor amplifier classification

Figure 5-44 shows the same relationship for a typical transistor dynamic characteristic while Fig. 5-45 shows the relationship between the bias point and class of operation for the collector family of curves.

Some tubes and transistors are manufactured to be used in a specific class of amplifier operation and give poor performance when used for other purposes.

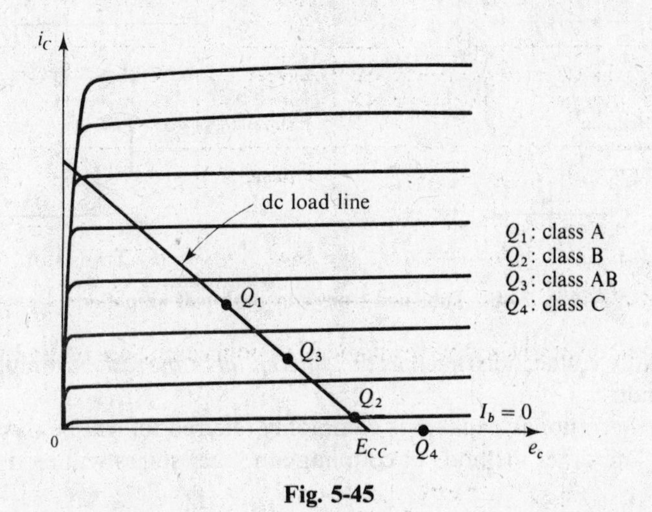

Fig. 5-45

Class A operation is mostly used in general and low-power applications, while class B and C operation is commonly encountered in radio frequency and high-power applications where the distortion can be either ignored or compensated for with filters.

Amplifiers may be connected in parallel if the power requirements are too great for a single amplifier. Figure 5-46 shows two amplifiers in parallel, feeding the same load.

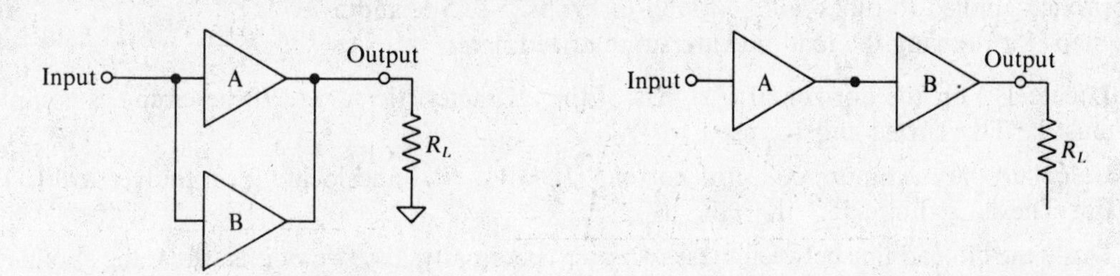

Fig. 5-46 Amplifiers in parallel **Fig. 5-47 Amplifiers in cascade (series)**

In order to amplify very small signals, the output of one amplifier is often fed into the input of another amplifier, as shown in Fig. 5-47. This is called *cascading*. While the gain of the overall system is obtained by adding the gain in decibels of the individual amplifiers, the bandwidth is decreased.

One of the most common connections of amplifiers that is used to increase the power-handling capabilities of a single circuit is the push-pull connection shown in Fig. 5-48. The usual approach is to let one tube or transistor handle the positive half of the input wave while the matching device handles the negative half. Some of the disadvantages of this connection include

- Possibility of "crossover" distortion
- Addition of extra circuitry to obtain phase inversion
- Difficulty of finding "matched" components

The advantages include

- Greater power handling
- Suppression of even harmonic distortion

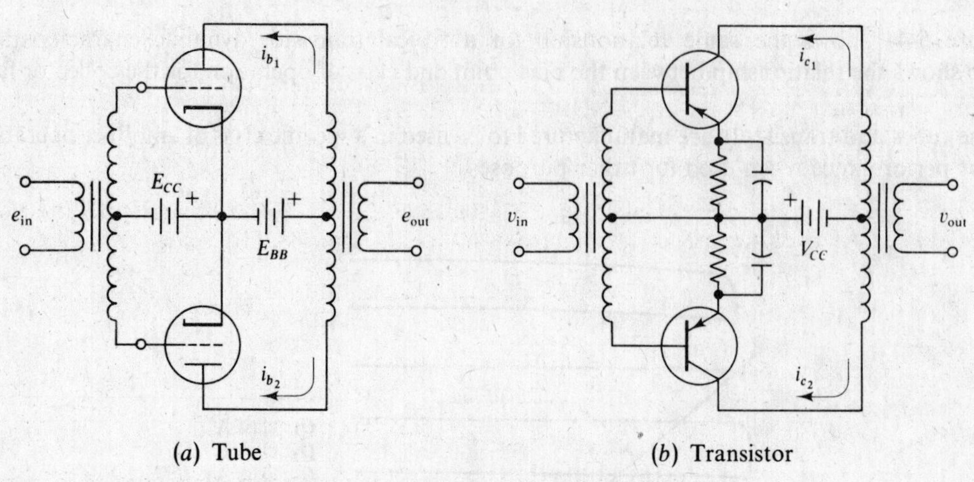

(*a*) Tube (*b*) Transistor

Fig. 5-48 Tube and transistor push-pull amplifier

Other special amplifier connections such as the *differential* or *operational* amplifier have found much use in the electronics industry.

The op amp, as the *op*erational *amp*lifier is commonly referred to, will be discussed in Chap. 7. The Darlington connection and other methods of coupling amplifier stages will be discussed in Chap. 6.

5.7 LOAD LINE AND BIASING

The criteria for determining the dc operating point and dc and ac load lines for a tube or transistor depend on the device's characteristic curves. The methods used in this section can be used for tubes as well as transistors and are an extension of the approach used in Chap. 4.

The descriptions in this section will make use of transistors and transistor characteristic curves. The same approach applied to tubes will be found in Probs. 5-5, 5-6, and 5-7.

The steps for drawing the load line are summarized here:

1. Locate V_{CC} on the horizontal (v_C) axis of the characteristic curves. If the circuit is complex, it must be Thevenized and V_{Th} used.
2. Calculate the maximum collector current ($I_N = V_{CC}/R_L$) and locate it on the vertical (i_C) axis. For Thevenized circuits, $I_N = V_{Th}/R_{Th}$.
3. Draw the dc load line between these two points. Actually, any two points where the actual i_C and v_C are known may be used.
4. Select the I_B curve for the desired class of operation and label Q, the operating point, as the intersection of this curve with the dc load line. For class A operation, this point is usually chosen as half the distance between the $I_B = 0$ and $I_B =$ maximum "safe" current points along the load line.
5. Draw the ac load line through point Q, but with a slope $= -1/R_{ac}$, where R_{ac} is the ac value of load resistance.
6. At point Q, construct a line at right angles to the ac load line and draw the signal voltage with this line as the zero axis.
7. Project the intersections of the signal waveform with the characteristic curves vertically and horizontally to obtain the voltage and current output waveform or values.

The distance from the operating point to the origin measured along the horizontal (voltage) axis equals the transistor collector-to-emitter voltage V_{CE}, while the distance from the operating point to V_{CC} or V_{Th} measured along the horizontal axis equals the voltage across the load resistor V_L.

The distance from the origin to the operating point measured along the vertical (current) axis equals the collector current I_C. All these values are known as the *quiescent* or *no-signal values.*

An example will demonstrate how the load line can be used to provide pertinent operating information.

Example 5.14 Using the amplifier circuit shown in Fig. 5-49*a* and the characteristic curves for the transistor given in Fig. 5-49*b*, develop the load line solution.

Figure 5-49*a* shows a coupled common-emitter NPN transistor amplifier. The capacitor C blocks the dc input to the next stage, so the circuit we use for drawing the dc load line is shown in Fig. 5-49*c*.

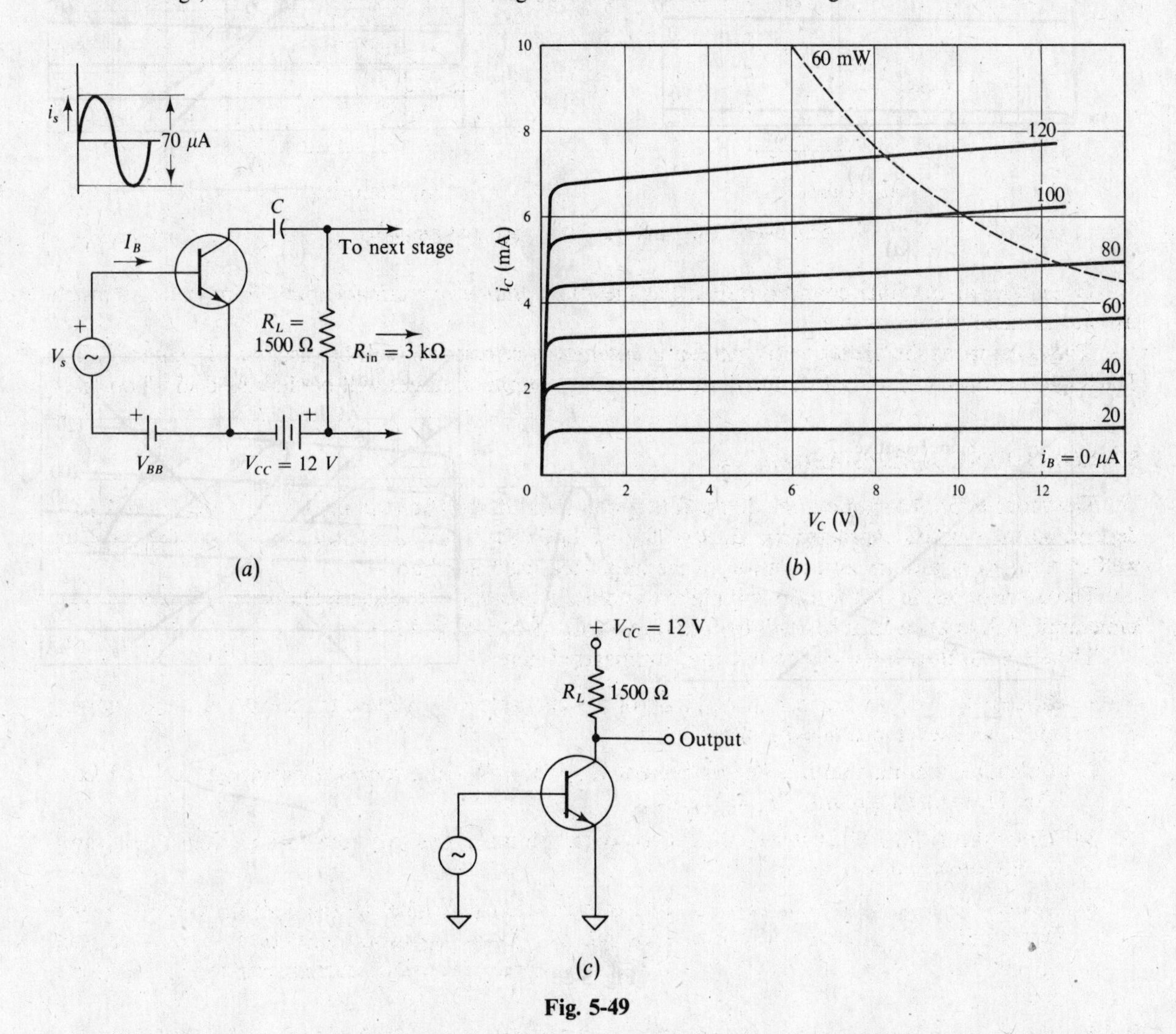

Fig. 5-49

If the circuit were more complex, it would have to be Thevenized by removing the transistor and applying Thevenin's theorem to the remaining circuit. This would result in a Thevenin generator V_{Th} in series with an R_{Th} resistor.

While this circuit does not need Thevenizing, it is a good idea to examine each individual case with Thevenizing in mind.

From Fig. 5-49*c*, $V_{CC} = 12$ V and $R_L = 1500\ \Omega$. Using Fig. 5-49*b*, we locate 12 V on the voltage axis and calculate I_N: According to Ohm's law,

$$I_N = \frac{V_{Th}}{R_{Th}} = \frac{V_{CC}}{R_L} = \frac{12}{1500} = 8.0 \times 10^{-3}\ \text{A} = 8.0\ \text{mA}$$

Locating 8.0 mA on the current axis, we now draw the dc load line between $I_N = 8.0$ and $V_{Th} = 12$, as shown in Fig. 5-50*b*.

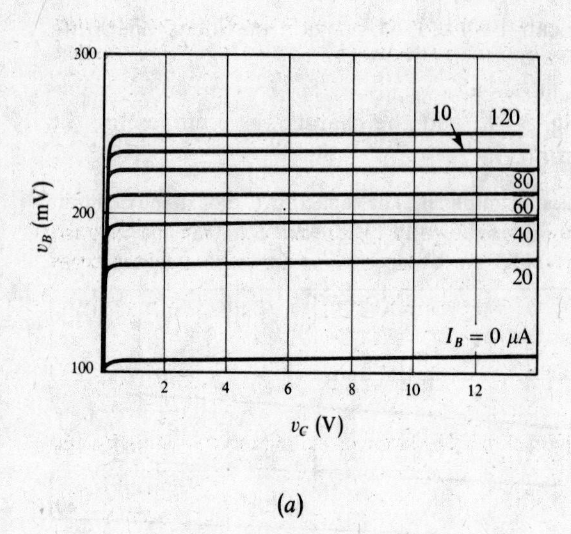

(a)

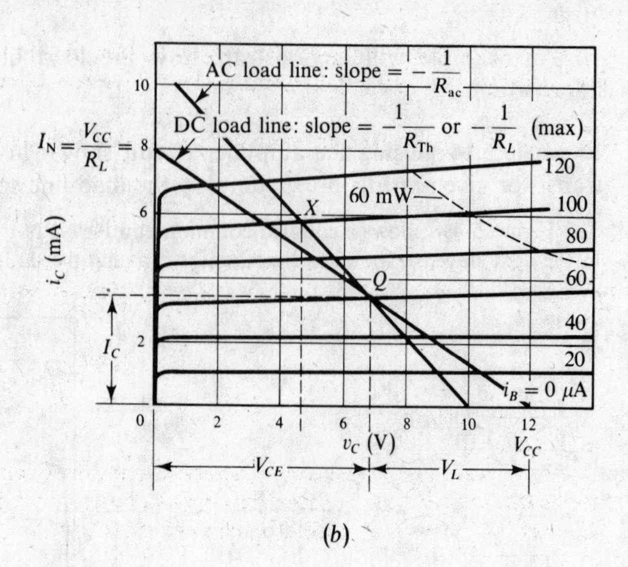

(b)

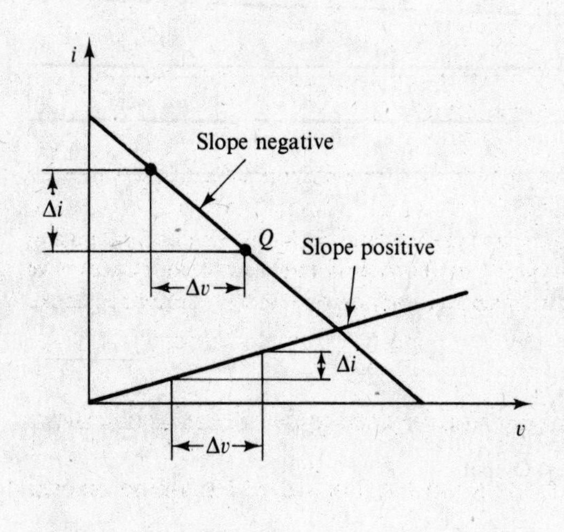

(c)

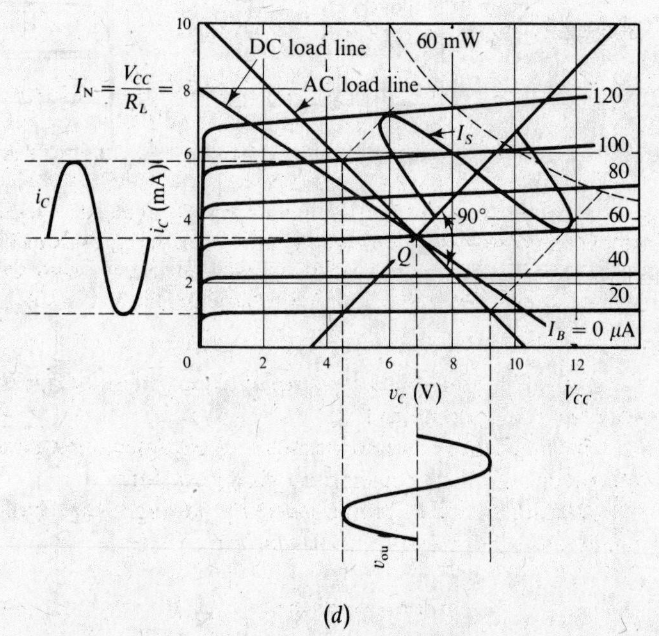

(d)

Fig. 5-50

Since the I_B range for the curves given in Fig. 5-49*a* is 0 to 120 μA, we select the desired base current as

$$I_B = \tfrac{1}{2}(120 - 0) = 60\ \mu\text{A}$$

Q is then located at the intersection of the load line with the 60-μA I_B curve.

From Q we draw horizontal and vertical dashed lines to the axes and label V_{CE}, V_L, and I_C as shown in Fig. 5-50*b*. So

$$V_{CE} = 6.9\ \text{V}$$

$$V_L = 12 - 6.9 = 5.1\ \text{V}$$

$$I_C = 3.4\ \text{mA}$$

We must now construct the ac load line.

For ac, the load resistance becomes R_L in parallel with the next stage input resistance since for ac the capacitor is essentially a short circuit. In Chap. 6, we will see how the value of C affects the frequency response of the coupled circuit.

From Fig. 5-49*a*, we may calculate the ac load resistance R_{ac} as [from (*1-7b*)]

$$R_{ac} = \frac{(R_L)(R_{in})}{R_L + R_{in}} = \frac{(1500)(3000)}{1500 + 3000} = 1000\ \Omega$$

To construct the ac load line through point Q, with a slope equal to $-1/R_{ac}$, we take a convenient distance along the voltage axis in the direction of decreasing voltage and make the distance along the current axis increase such that

$$\frac{\Delta i}{\Delta v} = \frac{1}{R_{ac}}$$

as shown in Fig. 5-50*c*.

The negative sign is taken into account by the direction of the slope. The slope of the load line is such that the horizontal and vertical distances are opposite in sign.

For $R_{ac} = 1000\ \Omega$ and a Δv chosen as 2.0 V,

$$\frac{\Delta i}{\Delta v} = \frac{1}{R_{ac}}$$

Solving for Δi,

$$\Delta i = \frac{\Delta v}{R_{ac}} = \frac{2.0}{1000} = 2.0 \times 10^{-3}\ \text{A} = 2.0\ \text{mA}$$

So from Q we move toward the origin 2.0 V and up 2.0 mA to point X. The ac load line is then drawn through points Q and X as shown in Fig. 5-50*b*.

Now we construct another line through Q at right angles to the ac load line and sketch our input signal using this line as an axis. Since the input signal is given in Fig. 5-49*a* as varying I_B by 70 μA peak to peak, we sketch the curve from $I_B = 25$ to 95 μA and project these intersections with the ac load line horizontally and vertically to get the output current and voltage variations as shown in Fig. 5-50*d*.

Once we establish the proper I_B for our desired operation point, we need to *bias* the original circuit to achieve this condition.

Biasing is not simple because of parameter temperature variations, but we will examine several approaches to the problem in a simplified form.

Biasing may be accomplished by adding a bias battery V_{BB} as shown in Fig. 5-51*a*, or by using a single tapped battery, as shown in Fig. 5-51*b*.

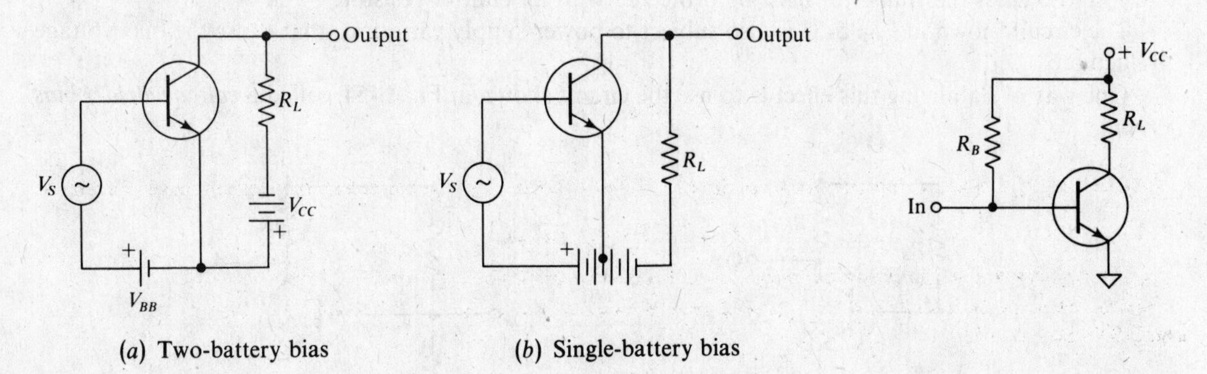

(*a*) Two-battery bias (*b*) Single-battery bias

Fig. 5-51

Fig. 5-52 Self-bias scheme

Usually a self-bias circuit is used to allow a single dc power supply to inexpensively provide both the collector and the bias voltages.

A simple circuit of this type is shown in Fig. 5-52.

The value of R_B can be easily calculated from Ohm's law as

$$R_B = \frac{V_{CC} - V_{BE}}{I_B} \tag{5-22}$$

where V_{BE} (the forward base-emitter drop) is 0.2 V for germanium and 0.7 V for silicon.

The circuit as shown in Fig. 5-52 is not protected against a phenomenon called *thermal runaway*, which can be described as follows: The operating point selected from the load line approach does not take the thermal effects of junction temperature into account. Actually, I_C causes the base-emitter junction to heat up, which causes the base current to increase since the heat makes more charge carriers available. The increase in base current shifts the operating point up the load line to a higher I_C. Although the V_{CE} is actually lowered by this shift, the increased I_C causes still more heating of the junction and a still higher I_C. This snowball effect occurs within a few milliseconds and can cause the transistor to exceed its maximum thermal ratings and be destroyed.

The simplest way to prevent thermal runaway is to divide the load resistance so that some portion of it is in the emitter circuit, as shown in Fig. 5-53.

Although there is a tradeoff between stability and gain, a good rule of thumb is to make

$$R_C = 0.9R_L \tag{5-23a}$$

$$R_E = 0.1R_L \tag{5-23b}$$

where

$$R_L = V_L/I_C(\text{max}) \tag{5-23c}$$

and

$$R_L = R_C + R_E \tag{5-23d}$$

With R_E in the emitter portion of the circuit, if the junction heats up or I_C increases for any reason, I_E will also increase and the $I_E R_E$ voltage drop will reduce V_{BE} sufficiently to ensure that I_C is held to a safe value.

The thermal effects may then cause some fluctuation in the operating point, but the safety of the transistor is assured.

For the circuit shown in Fig. 5-53, the dc load resistance is still R_L since the effect of the base current through R_E is usually very slight. For more precise calculations, I_B should be taken into account since it flows through R_E in addition to I_C.

As a general rule, thermal runaway is not likely to occur in amplifiers designed for voltage amplification. However, the amount of gain of a voltage amplifier can vary over a wide range of values when the ambient temperature is not constant. The emitter resistor stabilizes the amplifier gain against temperature variations.

Transistor power amplifiers, and certain switching transistors, are susceptible to thermal runaway, and in those cases the transistor *must* be protected with an emitter resistor.

The circuit shown in Fig. 5-53 is also subject to power-supply variations that cause the bias voltage to fluctuate.

One way of stabilizing this effect is to use the circuit shown in Fig. 5-54, called a *voltage divider bias circuit*.

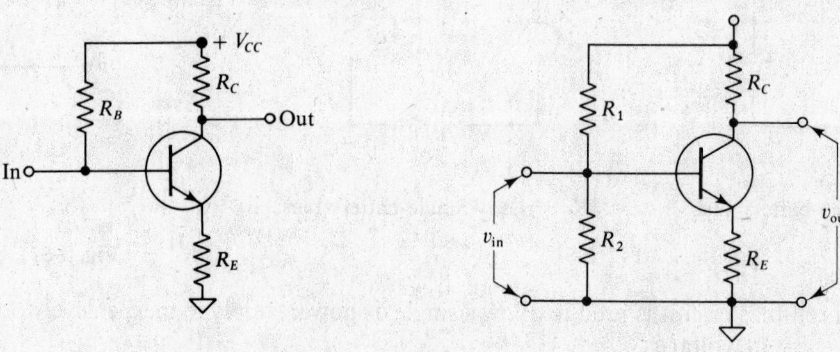

Fig. 5-53

Fig. 5-54

Since R_2 is also in the ac circuit and tends to shunt the input impedance of the transistor, there is another tradeoff between stability, gain, and input impedance. R_2 is usually made to range from R_L to about $10R_{in}$.

For the common-emitter configuration, R_{in} is on the order of 1 kΩ, so that a typical value for R_2 is around 10 kΩ.

We may then calculate R_1 from

$$R_1 = \frac{R_2(V_{CC} - V_B)}{I_B R_2 + V_B} \tag{5-24a}$$

where

$$V_B = V_{BE} + V_E \tag{5-24b}$$

and

$$R_2 = 10R_{in} \tag{5-24c}$$

Example 5.15 Find the values for R_1, R_2, R_C, and R_E for the amplifier circuit shown in Fig. 5-55*a*. $V_{CE} = 6.9$ V; $V_{BE} = 0.7$ V; $I_C = 3.4$ mA; $R_{in} = 850\ \Omega$; and $I_B = 60$ A.

Redrawing and labeling, as shown in Fig. 5-55*b*, $V_{CC} = 10$ V and $V_s(\text{max}) = 0.1$ V. Using Kirchhoff's voltage rule,

$$V_L = V_{CC} - V_{CE} = 10 - 6.9 = 3.1 \text{ V}$$

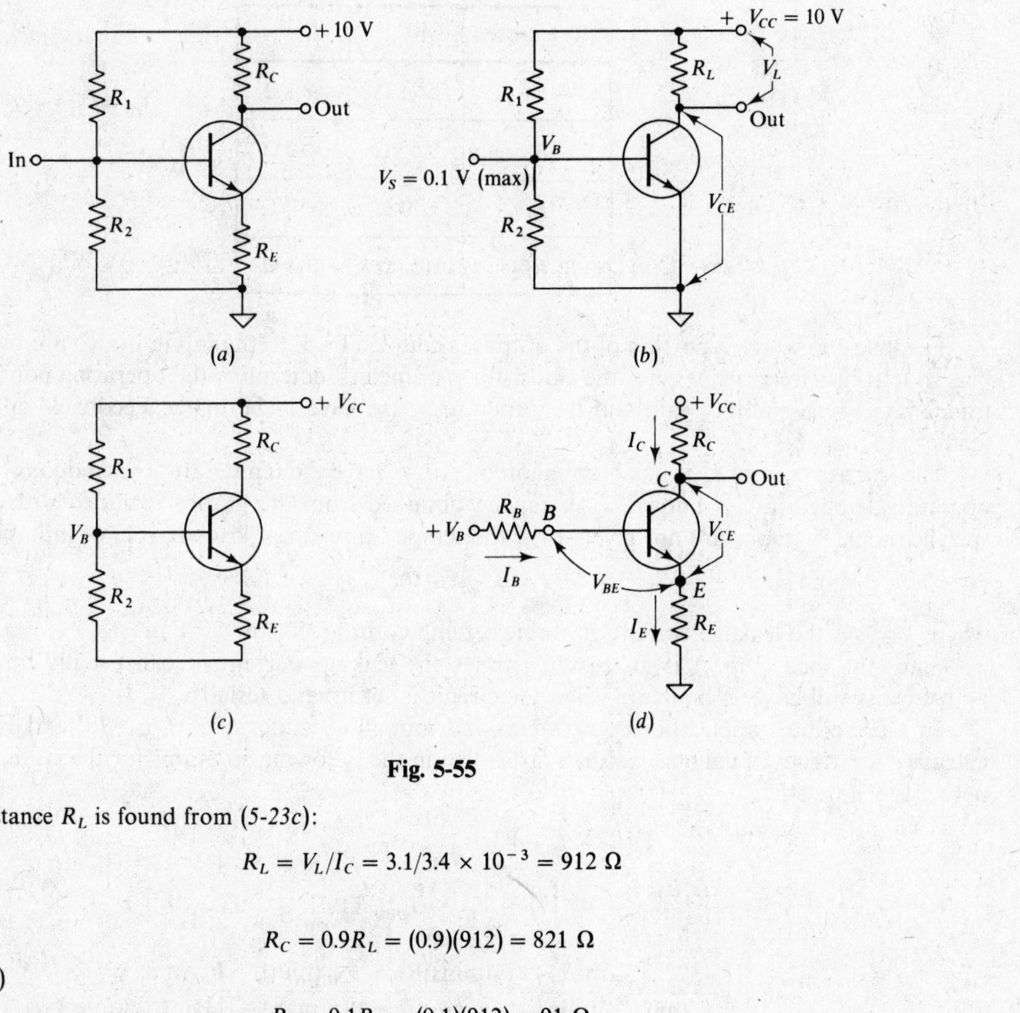

Fig. 5-55

The load resistance R_L is found from (5-23*c*):

$$R_L = V_L/I_C = 3.1/3.4 \times 10^{-3} = 912\ \Omega$$

By (5-23*a*)

$$R_C = 0.9R_L = (0.9)(912) = 821\ \Omega$$

and by (5-23*b*)

$$R_E = 0.1R_L = (0.1)(912) = 91\ \Omega$$

The R_{in} for this transistor is given as 850, and so by (5-24*c*)

$$R_2 = 10R_{in} = (10)(850) = 8500\ \Omega$$

Since by (*5-7a*)

$$I_E = I_B + I_C = (60 \times 10^{-6}) + (3.4 \times 10^{-3}) = 3.46 \times 10^{-3} \text{ A}$$

and by Ohm's law

$$V_E = I_E R_E = (3.46 \times 10^{-3})(91) = 0.31 \text{ V}$$

So by (*5-24b*)

$$V_B = V_{BE} + V_E = 0.7 + 0.31 = 1.01 \text{ V}$$

R_1 can now be found from (*5-24a*):

$$R_1 = \frac{R_2(V_{CC} - V_B)}{I_B R_2 + V_B} = \frac{(8500)(10 - 1.01)}{(60 \times 10^{-6})(8500) + 1.01} = 50.3 \text{ k}\Omega$$

Keep in mind that bias problems usually deal with a range of values and the calculated values are used as the starting point for a practical circuit. The circuit is then breadboarded and adjusted for desired operation. In addition, the resistors have a range of values, and standard resistors are normally specified.

In the previous example, we would specify as follows:

Tolerance	10%	20%
R_1	56 kΩ	68 kΩ
R_2	8.2 kΩ	10 kΩ
R_C	820 Ω	1.0 kΩ
R_E	100 Ω	100 Ω

Feedback, in which a portion of the output is coupled back to the input, is also used to stabilize the bias so that the circuit values, not the transistor parameters, determine the operating point. Once again, tradeoffs between stability gain and frequency response have to be made. Feedback will be discussed further in Chap. 6.

The leakage current I_{CBO} has been ignored here, as have the temperature variations of β. If the exact operating temperature is known, I_{CBO} can be obtained from the graphs included with the transistor specifications. If graphs are not given, or if the temperature is not known, rule of thumb specifies

$$I_{CBO} = 10 I_{CBOA}$$

where I_{CBOA} = the leakage current at ambient temperature (25°C or 77°F).

Since the load current is at least ten times the leakage current, it can usually be ignored since adjustments will have to be made when the circuit is set up and tested.

In more critical applications, maximum and minimum values of β, I_{CBO}, I_E, and V_{BE} are used to calculate the values of the bias resistors, as shown in the following solutions for the voltage divider bias network of Fig. 5-55*c*:

$$I_E = (\beta + 1)(I_B + I_{CBO}) \tag{5-25}$$

$$V_B = \left(\frac{R_B}{\beta + 1} + R_E\right) I_E + V_{BE} - I_{CBO} R_B \tag{5-26}$$

$$R_B = \frac{[I_E(\text{max}) - I_E(\text{min})]R_E + V_{BE}(\text{min}) - V_{BE}(\text{max})}{I_E(\text{min})/[\beta(\text{min}) + 1] - I_E(\text{max})/[\beta(\text{max}) + 1] + I_{CBO}(\text{max})} \tag{5-27}$$

Usually for silicon transistors,

$$I_{CBO} = 0 \tag{5-28}$$

Figure 5-55*d* shows the fundamental directions and nomenclature for all types of bias circuits.

One other point should be mentioned before we leave the load line solution, and that has to do with power dissipation.

The load line is normally drawn well below the maximum power dissipation curve to ensure that the transistor will always operate in the "safe" region. However, this criterion prevents the transistor from operating at its maximum capacity. In a voltage amplifier, this may be acceptable, but in a power amplifier, we would like to take advantage of the maximum power dissipation ability of the transistor. In such cases, the load line is drawn so that one end intersects the voltage axis at the V_{CC} point and the line itself is tangent to the maximum power dissipation curve, as shown in Fig. 5-56.

Since temperature considerations are important to bias and power conditions, it is necessary to ensure that any excess heat is properly dissipated from the transistor. This is done by a procedure called *heat sinking* and involves mounting the transistor on a chassis or heat sink that will efficiently radiate heat to the environment. Heat sinks are constructed of black aluminum with fins or other projections that maximize the surface area to aid in the dissipation of heat. Special mounts are made to provide good thermal conductivity to the heat sink or chassis and at the same time isolate the transistor electrically. Special thermal compounds, usually silicone greases, are applied to transistor contact surfaces to further increase the thermal conductivity.

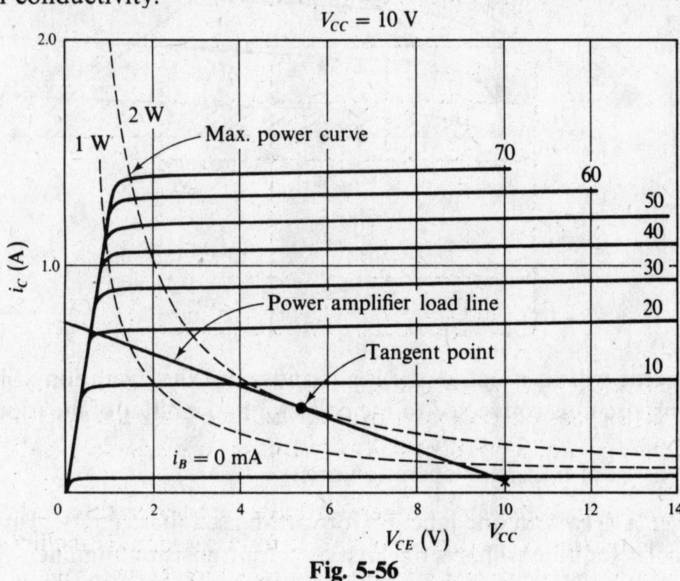

Fig. 5-56

5.8 SWITCHING

Transistors used for switching use a modified design approach to ensure good performance.

An ideal switch characteristic would look like Fig. 5-57. When the switch is off, the circuit is open, no current flows, and the full voltage is across the switch. When the switch is on, the circuit is closed, full current flows, and there is no voltage drop across the switch.

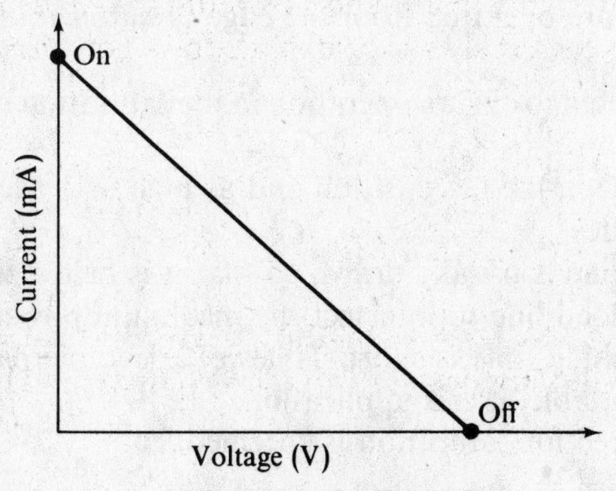

Fig. 5-57 Ideal switch

A transistor in the common-emitter configuration shown in Fig. 5-58*a* can approach this condition. Figure 5-58*b* shows the load line for this circuit. Points *A* and *B* approximate the ON and OFF states, respectively.

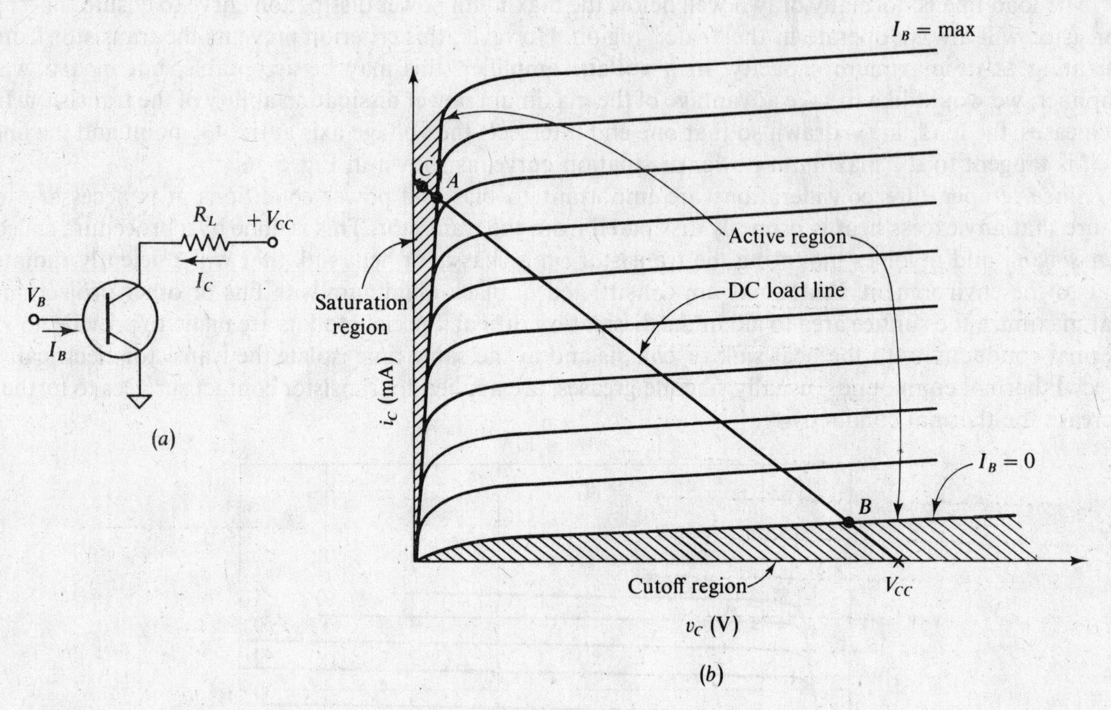

Fig. 5-58

Base-to-emitter current will then act as the input and move the operation point from *B* to *A* and back again. The working circuit is connected to the collector and emitter of the transistor as if they were the switch contacts.

The three possible areas of transistor operation are

- The normal or *active* area with one junction forward-biased (base emitter) and the other junction reverse-biased (collector base). This is the area used for transistor amplifiers and also is the region of fastest switch time.
- The *saturated* area with both junctions forward-biased.
- The *cutoff* area with both junctions reverse biased.

In order that the switch have a low resistance in the ON position, point *A* should be in the saturation region. Point *B* should be in the cutoff region so the switch will have a high resistance in the OFF position. For this reason, switches that are operated from the edge of cutoff to the edge of saturation are called *fully driven* as shown in Fig. 5-58*b*.

If the base current is increased to ensure operation in the saturation region like point *C* in Fig. 5-58*b*, the switch is called *overdriven*.

Resistances as low as 0.5 Ω in the ON position and as high as 1 MΩ in the OFF position have been obtained with transistor switches.

While the switching load line is usually drawn so that it is below the maximum power dissipation curve, it is acceptable for the load line to intersect the maximum power dissipation curve as shown in Fig. 5-59, so long as the switching time is short. Times of a few microseconds are typical of switching transistors and would be acceptable in this application.

The base current I_B required for saturation is specified from

$$I_B(\text{min}) = I_C/\beta = I_C/h_{FE} \tag{5-29}$$

The practical I_B is then made two or three times this value to account for temperature and other variations in β.

Equation (*5-29*) is derived from (*5-12*):

$$I_C = \beta I_B + (\beta + 1)I_{CBO}$$

with I_{CBO} taken as zero.

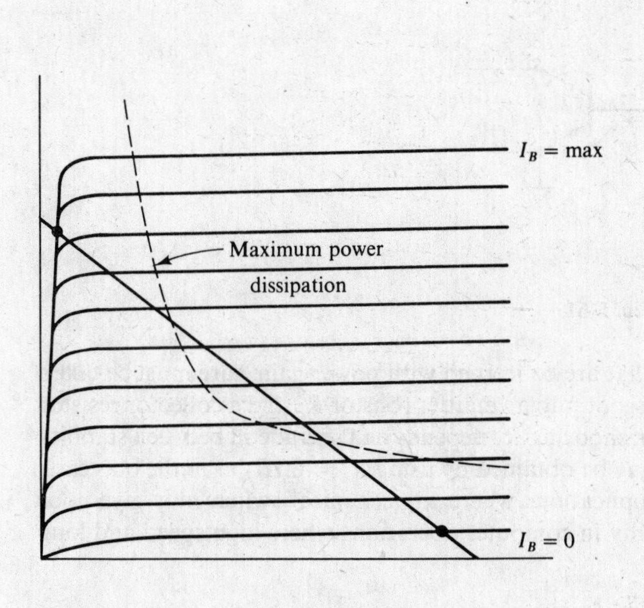

Fig. 5-59

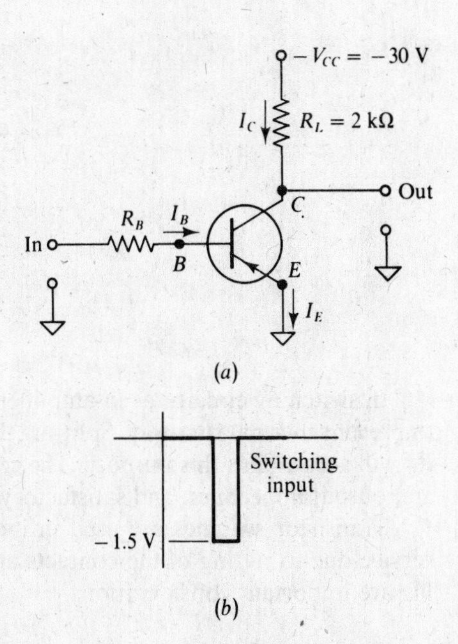

Fig. 5-60

Example 5.16 The transistor shown in Fig. 5-60*a* is to be used as a fully driven switch. The input voltage is shown in Fig. 5-60*b* and the transistor parameters are β (25°C) = 45; $V_{CE}(\text{sat}) = -0.15$ V; $V_{BE}(\text{sat}) = -0.65$ V. Find R_B.

The load voltage when the switch is on is

$$V_L = V_{CC} - V_{CE}(\text{sat}) = -30 - (-0.15) = -29.85 \text{ V}$$

The maximum load current is then calculated from Ohm's law:

$$I_C(\text{sat}) = V_L/R_L = -29.85/2000 = -14.92 \text{ mA}$$

where the negative sign indicating the direction of I_C is opposite to that drawn in Fig. 5-60*a*, as expected with a PNP transistor.

The minimum I_B can now be obtained from (*5-29*):

$$I_B(\text{min}) = I_C(\text{sat})/\beta = -14.92/45 = -0.332 \times 10^{-3} = -332 \times 10^{-6} \text{ A} = -332 \text{ } \mu\text{A}$$

To ensure operation in the saturation region, we use three times this value:

$$I_B = (3)(-332 \times 10^{-6}) = -996 \times 10^{-6} \text{ A} = -996 \text{ } \mu\text{A} \cong -1.0 \text{ mA}$$

Figure 5-60*b* shows that the maximum input voltage which triggers the switch is −1.5 V, and since the voltage V_{R_B} across R_B is

$$V_{R_B} = V_{\text{in}}(\text{max}) = V_{BE}(\text{sat}) = -1.5 - (-0.65) = -0.985 \text{ V}$$

R_B can be found from Ohm's law:

$$R_B = V_{R_B}/I_B = -0.985/(-1.0 \times 10^{-3}) = 985 \text{ } \Omega$$

Operation as an overdriven switch can be obtained by use of the circuit shown in Fig. 5-61. V_{BB} is made positive for the PNP connection shown and R_1 selected so that the current through R_1 is twice I_B(min). This protects against temperature variations and assures operation in the saturation region.

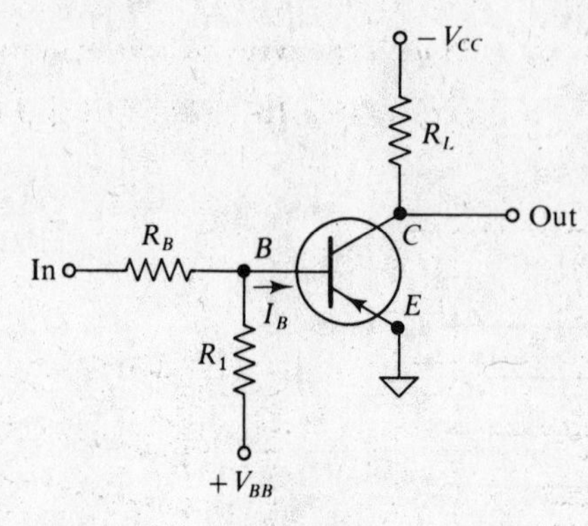

Fig. 5-61

In switching circuits, as in amplifier circuits that are concerned with power gain, care must be taken to prevent thermal runaway. Splitting the load resistor into an emitter resistor R_E and a collector resistor R_C will accomplish this purpose. The criterion for choosing R_E depends on the tradeoff between stability and output impedance, and satisfactory results may be obtained by using $R_E = 0.1R_L$ as in the bias case.

Transistor switches are used in industrial applications where a mechanical switch may give poor service due to pitting of the contacts and especially in computer operations where high speed and long life are important considerations.

Solved Problems

5.1 The value of α_{DC} for a certain transistor is 0.98. Find the collector current I_C when the emitter current I_E is 100 mA.

Using the definition of α_{DC} and substituting values, we have [from (5-8b)]

$$\alpha_{DC} = \frac{I_C}{I_E}$$

$$0.98 = \frac{I_C}{100 \times 10^{-3}}$$

or

$$I_C = 98 \times 10^{-3} \text{ A} = 98 \text{ mA}$$

5.2 Using the pentode plate characteristics shown in Fig. 5-62, find the amplification factor μ for a bias voltage of -2.50 V. The mutual conductance g_m for this tube is 75.0 μS.

We may obtain the plate resistance r_p from the characteristic curves of Fig. 5-62 by using (5-1):

$$r_p = \left.\frac{\Delta e_b}{\Delta i_b}\right|_{e_c=\text{constant}}$$

The bias voltage is given as -2.50 V, so we pick two convenient points along the $e_c = -2.50$ V curve shown in Fig. 5-62:

point A at $e_b = 100$ V, $i_b = 2.4$ mA

point B at $e_b = 300$ V, $i_b = 2.5$ mA

Then $$\Delta e_b = 300 - 100 = 200 \text{ V}$$

and $$\Delta i_b = (2.5 \times 10^{-3}) - (2.4 \times 10^{-3}) = 0.1 \times 10^{-3} \text{ A}$$

So $$r_p = \frac{200}{0.1 \times 10^{-3}} = 2000 \times 10^3 \ \Omega = 2.00 \text{ M}\Omega$$

The problem statement gives $g_m = 75\ \mu\text{S}$, and when we substitute into (*5-4*):

$$\mu = g_m r_p = (75 \times 10^{-6})(2.00 \times 10^6) = 150$$

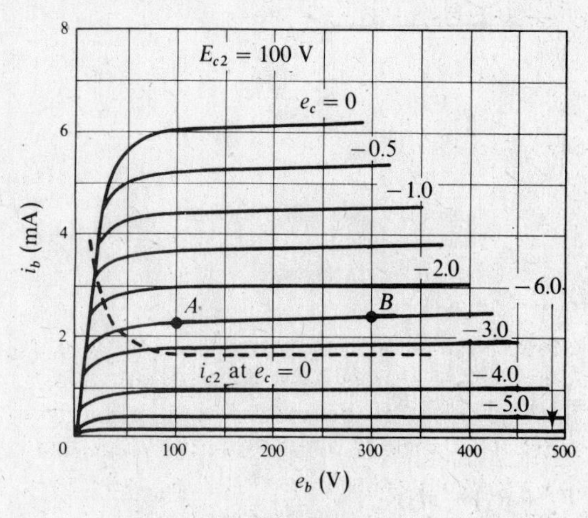

Fig. 5-62

Fig. 5-63

5.3 The common-emitter curves for a bipolar transistor are given in Fig. 5-63. Find h_{FE} for a collector voltage of 6.0 V.

h_{FE} or β_{DC} is defined as

$$\beta_{DC} = \left.\frac{\Delta I_C}{\Delta I_B}\right|_{V_C = \text{constant}}$$

Using the vertical line where $V_C = 6.0$ V, we pick two convenient I_B curves and read their corresponding I_C values:

When $I_B = 0.1$ mA, $$I_C = 16.0 \text{ mA}$$

and when $I_B = 0.2$ mA, $$I_C = 31.0 \text{ mA}$$

So $$\Delta I_B = (0.2 \times 10^{-3}) - (0.1 \times 10^{-3}) = 0.1 \times 10^{-3} \text{ A}$$

and $$\Delta I_C = (31.0 \times 10^{-3}) - (16.0 \times 10^{-3}) = 15.0 \times 10^{-3} \text{ A}$$

Therefore $$h_{FE} = \beta_{DC} = \frac{15.0 \times 10^{-3}}{0.1 \times 10^{-3}} = 150$$

5.4 Find the voltage gain A_v for the grounded grid amplifier shown in Fig. 5-64*a* if $r_p = 5.00$ kΩ, $\mu = 75.0$, $R_1 = 100\ \Omega$, and $R_L = 20$ kΩ.

First we draw the ac equivalent circuit as shown in Fig. 5-64*b* and develop an expression for A_v. Writing Kirchhoff's voltage rule for the input,

$$E_g - E_s = -I_p R_1$$

Solving for E_g,

$$E_g = E_s - I_p R_1$$

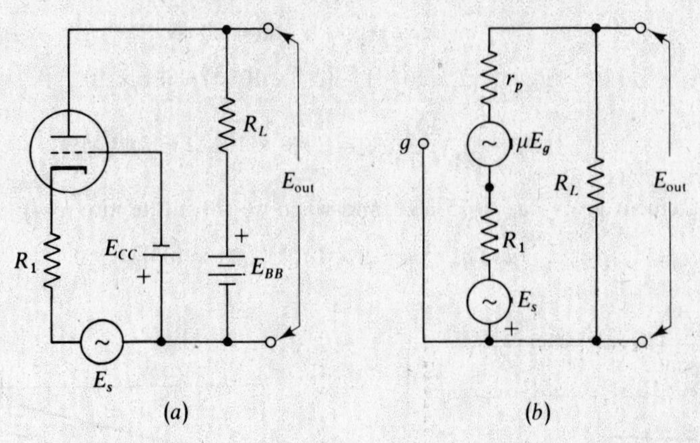

Fig. 5-64

Writing Kirchhoff's voltage rule for the output,

$$\mu E_g + E_s = I_p(r_p + R_L + R_1)$$

Substituting for E_g,

$$\mu(E_s - I_p R_1) + E_s = I_p(r_p + R_L + R_1)$$

Collecting terms,

$$\mu E_s + E_s = I_p(r_p + R_L + R_1) + \mu I_p R_1$$

Factoring,

$$(\mu + 1)E_s = I_p[r_p + R_L + (\mu + 1)R_1]$$

Dividing through by $(\mu + 1)$, we obtain

$$E_s = \frac{I_p[r_p + R_L + (\mu + 1)R_1]}{\mu + 1}$$

We also have from Ohm's law

$$E_{out} = I_p R_L$$

And since from (*5-19a*)

$$A_v = \frac{E_{out}}{E_s}$$

substitution yields

$$A_v = \frac{I_p R_L}{I_p[r_p + R_L + (\mu + 1)R_1]/(\mu + 1)} = \frac{(\mu + 1)R_L}{r_p + R_L + (\mu + 1)R_1}$$

If we define a new amplification factor $\mu' = \mu + 1$ and a new plate resistance $r'_p = r_p + \mu' R_1$, we may write the voltage gain in the familiar form shown in Table 5-10:

$$A_v = \frac{\mu' R_L}{r'_p + R_L}$$

Using the values given,

$$\mu' = \mu + 1 = 75 + 1 = 76$$

$$r'_p = r_p + \mu' R_1 = 5 \times 10^3 + (76)(100) = 12.6 \times 10^3\ \Omega = 12.6\ \text{k}\Omega$$

$$A_v = \frac{\mu' R_L}{r'_p + R_L} = \frac{(76)(20.0 \times 10^3)}{(12.6 \times 10^3) + (20.0 \times 10^3)} = 46.6$$

5.5 Find the current gain A_i for the common-base transistor ac equivalent circuit shown in Fig. 5-65 if $\alpha = 0.97$, $r_e = 12.0\ \Omega$, $r_b = 800\ \Omega$, and $r_c = 1.00\ \mathrm{M\Omega}$. Assume a load resistance $R_L = 10.0\ \mathrm{k\Omega}$.

Using Table 5-10, we see that the common-base circuit has a current gain A_i given by

$$\frac{\alpha r_c + r_b}{r_c + r_b + R_L} \cong \alpha$$

Substituting values,

$$A_i = \frac{(0.97)(1.00 \times 10^6) + 800}{1.00 \times 10^6 + 800 + 10.0 \times 10^3} = 0.9604$$

And we do not lose too much by making the assumption

$$A_i \cong \alpha \cong 0.97$$

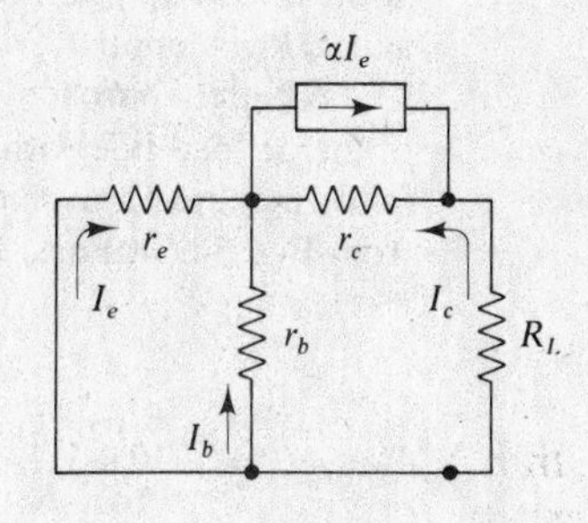

Fig. 5-65

5.6 Find the operating point, tube voltage, and plate current of the circuit shown in Fig. 5-66*a*. Figure 5-66*b* shows the tube plate characteristics. If the input voltage is 10.0 V peak to peak, what is the gain?

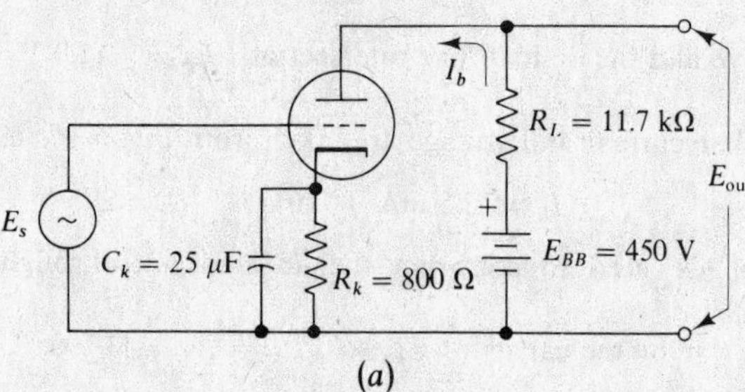

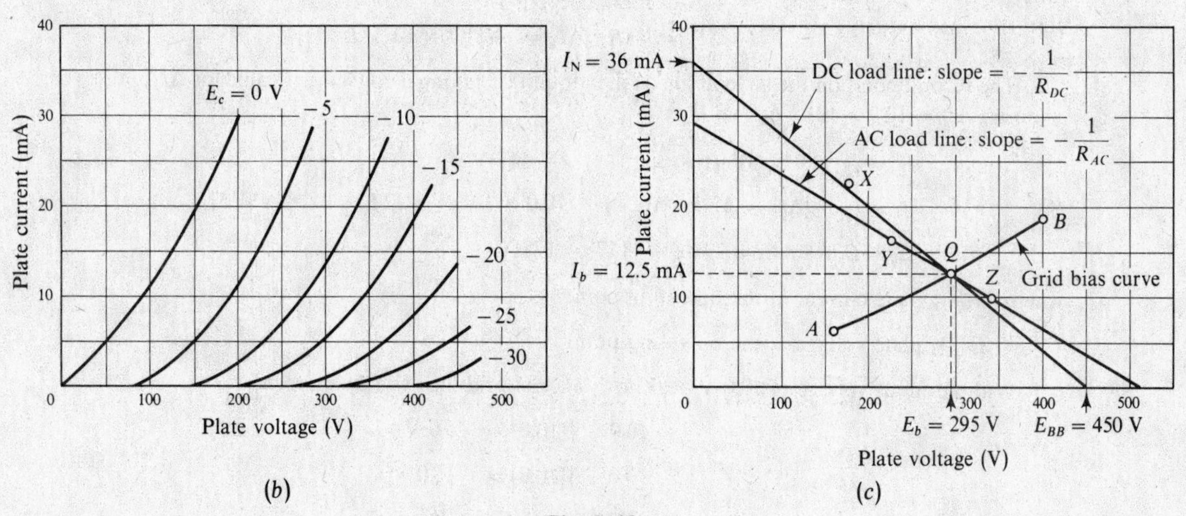

Fig. 5-66

To find the operating point, we first draw the dc load line by connecting $E_{BB} = 450$ V on the voltage axis and I_N on the current axis. By Ohm's law

$$I_N = \frac{E_{BB}}{R_{DC}}$$

and

$$R_{DC} = R_k + R_L = 800 + 11\,700 = 12\,500\ \Omega$$

So

$$I_N = \frac{450}{12.5 \times 10^3} = 36.0 \times 10^{-3}\ \mathrm{A} = 36.0\ \mathrm{mA}$$

Once the dc load line is drawn, we have to either pick an operating point and supply the necessary bias or find the self-bias from the circuit. Figure 5-66*a* shows that the tube current I_b through the cathode resistor R_k provides the self-bias. We also know that the operating point must lie on the load line. If we select an arbitrary operating point and calculate the bias voltage $I_b R_k$ at that point, we may compare it to the E_c value at the operating point. We keep adjusting our position on the load line until a reasonable match between E_c and $I_b R_k$ is found.

A better approach is to graph the grid bias curve by plotting E_c vs. I_b for several values. The intersection of the dc load line with the grid bias curve is then the operating point. Using this approach, we select the following three convenient E_c values and solve for I_b based on the equation $E_c = -I_b R_k$ with $R_k = 800$ from Fig. 5-66*a*. For $E_c = -5.0$ V,

$$I_b = -\frac{(-5.0)}{800} = 6.25 \text{ mA}$$

For $E_c = -10.0$ V,

$$I_b = -\frac{(-10.0)}{800} = 12.5 \text{ mA}$$

For $E_c = -15.0$ V,

$$I_b = -\frac{(-15.0)}{800} = 18.75 \text{ mA}$$

The grid bias curve and the dc load line intersect at $E_c = -10.0$ V and this point is labelled Q, as shown in Fig. 5-66*c*.

At point Q the plate current and voltage are taken from Fig. 5-66*c* as

$$I_b = 12.5 \text{ mA} \qquad \text{and} \qquad E_b = 295 \text{ V}$$

To find the gain, we need to construct the ac load line through point Q and having a slope $-1/R_{ac} = \Delta I_b/\Delta E_b$.

R_k is bypassed for ac by the capacitor C_k, so $R_{ac} = R_L = 11\,700\ \Omega$. Solving the slope equation for ΔE_b,

$$\Delta E_b = -(R_{ac})\Delta I_b = -(11.7 \times 10^3)\Delta I_b$$

The second point on the ac load line is then found by taking a convenient value for ΔI_b; in this case we choose $\Delta I_b = 10.0 \times 10^{-3}$ A, so

$$\Delta E_b = -(11.7 \times 10^3)(10.0 \times 10^{-3}) = -117 \text{ V}$$

So

$$I_b' = I_{bQ} + \Delta I_b = (12.5 \times 10^{-3}) + (10.0 \times 10^{-3}) = 22.5 \times 10^{-3} \text{ A} = 22.5 \text{ mA}$$

$$E_b' = E_{bQ} + \Delta E_b = 295 + (-117) = 178 \text{ V}$$

where I_{bQ} = plate current at the operating point

E_{bQ} = plate voltage at the operating point

With an input voltage of 10.0 V peak to peak, the grid bias will swing from

$$-10.0 + \tfrac{1}{2}(10.0) = -5.0 \text{ V}$$

to

$$-10.0 - \tfrac{1}{2}(10.0) = -15.0 \text{ V}$$

along the ac load line as shown by points Y and Z in Fig. 5-66*c*.

At point Y, $E_b = 226$ V, and at point Z, $E_b = 342$ V, from Fig. 5-66*c*, which is a difference of $342 - 226 = 116$ V.

Since the definition of gain for a tube is (*5-19a*)

$$A_v = \frac{E_{out}}{E_{in}}$$

substitution yields

$$A_v = \frac{116}{10.0} = 11.6$$

5.7 For the cathode bias circuit shown in Fig. 5-67*a*, find the values of R_k and R_L, so that Q is at $E_b = 250$ V and $I_b = 7.5$ mA. The triode has the average plate characteristics shown in Fig. 5-66*b*.

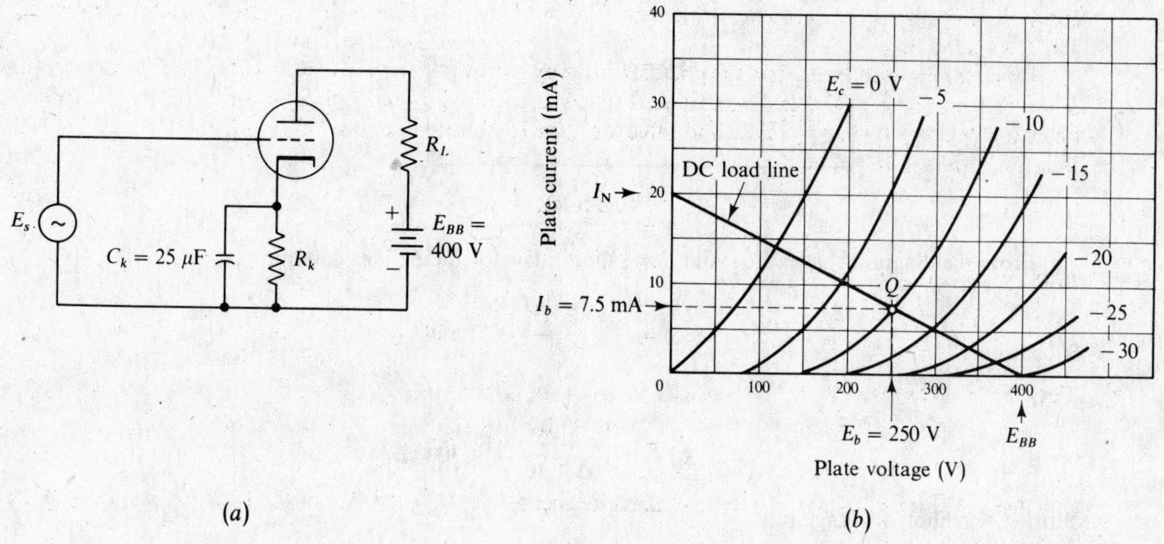

Fig. 5-67

First we locate the operating point Q on the plate characteristic curves as shown in Fig. 5-67*b*.

From Fig. 5-67*a*, $E_{BB} = 400$ V, so we may construct the dc load line from 400 V on the voltage axis through point Q.

As shown in Fig. 5-67*b*, the load line intersects the current axis at $I_N = 20$ mA. Now by Ohm's law

$$I_N = \frac{E_{BB}}{R_{DC}}$$

Solving for R_{DC},

$$R_{DC} = \frac{E_{BB}}{I_N} = \frac{400}{20 \times 10^{-3}} = 20 \times 10^3\ \Omega = 20\ \text{k}\Omega$$

Looking at Fig. 5-67*a*, we see that

$$R_{DC} = R_L + R_k$$

But we can solve for R_k since we know E_c and I_b: By Ohm's law

$$R_k = \frac{-E_c}{I_b} = \frac{-(-10.0)}{7.5 \times 10^{-3}} = 1.33 \times 10^3\ \Omega = 1.33\ \text{k}\Omega$$

Substituting yields

$$R_{DC} = R_L + R_k \qquad 20 \times 10^3 = R_L + (1.33 \times 10^3)$$

or

$$R_L = 18.67 \times 10^3\ \Omega = 18.7\ \text{k}\Omega$$

In practice, we would specify standard 10 percent resistors

$$R_k = 1.8\ \text{k}\Omega \qquad \text{and} \qquad R_L = 18\ \text{k}\Omega$$

which would put the operating point very close to the desired value.

5.8 For the FET circuit shown in Fig. 5-68, find the value of R_L and R_S that will give an operating point of $V_{GS} = -1.50$ V, $V_{DS} = 3.50$ V, and $I_D = 5.00$ mA.

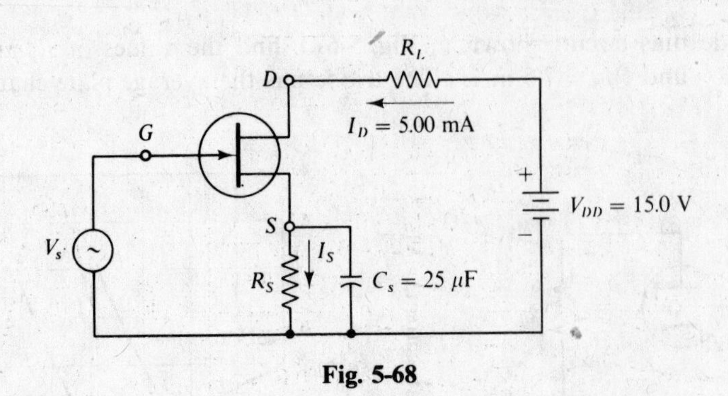

Fig. 5-68

From the diagram, $I_D = I_S$, and using Ohm's law for the dc bias condition

$$V_{GS} = -I_S R_S$$

Substituting,

$$-1.50 = -(5.00 \times 10^{-3})R_S$$

or

$$R_S = \frac{-1.50}{-5.00 \times 10^{-3}} = 300\ \Omega$$

Writing Kirchhoff's voltage rule,

$$V_{DD} = V_{R_L} + V_{DS} + V_{SG}$$

Substituting values and remembering that $V_{SG} = -V_{GS}$,

$$15.0 = V_{R_L} + 3.50 - (-1.50)$$

or

$$V_{R_L} = 10.0\ \text{V}$$

Using Ohm's law to solve for R_L, we have

$$R_L = \frac{V_{R_L}}{I_D} = \frac{10.0}{5.00 \times 10^{-3}} = 2.00 \times 10^3\ \Omega = 2.00\ \text{k}\Omega$$

5.9 The bipolar transistor amplifier shown in Fig. 5-69*a* is to be operated with an input signal of 0.30 mA peak to peak. Find V_{CE}, I_C, and the current gain A_i. Does distortion occur? Use the characteristic curves shown in Fig. 5-69*b*.

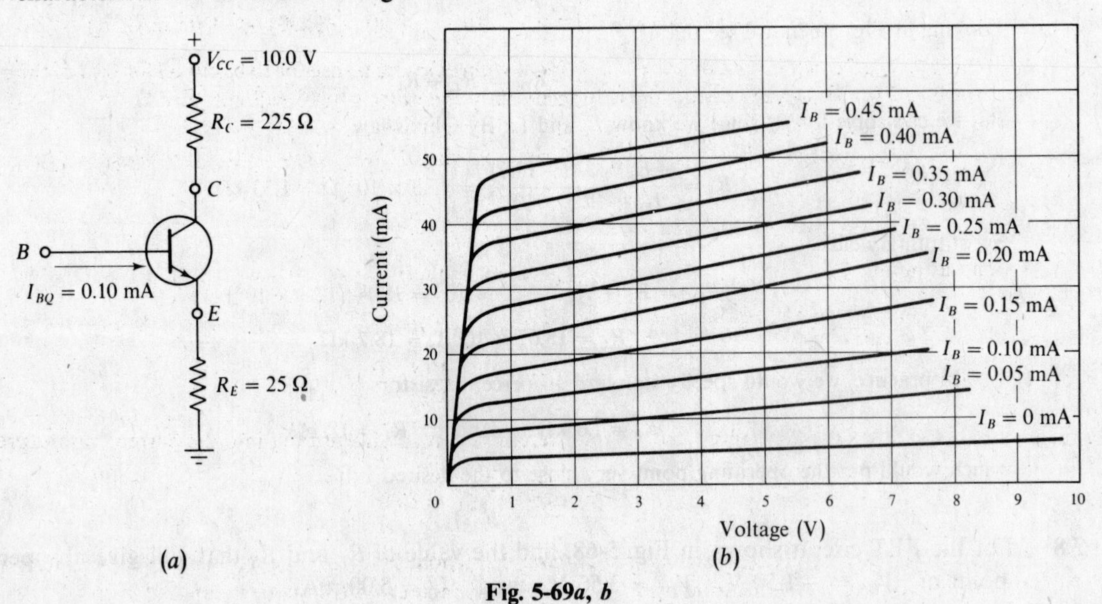

Fig. 5-69*a*, *b*

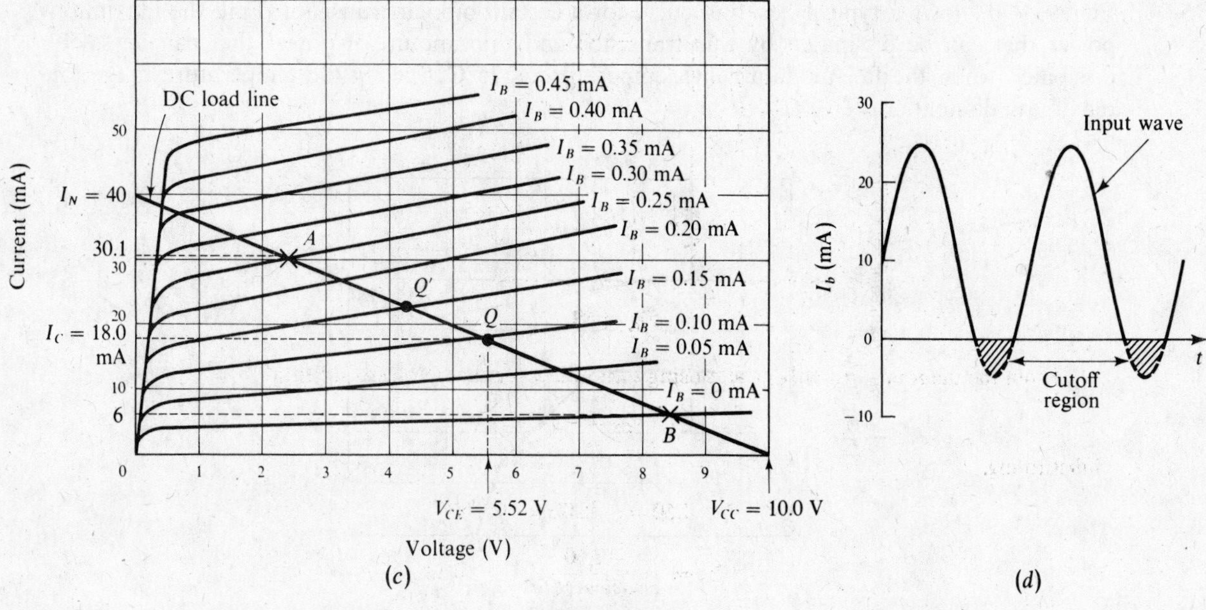

Fig. 5-69c, d

Again, the first step is to draw the dc load line and find the operating point. Then we can read the quiescent parameters V_{CE} and I_C. For this circuit,

$$R_{DC} = R_C + R_E = 225 + 25 = 250\ \Omega$$

So by Ohm's law

$$I_N = \frac{V_{CC}}{R_{DC}} = \frac{10.0}{250} = 40.0\ \text{mA}$$

We then connect $V_{CC} = 10.0$ V on the voltage axis with $I_N = 40.0$ mA on the current axis, as shown in Fig. 5-69*c*. Since Fig. 5-69*a* shows the quiescent base current (bias) as 0.10 mA, the intersection of the dc load line with the $I_B = 0.10$ mA bias curve is the operating point Q.

Q is shown on Fig. 5-69*c* and the values for V_{CE} and I_C are found to be

$$V_{CE} = 5.52\ \text{V} \quad \text{and} \quad I_C = 18.0\ \text{mA}$$

With an input signal swing of 0.30 mA, the bias current swings from

$$0.10 + \tfrac{1}{2}(0.30) = 0.25\ \text{mA} \quad \text{to} \quad 0.10 - \tfrac{1}{2}(0.30) = -0.05\ \text{mA}$$

But $I_B = 0$ is the *cutoff* point, so the input signal will cause the transistor to be cut off for part of the cycle, as shown in Fig. 5-69*d*, and distortion will occur since the output wave will not have the same shape as the input wave. This is an example of nonlinear distortion being introduced by clipping.

For this amplifier circuit, the collector current swing is from point A to B, as shown in Fig. 5-69*c*: At point A,

$$I_C = 30.1\ \text{mA}$$

and at point B,

$$I_C = 6.0\ \text{mA}$$

The current gain is defined as [from (*5-19b*)]

$$A_i = \frac{\Delta I_C}{\Delta I_B}$$

$\Delta I_C = 30.1 - 6.0 = 24.1$ mA and ΔI_B is given as 0.30 mA. Substituting into the current gain expression,

$$A_i = \frac{24.1}{0.30} = 80.0$$

If we wish to operate this amplifier without distortion, we must either reduce the input current swing or move the operating point to point Q' (Fig. 5-69*c*) by increasing the bias current to 0.15 mA.

5.10 Figure 5-70 shows a typical derating curve for a certain bipolar transistor. Find the maximum power that can be dissipated by this transistor and the amount of power that can be safely dissipated when the free-air (ambient) temperature is 75°C. Specify the temperature range for maximum dissipation.

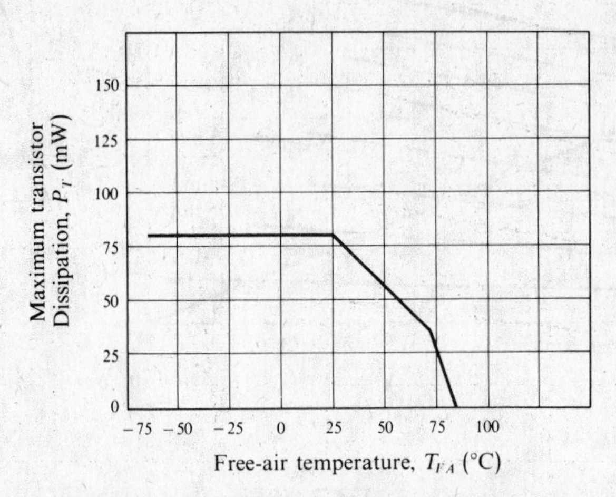

Fig. 5-70 Typical temperature derating curve

From Fig. 5-70, we see that the maximum value of the dissipation power P_T is about 80 mW and occurs throughout the range of −70 to 25°C after which it begins to fall off.

To be on the safe side, we would probably specify a temperature range of about −50 to +20°C:

$$-50°C < T_{FA} < 20°C$$

At 75°C we see from Fig. 5-70 that the P_T has fallen off to about 25 mW. This means that we can only use this transistor in circuits where it dissipates 25 mW or less if the free-air temperature is 75°C.

5.11 Design a bias circuit for the silicon NPN power transistor whose family of characteristic curves is shown in Fig. 5-71*a*. The maximum allowable power dissipation for this transistor is 15.0 W at the operating temperature. Use a power supply voltage of 4.00 V.

We need to find the operating point which will allow us to take advantage of the transistor's full power rating of 15.0 W. Since this value is given at the operating temperature, no derating curves need to be used.

We first superimpose the maximum power dissipation curve on the family of curves, as shown in Fig. 5-71*b*.

It is usually easier to obtain a few points using the power formula (*5-15*):

$$P_T = V_{CE} I_C$$

So $V_{CE} I_C = 15.0$ and we construct a table before placing the points on the characteristic curves:

V_{CE} (V)	1.00	2.00	3.00	4.00	5.00
I_C (A)	15.00	7.50	5.00	3.75	3.00

The maximum power dissipation curve is then drawn with a dashed line, as shown in Fig. 5-71*b*.

We then draw the load line *tangent* to the maximum power dissipation curve through the V_{CC} point on the voltage axis. By making the load line tangent to the power curve, we ensure operation close to maximum power without the danger of exceeding the transistor ratings on the input swing.

The operating point is then chosen to be the point of tangency and is labeled Q in Fig. 5-71*b*. At Q, $V_{CE} = 2.00$ V, $I_C = 7.5$ A, and $I_B = 200$ mA.

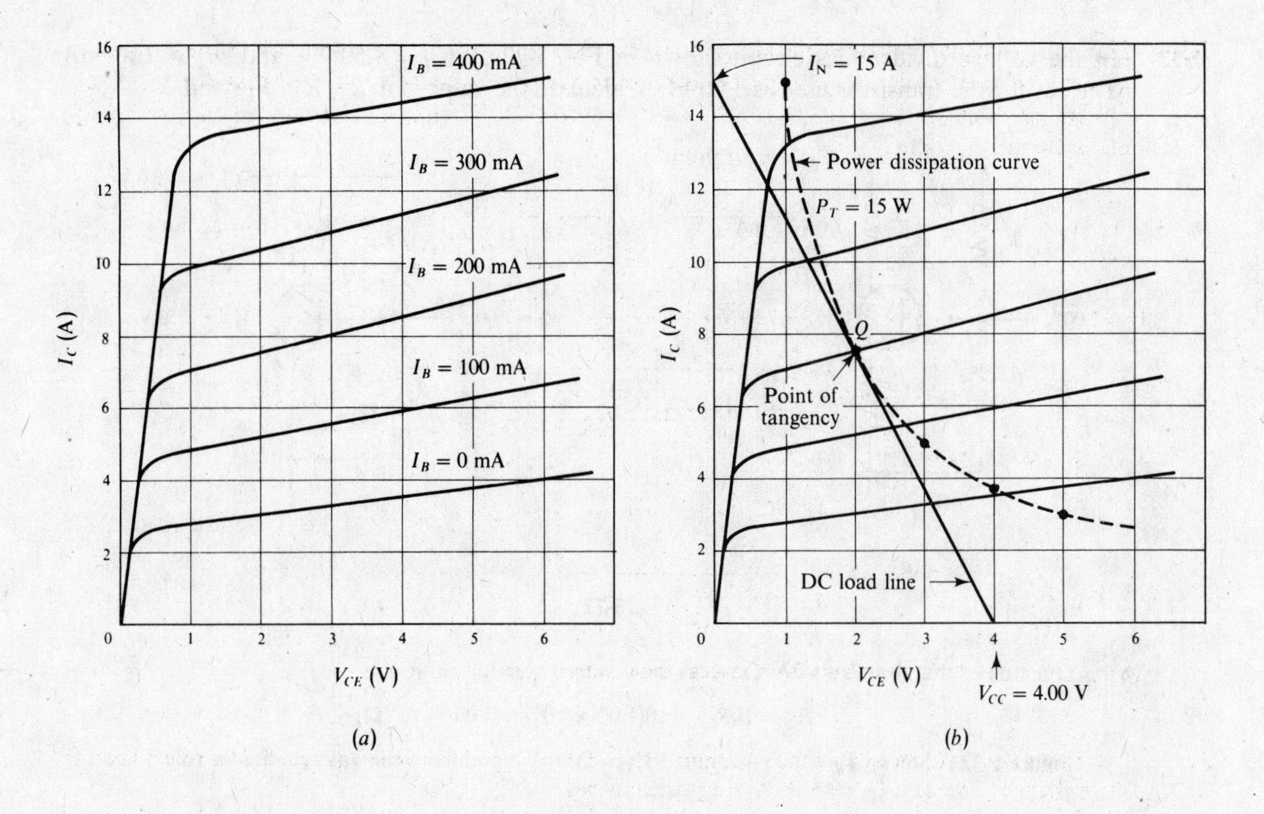

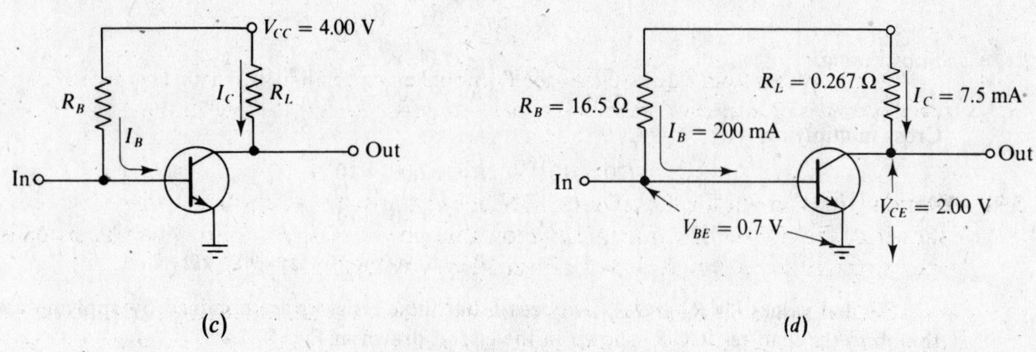

Fig. 5-71

If we use a simple bias circuit as shown in Fig. 5-71*c*, we need to find R_L and R_B. R_L is the dc resistance, so

$$I_N = \frac{V_{CC}}{R_L}$$

I_N for the dc load line is 15.0 A as shown in Fig. 5-71*b*, so

$$R_L = \frac{4.00}{15.0} = 0.267\ \Omega$$

This low value could be the dc resistance of a transformer primary winding.

R_B is found from Ohm's law, since we know I_B and V_{CC}, and $V_{BE} = 0.7$ V for silicon:

$$R_B = \frac{V_{CC} - V_{BE}}{I_B} = \frac{4.00 - 0.7}{200 \times 10^{-3}} = 16.5\ \Omega$$

The final circuit is shown in Fig. 5-71*d*.

5.12 In the voltage divider bias circuit shown in Fig. 5-72*a*, $V_{CE} = 7.50$ V and $I_C = 2.50$ mA. The R_{in} for the transistor used is 1.00 kΩ. Calculate the values for R_1, R_2, R_C, and R_E.

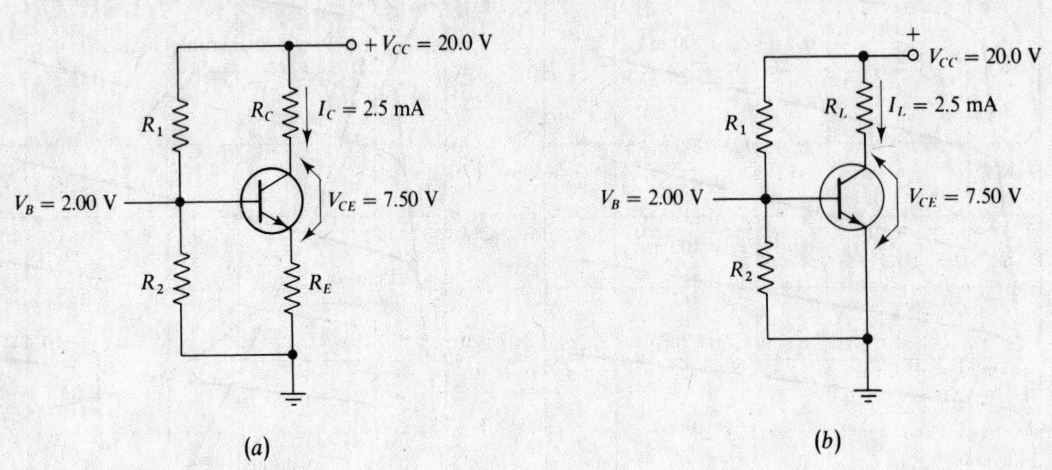

Fig. 5-72

Our rule of thumb tells us from (*5-24c*) that

$$R_2 = 10R_{in} = 10(1.00 \times 10^3) = 1.00 \times 10^4\ \Omega$$

Figure 5-72*a* shows $V_B = 2.00$ V and $V_{CC} = 20.0$ V and from the voltage divider rule

$$V_B = \frac{R_2}{R_1 + R_2} V_{CC}$$

Substituting,

$$2.00 = \left(\frac{1.00 \times 10^4}{R_1 + 1.00 \times 10^4}\right)(20.0)$$

Cross multiplying,

$$(20 \times 10^3) + 2R_1 = 2.00 \times 10^5$$

$$2R_1 = 1.80 \times 10^5$$

$$R_1 = 0.900 \times 10^5\ \Omega = 90.0\ \text{k}\Omega$$

To find values for R_C and R_E, we recall that these resistors are obtained by applying another rule of thumb to the load resistor R_L shown in the circuit drawn in Fig. 5-72*b*.

First we obtain R_L using Ohm's law:

$$R_L = \frac{V_{CC} - V_{CE}}{I_C} = \frac{20.0 - 7.50}{2.5 \times 10^{-3}} = 5.00 \times 10^3\ \Omega = 5.00\ \text{k}\Omega$$

Now we apply the rule of thumb (*5-23a*) that says

$$R_C = 0.9R_L = (0.9)(5.00 \times 10^3) = 4.50\ \text{K}\Omega$$

and (*5-23b*)

$$R_E = 0.1R_L = (0.1)(5.00 \times 10^3) = 500\ \Omega$$

Some texts use different criteria for determining R_1 and R_2, and one commonly seen set is given below [see Eqs. (*5-22*) and (*5-24a*)–(*5-24c*)]:

$$R_B = \frac{V_{CC} - V_B}{I_B} \qquad (5\text{-}22a)$$

$$R_1 = R_B \frac{V_{CC}}{V_B} \qquad (5\text{-}24d)$$

$$R_2 = R_1 \frac{V_B}{V_{CC} - V_B} \qquad (5\text{-}24c)$$

For our problem, and with an I_B of 250 μA, substitution gives

$$R_B = \frac{20.0 - 2.00}{250 \times 10^{-6}} = 72.0 \times 10^3\ \Omega$$

$$R_1 = \frac{(72.0 \times 10^3)(20.0)}{2.00} = 7.20 \times 10^5 = 720 \times 10^3\ \Omega = 720\ \text{k}\Omega$$

and

$$R_2 = \frac{(7.20 \times 10^5)(2.00)}{20.0 - 2.00} = 8.00 \times 10^4 = 80.0 \times 10^3\ \Omega = 80.0\ \text{k}\Omega$$

Comparing results, the second criterion gives values eight times as large and is a more conservative method than the criterion we used.

5.13 Find the value of R_B for the silicon transistor amplifier shown in Fig. 5-73*a* if we wish class B operation. The transfer characteristic is shown in Fig. 5-73*b*.

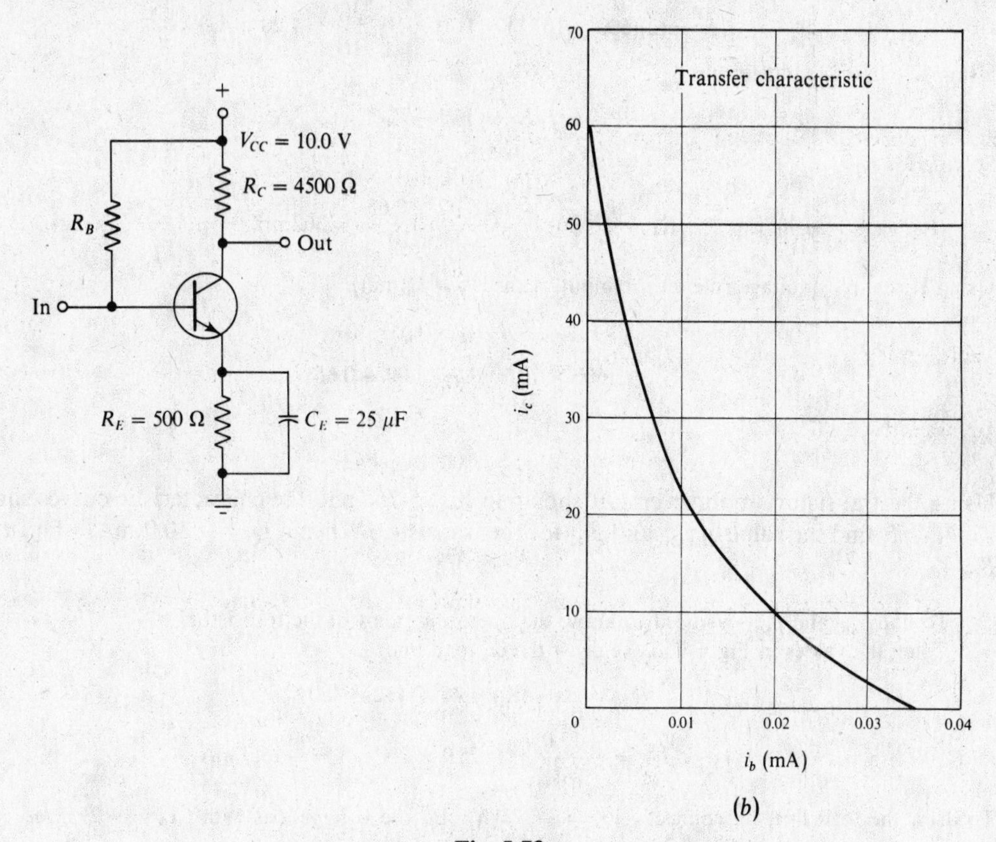

Fig. 5-73

Class B operation means we wish the operating point to be located at the cutoff point.

From Fig. 5-73*b* we see that the cutoff point, or the point at which the collector current equals zero, occurs when

$$I_B = 0.035\ \text{mA} = 35.0\ \mu\text{A}$$

From Fig. 5-73*a*, $V_{CC} = 10.0$ V and remembering that $V_{BE} = 0.7$ V for a silicon transistor, we have from Ohm's law

$$R_B = \frac{V_{CC} - V_{BE}}{I_B} = \frac{10.0 - 0.7}{35 \times 10^{-6}} = 266 \times 10^3\ \Omega = 266\ \text{k}\Omega$$

5.14 The switching transistor circuit shown in Fig. 5-74 is to be operated as an overdriven switch with a trigger voltage of 1.00 V. If $\beta = 50$ and $V_{BE} = 0.65$ V for this transistor, find R_1 and R_B.

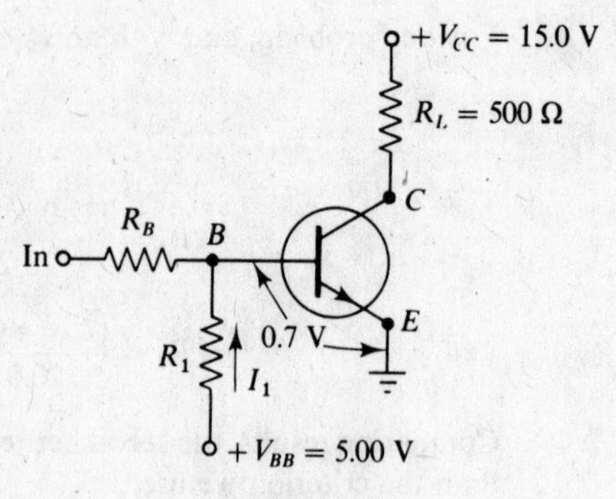

Fig. 5-74

From Fig. 5-74, $V_{CC} = 15.0$ V and $R_L = 500\ \Omega$, so by Ohm's law

$$I_C = \frac{V_{CC}}{R_L} = \frac{15.0}{500} = 30.0 \times 10^{-3}\ \text{A} = 30.0\ \text{mA}$$

Ignoring any leakage current I_{CBO},

$$I_B(\text{min}) = \frac{I_C}{\beta} = \frac{30.0 \times 10^{-3}}{50} = 6.00 \times 10^{-4}\ \text{A} = 600\ \mu\text{A}$$

The current through R_1 necessary to ensure cutoff for an overdriven switch is (by rule of thumb) equal to $2I_B(\text{min})$, so

$$I_1 = (2)(600 \times 10^{-6}) = 1.20 \times 10^{-3}\ \text{A} = 1.20\ \text{mA}$$

Using Kirchhoff's voltage rule,

$$I_1 R_1 = V_{BB} - V_{BE}$$

Solving for R_1,

$$R_1 = \frac{V_{BB} - V_{BE}}{I_1} = \frac{5.00 - 0.65}{1.20 \times 10^{-3}} = 3625\ \Omega$$

Using Kirchhoff's voltage rule on the input, with $I_B = I_B(\text{min})$,

$$I_B R_B = V_{\text{in}} - V_{BE}$$

$$(600 \times 10^{-6})R_B = 1.00 - 0.65$$

$$R_B = 583\ \Omega$$

5.15 Using the transistor amplifier circuit shown in Fig. 5-75*a* and the characteristic curves shown in Fig. 5-75*b*, find the value of α_{DC} and β_{DC} for the transistor. When $I_C = -10.0$ mA, find α_{DC} and β_{DC}.

To find α_{DC} and β_{DC}, we need to know the operating point of the transistor. Using the values in Fig. 5-75*a*, we draw the dc load line:

$$R_{DC} = R_E + R_C = 48 + 432 = 480\ \Omega$$

and

$$I_N = \frac{V_{CC}}{R_{DC}} = \frac{-12.0}{480} = -25.0 \times 10^{-3}\ \text{A} = -25.0\ \text{mA}$$

To draw the load line, we connect $V_{CC} = -12.0$ V on the voltage axis with $I_N = -25.0$ mA on the current axis, as shown in Fig. 5-75*c*.

The operating point is selected by finding the bias current I_B from

$$I_B = \frac{V_{CC} - V_{BE}}{R_B}$$

With (from Fig. 5-75*a*) $R_B = 226\ \text{k}\Omega$, $V_{CC} = -12.0$ V, $V_{BE} = -0.7$ V.

So

$$I_B = \frac{(-12.0) - (-0.7)}{226 \times 10^3} = -5.00 \times 10^{-5}\ \text{A} = -50.0\ \mu\text{A}$$

The intersection of the load line with the $I_B = -50.0\ \mu$A curve is Q and is shown in Fig. 5-75*c*. From Fig. 5-75*c*, we also note that when $I_B = -50.0\ \mu$A, $I_C = -15.0$ mA. Since by (5-7*a*)

$$I_E = I_B + I_C$$

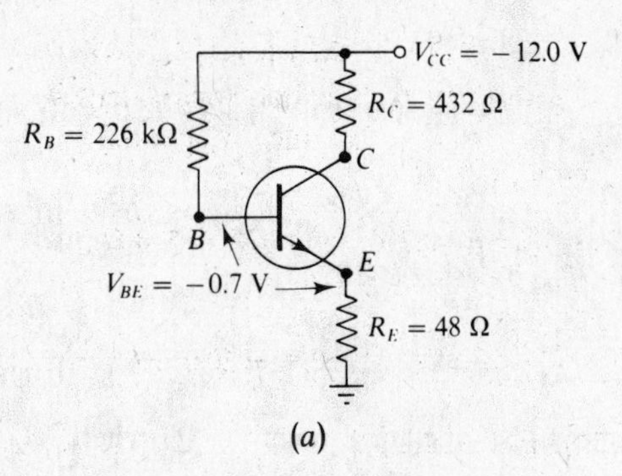
$V_{CC} = -12.0$ V
$R_C = 432\ \Omega$
$R_B = 226$ kΩ
C
B
E
$V_{BE} = -0.7$ V
$R_E = 48\ \Omega$

(a)

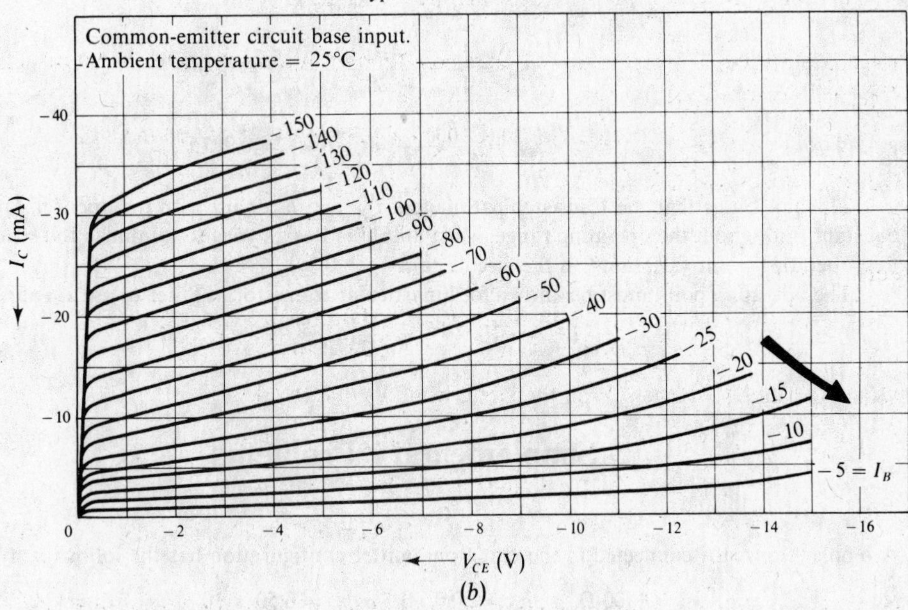
Typical collector characteristics
Common-emitter circuit base input.
Ambient temperature = 25°C
I_C (mA)
V_{CE} (V)
$-5 = I_B$

(b)

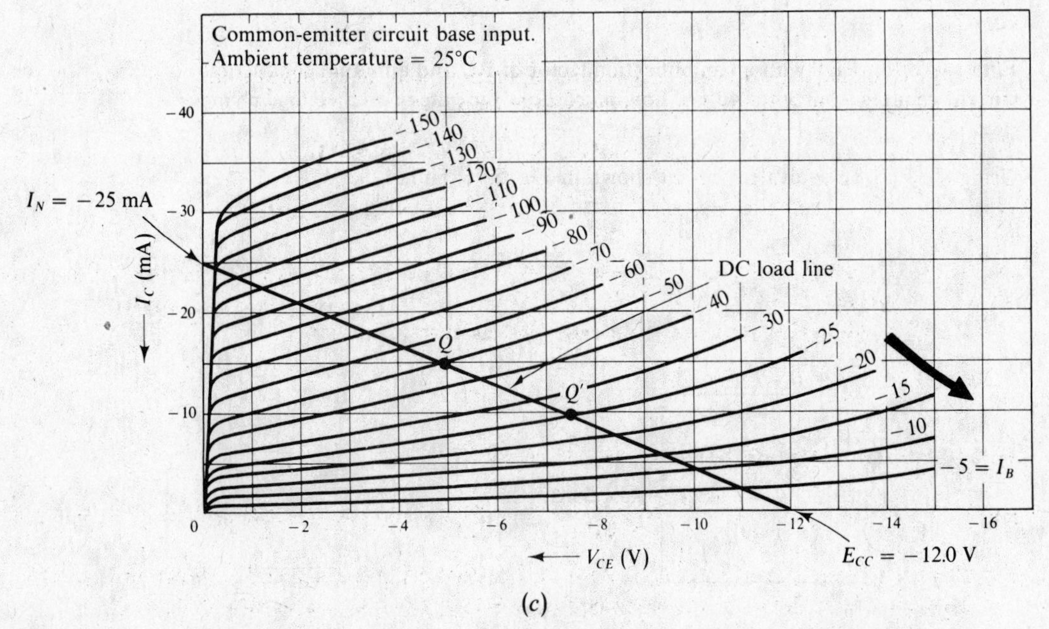
Typical collector characteristics
Common-emitter circuit base input.
Ambient temperature = 25°C
$I_N = -25$ mA
I_C (mA)
DC load line
Q
Q'
V_{CE} (V)
$E_{CC} = -12.0$ V
$-5 = I_B$

(c)

Fig. 5-75

substitution gives

$$I_E = (-50.0 \times 10^{-6}) + (-15.0 \times 10^{-3}) = -15.05 \times 10^{-3} \text{ A}$$

Now from (*5-8b*)

$$\alpha_{DC} = \frac{I_C}{I_E} = \frac{-15.0 \times 10^{-3}}{-15.05 \times 10^{-3}} = 0.9967$$

We may solve for β_{DC} by using (*5-10*):

$$\beta_{DC} = \frac{\alpha_{DC}}{1 - \alpha_{DC}} = \frac{0.9967}{1 - 0.9967} = 302$$

When we move the operating point to Q', where $I_C = -10$ mA as in the problem statement, Fig. 5-75*c* shows

$$I_B = -25.0\ \mu\text{A}$$

Then (*5-9b*)

$$\beta_{DC} = \frac{I_C}{I_B} = \frac{-10.0 \times 10^{-3}}{-25.0 \times 10^{-6}} = 400$$

and (*5-10a*)

$$\alpha_{DC} = \frac{\beta_{DC}}{\beta_{DC} + 1} = \frac{400}{400 + 1} = 0.9975$$

This illustrates that the transistor parameters α_{DC} and β_{DC} vary with the operating point and are not constant throughout the operating range. Most amplifier design tries to minimize the effect of temperature and operating point variations on the circuit gain.

The operating point must be known for a particular transistor in order to use α_{DC} and β_{DC} as constants.

Supplementary Problems

5.16 A bipolar transistor connected in the common-emitter configuration has the following hybrid parameters:

$$h_{ie} = 1500\ \Omega \qquad h_{oe} = 30.0\ \mu\text{S} \qquad h_{re} = 3.50 \times 10^{-4} \qquad h_{fe} = 49.0$$

Find r_m, r_c, r_e, r_b, and α.

5.17 Find r_{DS} for an FET with an amplification factor of 100 and a mutual conductance of 5000 μS. If the drain current changes from 30 to 40 μA, how much does the gate-source voltage change?

5.18 Using the FET ac equivalent circuit shown in Fig. 5-76, find A_v.

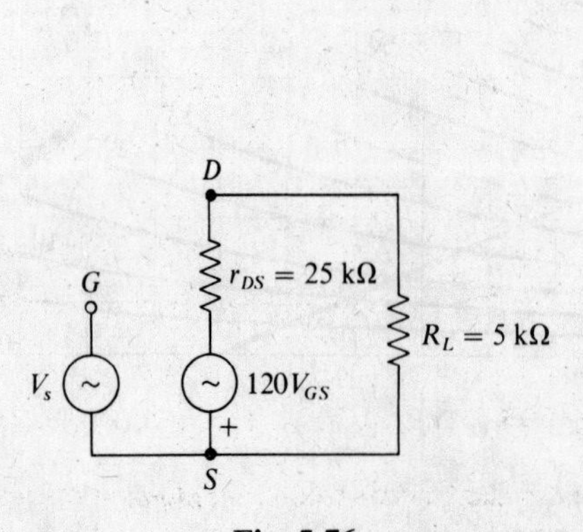

Fig. 5-76

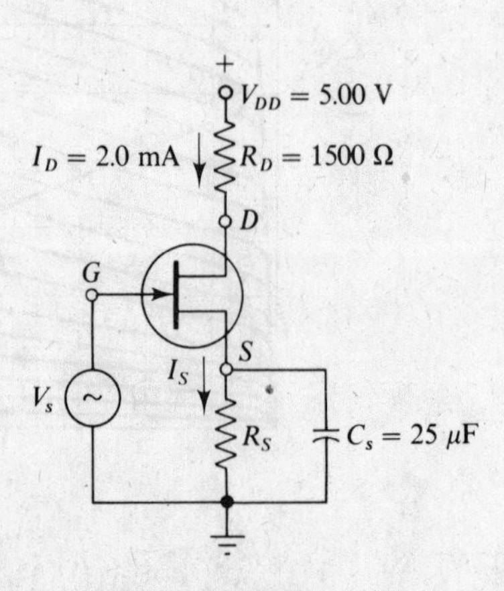

Fig. 5-77

5.19 A bipolar transistor with the transfer characteristic given in Fig. 5-73*b* is to be operated as a class C amplifier. Find the quiescent bias current for this type of operation.

5.20 Find the value of the source bias resistor R_S in the circuit shown in Fig. 5-77 if the FET is to be operated at $V_{GS} = -1.50$ V.

5.21 Output measurements for an amplifier are shown in Fig. 5-78. If the input is a perfect square wave as shown, find the rise time and the frequency of the input wave.

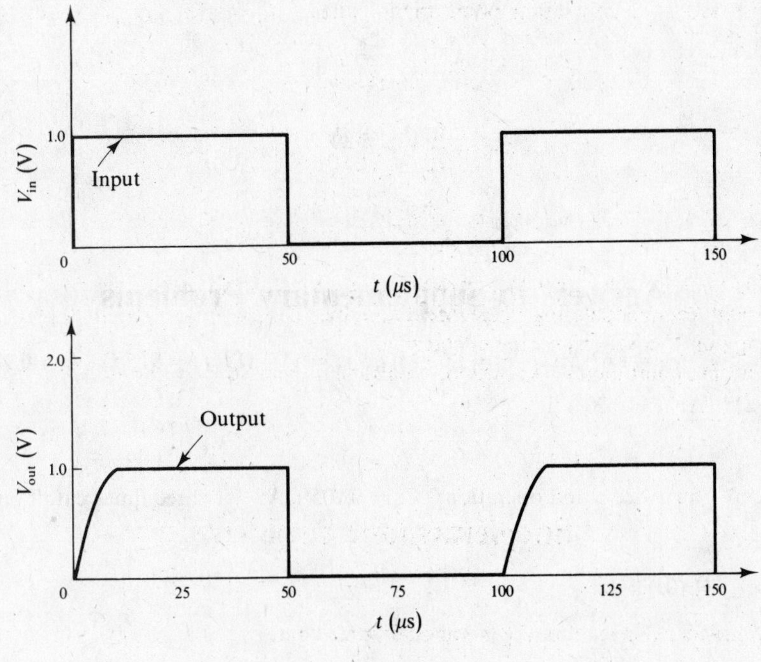

Fig. 5-78

5.22 Find A_v for a common-cathode tube circuit with $r_p = 3.00$ kΩ, $R_L = 5.00$ kΩ, and $\mu = 80$.

5.23 For the FET source follower circuit shown in Fig. 5-79*a* and its ac equivalent circuit drawn in Fig. 5-79*b*, find the current gain A_i.

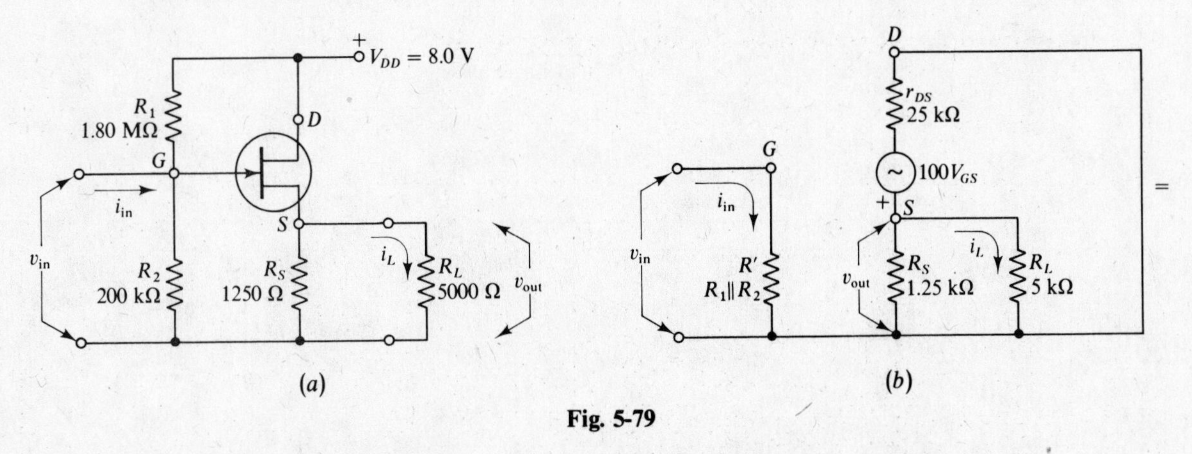

Fig. 5-79

5.24 Find the h parameters and α for a transistor in the common-base configuration if $r_b = 1.00$ kΩ, $r_e = 10.0$ Ω, and $r_c = 2.00$ MΩ.

5.25 What is the minimum switching voltage necessary to drive the transistor shown in Fig. 5-80 into saturation? Assume the β for this transistor to be 100. What switching voltage would you specify to ensure fully driven operation?

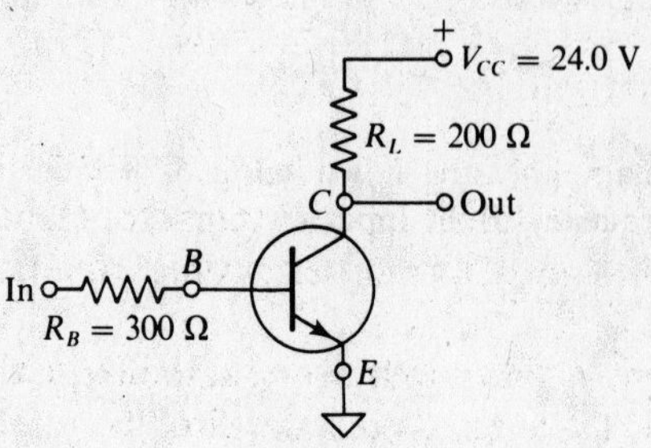

Fig. 5-80

Answers to Supplementary Problems

5.16 Using Table 5-4, $r_m = 1.63$ MΩ, $r_c = 1.67$ MΩ, $r_e = 11.7\ \Omega$, $r_b = 915\ \Omega$, $\alpha = 0.98$.

5.17 $r_{DS} = 20.0\ k\Omega$; $\Delta V_{GS} = 0.20$ V.

5.18 $A_v = 20.0$.

5.19 $I_B = 0.070$ mA for twice cutoff operation; $I_B = 0.105$ mA for three times cutoff operation.

5.20 $R_S = 750\ \Omega$.

5.21 $t_r = 7.50\ \mu s$; $f = 10.0$ kHz.

5.22 $A_v = 50.0$.

5.23 $A_i = 28.6$.

5.24 $h_{ib} = 30.0\ \Omega$, $h_{rb} = 5 \times 10^{-4}$, $h_{fb} = -0.98$, $h_{ob} = 0.5\ \mu S$, $\alpha = -h_{fb} = 0.98$.

5.25 0.36 V; about 1 V.

Chapter 6

Coupling, Oscillation, and Filtering

6.1 INTRODUCTION

In Chap. 5, we explored the use of tubes, bipolar transistors, and FETs as amplifying devices.

Because the signals that are developed by most electronic circuits are low in value, we normally need to use more than one amplifier.

The methods by which we transfer the output of one amplifier to the input of the next are classified in terms of the components used to *couple* or *cascade* the circuits.

Tube amplifiers generally use *RC* (resistor-capacitor) *coupling* and transformer coupling. For specific purposes that will be explained in this chapter, *LC* (inductor-capacitor) *coupling* and *dc* (or *direct coupling*) are also used.

When transistor amplifiers were developed, these same coupling methods were employed. Since the tube is essentially a voltage device and the transistor a current device, the coupling methods have slightly different design criteria for transistors and tubes.

Of special interest is *tuned coupling*, or *tank-circuit coupling*, which is used for tuned amplifiers and employs *LC* resonant circuits to transfer the energy from one part of the circuit to another.

When the coupling is from the output of an amplifier stage to the input of the same stage, as shown in Fig. 6-1, we call the connection a *feedback loop*.

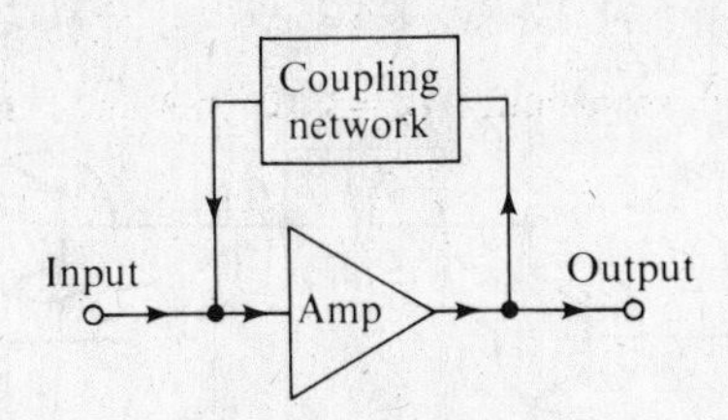

Fig. 6-1 Feedback

Under proper conditions, the feedback from the output to the input can cause the circuit to *ring* or *oscillate*, and the frequency of oscillation can then be adjusted by the circuit parameters.

Another way to achieve oscillation is to use a switch or *multivibrator*. This type of oscillator may have a more complex waveform than the tank or feedback circuit since pulses, rather than sine waves, result from the switching action.

Frequency considerations become important in design of electronic circuitry, and we will summarize the ways of controlling the frequency and waveshape of a signal by the use of *filters*.

6.2 *RC* COUPLING

One of the more widely used ways of coupling the signal output voltage from the load of one amplifier to the input of the next is the *RC* network shown in Fig. 6-2*a*.

The coupling capacitor C_1 couples the ac signal to the second stage and blocks the dc from affecting the bias of the second amplifier.

The *grid leak resistor* R_g provides a leakage path for the small number of electrons that strike the grid and also serves as a means of providing bias for the second grid.

The value of R_g should be high enough so that it does not materially lower the first tube's load resistance and low enough to provide a leakage path for the electrons that strike the grid.

Usual values of R_g for tubes range from 250 kΩ to 1 MΩ. A tube *RC* coupled amplifier is shown in Fig. 6-2*b*.

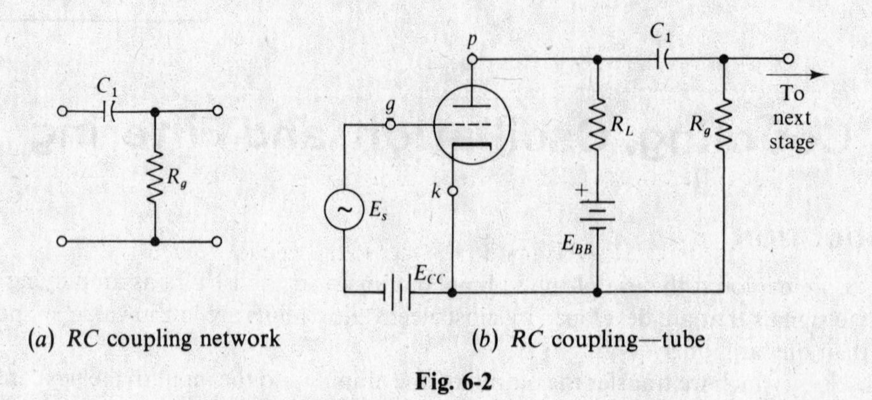

(*a*) *RC* coupling network (*b*) *RC* coupling—tube

Fig. 6-2

For *RC* coupled transistor circuits, the value of R_g (or its counterpart) is usually selected in terms of the bias conditions mentioned in Chap. 5.

As the circuit for the common emitter amplifier in Fig. 6-3 shows, R_2 and R_6 have been selected as 10 kΩ, in keeping with the bias considerations.

The coupling network in Fig. 6-3 consists of C_3 and R_6 and is enclosed with dashed lines.

The effectiveness of all coupled circuits is usually determined by the amplifier stage gain, which is frequency-dependent.

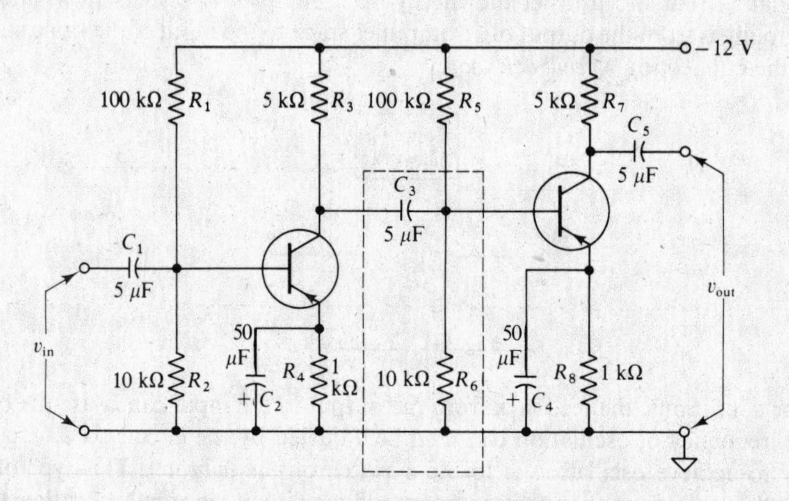

Fig. 6-3 ***RC* coupled two-stage transistor amplifier**

At this point it is important to return to the tube and transistor ac equivalent circuits and redraw them including the capacitance values.

Figure 6-4*a* shows a two-stage *RC* coupled tube amplifier. Figure 6-4*b* is the all-frequency ac equivalent circuit for the first stage.

At low frequencies, C_{pk} and C_{in} in Fig. 6-4*b* have a large enough impedance to be ignored, and the equivalent circuit becomes dependent on C_1 for gain, as is shown in Fig. 6-4*c*.

At midrange frequencies, the impedances of C_{pk} and C_{in} are still large enough to ignore, but the impedance of C_1 has dropped to a point where it can be considered to be a short. The equivalent circuit then becomes the one discussed in Chap. 5. It is drawn again as Fig. 6-4*d*.

At high frequencies, C_1 is still a short, but C_{pk} and C_{in} are added together and called C_g. The equivalent circuit becomes that shown in Fig. 6-4*e* with the gain dependent on C_g.

The gain **A** is a complex number, and our use of the term gain as *A* describes the magnitude of **A**, while the angle of **A** is denoted as phase, ϕ.

A typical gain-frequency curve is shown in Fig. 6-5. For the midband the gain is 1.0 and the phase angle is 180°.

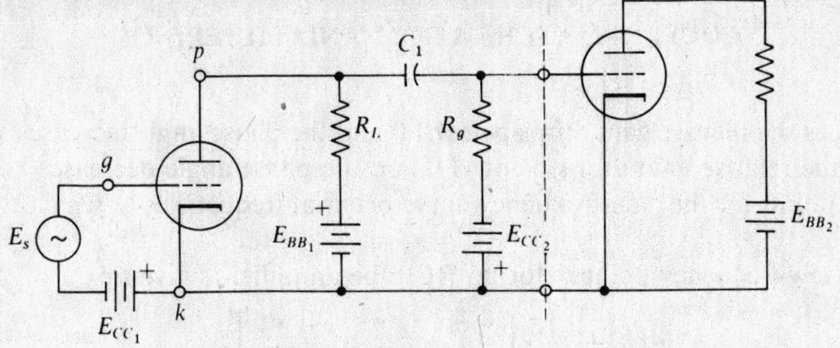

(*a*) *RC* coupled two-stage tube amplifier

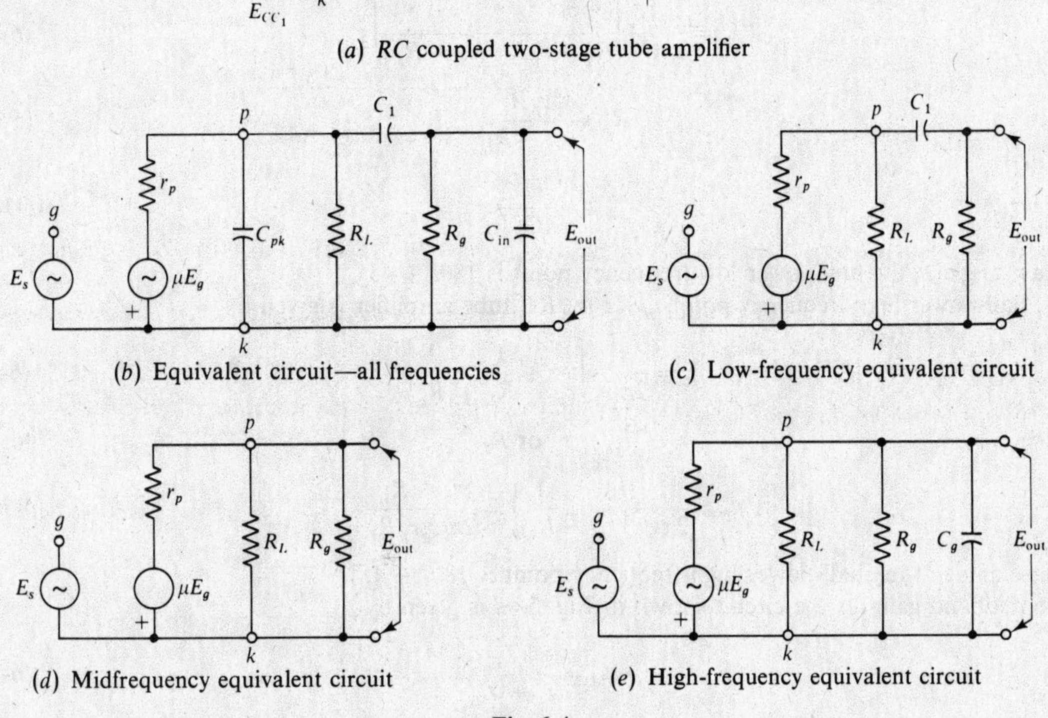

(*b*) Equivalent circuit—all frequencies

(*c*) Low-frequency equivalent circuit

(*d*) Midfrequency equivalent circuit

(*e*) High-frequency equivalent circuit

Fig. 6-4

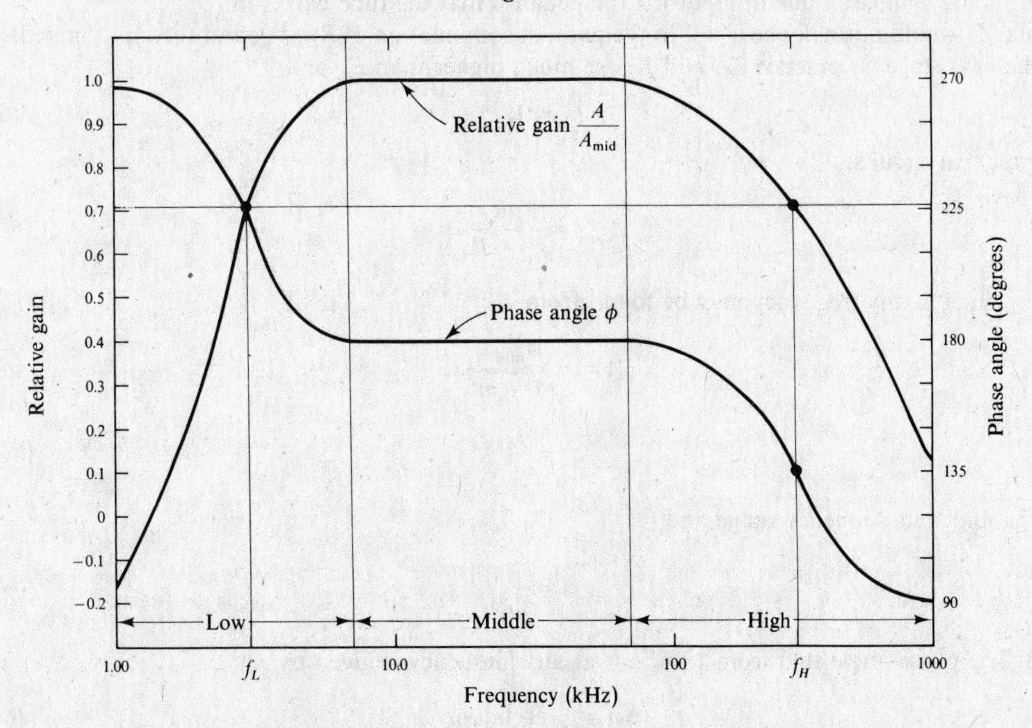

Fig. 6-5 Gain and phase angle vs. frequency

At low frequencies the relative gain drops below 1.0 and the phase angle increases above 180°, while at high frequencies the relative gain drops below 1.0 and the phase angle decreases below 180°.

The half-power points for the gain-frequency curve occur at frequencies at which the relative gain is 0.707.

The half-power low-frequency point f_L for an RC tube amplifier is given by

$$f_L = \frac{1}{2\pi C_1}\left[\frac{1}{R_g + r_p R_L/(r_p + R_L)}\right] \tag{6-1}$$

but since

$$R_g \gg \frac{r_p R_L}{r_p + R_L} \tag{6-2}$$

we can write

$$f_L \cong \frac{1}{2\pi C_1 R_g} \tag{6-1a}$$

The phase angle at the half-power low-frequency point is 180° + 45°.

The half-power high-frequency point f_H for an RC tube amplifier is given by

$$f_H = \frac{1}{2\pi C_g}\left[\frac{1}{r_p} + \frac{1}{R_L} + \frac{1}{R_g}\right] \tag{6-3}$$

and since

$$R_g \gg r_p \text{ or } R_L \tag{6-4}$$

$$f_H \cong \frac{1}{2\pi C_g}\left(\frac{1}{r_p} + \frac{1}{R_L}\right) = \frac{r_p + R_L}{2\pi C_g r_p R_L} \tag{6-3a}$$

The phase angle at the half-power high-frequency point is 180° − 45°.

The midband gain for the circuit shown in Fig. 6-4*d* is given by

$$A_{\text{mid}} = \frac{\mu \mathbf{Z}_L}{r_p + \mathbf{Z}_L} \tag{6-5}$$

where $\mathbf{Z}_L$ is the complex value of input load impedance that the tube works into.

While $\mathbf{Z}_L$ would normally be equal to the parallel combination of R_L, R_g, and the input impedance $\mathbf{Z}_{\text{in}}$ of the next stage, in practice $\mathbf{Z}_{\text{in}}$ and R_g are much higher than R_L and

$$\mathbf{Z}_L = R_L \tag{6-6}$$

So for practical circuits,

$$A_{\text{mid}} = \frac{\mu R_L}{r_p + R_L} \tag{6-5a}$$

The gain A at any frequency may be found from

$$A = \frac{A_{\text{mid}}}{\sqrt{1 + m^2}} \tag{6-7}$$

where

$$m = \frac{f_L}{f} \tag{6-8a}$$

below the midband frequency range and

$$m = \frac{f}{f_H} \tag{6-8b}$$

above it. The phase angle shift from 180°, $\Delta\phi$, at any frequency is given by

$$\Delta\phi = \pm \arctan m \tag{6-9}$$

where the negative sign is only used above the midband frequency range and m is defined as before.

The midband frequency range is approximately determined by

$$10f_L < f_m < \frac{f_H}{10} \tag{6-10}$$

In the tube amplifier, selection of tubes with low values of C_{pk} and C_{gk} extends the high-frequency range. For the common-emitter amplifier, selection of transistors with low C_c and low R_{ie} will extend the high-frequency range. To extend the low-frequency range in both cases, increase the value of C_1.

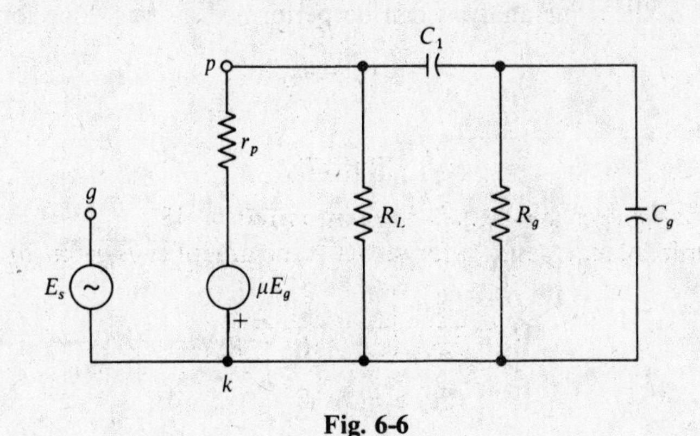

Fig. 6-6

Example 6.1 A tube amplifier equivalent circuit shown in Fig. 6-6 has the following values:

$$r_p = 2\ \text{k}\Omega \qquad \mu = 100$$

$$R_g = 255\ \text{k}\Omega \qquad C_1 = 0.025\ \mu\text{F}$$

$$R_L = 500\ \Omega \qquad C_g = 0.133\ \mu\text{F}$$

Find A_{mid}, f_m, A at 100 Hz, and A at 2 kHz.

Using the given circuit parameters, $R_g \gg R_L$ and the midband gain A_{mid} is calculated directly from (*6-5a*):

$$A_{\text{mid}} = \frac{\mu R_L}{r_p + R_L} = \frac{(100)(500)}{2000 + 500} = 20$$

The midband frequency range f_M can be found from Eq. (*6-10*) once f_L and f_H are known. Using (*6-1a*),

$$f_L = \frac{1}{2\pi C_1 R_g} = \frac{1}{2\pi(0.025 \times 10^{-6})(255 \times 10^3)} = 25.0\ \text{Hz}$$

Using (*6-3a*),

$$f_H = \frac{r_p + R_L}{2\pi C_g r_p R_L} = \frac{2000 + 500}{2\pi(0.133 \times 10^{-6})(2000)(500)} = \frac{2.5 \times 10^3}{8.36 \times 10^{-1}} = 2.99 \times 10^3\ \text{Hz} = 2.99\ \text{kHz}$$

Midband frequency range f_m is then given by (*6-10*):

$$10f_L < f_m < \frac{f_H}{10}$$

So

$$250 < f_m < 299\ \text{Hz}$$

For 100 Hz, using Eq. (*6-8a*),

$$m = \frac{f_L}{f} = \frac{25}{100} = 0.25$$

and from (6-7)

$$A_{100} = \frac{A_{\text{mid}}}{\sqrt{1 + m^2}} = \frac{20}{\sqrt{1 + (0.25)^2}} = 19.4$$

For 2 kHz, using Eq. (*6-8b*),

$$m = \frac{f}{f_H} = \frac{2.00 \times 10^3}{2.99 \times 10^3} = 0.669$$

and from (*6-7*)

$$A_{2000} = \frac{A_{\text{mid}}}{\sqrt{1 + m^2}} = \frac{20}{\sqrt{1 + (0.669)^2}} = 16.6$$

For the transistor common-emitter *RC* coupled amplifier and its all-frequency equivalent circuit shown in Fig. 6-7*a* and *b*, the same analysis can be performed as was done for the *RC* coupled tube amplifier.

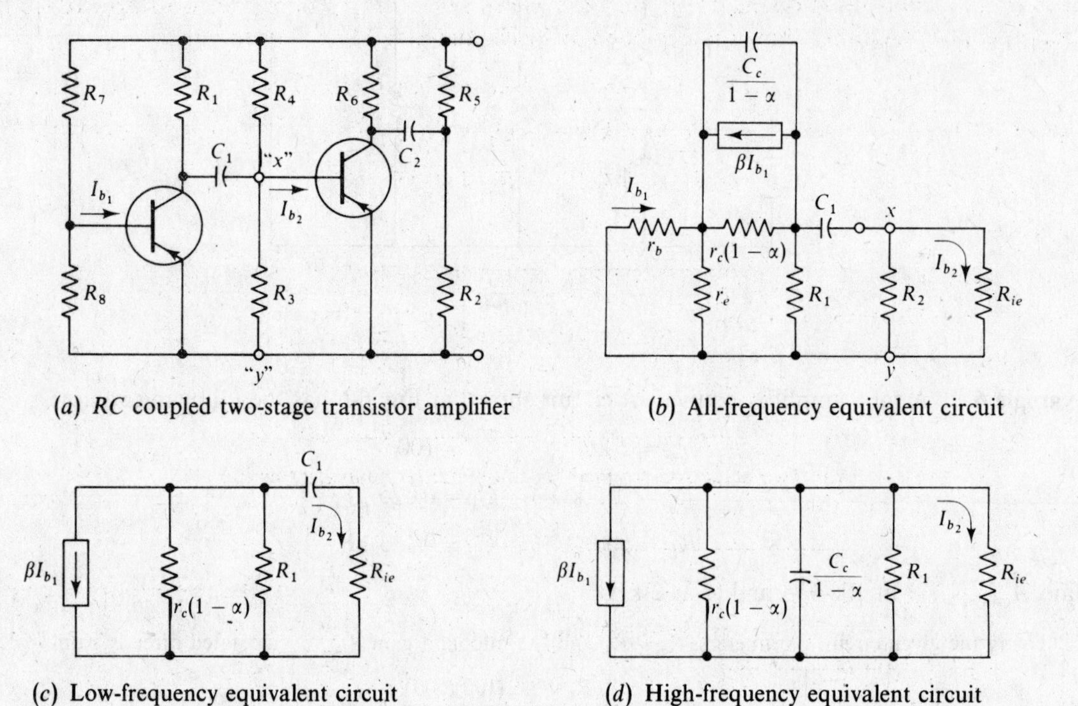

(*a*) *RC* coupled two-stage transistor amplifier

(*b*) All-frequency equivalent circuit

(*c*) Low-frequency equivalent circuit

(*d*) High-frequency equivalent circuit

Fig. 6-7

For low frequencies, r_b and r_e are small and may be ignored. $R_2 \gg R_{ie}$ as a rule, and since they are in parallel, R_2 can be removed from the circuit. Since β is normally greater than 10, I_{b_1} can be neglected with respect to βI_{b_1}, and the circuit becomes the one shown in Fig. 6-7*c* with the gain dependent on C_1.

For midrange frequencies, in addition to the low-frequency assumptions, we can replace C_1 with a short since its impedance becomes low with respect to the other values and the circuit reduces to the ac equivalent circuit discussed in Chap. 5.

For high frequencies, C_1 may be replaced by a short but the reflected collector capacitance $C_c/1 - \alpha$ shunts the reflected collector resistance $r_c(1 - \alpha)$, and the circuit becomes that shown in Fig. 6-7*d* with the gain dependent on α and C_c.

C_1 and R_{ie} in series control the low-frequency limit, while C_c and R_{ie} in shunt control the high-frequency limit.

The half-power low-frequency point for an *RC* coupled common-emitter amplifier is given by Eq. (*6-1*) with R_1 replacing R_g, R_{ie} replacing $r_p R_L/(r_p + R_L)$, and assuming that $r_c(1 - \alpha) \gg R_1$.

Equation (*6-1*) can then be written

$$f_L = \frac{1}{2\pi C_1(R_{ie} + R_1)} \tag{6-1b}$$

The half-power high-frequency point for an RC coupled common-emitter amplifier can be obtained from Eq. (6-3a) with R_1 replacing R_L, R_{ie} replacing r_p, $(\beta + 1)C_c$ replacing C_g, and assuming that $r_c(1 - \alpha) \gg R_1$. Equation (6-3a) can then be written

$$f_H = \frac{R_{ie} + R_1}{2\pi(\beta + 1)C_c R_{ie} R_1} \tag{6-3b}$$

The gain and phase shift formulas given for the tube case can then be used with $A_{mid} \cong \beta$. For the common-emitter amplifier, we are usually interested in the current gain β. If voltage gain is desired, the appropriate formula from Chap. 5 should be used for A_{mid}.

Example 6.2 Find the midband range and the phase shift from 180° for the two-stage RC coupled common-emitter amplifier shown in Fig. 6-8a at 1 kHz and at 1 MHz. Given:

$$\beta = 49 \qquad C_c = 7.78 \text{ pF}$$
$$R_1 = 5 \text{ k}\Omega \qquad C_1 = 0.032\ \mu\text{F}$$
$$R_{ie} = 500\ \Omega \qquad R_2 = 10 \text{ k}\Omega$$

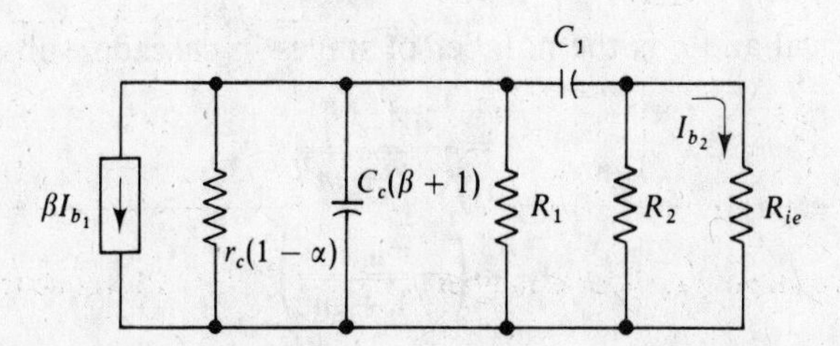

(a) Two-stage RC coupled common emitter equivalent circuit

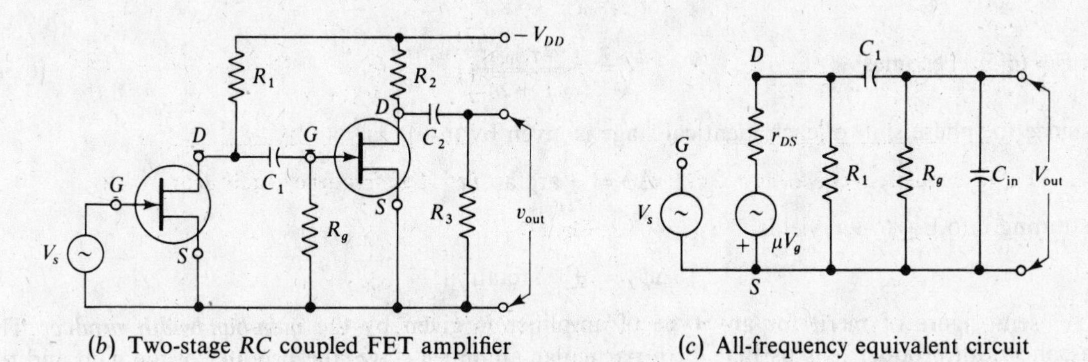

(b) Two-stage RC coupled FET amplifier

(c) All-frequency equivalent circuit

Fig. 6-8

Using the values given in the problem statement, we calculate f_L and f_H. Using (6-1b),

$$f_L = \frac{1}{2\pi C_1(R_{ie} + R_1)} = \frac{1}{2\pi(0.032 \times 10^{-6})(500 + 5000)} = 904 \text{ Hz}$$

Using Eq. (6-3b),

$$f_H = \frac{R_{ie} + R_1}{2\pi(\beta + 1)C_c R_{ie} R_1} = \frac{(500 + 5000)}{2\pi(49 + 1)(7.78 \times 10^{-12})(500)(5000)} = 9.00 \times 10^4 \text{ Hz} = 900 \text{ kHz}$$

The range of f_m is given by (6-10):

$$10 f_L < f_m < \frac{f_H}{10}$$

or

$$9.04 \text{ kHz} < f_m < 90.0 \text{ kHz}$$

For 1 kHz, using (*6-8a*),

$$m = \frac{f_L}{f} = \frac{904}{1000} = 0.904$$

and using (*6-9*) $\Delta\phi = \arctan m = \arctan(0.904) = 42.1°$

For 1 MHz, using (*6-8b*), $m = \frac{f}{f_H} = \frac{1 \times 10^6}{900 \times 10^3} = 1.11$

and using (*6-9*) $\Delta\phi = -\arctan m = -\arctan(1.11) = -48.0°$

For FET amplifiers, the tube formulas may be used with r_{DS} replacing r_p.

An FET *RC* coupled amplifier and its all-frequency equivalent circuit are shown in Fig. 6-8*b* and *c*.

If nonidentical *RC* amplifiers are cascaded, the overall gain A_T is the product of the individual gains:

$$A_T = A_1 A_2 A_3 \cdots A_n \tag{6-5b}$$

The overall phase shift $\Delta\phi_T$ is the algebraic sum of the individual phase shifts:

$$\Delta\phi_T = \Delta\phi_1 + \Delta\phi_2 + \Delta\phi_3 + \cdots + \Delta\phi_n \tag{6-9a}$$

If the stages are identical and n is the number of stages in cascade, substituting (*6-7*)

$$A = \frac{A_{\text{mid}}}{\sqrt{1+m^2}}$$

into Eq. (*6-5b*) yields

$$A_T = \left(\frac{A_{\text{mid}}}{\sqrt{1+m^2}}\right)^n \tag{6-5c}$$

If we define the overall midband gain of the cascaded amplifier as

$$A_{T,\,\text{mid}} = (A_{\text{mid}})^n$$

then Eq. (*6-5c*) becomes

$$A_T = \frac{A_{T,\,\text{mid}}}{(\sqrt{1+m^2})^n} \tag{6-5d}$$

and since the phase shift of each identical stage is given by (*6-9*)

$$\Delta\phi = \pm\arctan m$$

substituting into Eq. (*6-9a*) yields

$$\Delta\phi_T = \pm n \arctan m \tag{6-9b}$$

A useful figure of merit for any type of amplifier is given by the *gain-bandwidth product*. The gain-bandwidth product tells us that for a particular amplifier stage, the product of the gain and the bandwidth is constant at all frequencies.

In general this means that if we wish to improve the bandwidth of an amplifier we must sacrifice gain and vice versa.

In the case of cascaded amplifiers, for the overall gain-bandwidth product to remain the same as that of an individual stage, the overall bandwidth would decrease since the overall gain is larger than the gain per stage.

However, if the gain-bandwidth product of each stage were kept constant, increasing the number of stages would actually increase the bandwidth up to a maximum, after which it would decrease.

For the case of the cascaded *RC* amplifier, the gain per stage is optimized at $A_v = 1.6487$ or 4.343 dB.

6.3 DIRECT COUPLING

For tube amplifiers, the low-frequency limit is controlled by the size of the coupling capacitor. If the coupling capacitor is eliminated, we have a *direct-coupled* or *dc* amplifier.

An example of such a direct-coupled tube amplifier is shown in Fig. 6-9. Note the low-resistance, high-current voltage divider R_1 to R_4 necessary to supply proper bias voltages for the tubes. Since the tube grids cannot distinguish between a change in signal level or a change in bias supply level, high stability is required in the bias supply. This is one of the major disadvantages of direct coupling, both with tubes and transistors, although in the case of transistorized circuits a well-regulated power supply is relatively inexpensive and direct coupling is common.

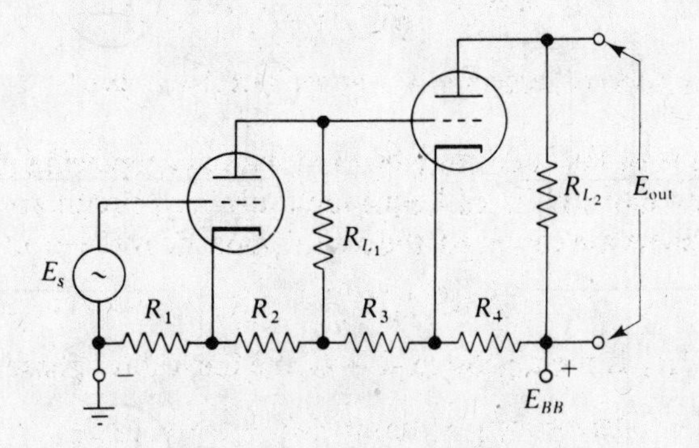

Fig. 6-9 ***DC* tube amplifier**

For transistor *dc* amplifiers, the number of components is reduced for bias stabilization and there is less power-supply current drawn than with *RC* coupled amplifiers with the same bias stability. An example of a *dc* coupled bipolar transistor amplifier is shown in Fig. 6-10*a*.

Direct coupling is extensively used in integrated circuits, since large-value coupling capacitors are difficult to produce in a monolithic format.

One method of isolating different voltage levels is to use a zener diode, as shown in Fig. 6-10*b*.

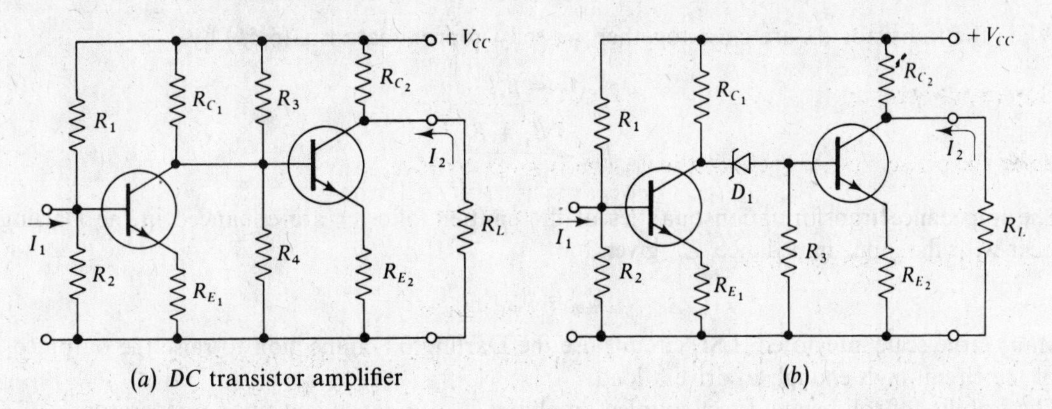

(*a*) *DC* transistor amplifier (*b*)

Fig. 6-10

Direct coupling may also be used to couple an FET to an FET, as shown in Fig. 6-11*a*, or to a bipolar transistor, as shown in Fig. 6-11*b*.

Another commonly seen direct-coupled configuration is the Darlington pair shown in Fig. 6-11*c*. A more practical Darlington circuit is shown in Fig. 6-11*d*.

The Darlington connection is essentially two identical emitter follower stages cascaded directly. It is sometimes referred to as a "bootstrap connection" since the output of the first stage directly feeds the input to the second stage.

The total gain of the Darlington circuit is the product of the individual A_i's or

$$A_T = (A_{i1})(A_{i2}) \tag{6-5e}$$

For the common-emitter follower $A_i \cong \beta$, so

$$A_T = \beta_1 \beta_2 \tag{6-5f}$$

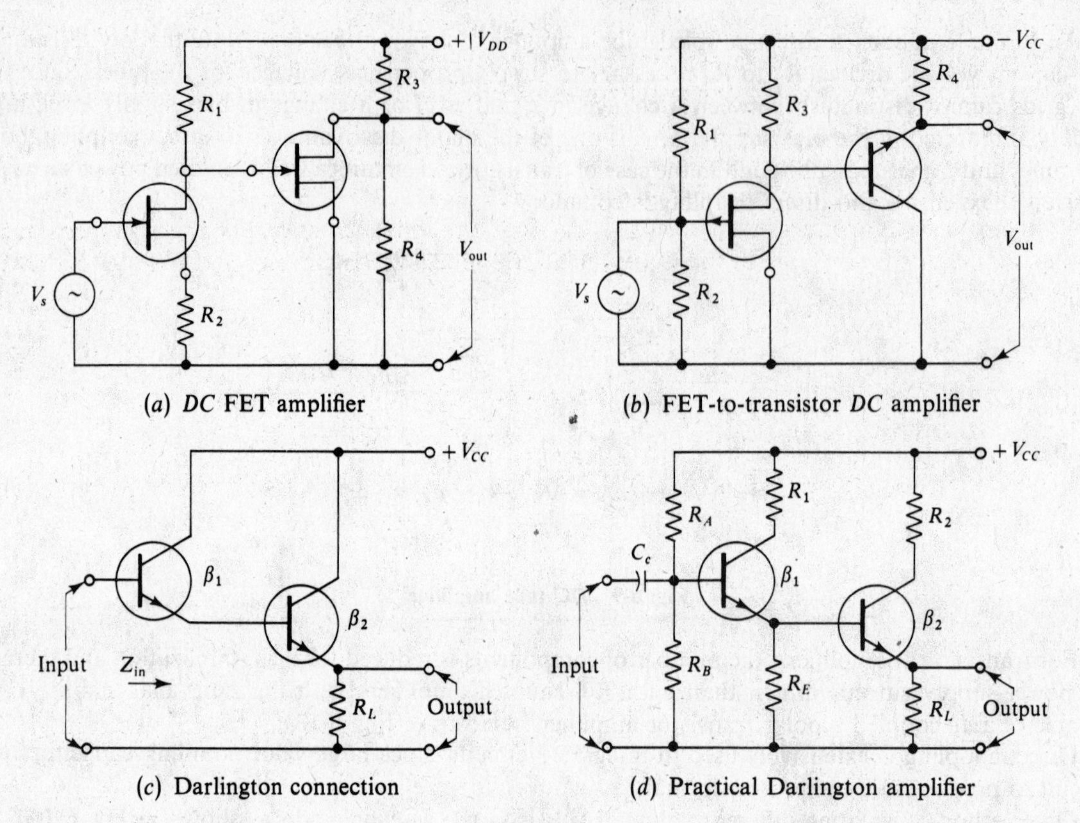

Fig. 6-11

When the transistor β's are close together, we may approximate Eq. (6-5f) by

$$A_T = \beta_{av}^2 \tag{6-5g}$$

with

$$\beta_{av} = \frac{\beta_1 + \beta_2}{2}$$

The impedance transformation qualities of the emitter follower are enhanced in the Darlington amplifier with the input impedance, Z_{in} given by

$$Z_{in} = \beta_{av}^2 R_L$$

Many large-scale integrated (LSI) circuits use the Darlington connection to raise the output of the rest of the circuit high enough to drive a load.

The stability of solid state direct-coupled amplifiers is also dependent upon temperature.

Balanced amplifiers with two active elements, as shown in Fig. 6-12a and b, can be used to cancel out the thermal variations as long as the elements used are nearly identical in performance and are mounted

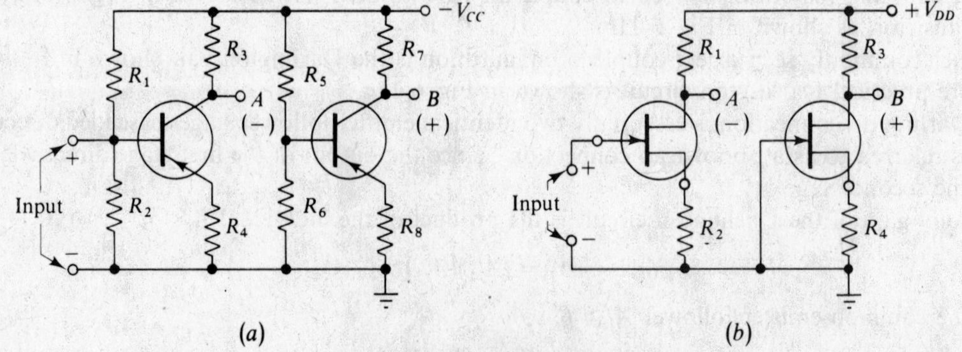

Fig. 6-12 Balanced amplifiers

closely together. The succeeding stages are also balanced and the input points connected to points A and B of the preceding stage. Temperature-compensated amplifiers using elements that have a temperature coefficient opposite to that of the primary transistor or FET can also eliminate temperature variations.

6.4 TRANSFORMER COUPLING

Transformer coupling is shown in Fig. 6-13*a*, *b*, and *c*.

A transformer can perform many tasks in addition to coupling the output of one amplifier stage to the input of the next—it can be used for isolation, impedance matching, and voltage or current multiplication.

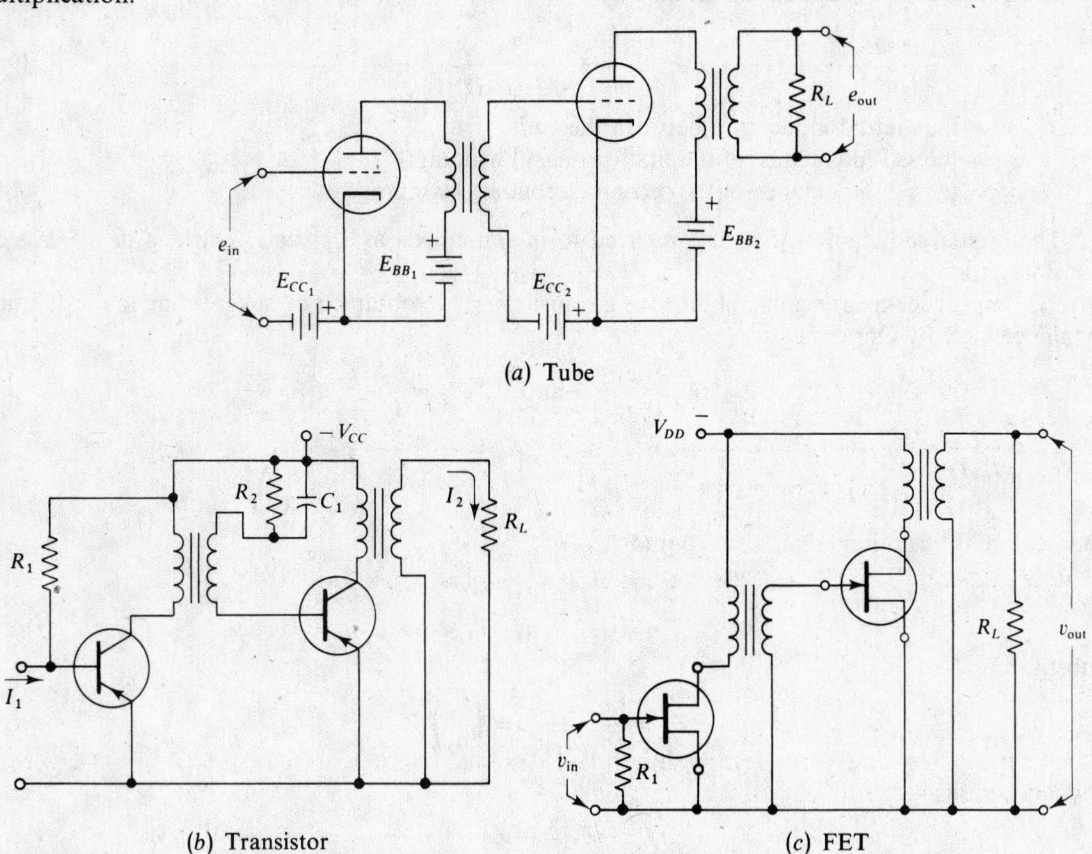

Fig. 6-13 Transformer coupling

The principal disadvantage of transformer coupling is the bulk and expense of the transformer itself along with the fact that the frequency response of the transformer is poorer than with RC coupling.

An ideal transformer and its schematic representation are shown in Fig. 6-14*a* and *b*. The dots indicate polarity and mean that, if the current increases constantly into one dotted terminal, the other dotted terminal is positive.

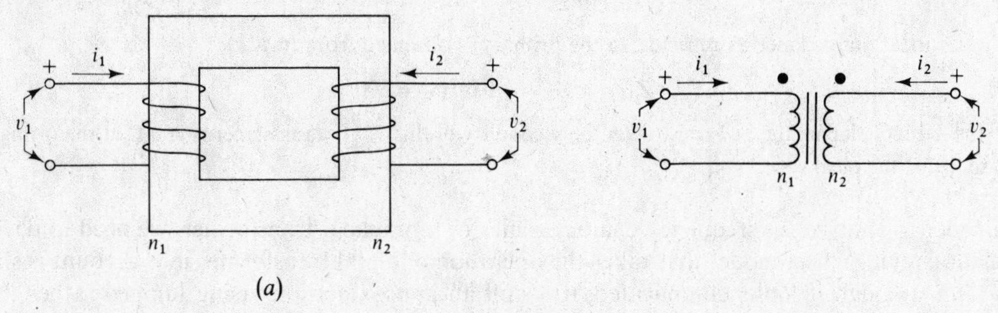

Fig. 6-14

Since the transformer operates by the magnetic flux linking primary and secondary, we have the following relationships:

$$\frac{n_1}{n_2} = \frac{v_1}{v_2} = -\frac{i_2}{i_1} \tag{6-11}$$

where the negative sign indicates that the currents are entering the dotted ends.

The turns ratio n_1/n_2 is called a and an $n_1 : n_2$ transformer is sometimes designated as an $a : 1$ transformer.

The coefficient of coupling k is defined by

$$k = \frac{M}{\sqrt{L_{11}L_{12}}} \tag{6-12}$$

where M = the mutual inductance between the coils
L_{11} = the self-inductance of the first (primary) winding
L_{22} = the self-inductance of the second (secondary) winding

The mutual inductance M is also referred to in some texts as L_{12} or L_{21}, which in the case of a transformer are identical.

The impedance-transforming abilities of a transformer are obtained by substituting in the definition for impedance. By Ohm's law,

$$Z_1 = \frac{v_1}{i_1} \quad \text{and} \quad Z_2 = \frac{v_2}{i_2}$$

But from (*6-11*)

$$v_1 = \frac{n_1}{n_2} v_2$$

and neglecting the minus sign, also from (*6-11*),

$$i_1 = \frac{n_2}{n_1} i_2$$

Substituting,

$$Z_1 = \frac{(n_1/n_2)v_2}{(n_2/n_1)i_2} = \left(\frac{n_1}{n_2}\right)^2 Z_2$$

and since $n_1/n_2 = a$,

$$Z_1 = a^2 Z_2 \tag{6-11a}$$

Example 6.3 A transformer secondary is connected to an 8-Ω resistor. Find the impedance as reflected to the primary side if 120 V on the primary is stepped down to 12 V at the secondary.

The turns ratio may be found first from the given voltage ratios. From (*6-11*),

$$a = \frac{n_1}{n_2} = \frac{v_1}{v_2} = \frac{120}{12} = 10.0$$

The secondary impedance as reflected to the primary is obtained from (*6-11a*):

$$Z_1 = a^2 Z_2 = (10.0)^2(8) = 800\ \Omega$$

So the effect of placing the 8-Ω resistor on the secondary of this 10 : 1 transformer is 800 Ω at the primary. This, of course, is for the ideal case.

In order to analyze the frequency characteristics of a practical transformer, we need to develop an equivalent circuit. A true model that takes the operation of a real transformer into account is shown in Fig. 6-15*a*. Although it looks complicated, it is still an approximation, using lumped rather than distributed elements.

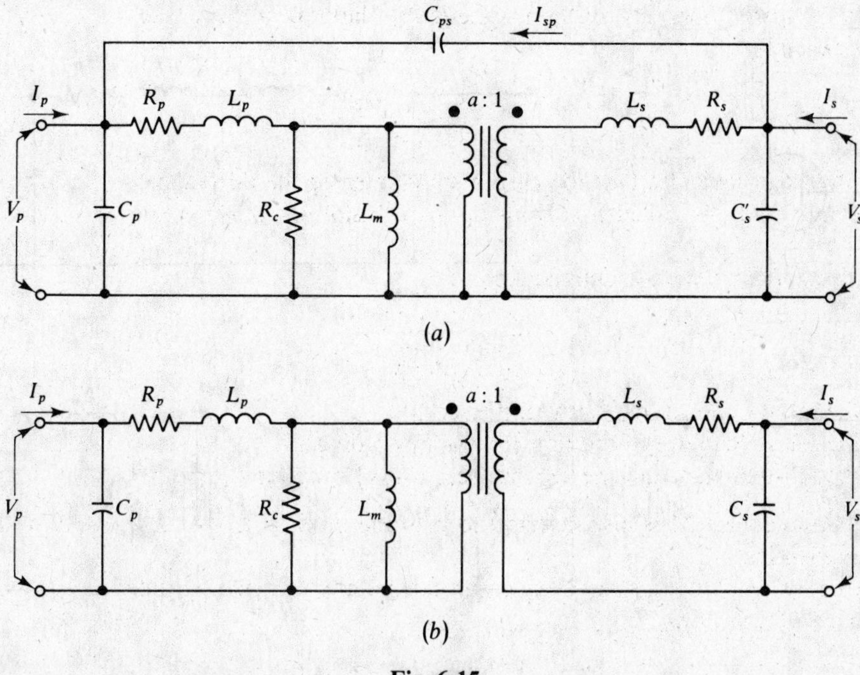

Fig. 6-15

We will further modify this circuit to get some usable results.

L_m is the shunt-magnetizing inductance added because the primary inductance is not infinite.

R_c is the core-loss equivalent shunt resistance that takes into account *hysteresis* and *eddy current* losses. The higher the losses, the lower R_c becomes.

L_p and L_s are leakage inductances that account for the fact that the primary and secondary coils set up self-induced voltage drops, while R_p and R_s are the resistances of the primary and secondary windings. As L_p, L_s, R_p, and R_s are decreased, the transformer becomes better.

C_p and C_s' are the values of capacitance between the successive turns of the primary and secondary. C_{ps} is the capacitive coupling between the primary and secondary windings.

We may approximately account for C_{ps} by increasing C_s' by the factor $(1 \mp a)C_{ps}$, where the minus sign is used if the dots are as shown in Fig. 6-15*a*.

Using $C_s = C_s' + (1 \mp a)C_{ps}$, we then have Fig. 6-15*b*.

We may then further simplify the transformer equivalent circuit by reflecting all impedances to the primary or secondary side, as shown in Fig. 6-16*a* and *b*. The primary-side model of Fig. 6-16*a* is simpler and more commonly used.

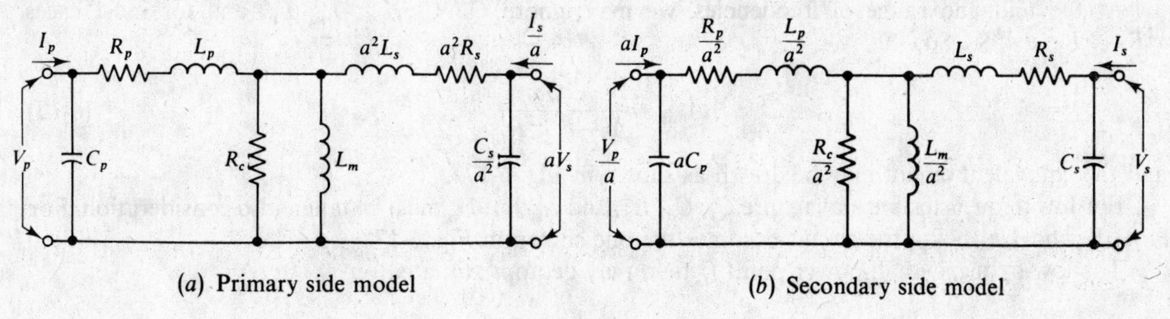

(*a*) Primary side model (*b*) Secondary side model

Fig. 6-16

For the transformer-coupled bipolar transistor amplifier shown in Fig. 6-17*a*, the all-frequency primary-side equivalent circuit is shown in Fig. 6-17*b*.

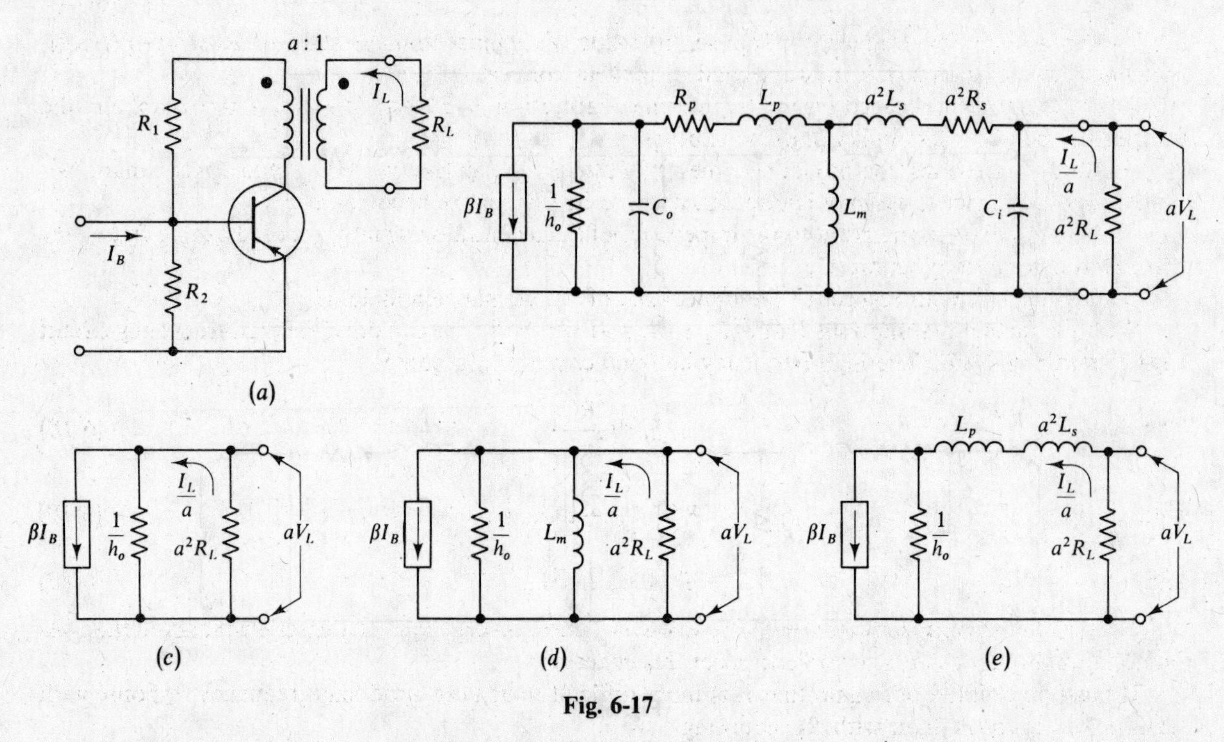

Fig. 6-17

C_o includes the transistor output capacitance plus the primary capacitance of the transformer, while C_i includes the reflected input capacitance of the load element (usually the input capacitance of the next stage) plus the reflected capacitance of C_s for the transformer:

$$C_o = C_{cb} + C_p \tag{6-13}$$

$$C_i = \frac{C_{be}}{a^2} + \frac{C_s}{a^2} \tag{6-14}$$

where C_o = total output capacitance
C_{cb} = transistor collector-to-base capacitance
C_p = transformer primary capacitance
C_i = total input capacitance
C_{be} = transistor base-to-emitter capacitance
C_s = transformer secondary capacitance
a = transformer turns ratio

R_c is usually much larger than the impedance of L_m at the frequencies of interest, so it is omitted from Fig. 6-17*b*.

We find that we may describe the gain, frequency shift, and frequency range the same way we did for the *RC* coupled amplifier with the following amendments.

At the midband range of frequencies, we may ignore C_i, C_o, L_m, L_s, L_p, and for most cases $a^2R_L \gg R_p + a^2R_s$. So

$$A_{\text{mid}} = \frac{a\beta}{1 + h_o a^2 R_L} \tag{6-15}$$

and the equivalent circuit may be drawn as shown in Fig. 6-17*c*.

For low frequencies, we may ignore C_i, C_o, L_p, and L_s, but L_m must be taken into consideration. For $R_L \gg R_s$ and $1/h_o \gg R_p$, the circuit becomes the one shown in Fig. 6-17*d*.

The low-frequency half-power point f_L then may be approximated by

$$f_L = \frac{R'}{2\pi L_m} \tag{6-16}$$

with

$$R' = \frac{a^2 R_L}{1 + h_o a^2 R_L} \tag{6-17}$$

To improve the low-frequency response, the value of L_m must be made high, an expensive proposition since large amounts of iron are needed to prevent core saturation.

When no dc bias is placed on the transformer, saturation is reduced, less iron is necessary, and the transformer becomes less bulky and less expensive.

For high frequencies, the model becomes difficult to solve unless we make some approximations.

At high frequencies, we may ignore L_m, because its impedance becomes large.

If $1/h_o$ is small in comparison to the impedance of C_o, a condition usually imposed by low-frequency considerations, we may ignore C_o.

If a^2R_L is small with respect to the impedance of C_i, we may eliminate C_i.

With the further assumptions that $a^2R_L \gg R_s$ and $1/h_o \gg R_p$, we can draw the high-frequency circuit as shown in Fig. 6-17*e*. The high-frequency half-power point f_H becomes

$$f_H = \frac{R''}{2\pi L''} \tag{6-18}$$

with

$$R'' = \frac{1}{h_o} + a^2R_L \tag{6-19}$$

and

$$L'' = L_p + a^2L_s \tag{6-20}$$

For good high-frequency response, the value of L'' must be kept small. By careful winding of the coils on good quality cores, it is possible to meet this criterion.

It must be pointed out again that it is more difficult to achieve good high-frequency response with transformer coupling than with *RC* coupling.

Example 6.4 Find the midband range and the gain at 200 Hz and 120 kHz for the transformer-coupled amplifier with the following parameters:

$$h_o = 12.5\ \mu S \qquad L_s = 200\ \mu H$$
$$\beta = 50 \qquad R_L = 16\ \Omega$$
$$L_m = 15\ H \qquad a = 25$$
$$L_p = 20\ mH$$

First we solve for the midband gain. Using (*6-15*),

$$A_{mid} = \frac{a\beta}{1 + h_o a^2 R_L} = \frac{(25)(50)}{1 + (12.5 \times 10^{-6})(25)^2(16)} = 1.11 \times 10^3$$

To get the midband range, we need to solve for f_L and f_H. Using (*6-17*),

$$R' = \frac{a^2R_L}{1 + h_o a^2 R_L} = \frac{(25)^2(16)}{1 + (12.5 \times 10^{-6})(25)^2(16)} = 8.89 \times 10^3\ \Omega = 8.89\ k\Omega$$

Therefore, using (*6-16*)

$$f_L = \frac{R'}{2\pi L_m} = \frac{8.89 \times 10^3}{2\pi(15)} = 94.3\ Hz$$

Using (*6-19*),

$$R'' = \frac{1}{h_o} + a^2R_L = \frac{1}{12.5 \times 10^{-6}} + (25)^2(16) = 9.00 \times 10^4\ \Omega = 90.0\ k\Omega$$

and using (*6-20*),

$$L'' = L_p + a^2L_s = 20 \times 10^{-3} + (25)^2(200 \times 10^{-6}) = 1.45 \times 10^{-1}\ H = 145\ mH$$

Therefore, using (*6-18*),

$$f_H = \frac{R''}{2\pi L''} = \frac{90.0 \times 10^3}{2\pi(145 \times 10^{-3})} = 9.88 \times 10^4\ Hz = 98.8\ kHz$$

The midband range is then, using (6-10),

$$10f_L < f_m < \frac{f_H}{10} \quad \text{or} \quad 943 < f_m < 9890 \text{ Hz}$$

The gain at 200 Hz becomes [from (6-5a)]

$$A_{200} = \frac{A_{\text{mid}}}{\sqrt{1 + m^2}}$$

with [using (6-8a)]

$$m = \frac{f_L}{f}$$

So

$$m = \frac{94.3}{200} = 0.472$$

and

$$A_{200} = \frac{1.11 \times 10^3}{\sqrt{1 + (0.472)^2}} = 1.00 \times 10^3 = 1000$$

For 120 kHz, using (6-8b),

$$m = \frac{f}{f_H} = \frac{120 \times 10^3}{98.9 \times 10^3} = 1.21$$

and

$$A_{120\text{k}} = \frac{1.11 \times 10^3}{\sqrt{1 + (1.21)^2}} = 706$$

For transformer-coupled FET and tube circuits, we replace $1/h_o$, with r_{DS} for FETs or r_p for tubes and βI_b with $g_m V_g$ in the equivalent circuits. For FET and tube circuits, we also may wish to change the current source equivalent circuit to a voltage source. The low- and high-frequency approximate equivalent circuits for FETs and tubes are shown in Fig. 6-18*a* and *b*.

For the midfrequency range, the circuit is the same as discussed in Chap. 5 with the addition of the a^2 multiplier on the load resistance R_L. The midfrequency approximate circuit is shown in Fig. 6-18*c*.

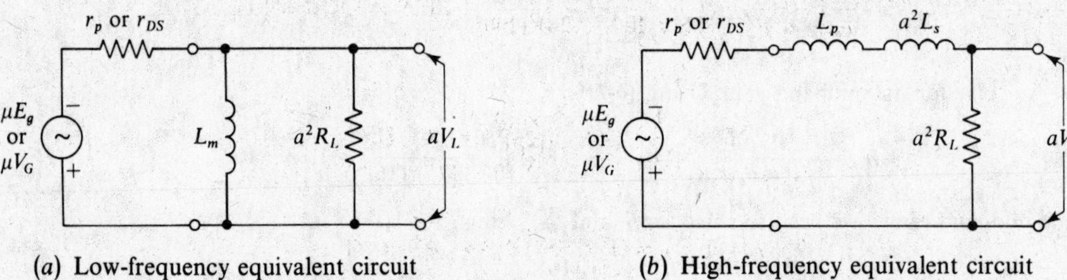

(*a*) Low-frequency equivalent circuit (*b*) High-frequency equivalent circuit

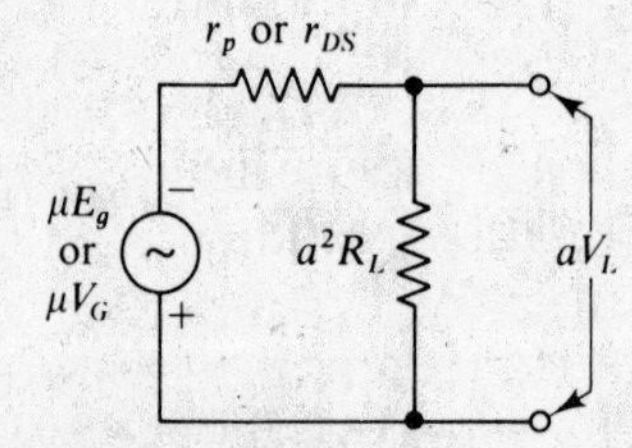

(*c*) Midfrequency equivalent circuit

Fig. 6-18

Equation (6-15) becomes

$$A_{\text{mid}} = \frac{\mu a R_L}{r_p + a^2 R_L} \tag{6-15a}$$

for tubes and

$$A_{\text{mid}} = \frac{\mu a R_L}{r_{DS} + a^2 R_L} \tag{6-15b}$$

for FETs. Equation (*6-17*) becomes

$$R' = \frac{(r_p)(a^2 R_L)}{r_p + a^2 R_L} \tag{6-17a}$$

for tubes and

$$R' = \frac{(r_{DS})(a^2 R_L)}{r_{DS} + a^2 R_L} \tag{6-17b}$$

for FETs. Equation (*6-19*) becomes

$$R'' = r_p + a^2 R_L \tag{6-19a}$$

for tubes and

$$R'' = r_{DS} + a^2 R_L \tag{6-19b}$$

for FETs.

Example 6.5 Find the gain and phase shift at 120 Hz for the FET transformer-coupled amplifier characterized by the following parameters:

$$g_m = 1 \times 10^{-2}\ \text{S} \qquad a = 20$$
$$r_{DS} = 15\ \text{k}\Omega \qquad R_L = 100\ \Omega$$
$$L_m = 20\ \text{H}$$

To find the low-frequency half-power point f_L, we use (*6-17b*),

$$R' = \frac{(r_{DS})(a^2 R_L)}{r_{DS} + a^2 R_L} = \frac{(15 \times 10^3)(20)^2(100)}{15 \times 10^3 + (20)^2(100)} = 10.9 \times 10^3\ \Omega = 10.9\ \text{k}\Omega$$

and (*6-16*),

$$f_L = \frac{R'}{2\pi L_m} = \frac{10.9 \times 10^3}{2\pi(20)} = 86.8\ \text{Hz}$$

Using (*5-4*),

$$\mu = g_m r_{DS} = (1 \times 10^{-2})(15 \times 10^3) = 150$$

and (*6-15b*),

$$A_{\text{mid}} = \frac{\mu a R_L}{r_{DS} + a^2 R_L} = \frac{(150)(20)(100)}{15 \times 10^3 + (20)^2(100)} = 5.45$$

with [from (*6-8a*)]

$$m = \frac{f_L}{f} = \frac{86.8}{120} = 0.723$$

we have from (*6-6a*)

$$A_{120} = \frac{A_{\text{mid}}}{\sqrt{1 + m^2}} = \frac{5.45}{\sqrt{1 + (0.723)^2}} = 4.42$$

and from (*6-9*)

$$\Delta\phi_{120} = \arctan m = \arctan (0.723) = 35.9°$$

This is the phase shift from either 180° or the midband phase angle. In some amplifiers the phase angle at midband is 0°

6.5 *LC* COUPLING AND TUNED AMPLIFIERS

When the coupling network that transfers a signal from the output of one circuit to the input of the next consists of a choke and a capacitor instead of a resistor and a capacitor, *RC* coupling becomes *LC* coupling.

This type of circuit applied to a tube has some limited use in radio circuitry where the choke is used to block the high-frequency components of the output signal from being grounded through the power supply.

The choke is frequency-sensitive while the blocking resistor in an *RC* coupled circuit is not. The network is a high-pass filter and exhibits some resonance phenomena which will be examined in the section on filters. As a coupling system, it is not as useful as the tuned or tank circuit.

A simple *LC* coupled tube amplifier is shown in Fig. 6-19*a* and a singly tuned tube amplifier in Fig. 6-19*b*.

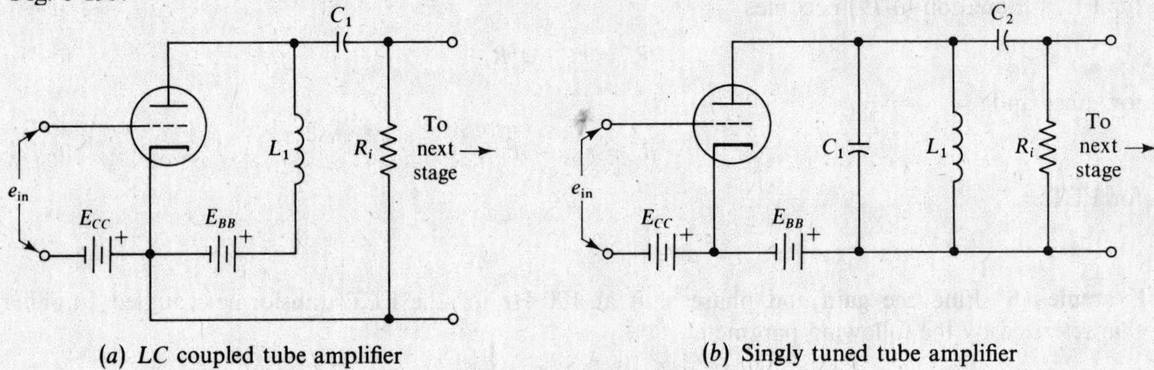

(*a*) *LC* coupled tube amplifier (*b*) Singly tuned tube amplifier

Fig. 6-19

The major advantage of a tuned amplifier is its ability to amplify only a range or band of frequencies. This makes it useful in radio circuitry where selectivity is important.

A transistor, singly tuned, *RC* coupled amplifier and its equivalent circuit are shown in Fig. 6-20*a* and *b*. In Fig. 6-20*b*, C_1 is the total shunt capacitance made up of the sum of C_{CE} of the first stage, C_{BE} of the next stage, and C.

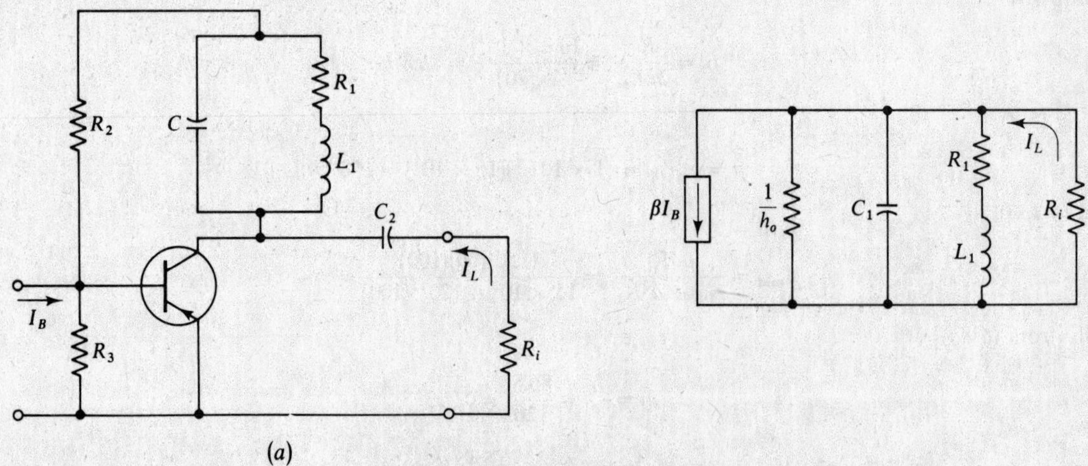

Fig. 6-20 Transistor tuned amplifier

The tank circuit consisting of C_1, L_1, and R_1 exhibits parallel resonance phenomena, and the resonant frequency ω_o is approximately given by

$$\omega_o = \frac{1}{\sqrt{L_1 C_1}} \tag{6-20a}$$

$$f_o = \frac{\omega_o}{2\pi} = \frac{1}{2\pi\sqrt{L_1 C_1}} \tag{6-20b}$$

and the quality factor Q_o of the coil is given by

$$Q_o = \frac{\omega_o L_1}{R_1} = \frac{1}{\omega_o R_1 C_1} \tag{6-21a}$$

or

$$Q_o = \frac{1}{R_1}\sqrt{\frac{L_1}{C_1}} \tag{6-21b}$$

Actually, ω_o is multiplied by a correction factor $\sqrt{Q_o^2 - 1/Q_o}$ to get the parallel resonant frequency ω'_o; but if $Q_o \gg 1$, as is the normal situation,

$$\omega'_o \cong \omega_o$$

We define the maximum value of impedance R_o, which is a pure resistance occurring at the resonant frequency ω_o, by the expression

$$R_o = Q_o^2 R_1 = \omega_o L_1 Q_o \tag{6-21c}$$

The half-power frequency points for the resonant circuit are, for $Q_o \gg 1$, approximated by

$$f_L = f_o\left(1 - \frac{1}{2Q_o}\right) \tag{6-22a}$$

$$f_H = f_o\left(1 + \frac{1}{2Q_o}\right) \tag{6-22b}$$

The *bandwidth B* is then given as the difference between the two half-power frequencies, although it may be defined differently for precise applications:

$$B = f_H - f_L \tag{6-23}$$

Substituting for f_H and f_L, we may also arrive at

$$B = f_o\left(1 + \frac{1}{2Q_o}\right) - f_o\left(1 - \frac{1}{2Q_o}\right) = \not{f_o} + \frac{f_o}{2Q_o} - \not{f_o} + \frac{f_o}{2Q_o} = \frac{f_o}{Q_o} \tag{6-23a}$$

Notice that if we multiply Eq. (*6-22a*) by (*6-22b*) we obtain

$$f_L f_H = f_o^2\left(1 - \frac{1}{2Q_o}\right)\left(1 + \frac{1}{2Q_o}\right) = f_o^2\left(1 - \frac{1}{4Q_o^2}\right)$$

and if $Q_o \gg 1$, the second term in the parentheses drops out and

$$f_L f_H = f_o^2 \tag{6-24}$$

Although Eqs. (*6-22a*) and (*6-22b*) are approximations that depend on Q_o, Eq. (*6-24*) can be derived without any approximations and is always true.

Another expression that is sometimes used in parallel resonant circuit analysis is the normalized frequency variable δ:

$$\delta = \frac{f - f_o}{f_o} = \frac{f}{f_o} - 1 = \frac{\omega - \omega_o}{\omega_o} = \frac{\omega}{\omega_o} - 1 \tag{6-25}$$

If we look at the transistor tuned amplifier equivalent circuit, we see that at resonance the tank formed by L_1, C_1, and R_1 can be replaced by R_o in parallel with $1/h_o$ and R_i.

If we define an effective Q for this circuit using

$$\frac{1}{R_{Sh}} = h_o + \frac{1}{R_o} + \frac{1}{R_i} \tag{6-26}$$

then

$$Q_{\text{eff}} = Q_o \frac{R_{\text{Sh}}}{R_o} \tag{6-27}$$

We may then obtain useful relationships for the single-tuned amplifier as follows:

$$A_o = \beta \frac{R_{\text{Sh}}}{R_i} \tag{6-28}$$

(with A_o = gain at resonance)

$$B = \frac{f_o}{Q_{\text{eff}}} \tag{6-23b}$$

$$f_L \cong f_o\left(1 - \frac{1}{2Q_{\text{eff}}}\right) \tag{6-29a}$$

$$f_H \cong f_0\left(1 + \frac{1}{2Q_{\text{eff}}}\right) \tag{6-29b}$$

$$A = \frac{A_o}{\sqrt{1 + (m_o Q_{\text{eff}})^2}} \tag{6-30}$$

where

$$m_o = \frac{f}{f_o} - \frac{f_o}{f} \tag{6-31a}$$

or

$$m_o = \frac{\omega}{\omega_o} - \frac{\omega_o}{\omega} \tag{6-31b}$$

and

$$\Delta\phi = \arctan m_o Q_{\text{eff}} \tag{6-32}$$

This is close to the form we used for the RC amplifier, with $m_o Q_{\text{eff}}$ taking the place of m.
A graph of the gain-frequency curve for a single-tuned transistor amplifier is shown in Fig. 6-21 for different Q's.

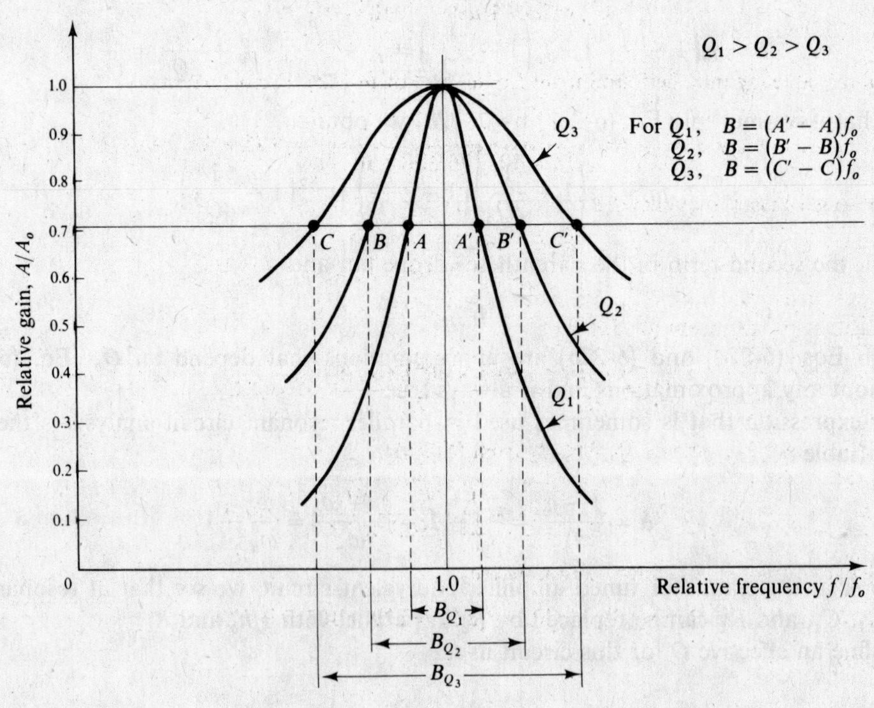

Fig. 6-21

It is possible to increase the flexibility of the gain-frequency curve by using more than one tuned circuit in the coupling network. One of the most useful forms of this approach is the doubly tuned coupling circuit utilizing a transformer shown in Fig. 6-22*a*. This type of circuit is used extensively in communications products. As shown by Fig. 6-22*b* it can give a broad response curve, shown dashed, or a double humped curve, shown solid, depending on the adjustment of the values of each tuned circuit.

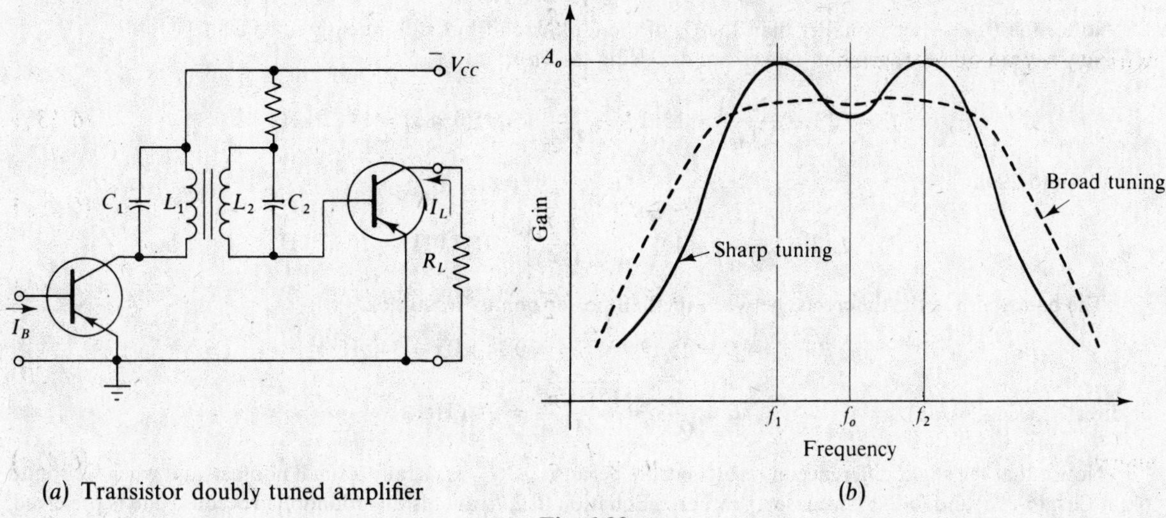

(*a*) Transistor doubly tuned amplifier (*b*)

Fig. 6-22

Example 6.6 Find the resonant frequency, the half-power points, and the bandwidth for the circuit shown in Fig. 6-20*b* that has the following values:

$$\beta = 50 \qquad L_1 = 8 \text{ mH}$$
$$h_o = 13.5\ \mu\text{S} \qquad R_1 = 2.5\ \Omega$$
$$C_1 = 0.014\ \mu\text{F} \qquad R_i = 80\text{ k}\Omega$$

The resonant frequency may be found from Eq. (*6-20a*) or (*6-20b*). From (*6-20a*):

$$\omega_o = \frac{1}{\sqrt{L_1 C_1}} = \frac{1}{\sqrt{(8 \times 10^{-3})(0.014 \times 10^{-6})}} = 9.45 \times 10^4 \text{ rad/s}$$

Dividing by 2π to get linear frequency f_o:

$$f_o = \frac{\omega_o}{2\pi} = \frac{9.45 \times 10^4}{2\pi} = 1.5 \times 10^4 \text{ Hz} = 15 \text{ kHz}$$

Since we have calculated ω_o, from (*6-21a*):

$$Q_o = \frac{\omega_o L_1}{R_1} = \frac{(9.45 \times 10^4)(8 \times 10^{-3})}{2.5} = 302.4$$

As a check we may wish to calculate Q_o directly from (*6-21b*):

$$Q_o = \frac{1}{R_1}\sqrt{\frac{L_1}{C_1}} = \frac{1}{2.5}\sqrt{\frac{8 \times 10^{-3}}{0.14 \times 10^{-6}}} = 302.4$$

Q_{eff} can be calculated once R_o is known. From (*6-21c*)

$$R_o = Q_o^2 R_1 = (302.4)^2(2.5) = 2.29 \times 10^5\ \Omega = 229\text{ k}\Omega$$

Then, using (*6-26*),

$$\frac{1}{R_{\text{Sh}}} = h_o + \frac{1}{R_o} + \frac{1}{R_i} = 13.5 \times 10^{-6} + \frac{1}{2.29 \times 10^3} + \frac{1}{80 \times 10^3} = 3.04 \times 10^{-5}\text{ S}$$

Inverting,

$$R_{\text{Sh}} = \frac{1}{3.04 \times 10^{-5}} = 32.9 \times 10^3 \ \Omega = 32.9 \text{ k}\Omega$$

So, according to (*6-27*),

$$Q_{\text{eff}} = Q_o \frac{R_{\text{Sh}}}{R_o} = \frac{(302.4)(32.9 \times 10^3)}{229 \times 10^3} = 43.4$$

Notice the Q_{eff} is much smaller than the Q_o of the tank circuit but still much greater than 1.
Now we can calculate the half-power points. Using (*6-29a*),

$$f_L = f_o\left(1 - \frac{1}{2Q_{\text{eff}}}\right) = 15\left(1 - \frac{1}{2(43.4)}\right) = 15(0.988) = 14.83 \text{ kHz}$$

and using (*6-29b*),

$$f_H = f_o\left(1 + \frac{1}{2Q_{\text{eff}}}\right) = 15\left(1 + \frac{1}{2(43.4)}\right) = 15(1.011) = 15.17 \text{ kHz}$$

The bandwidth is the difference between the half-power points. Using (*6-23*),

$$B = f_H - f_L = 15.17 - 14.83 = 0.34 \text{ kHz} = 340 \text{ Hz}$$

or directly, using (*6-23b*),

$$B = \frac{f_o}{Q_{\text{eff}}} = \frac{15 \times 10^3}{43.4} = 346 \text{ Hz}$$

Notice that the slight difference in results occurs because $1/2Q_{\text{eff}}$ is usually a small number, and when we round off in Eqs. (*6-29a*) and (*6-29b*), we lose a few hertz. Equation (*6-23b*) avoids this roundoff procedure and is preferred, if the information is available.

Tube and FET tuned amplifiers have the same gain-frequency curves; however for tubes

$$\frac{1}{R_{\text{Sh}}} = \frac{1}{r_p} + \frac{1}{R_o} + \frac{1}{R_i} \qquad (6\text{-}26a)$$

and for FETs

$$\frac{1}{R_{\text{Sh}}} = \frac{1}{r_{DS}} + \frac{1}{R_o} + \frac{1}{R_i} \qquad (6\text{-}26b)$$

For both tubes and FETs

$$A_o = \mu R_{\text{Sh}} \qquad (6\text{-}28a)$$

For tube and FET circuits, the rest of the calculations are the same.

An example of an FET single-tuned amplifier and equivalent circuit is shown in Fig. 6-23*a* and *b*. As in the transistor circuit, C' is the sum of all the shunt capacitances, the output capacitance of the first stage, the input capacitance of the second stage, and C_1. R_i is the input resistance of the next stage.

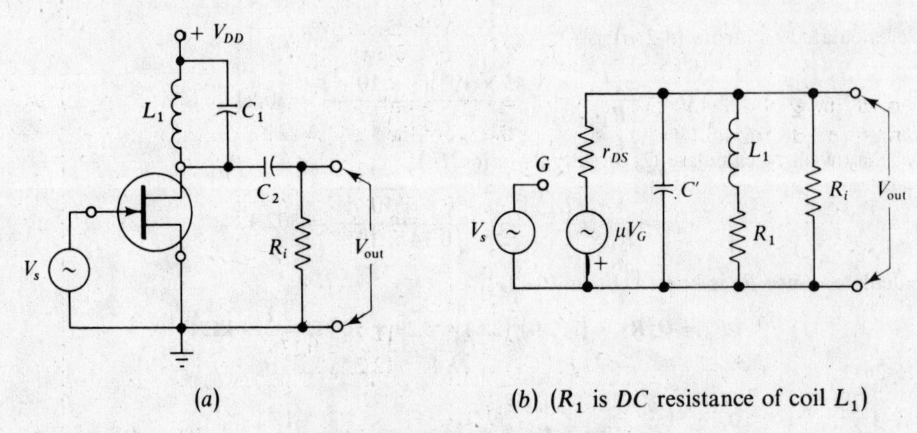

(*a*) (*b*) (R_1 is *DC* resistance of coil L_1)

Fig. 6-23 FET tuned amplifier

Example 6.7 In the FET single-tuned amplifier equivalent circuit shown in Fig. 6-24*a*, find the Q_{eff}, bandwidth, and phase shift at 12.2 kHz.

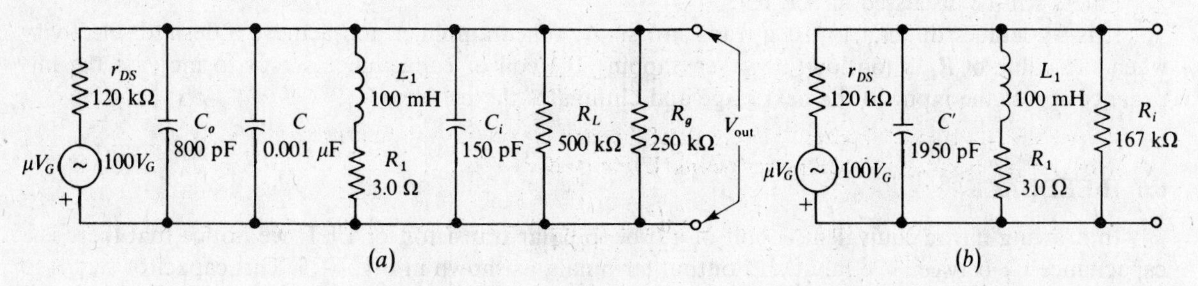

Fig. 6-24

Figure 6-24*a* is slightly different than Fig. 6-23*b* since C' and R_i are not given in that form. We may change to the form shown in Fig. 6-24*b* by the following method, using the values given in Fig. 6-24*a*:

$$C' = C_o + C + C_i = (800 \times 10^{-12}) + (0.001 \times 10^{-6}) + (150 \times 10^{-12}) = 1.95 \times 10^{-9}\ \text{F} = 1950\ \text{pF}$$

and
$$R_i = \frac{R_L R_g}{R_L + R_g} = \frac{(500 \times 10^3)(250 \times 10^3)}{500 \times 10^3 + 250 \times 10^3} = 1.67 \times 10^5\ \Omega = 167\ \text{k}\Omega$$

We need to calculate f_o, Q_o, R_o, and R_{Sh} to solve for Q_{eff} and B. Using the values given in Fig. 6-24*a* and *b*, and using (*6-20b*),

$$f_o = \frac{1}{2\pi\sqrt{L_1 C_1}} = \frac{1}{2\pi\sqrt{(100 \times 10^{-3})(1950 \times 10^{-12})}} = 11.4 \times 10^3\ \text{Hz} = 11.4\ \text{kHz}$$

Using (*6-21b*),

$$Q_o = \frac{1}{R_1}\sqrt{\frac{L_1}{C_1}} = \frac{1}{3.0}\sqrt{\frac{100 \times 10^{-3}}{1950 \times 10^{-12}}} = 2387$$

Using (*6-21c*),

$$R_o = Q_o^2 R_1 = (2387)^2(3.0) = 1.71 \times 10^7\ \Omega = 17.1\ \text{M}\Omega$$

And using (*6-26b*),

$$\frac{1}{R_{Sh}} = \frac{1}{r_{DS}} + \frac{1}{R_o} + \frac{1}{R_i} = \frac{1}{120 \times 10^3} + \frac{1}{17.1 \times 10^6} + \frac{1}{167 \times 10^3} = 1.44 \times 10^{-5}\ \text{S}$$

Inverting gives
$$R_{Sh} = \frac{1}{1.44 \times 10^{-5}} = 6.95 \times 10^4\ \Omega = 69.5\ \text{k}\Omega$$

Now, using (*6-27*),

$$Q_{eff} = Q_o \frac{R_{Sh}}{R_o} = \frac{(2387)(69.5 \times 10^3)}{17.1 \times 10^6} = 9.71$$

Even though the Q of the tank circuit is 2387, the Q_{eff} is only 9.71. This illustrates the effect of amplifier loading on the performance of the circuit. If either the R_i or the r_{DS} of the FET were made larger, the Q_{eff} could be increased.

The bandwidth is now obtained from (*6-23b*):

$$B = \frac{f_o}{Q_{eff}} = \frac{11.4 \times 10^3}{9.71} = 1174\ \text{Hz} = 1.17\ \text{kHz}$$

Using (*6-31a*),

$$m_o = \frac{f}{f_o} - \frac{f_o}{f} = \frac{12.2}{11.4} - \frac{11.4}{12.2} = 0.136$$

And the phase shift may now be calculated from (*6-32*):

$$\Delta\phi_{12.2k} = \arctan m_o Q_{eff} = \arctan (0.136)(9.71) = 52.8°$$

The tuned amplifier is also used as an *active filter* since it amplifies only a selected frequency range dependent on its parameters.

Filters will be discussed in Sec. 6.9.

It is sometimes difficult to use a transistor singly tuned amplifier and achieve a desired selectivity when the value of R_i is too low; however, tapping the coil or capacitance serves to increase the impedance across the input to the next stage and eliminates the problem.

6.6 FEEDBACK

In drawing the ac equivalent circuit of a tube, bipolar transistor, or FET, we notice that there is a capacitance C_F between the input and output terminals, as shown in Fig. 6-25. This capacitor serves to couple part of the output back to the input and to divide the feedback impedance by the factor $(1 - A_v)$.

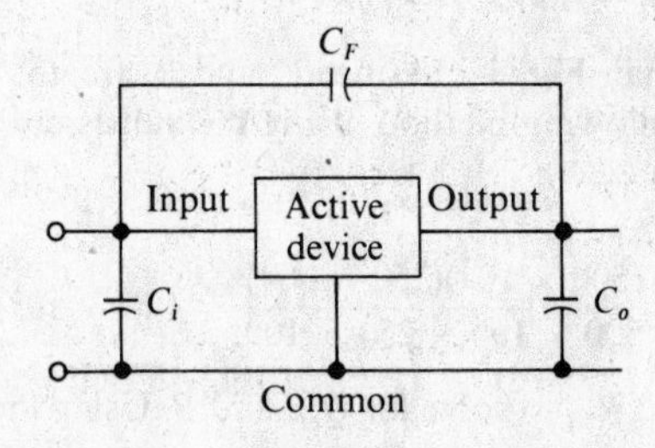

Fig. 6-25

The effect of this coupling causes the coupling capacitor to appear in the input circuit, so that

$$C_{\text{in}} = C_i + C_F(1 - A_v) \tag{6-33}$$

This is known as the *Miller effect* and is usually detrimental to the high-frequency response of the amplifier. Equation (*6-33*) takes the following forms for tube, common-emitter transistor, and FET, respectively:

$$C_{\text{in}} = C_{gk} + C_{gp}(1 - A_v) \tag{6-33a}$$

$$C_{\text{in}} = C_{be} + C_{cb}(1 - A_v) \tag{6-33b}$$

$$C_{\text{in}} = C_{GS} + C_{GD}(1 - A_v) \tag{6-33c}$$

where A_v = magnitude of the voltage gain $\mathbf{A}_v$ and is usually negative
C_{in} = total input capacitance
C_{gk} = grid-to-cathode capacitance
C_{gp} = grid-to-plate capacitance
C_{be} = base-to-emitter capacitance
C_{cb} = collector-to-base capacitance
C_{GS} = gate-to-source capacitance
C_{GD} = gate-to-drain capacitance

If the gain is high, what looks to be only a negligible feedback capacitance becomes multiplied by a large number and must be considered in the design.

Any network, active or passive, may be used to couple part of the output signal back to the input; however, we will limit our discussion to passive networks.

The block diagram of an amplifier with feedback is shown in Fig. 6-26.

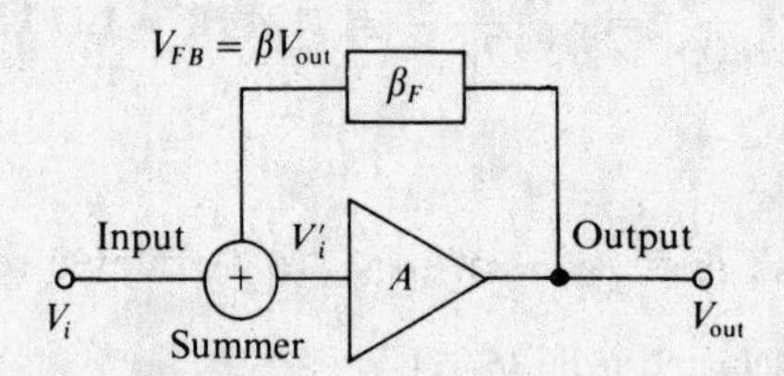

Fig. 6-26

The β_F used in the feedback loop should not be confused with the transistor $\beta = h_{fe}$. To distinguish between the two, the feedback β_F will be defined by any one of the following relations:

$$V_{FB} = \beta_F V_{out} \tag{6-34a}$$

$$V_{FB} = \beta_F I_{out} \tag{6-34b}$$

$$I_{FB} = \beta_F I_{out} \tag{6-34c}$$

$$I_{FB} = \beta_F V_{out} \tag{6-34d}$$

Equations (*6-34a*) and (*6-34d*) illustrate *voltage feedback*, in which the feedback voltage or current is proportional to the output voltage. Voltage feedback serves to isolate the output voltage from changes and is a practical approach to a constant-voltage source.

Equations (*6-34b*) and (*6-34c*) illustrate *current feedback*, in which the feedback voltage or current is proportional to the output current. Current feedback tends to stabilize the output current and is a practical approach to a constant-current source.

Feedback amplifiers with unity gain are used as operational amplifiers (op amps).

The gain of the amplifier without feedback may also be a voltage gain A_v or current gain A_i, and the gain of the amplifier with feedback becomes

$$A_{FB} = \frac{A}{1 + \beta_F A} \tag{6-35}$$

For voltage gain,

$$A_{vFB} = \frac{A_v}{1 + \beta_F A_v} \tag{6-35a}$$

For current gain,

$$A_{iFB} = \frac{A_i}{1 + \beta_F A_i} \tag{6-35b}$$

For power gain,

$$G_{FB} = \frac{G}{1 + \beta_F G} \tag{6-35c}$$

Example 6.8 Derive the expression for A_{FB} as shown in Eq. (*6-35*). Use Fig. 6-26.

Using Fig. 6-26,

$$V_{out} = AV_i$$

By (*6-34a*),

$$V_{FB} = \beta_F V_{out}$$

$$V_i' = V_i + V_{FB} = V_i + \beta_F V_{out} = V_i + \beta_F AV_i = V_i(1 + \beta_F A)$$

But the basic definition of gain with feedback is the output voltage or current over the amplifier input voltage or current. In this case,

$$A_{FB} = \frac{V_{out}}{V_i'} = \frac{V_{out}}{V_i(1 + \beta_F A)}$$

But V_{out}/V_i is equal to the gain without feedback, so

$$A_{FB} = \frac{A}{1 + \beta_F A}$$

which is Eq. (*6-35*).

If the gain A is negative, A_{FB} is negative or positive depending on the polarity of β_F:

$$A_{FB} = -\frac{A}{1 - \beta_F A} \tag{6-35d}$$

This form is used in some texts where the denominator $(1 - \beta_F A)$ assumes a negative A or an **A** with 180° phase angle.

If β is positive, this is sometimes written

$$A_{FB} = \frac{A}{1 - \beta_F A} \tag{6-35e}$$

with the understanding that A and A_{FB} are negative or 180° phase-shifted quantities and $\beta_F A < 1$.

If $\beta_F A$ is much greater than 1, then

$$A_{FB} \cong \frac{1}{\beta_F} \tag{6-35f}$$

and the gain with feedback becomes independent of the active elements.

Distortion is also reduced but the overall gain is less than 1. So stability and active-element independence are improved at the expense of gain.

This type of feedback is known as *negative* or *degenerative* feedback and is used to increase bandwidth and stabilize amplifier circuits. Negative feedback may also be used to stabilize bias circuitry.

The bandwidth of an amplifier with feedback is given by

$$B_{FB} = (1 + \beta_F A_{\text{mid}})B \tag{6-36}$$

where B is the bandwidth of the amplifier without feedback and A_{mid} is the midband gain without feedback.

If $\beta_F A$ is negative and less than 1, then A_{FB} is larger than A and we have *positive* or *regenerative* feedback. Positive feedback in amplifiers is usually undesirable since distortion is increased, and the dependence of the gain on the active circuit elements is greater. In addition, positive feedback tends to affect amplifier stability since each increase in output reappears at the input and the amplifier may become an *oscillator*.

The terms *gain margin* and *phase margin* are sometimes encountered in determining stability, and can be explained as follows:

If we plot the magnitude of $\beta_F A$ in decibels, against frequency, for high values of frequency, the gain will drop off as shown in Fig. 6-27. We also plot the phase angle against the same frequency range.

The value of gain in decibels when the phase angle is zero is the gain margin, while the value of phase when the gain is unity (or 0 dB) is the phase margin.

If the gain margin is negative, it indicates the rise in gain possible without oscillation. If the gain margin is positive, the amplifier is unstable and becomes an oscillator.

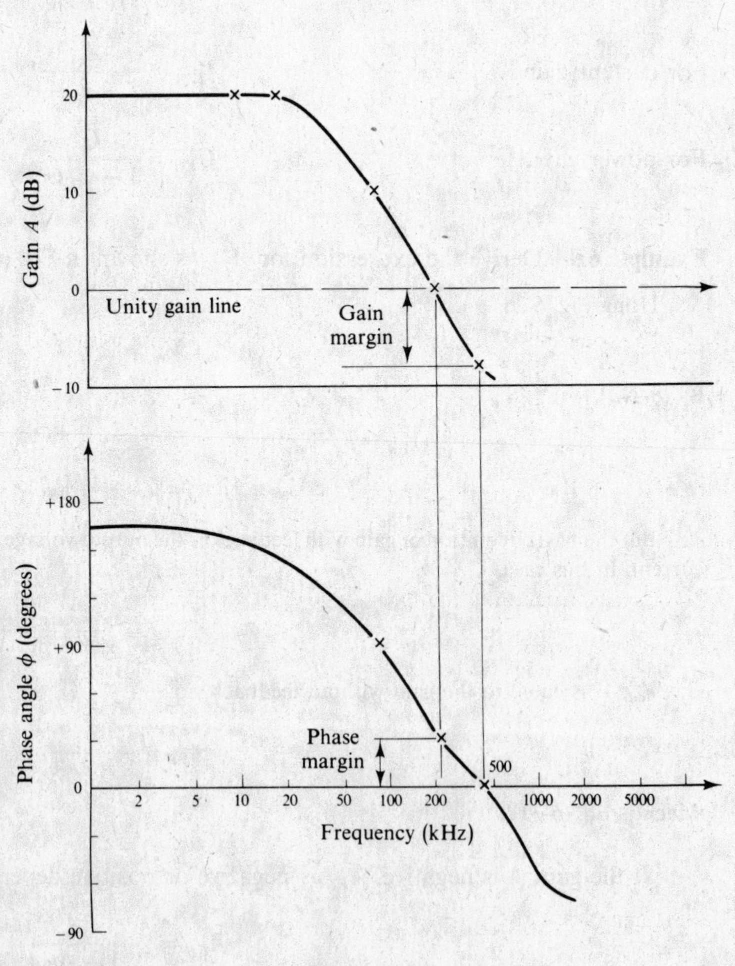

Fig. 6-27

If the phase margin is positive, it indicates the shift in phase allowable without oscillation. If the phase margin is negative, we have an oscillator.

Normal design requires a gain margin of -10 dB and a phase margin of 30° for a stable amplifier.

It must be pointed out that, in practical feedback amplifiers, the gain may not always be in the form shown in Eq. (*6-35*). In only the simplest cases will that equation describe A_{FB}, but it is still a useful relationship for general study of feedback.

The calculations for most real feedback amplifiers involve using the equivalent circuits and writing complete expressions for V_i or I_i, V_{FB} or I_{FB}, and V_{out} or I_{out}.

Example 6.9 The FET feedback amplifier shown in Fig. 6-28*a* and *b* has the following values:

$$r_{DS} = 25\ \text{k}\Omega \qquad R_1 = 5\ \text{k}\Omega \qquad R_L = 10\ \text{k}\Omega \qquad \mu = 150$$

Assume C_s is a short at the frequencies under consideration and that Eq. (*6-35*) holds. Find the gain without feedback, A; β_F; and the gain with feedback, A_{FB}.

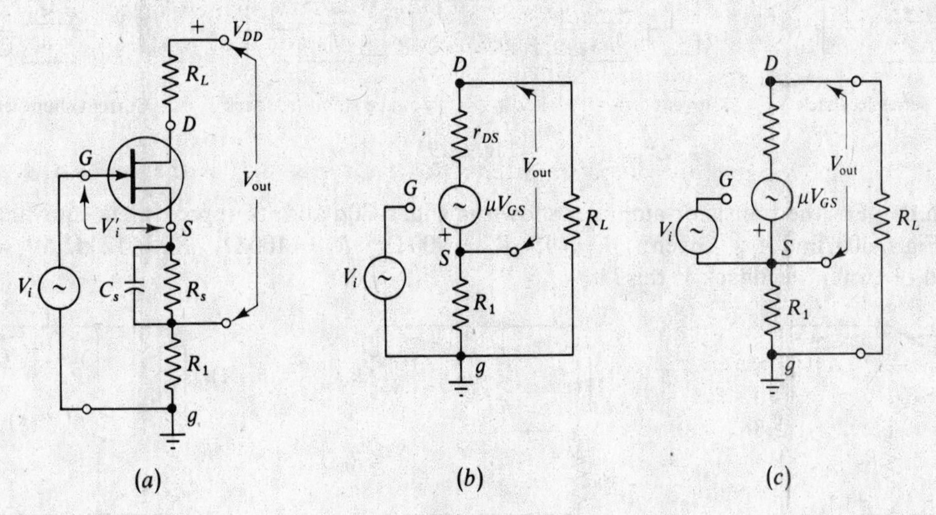

Fig. 6-28

The gain without feedback would be obtained by breaking the wire from V_i to g and replacing it with a wire from V_i to S, as shown in Fig. 6-28*c*.

Since $V_{GS} = V_i$ for the no-feedback case, we have, from the voltage divider rule,

$$V_{out} = -\frac{\mu V_i(R_1 + R_L)}{r_{DS} + R_1 + R_L}$$

or

$$A = -\frac{\mu(R_1 + R_L)}{r_{DS} + R_1 + R_L}$$

Substituting values into the second equation,

$$A = -\frac{(150)(5 + 10) \times 10^3}{(25 + 5 + 10) \times 10^3} = -56.25$$

With feedback,

$$V_{GS} = V_i + V_{FB}$$

and from the voltage divider rule,

$$V_{FB} = -\frac{R_1}{R_1 + R_L} V_{out}$$

Comparing this with Eq. (*6-34a*), we see that

$$\beta_F = -\frac{R_1}{R_1 + R_L} = -\frac{5 \times 10^3}{(5 + 10) \times 10^3} = -0.333$$

So, using (*6-35*),

$$A_{FB} = \frac{A}{1 + \beta_F A} = \frac{-56.25}{1 + (-0.333)(-56.25)} = -2.85$$

The gain has been lowered with this negative voltage feedback.

Note that if the source bypass capacitor C_s were not a short circuit, the impedance of C_s and R_s in parallel would introduce negative current feedback in addition to the negative voltage feedback of R_1. The problem would become more difficult to solve since β_F would consist of two parts.

Feedback types can also be classified as series or shunt, depending on the method of connecting the feedback component to the input circuit. The various possibilities are shown in Fig. 6-29.

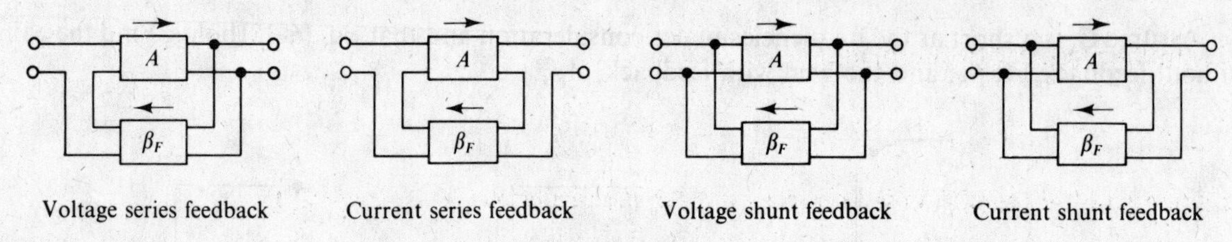

Fig. 6-29

Example 6.10 For the transistor amplifier shown in Fig. 6-30*a* and its approximate equivalent circuit shown in Fig. 6-30*b*, find A_{FB}. Given: $\beta = 49$; $R_C = 900\ \Omega$; $R_E = 100\ \Omega$; $R_B = 12\ \text{k}\Omega$; $r_c = 120\ \text{k}\Omega$. What kind of current feedback is this?

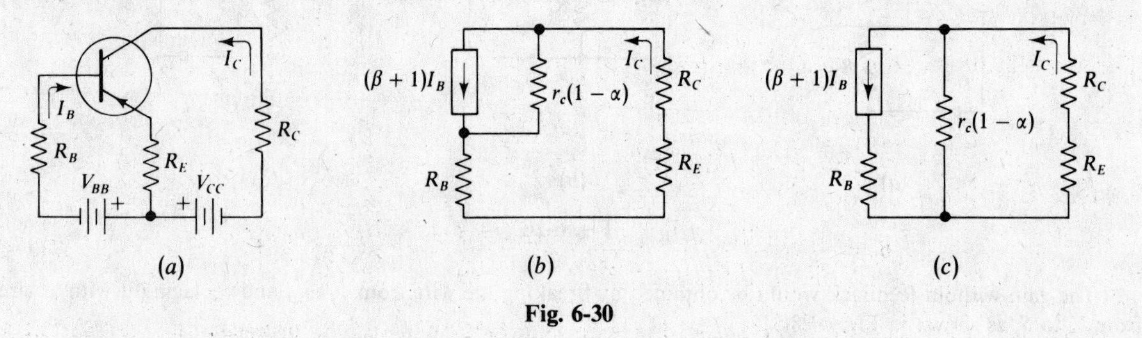

Fig. 6-30

The β_F is not easy to find in this case of current feedback, so we solve for A_{FB} directly:

$$\alpha = \frac{\beta}{\beta + 1} = \frac{49}{49 + 1} = 0.98$$

The current gain is given by I_C/I_B for this amplifier, and by the current divider rule in Fig. 6-30*b*,

$$I_C = (\beta + 1)I_B \frac{r_c(1 - \alpha)}{r_c(1 - \alpha) + R_B + R_C + R_E} = (49 + 1)I_B \frac{(1.2 \times 10^5)(1 - 0.98)}{(1.2 \times 10^5)(1 - 0.98) + 1.2 \times 10^4 + 900 + 100}$$

$$= 50I_B \frac{2.4 \times 10^3}{15.4 \times 10^3} = 7.79I_B$$

So

$$A_{FB} = \frac{I_C}{I_B} = \frac{7.79\not{I}_B}{\not{I}_B} = 7.79$$

To check the type of feedback, we redraw Fig. 6-30*b* without the current feedback by moving the bottom of the $r_c(1 - \alpha)$ resistor to the junction of R_B amd R_E, as shown in Fig. 6-30*c*.

The gain is still given by I_C/I_B but now I_C is written by the current divider rule as

$$I_C = (\beta + 1)I_B \frac{r_c(1 - \alpha)}{r_c(1 - \alpha) + R_C + R_E} = (49 + 1)I_B \frac{1.2 \times 10^5(1 - 0.98)}{1.2 \times 10^5(1 - 0.98) + 900 + 100} = 50I_B \frac{2.4 \times \not{10^3}}{3.4 \times \not{10^3}} = 35.3I_B$$

The gain A_i is then

$$A_i = \frac{I_C}{I_B} = \frac{35.3 I_B}{I_B} = 35.3$$

This is higher than the gain with feedback, so the feedback must be negative.

6.7 OSCILLATORS

In order for an amplifier to become an oscillator, it must be designed for that purpose and have

$$\beta'_F = \frac{1}{A} \tag{6-37}$$

For a tube amplifier [from (*6-5a*)]

$$A = \frac{-\mu Z_L}{r_p + Z_L}$$

and applying this condition is known as the *Barkhausen criterion*. Using (*6-37*),

$$\beta'_F = \frac{1}{A} = \frac{1}{-\mu Z_L/(r_p + Z_L)} = -\frac{r_p + Z_L}{\mu Z_L} = -\frac{1}{\mu}\left[\frac{r_p}{Z_L} + 1\right] \tag{6-37a}$$

For a transistor amplifier in which $A_i = -\beta$

$$\beta'_F = \frac{1}{A_i} = -\frac{1}{\beta} \tag{6-37b}$$

For an FET amplifier in which [from (*6-5a*)]

$$A = \frac{-\mu Z_L}{r_{DS} + Z_L}$$

using (*6-37*),

$$\beta'_F = \frac{1}{A} = \frac{1}{-\mu Z_L/(r_{DS} + Z_L)} = \frac{r_{DS} + Z_L}{-\mu Z_L} = -\frac{1}{\mu}\left[\frac{r_{DS}}{Z_L} + 1\right] \tag{6-37c}$$

Example 6.11 For the FET oscillator shown in Fig. 6-31, find the β'_F necessary for oscillation at 100 kHz.

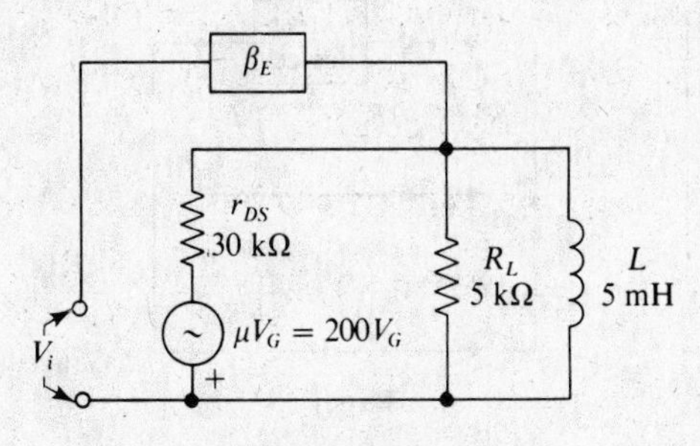

Fig. 6-31

For this circuit, $Z_L = R_L$ in parallel with L, $\mu = 200$, and $r_{DS} = 30 \times 10^3\ \Omega$.

So if

$$f = 100 \times 10^3 \text{ Hz}$$

$$\omega = 2\pi f = 2\pi(100 \times 10^3) = 6.28 \times 10^5 \text{ rad/s}$$

and

$$Z_L = \frac{(R_L)(j\omega L)}{R_L + j\omega L} = \frac{(5 \times 10^3)(j)(6.28 \times 10^5)(5.0 \times 10^{-3})}{5 \times 10^3 + (j)(6.28 \times 10^5)(5.0 \times 10^{-3})} = \frac{j(1.57 \times 10^8)}{5 \times 10^3 + j(3.14 \times 10^3)}$$

Converting to polar form, $$Z_L = \frac{1.57 \times 10^8 \angle 90^\circ}{5.90 \times 10^3 \angle 32.1^\circ} = 2.66 \times 10^4 \angle 57.9^\circ$$

The Barkhausen criterion for an FET oscillator is given by (*6-37c*):

$$\beta'_F = -\frac{1}{\mu}\left(\frac{r_{DS}}{Z_L} + 1\right)$$

Substituting,

$$\beta'_F = -\frac{1}{200}\left(\frac{30 \times 10^3 \angle 0^\circ}{2.66 \times 10^4 \angle 57.9^\circ} + 1\right) = -\frac{1}{200}(1.13 \angle -57.9^\circ + 1)$$

Converting to rectangular form,

$$\beta'_F = -\tfrac{1}{200}[(0.6 - j0.955) + 1] = -\tfrac{1}{200}(1.6 - j0.955) = -8.00 \times 10^{-3} + j4.78 \times 10^{-3}$$

So the feedback has to provide a β'_F with a real negative part of 8.00×10^{-3} and a j positive part of 4.78×10^{-3}.

Oscillators have different ways of coupling the output to the input and are classified accordingly. The basic oscillator types are shown in Fig. 6-32 as developed for a general active device which may be a tube, bipolar transistor, or FET.

The feedback in the tuned-input, tuned-output, or combination tuned-input-output oscillators is the mutual inductance or inductive coupling that exists between L_1 and L_2. The tuning capacitor C_1 selects the frequency at which the tank resonates, and the adjustment of the coupling between L_1 and L_2 determines if the circuit will oscillate under the Barkhausen criterion. If

$$M \geq -\frac{L_1}{\mu}\left(1 + \frac{r_p}{R_L}\right) \tag{6-38}$$

the circuit will oscillate.

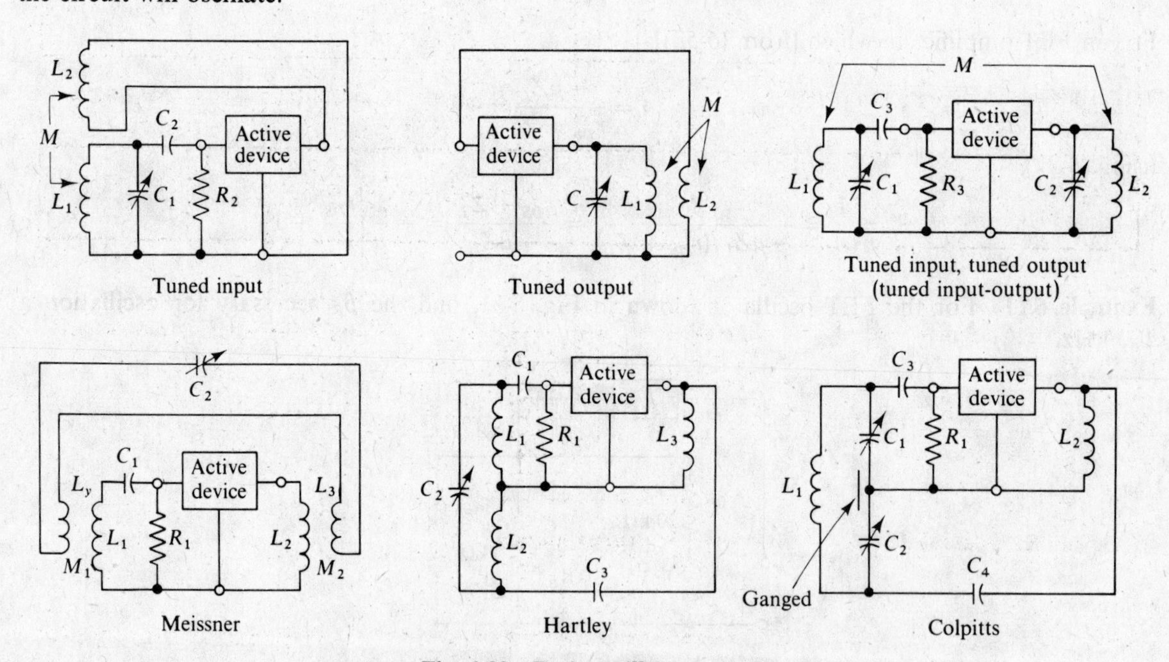

Fig. 6-32 Basic oscillator types

The feedback in the Meissner oscillator is governed by the LC tuned circuit consisting of L_4, L_3, and C_2. Varying C_2 adjusts the frequency of oscillation and determines whether the circuit meets the oscillation criterion. Because the feedback circuit is tuned, oscillation normally occurs at only one frequency.

Both the Hartley and Colpitts oscillators use capacitor C_1 to couple the output to the input. The Hartley oscillator uses a tapped coil to determine the amount of feedback and the tuning capacitor C_2 to adjust the frequency.

The Colpitts oscillator uses the tapped capacitor to determine the amount of feedback and is usually ganged, so that once the circuit is oscillating, the feedback ratio is maintained as the circuit is tuned to resonance.

The analysis of oscillation criteria involves a great deal of algebra and gives two conditions: One is the value of voltage gain necessary to start the circuit into oscillation and keep it oscillating, while the other is the circuit conditions that control the frequency of oscillation.

These results are found either by solving for the open-loop gain and setting it equal to unity or solving for the impedance of any loop and setting it equal to zero.

The real part of the result gives the gain requirements of the active device while the j (imaginary) part gives the frequency of oscillation.

Some of the results of these calculations for common oscillators are listed in Table 6-1.

Table 6.1 Oscillator design criteria

Oscillator type	Gain requirements	Frequency	Diagram
RC phase shift	$A_v \geq -\dfrac{5-(\omega_o RC)^2}{(\omega_o RC)^2} \geq -29$	$\omega_o = 2\pi f_o = \dfrac{1}{\sqrt{6}RC}$	Fig. 6-33*a*
Tuned output	$A_v \leq M$	$\omega_o = 2\pi f_o = \sqrt{\dfrac{1}{L_1 C_1}}$	Fig. 6-33*b*
Colpitts	$A_v \geq \dfrac{C_2}{C_1}$	$\omega_o = 2\pi f_o = \sqrt{\dfrac{C_1 + C_2}{LC_1C_2}}$	Fig. 6-33*c*
Clapp	$A_v \geq \dfrac{C_2}{C_1}$	$\omega_o = 2\pi f_o = \dfrac{1}{\sqrt{LC_3}}$	Fig. 6-33*d*
Hartley	$A_v \geq \dfrac{L_1 + L_2 + 2M}{L_2}$	$\omega_o = 2\pi f_o = \dfrac{1}{\sqrt{(L_1 + L_2 + 2M)C}}$	Fig. 6-33*e*
Pierce	$A_v \geq 1$	$\omega_o = 2\pi f_o = \dfrac{1}{\sqrt{LC}}$	Fig. 6-33*f*

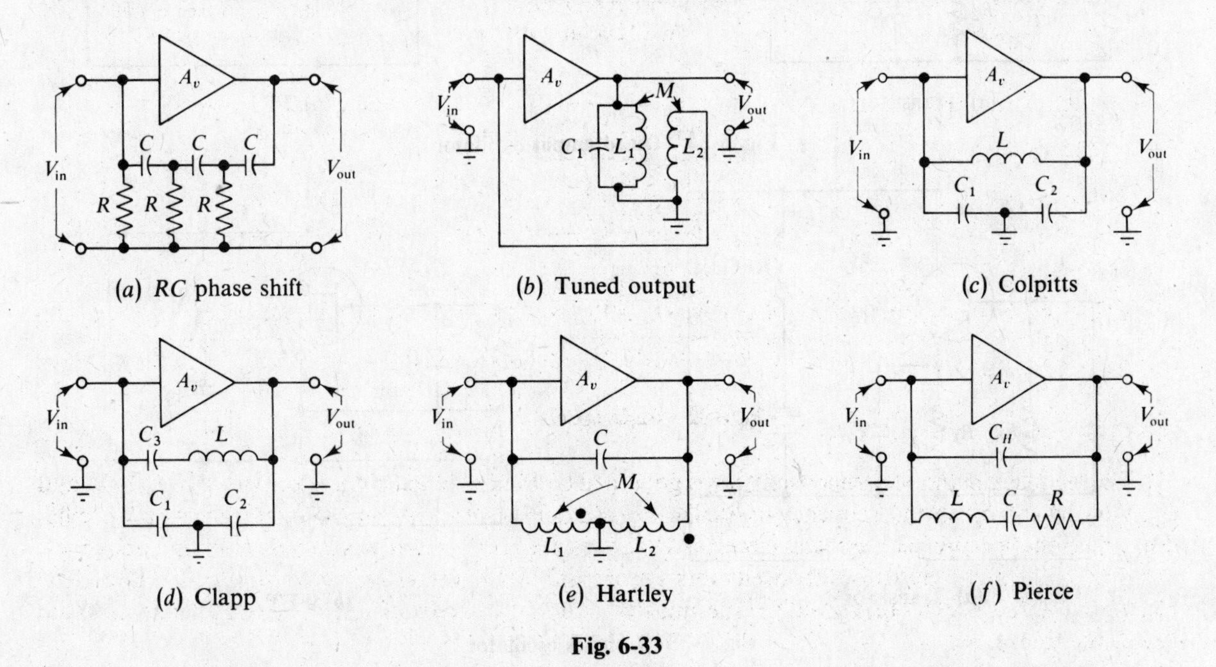

(*a*) *RC* phase shift (*b*) Tuned output (*c*) Colpitts

(*d*) Clapp (*e*) Hartley (*f*) Pierce

Fig. 6-33

Example 6.12 Find the minimum voltage gain and the frequency of oscillation for a Colpitts oscillator with $C_1 = 0.004\ \mu F$, $C_2 = 0.03\ \mu F$, and $L = 4.0$ mH.

From Table 6-1,

$$A_v \geq \frac{C_2}{C_1}$$

So

$$A_v \geq \frac{0.03 \times 10^{-6}}{0.004 \times 10^{-6}} \geq 7.5$$

Also from Table 6-1,

$$2\pi f_o = \sqrt{\frac{C_1 + C_2}{LC_1C_2}}$$

So

$$f_o = \frac{1}{2\pi}\sqrt{\frac{0.004 \times 10^{-6} + 0.03 \times 10^{-6}}{(4 \times 10^{-3})(0.004 \times 10^{-6})(0.03 \times 10^{-6})}} = \frac{1}{2\pi}\sqrt{7.08 \times 10^{10}} = 42.4 \times 10^3 \text{ Hz} = 42.4 \text{ kHz}$$

The tuned-output oscillator, Fig. 6-33*b*, is shown for transistor and FET active elements in Fig. 6-34*a* and *b*. The Colpitts oscillator, Fig. 6-33*c*, is shown in transistor and FET forms in Fig. 6-35*a* and *b*. The Hartley oscillator, Fig. 6-33*e*, is shown in transistor and FET forms in Fig. 6-36*a* and *b*.

The Pierce oscillator shown in Fig. 6-33*f* is redrawn in Fig. 6-37*a*, *b*, and *c* in tube, transistor, and FET format. It is an example of a *crystal controlled oscillator*. Capacitors C_A and C_B are dc blocking capacitors and the *RF* choke prevents the high-frequency signals from grounding through the power supply.

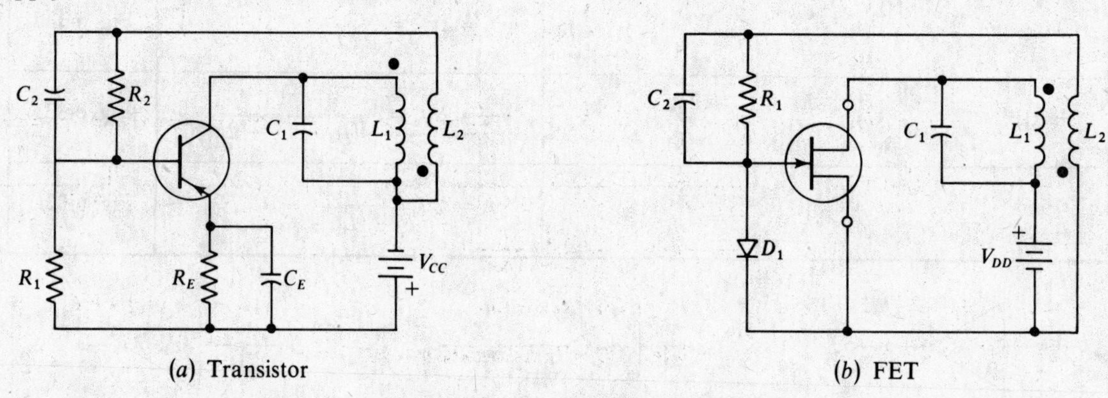

(*a*) Transistor (*b*) FET

Fig. 6-34 Tuned output oscillator

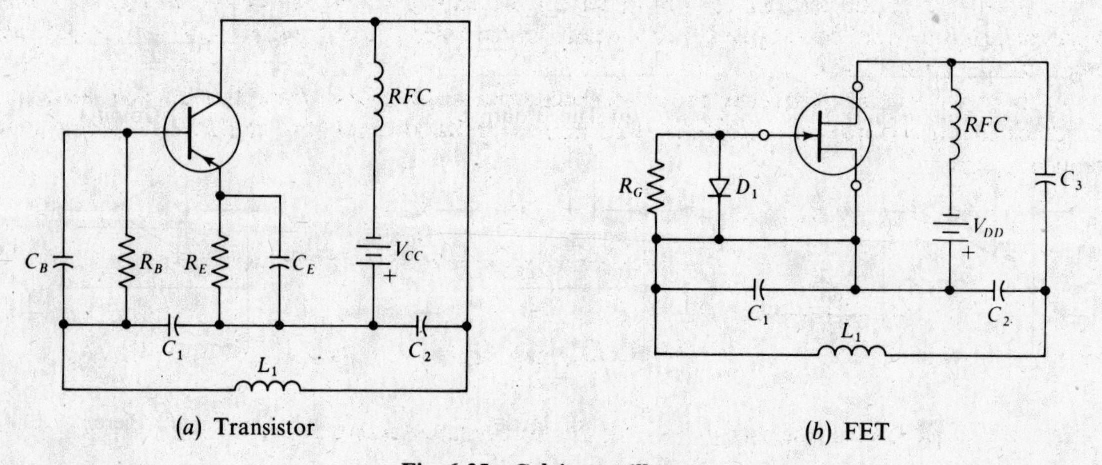

(*a*) Transistor (*b*) FET

Fig. 6-35 Colpitts oscillator

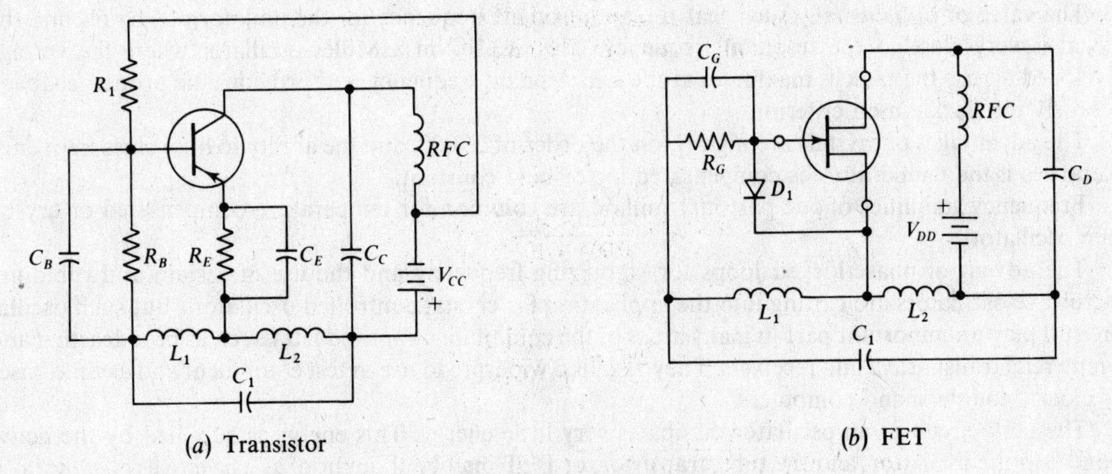

(*a*) Transistor (*b*) FET

Fig. 6-36 Hartley oscillator

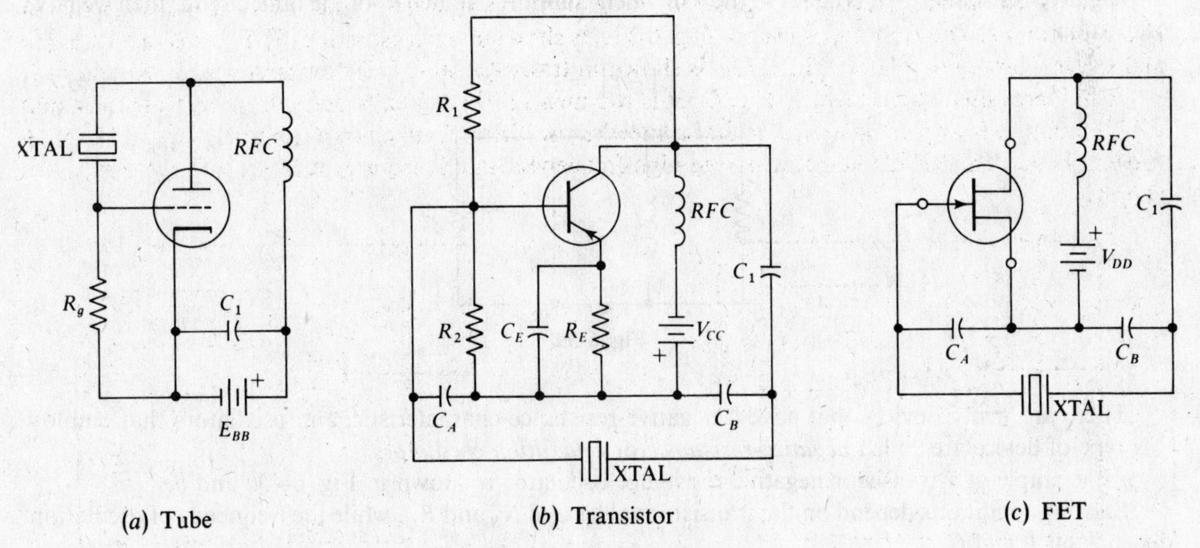

(*a*) Tube (*b*) Transistor (*c*) FET

Fig. 6-37 Pierce oscillator

A crystal schematic and an equivalent circuit are shown in Fig. 6-38*a* and *b*. C_H is the shunt capacitance of the input circuit plus the capacitance of the holder. L_1, C_1, and R_1 make up the equivalent circuit for the crystal itself.

At the series resonant frequency, the crystal becomes a small resistance and this is made use of in the Pierce oscillator to provide a feedback that meets the Barkhausen criterion at the series resonant frequency only.

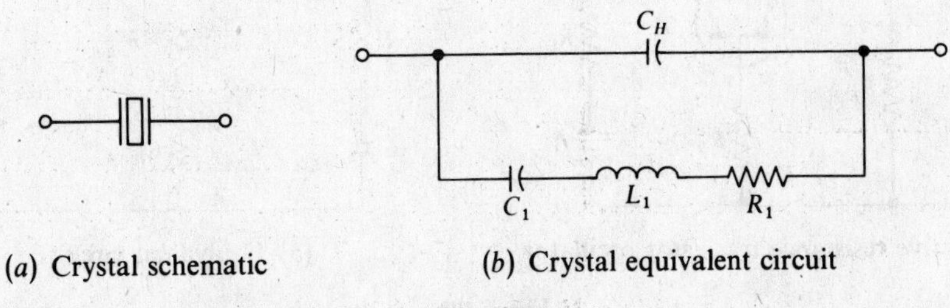

(*a*) Crystal schematic (*b*) Crystal equivalent circuit

Fig. 6-38

The value of C_H is usually such that the antiresonant frequency for the tank formed by C_H and the crystal is very close to the resonant frequency. This is used in a Miller oscillator where the voltage developed across the tank is maximum at the antiresonant frequency and provides the proper feedback to satisfy the Barkhausen criterion.

The advantages of crystals are high Q, on the order of 25 000, and the ability to hold close frequency tolerances if the temperature is compensated for or held constant.

Frequency stabilities of one part in 1 million are common for temperature-compensated or crystal oven oscillators.

The advent of phase-locked loops for stabilizing frequency and the use of cesium and rubidium microwave oscillators are cutting into the application for crystal controlled oscillators, but such oscillators still play an important part in many areas of the communications industry such as broadcasting and commercial transmitters and receivers. They also find widespread use in test equipment and as time bases for clocks, counters, and computers.

The tank circuit in an oscillator dissipates very little energy. This energy is supplied by the active element in the oscillator, and the tube, transistor, or FET may be thought of as a *negative resistance* $-r$, which supplies rather than dissipates energy.

A negative resistance $-r$ is placed across the tank circuit, as shown in Fig. 6-39. If the magnitude of the negative resistance, $-r$, is equal to the equivalent shunt resistance R' of the tank circuit, then we have oscillation:

$$|-r| = R' \tag{6-39}$$

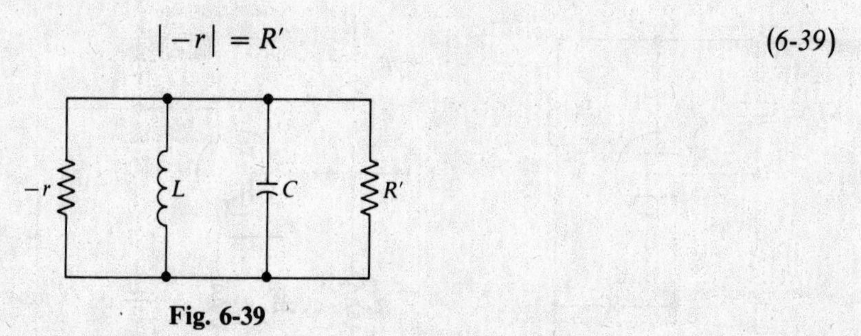

Fig. 6-39

There are many devices that have a negative-resistance characteristic, and oscillators that employ this type of device are called *negative-resistance* or *relaxation oscillators.*

An example of a transistor negative-resistance oscillator is shown in Fig. 6-40*a* and *b*.

Starting conditions depend on the transistor values and R_B and R_C, while the frequency of oscillation depends on L and C.

A special configuration of relaxation oscillator, known as a *multivibrator*, is shown in Fig. 6-41*a*, *b*, and *c* for tube, transistor, and FET, respectively. The multivibrator circuit has many applications when the waveform is not required to be sinusoidal.

We classify multivibrators into three types; *astable* or *free-running*, *monostable* or *one-shot*, and *bistable* or *flip-flop*. These are shown in Fig. 6-42*a*, *b*, and *c*. Most multivibrator circuits are built as IC devices and have their greatest application in digital work. We will look at IC multivibrators more closely in Chap. 7.

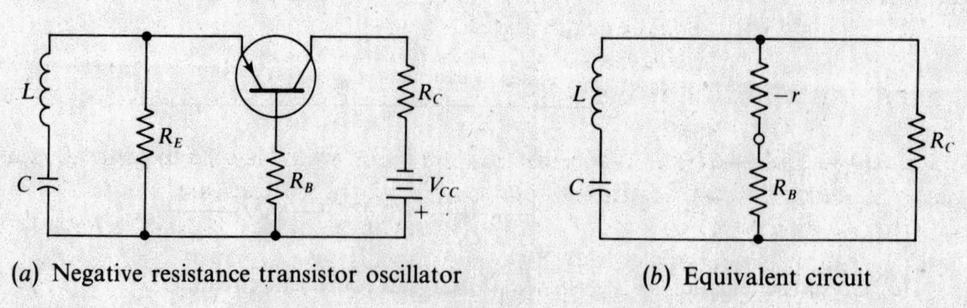

(*a*) Negative resistance transistor oscillator (*b*) Equivalent circuit

Fig. 6-40

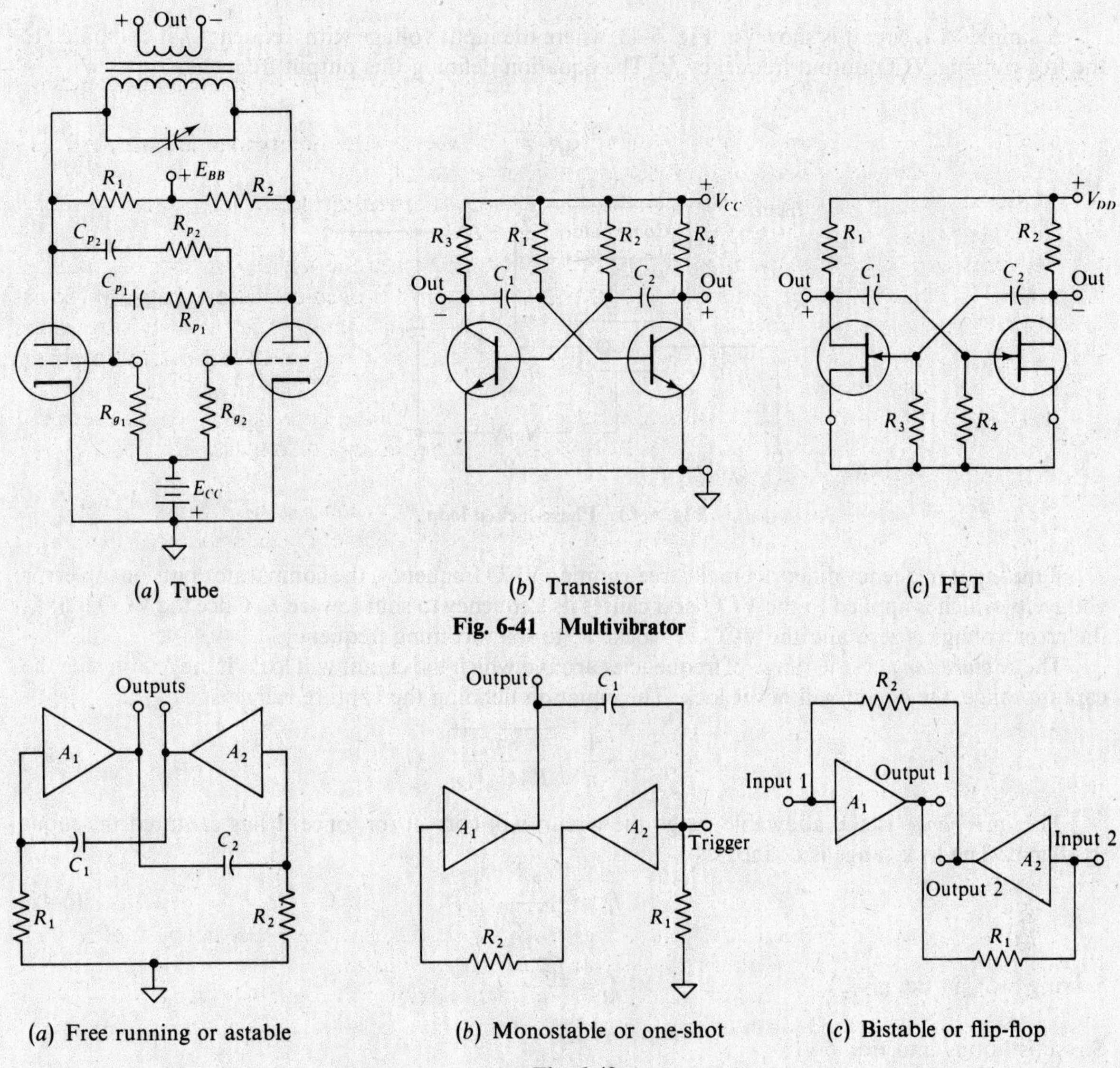

(a) Tube (b) Transistor (c) FET

Fig. 6-41 Multivibrator

(a) Free running or astable (b) Monostable or one-shot (c) Bistable or flip-flop

Fig. 6-42

6.8 CONTROLLED OSCILLATORS AND PHASE-LOCKED LOOPS

In our studies of the oscillator, we found that the frequency of oscillation was determined by the feedback components, and sometimes the active-element parameters.

It is possible to construct an oscillator which changes frequency with temperature or voltage. If one of the elements in the feedback frequency-controlling loop changes linearly with temperature, the frequency of oscillation will depend linearly on temperature, and we have a *temperature-controlled oscillator*, or TCO. It is also possible to have *gain-controlled oscillators*, or GCOs; *current-controlled oscillators*, or ICOs; *voltage-controlled oscillators*, or VCOs; and crystal controlled oscillators, or COs. Voltage controlled crystal oscillators, or VCXOs, have many applications in commercial modulator-demodulator (MODEM) circuits.

The most useful of these specialized oscillators is the VCO, which is used to generate musical tones for electronic instruments and to provide modulation and frequency tracking.

A circuit which uses a VCO to "lock" onto a particular frequency is called a *locked loop circuit*. When the locking parameters depend on the phase, the circuit is known as a *phase-locked loop circuit*, or PLL. PLLs have become widely used in IC form and are the basis for many receiver and frequency synthesizer circuits.

A simple PLL circuit is shown in Fig. 6-43, where the input voltage with frequency f_i is compared to the free-running VCO output frequency f_o. The equation defining this output frequency is

$$f_o \cong \frac{1}{4R_1C_1} \tag{6-40}$$

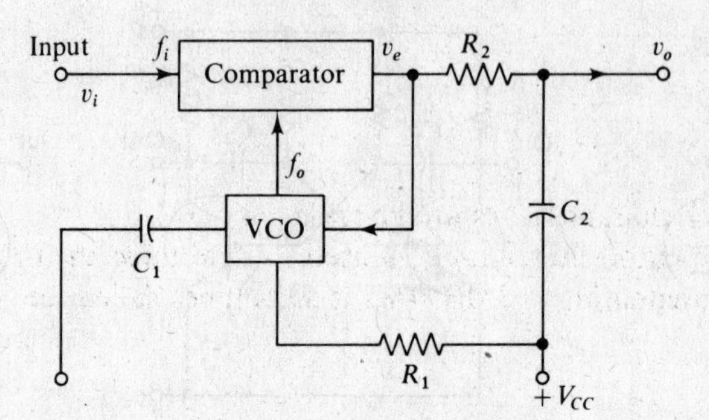

Fig. 6-43 Phase-locked loop

If the input frequency differs from the free-running VCO frequency, the comparator puts out an error voltage v_e which is applied to the VCO and causes its frequency to shift toward f_i. Once the VCO is at f_i, the error voltage is zero and the VCO is "locked" to the incoming frequency.

The *capture range* is the range of frequencies around which the circuit will lock. If the f_i is outside the capture range, the circuit will never lock. The equation defining the capture range is

$$f_{cp} \cong \pm\frac{1}{\pi}\sqrt{\frac{8\pi f_o}{R_2C_2V_{CC}}} \tag{6-41}$$

The *lock range* is the allowable range the circuit will correct for, once it has captured the input frequency. The lock range is defined as

$$f_l \cong \pm\frac{8f_o}{V_{CC}} \tag{6-42}$$

Solving for f_o in Eq. (*6-42*),

$$f_o = \frac{f_l V_{CC}}{8}$$

So, substituting into Eq. (*6-41*),

$$f_{cp} \cong \pm\frac{1}{\pi}\sqrt{\left(\frac{8\pi}{R_2C_2V_{CC}}\right)\left(\frac{f_l V_{CC}}{8}\right)} \cong \pm\frac{1}{\pi}\sqrt{\frac{\pi f_l}{R_2C_2}} \tag{6-41a}$$

Example 6.13 Find the free-running VCO frequency, the lock range, and the capture range for a PLL circuit with the following values:

$$R_1 = 4000\ \Omega,\quad C_1 = 0.0047\ \mu\text{F},\quad V_{CC} = 10.0\ \text{V},\quad R_2 = 3.2\ \text{k}\Omega,\quad \text{and}\quad C_2 = 0.08\ \mu\text{F}.$$

Using (*6-40*), $$f_o \cong \frac{1}{4R_1C_1} = \frac{1}{4(4\times10^3)(0.0047\times10^{-6})} = 13\,298\ \text{Hz} = 13.3\ \text{kHz}$$

which is the free-running VCO frequency. Then, using (*6-42*),

$$f_l \cong \pm\frac{8f_o}{V_{CC}} = \pm\frac{(8)(13.3\times10^3)}{(10.0)} = \pm 10\,638\ \text{Hz} = \pm 10.6\ \text{kHz}$$

from the desired frequency, or a range of 21.2 kHz, and using (*6-41a*),

$$f_{cp} \cong \pm\frac{1}{\pi}\sqrt{\frac{\pi f_l}{R_2C_2}} = \pm\frac{1}{\pi}\sqrt{\frac{(\pi)(10.6\times10^3)}{(3.2\times10^3)(0.08\times10^{-6})}} = \pm 3637\ \text{Hz} = \pm 3.64\ \text{kHz}$$

from the desired frequency, or a range of 7.28 kHz.

6.9 FILTERS

Any device that has varying output with frequency can be thought of as a filter.

Inductors or capacitors are examples of simple *passive* filters, while a tuned amplifier is an example of an *active* filter. In addition, filters can be grouped into categories as follows:

- Low pass
- High pass
- Bandpass
- Bandstop

Examples of ideal filter characteristics are given in Fig. 6-44.

In practical filters, the cutoff frequency f_c is used to determine the pass point limit and is usually selected to be where the attenuation is 3 dB. This is sometimes called the *half-power cutoff point.*

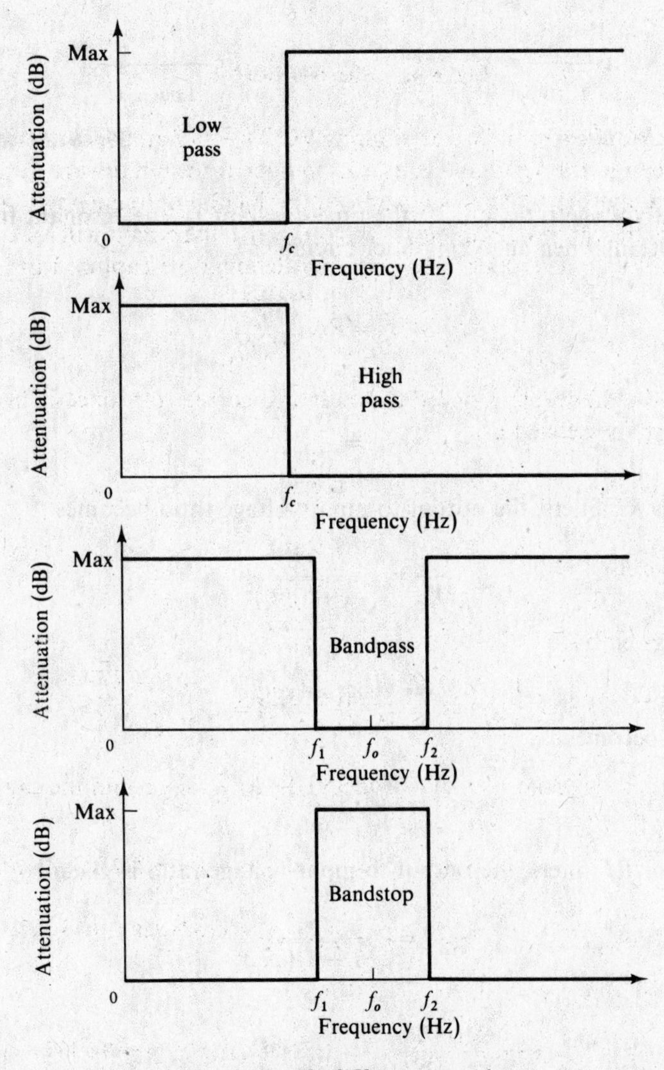

Fig. 6-44 Ideal filter attenuation

The center frequency f_o is the geometric mean of the high and low cutoff frequencies. From (*6-24*),

$$f_o = \sqrt{f_L f_H}$$

When f_L and f_H are close together, so that $f_H/f_L \leq 1.2$, we may approximate f_o by the average

$$f_o = \frac{f_L + f_H}{2} \tag{6-24a}$$

The simplest filters are passive *RC* or *RL* filters, as shown in Figs. 6-45 and 6-46 in both the low-pass and high-pass forms.

In building analog computer blocks or simulators, the *RC* or *RL* low-pass filter is used as an *integrator* while the low-pass *RC* or *RL* filter is used as a *differentiator*.

Fig. 6-45 *RC* filters

Fig. 6-46 *RL* filters

The output-to-input voltage ratio V_{out}/V_i, the time constant t_c, the resonant frequency f_o, and the phase angle ϕ are important when analyzing filter circuits.

For *RC* filters,

$$t_c = RC \tag{6-43a}$$

For *RL* filters,

$$t_c = \frac{L}{R} \tag{6-43b}$$

For low-pass *RC* or *RL* filters, the output-to-input voltage ratio becomes

$$\frac{V_{out}}{V_i} = \frac{1}{\sqrt{1 + (\omega t_c)^2}} \tag{6-44}$$

and the phase angle is given by

$$\phi = -\arctan \omega t_c \tag{6-45}$$

For $\omega t_c \gg 1$, Eq. (*6-44*) becomes

$$\frac{V_{out}}{V_i} \cong \frac{1}{\omega t_c} \tag{6-44a}$$

For high-pass *RC* or *RL* filters, the output-to-input voltage ratio is given by

$$\frac{V_{out}}{V_i} = \frac{1}{\sqrt{1 + 1/(\omega t_c)^2}} \tag{6-46}$$

with the phase angle

$$\phi = \arctan \frac{1}{\omega t_c} \tag{6-47}$$

For $\omega t_c^2 \ll 1$, Eq. (*6-46*) becomes

$$\frac{V_{out}}{V_i} \cong \omega t_c \tag{6-46a}$$

Since V_{out}/V_i is less than unity, the attenuation α in decibels is given by

$$\alpha = -20 \log \frac{V_{out}}{V_i}$$

Setting either Eq. (*6-44*) or (*6-46*) equal to 0.707 to obtain the half-power cutoff points, we find

$$f_c = \frac{1}{2\pi t_c} \tag{6-48}$$

A plot of frequency vs. attenuation and phase angle is shown for the low-pass *RC* or *RL* filters in Fig. 6-47 and for the high-pass *RC* or *RL* filters in Fig. 6-48.

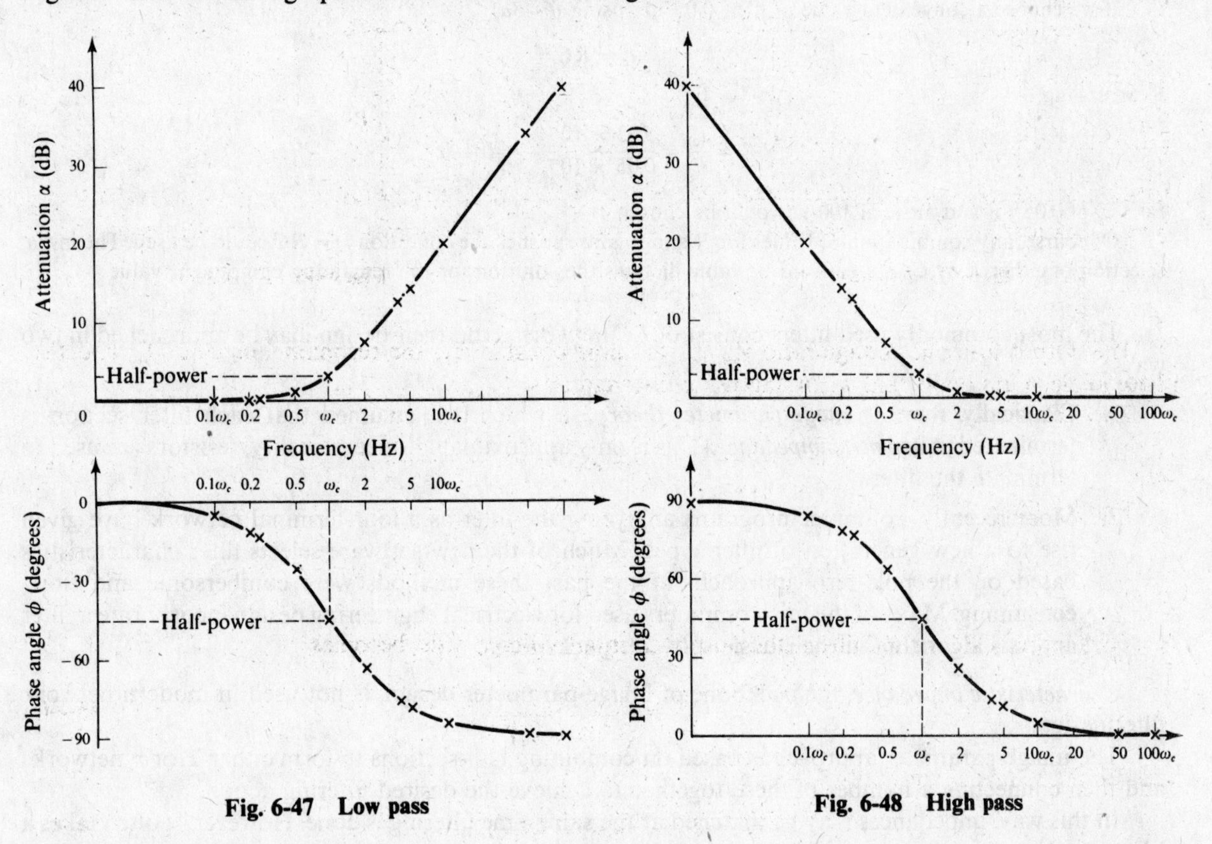

Fig. 6-47 Low pass **Fig. 6-48 High pass**

Example 6.14 Find the values of *R* and *C* for a high-pass filter that has an attenuation of 10 dB at 10 kHz.

For a high-pass *RC* filter, we use (*6-46a*):

$$\frac{V_{out}}{V_i} \cong \omega t_c$$

We are given the attenuation α in decibels, so

$$\alpha = -20 \log \frac{V_{out}}{V_i}$$

and the desired V_{out}/V_i may be calculated from

$$10 = -20 \log \frac{V_{out}}{V_i}$$

$$-0.5 = \log \frac{V_{out}}{V_i}$$

Taking the inverse log of both sides;

$$0.316 = \frac{V_{\text{out}}}{V_i}$$

So at $f = 10$ kHz,

$$0.316 = \omega = (2\pi)(10 \times 10^3)t_c$$

and

$$t_c = \frac{0.316}{(2\pi)(10 \times 10^3)} = 5.03 \times 10^{-6}\,\text{s} = 5.03\ \mu\text{s}$$

If we choose a convenient value of C as 0.05 μF, using (*6-43a*),

$$t_c = RC$$

Rearranging,

$$R = \frac{t_c}{C} = \frac{5.03 \times 10^{-6}}{0.05 \times 10^{-6}} = 101\ \Omega$$

So a C of 0.05 μF and an R of 100 Ω would be chosen.

Of course, any combination of values for R and C that satisfies the condition $t_c = RC$ could be used. The initial selection of either R or C as a convenient value dictates the solution for the remaining component value.

The most commonly used filters consist of LC networks, and their design may be approached in two ways:

1. Classically, there is *image-parameter theory*, in which it is assumed that each filter section is terminated in its *image impedance*. This is only approximated since in reality, resistors are used to terminate the filter.
2. More recently, computer programs analyzing the filter as a four-terminal network have given rise to a new generation of filter types. Much of the new software selects filter characteristics based on the pole-zero approach. In the past these methods were cumbersome and time-consuming. Many of the minicomputers used for electrical engineering design incorporate a filter analysis algorithm in their basic software package.

Characteristic impedance, the backbone of image-parameter design, is not used in modern network filter design.

The image-parameter approach is based on combining half-sections to form either T or π networks and then connecting a number of these together to achieve the desired filtering action.

In this way, impedances may be matched at the same time filtering is done. However, it often takes a large number of sections to achieve the desired result and not every requirement may be met.

Figure 6-49*a* shows the basic half-section and the image impedance looking into each side. Figure 6-49*b* and *c* show how the half-sections are combined to make a full T and a full π section.

The image impedance Z_T is the impedance seen looking into terminals 1 and 2 with Z_π connected across terminals 3 and 4. Z_T is sometimes termed the *midseries image impedance*.

The image impedance Z_π is the impedance seen looking into terminals 3 and 4 with Z_T connected across terminals 1 and 2. Z_π is sometimes termed the *midshunt image impedance*.

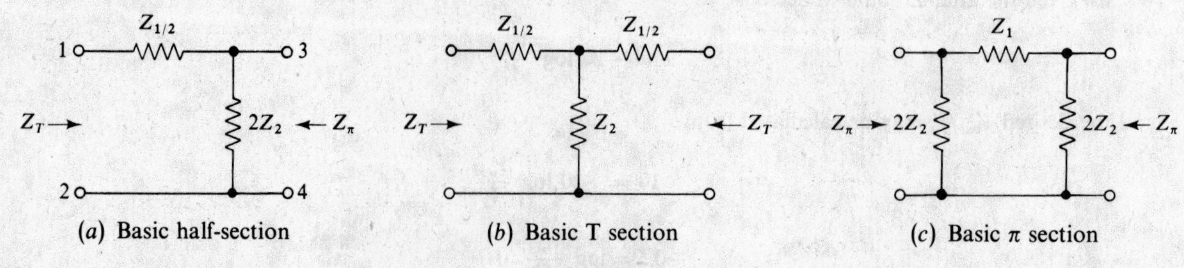

(*a*) Basic half-section (*b*) Basic T section (*c*) Basic π section

Fig. 6-49

The basic relationships are given by

$$Z_T Z_\pi = Z_1 Z_2 \tag{6-49}$$

$$Z_T^2 = Z_1 Z_2 + 0.25 Z_1^2 \tag{6-50}$$

$$Z_\pi^2 = \frac{(Z_1 Z_2)^2}{Z_1 Z_2 + 0.25 Z_1^2} \tag{6-51}$$

If we terminate the sections with an impedance Z_o such that

$$Z_T Z_\pi = Z_o^2 \tag{6-52}$$

we call Z_o the characteristic impedance.

If Z_o is a pure resistance, we call it the characteristic resistance k, and Eq. (*6-52*) becomes

$$Z_T Z_\pi = k^2 \tag{6-53}$$

Many filters are used with this condition imposed and are known as *constant-k filters.*

Examples of constant-k half-sections are shown in Fig. 6-50, while Table 6-2 gives the impedance and design criteria for the constant-k full sections shown in Figs. 6-51 and 6-52.

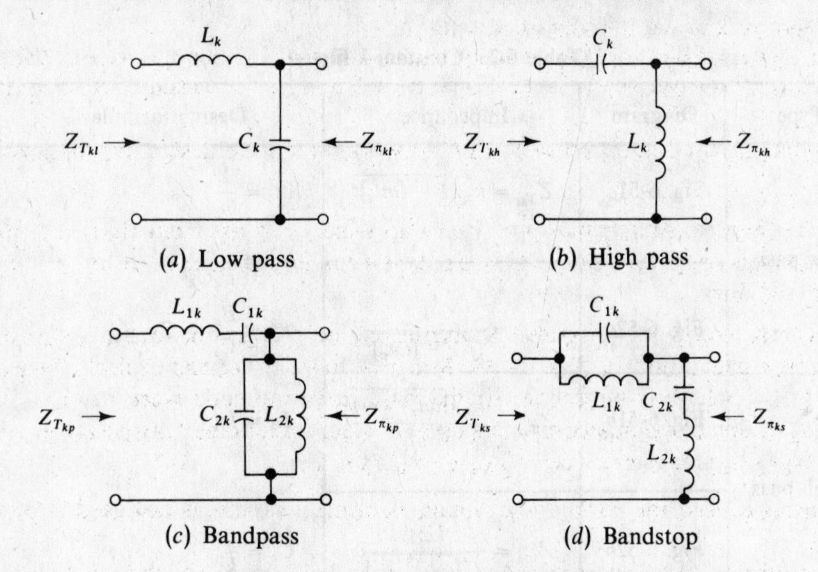

Fig. 6-50 Constant-*k* half-sections

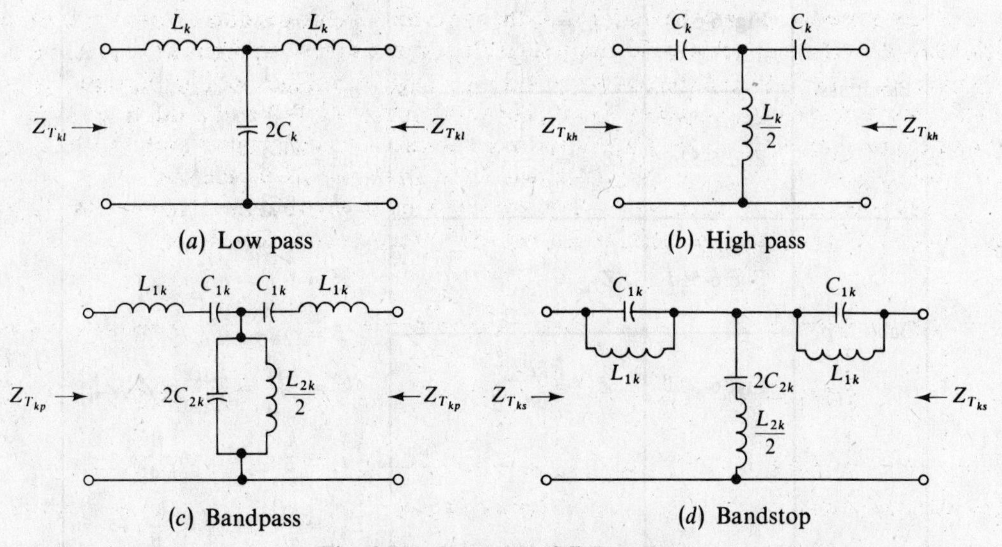

Fig. 6-51 Constant-*k* full T sections

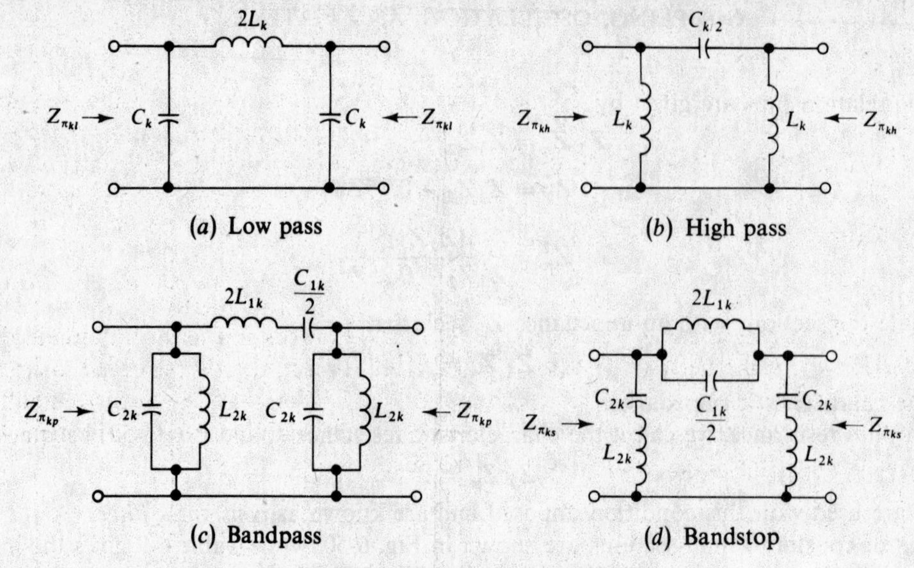

Fig. 6-52 Constant-*k* full π sections

Table 6-2 Constant-*k* filters

Type	Diagram	Impedance	Design formulas
Low pass	Fig. 6-51*a*	$Z_{T_{kl}} = k\sqrt{1-(\omega t_c)^2}$	$t_c = \dfrac{1}{\omega_c}$
	Fig. 6-52*a*	$Z_{\pi kl} = \dfrac{k}{\sqrt{1-(\omega t_c)^2}}$	$\omega_c^2 = \dfrac{1}{L_k C_k}$ $k^2 = \dfrac{L_k}{C_k} = Z_{T_k} Z_{\pi k}$
High pass	Fig. 6-51*b*	$Z_{T_{kh}} = \dfrac{k\sqrt{(\omega t_c)^2 - 1}}{\omega t_c}$	$L_k = kt_c$
	Fig. 6-52*b*	$Z_{\pi kh} = \dfrac{k\omega t_c}{\sqrt{(\omega t_c)^2 - 1}}$	$C_k = \dfrac{t_c}{k}$
Bandpass	Fig. 6-51*c*	$Z_{T_{kp}} = \dfrac{kad}{B\omega}$	$\omega_o^2 = \omega_1 \omega_2$ $\omega_o^2 = \dfrac{1}{L_{1k}C_{1k}} = \dfrac{1}{L_{2k}C_{2k}}$
	Fig. 6-52*c*	$Z_{\pi kp} = \dfrac{kB\omega}{ad}$	$a^2 = \lvert \omega_2^2 - \omega^2 \rvert$ $c^2 = \lvert \omega_o^2 - \omega^2 \rvert$
Bandstop	Fig. 6-51*d*	$Z_{T_{ks}} = \dfrac{kad}{c}$	$d^2 = \lvert \omega^2 - \omega_1^2 \rvert$ $B = \omega_2 - \omega_1$
	Fig. 6-52*d*	$Z_{\pi ks} = \dfrac{kc}{ad}$	$k^2 = \dfrac{L_{1k}}{C_{2k}} = \dfrac{L_{2k}}{C_{1k}} = Z_{T_k} Z_{\pi k}$ $L_{1k} = \dfrac{k}{B} \quad C_{1k} = \dfrac{B}{k\omega_o^2}$ $L_{2k} = \dfrac{kB}{\omega_o^2} \quad C_{2k} = \dfrac{1}{kB}$

Constant-k filters approach the cutoff frequencies slowly and therefore are not very selective. To improve the selectivity, we may modify either leg of the filter by putting in a series or parallel resonant LC circuit. This causes the filter response to become more selective and introduces a *peak attenuation frequency* f_∞.

We then define

$$m = \sqrt{1 - \left(\frac{f_c}{f_\infty}\right)^2} \tag{6-54}$$

and call filters based on this approach m-derived. If the added network is a series LC circuit, the filter is called *series m-derived*. If the network added is a parallel LC circuit, the filter is called *shunt m-derived*.

Figure 6-53 shows series m-derived half-sections and Fig. 6-54 shows shunt m-derived half-sections. Table 6-3 gives the impedance and design criteria for series m-derived full sections and Table 6-4 gives the impedance and design criteria for shunt m-derived full sections.

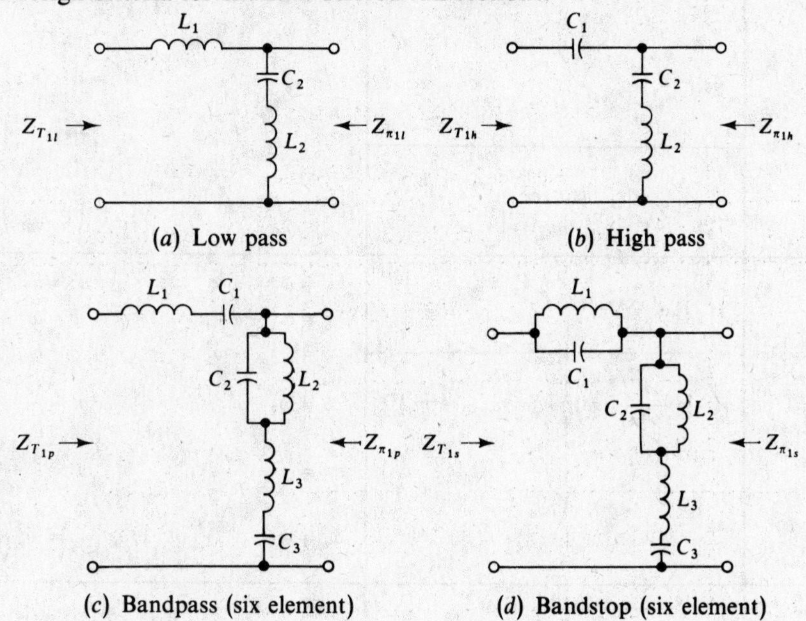

Fig. 6-53

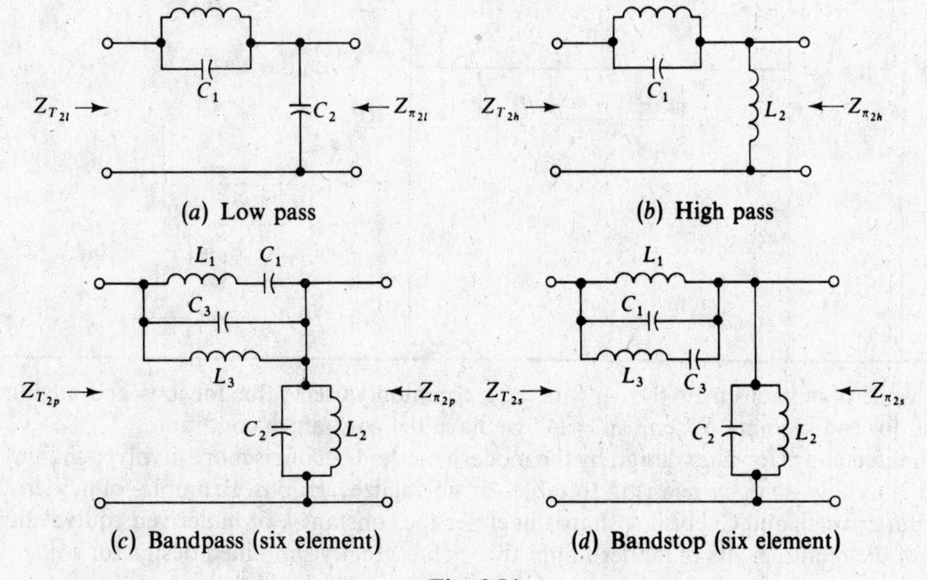

Fig. 6-54

Table 6-3 Series *m*-derived filters

Type	Diagram	Impedances	Design formulas
Low pass	Fig. 6-53*a*	$Z_{T_{1l}} = Z_{T_{kl}}$ $Z_{\pi_{1l}} = Z_{\pi_{kl}}[1 - (\omega t_c)^2(1 - m^2)]$	$t_c = \frac{1}{\omega_c}$ $\quad k^2 = \frac{L_k}{C_k}$ $\omega_c^2 = \frac{1}{L_k C_k}$ $\quad m^2 = 1 - \frac{1}{(t_c \omega_\infty)^2}$ $L_1 = mL_k$ $\quad L_2 = \left(\frac{1 - m^2}{m}\right)L_k = \frac{L_1}{m^2} - L_1$ $C_2 = mC_k$
High pass	Fig. 6-53*b*	$Z_{T_{1h}} = Z_{T_{kh}}$ $Z_{\pi_{1h}} = Z_{\pi_{kh}}\left[1 - \frac{(1 - m^2)}{(\omega t_c)^2}\right]$	$t_c = \frac{1}{\omega_c}$ $\quad k^2 = \frac{L_k}{C_k}$ $\omega_c^2 = \frac{1}{L_k C_k}$ $\quad m^2 = 1 - (t_c \omega_\infty)^2$ $L_2 = \frac{L_k}{m}$ $C_1 = \frac{C_k}{m}$ $\quad C_2 = \left(\frac{m}{1 - m^2}\right)C_k = \left(\frac{m^2}{1 - m^2}\right)C_1$
Bandpass	Fig. 6-53*c*	$Z_{T_{1p}} = Z_{T_{kp}}$ $Z_{\pi_{1p}} = \frac{k}{\omega Bad}[(ad)^2 + (\omega_o^2 m_2 - \omega^2 m_1)^2]$	$a^2 = \omega_2^2 - \omega^2;$ $\quad d^2 = \omega^2 - \omega_1^2$ $g = \omega_1^2 - \omega_{1\infty}^2;$ $\quad h = \omega_2^2 - \omega_{1\infty}^2$ $p = \omega_{2\infty}^2 - \omega_1^2;$ $\quad q = \omega_{2\infty}^2 - \omega_2^2$ $B = \omega_2 - \omega_1;$ $\quad m' = \frac{(m_1 - m_2)^2}{m_1 m_2}$ $m_1 = \frac{gb + pq}{g + p};$ $\quad m_2 = \frac{gh\omega_{2\infty}^2 + pq\omega_{1\infty}^2}{\omega_o^2(g + p)}$ $\omega_o^2 = \omega_1 \omega_2;$ $\quad \omega_{1\infty}\omega_{2\infty} = \omega_o^2 \sqrt{\frac{1 - m_2^2}{1 - m_1^2}}$
Bandstop	Fig. 6-53*d*	$Z_{T_{1s}} = Z_{T_{ks}}$ $Z_{\pi_{1s}} = \frac{k}{c^2}\left[\frac{c^4 - (1 - m^2)(\omega B)^2}{\sqrt{c^4 - (\omega B)^2}}\right]$	$L_1 = m_1 L_k;$ $L_2 = \frac{L_{1k}}{m_2}\left[\left(\frac{B}{\omega_o}\right)^2 - m'\right];$ $L_3 = \left(\frac{1 - m_1^2}{m_1}\right)L_{1k}$ $C_1 = \frac{C_{1k}}{m}$ $\quad C_2 = \frac{m_1 C_{1k}}{(B/\omega_o)^2 - m'}$ $\quad C_3 = \left(\frac{m_2}{1 - m_2^2}\right)C_{1k}$ $B = \omega_2 - \omega_1$ $\quad \omega_o^2 = \omega_{1\infty}\omega_{2\infty}$ $C^2 = \omega_o^2 - \omega^2$ $\quad k^2 = \frac{L_{1k}}{C_{2k}} = \frac{L_{2k}}{C_{1k}}$ $m^2 = 1 - \frac{(\omega_{2\infty} - \omega_{1\infty})^2}{B^2}$ $L_1 = mL_{1k}$ $\quad L_2 = \left(\frac{1 - m^2}{m}\right)L_{1k}$ $\quad L_3 = \frac{L_{2k}}{m}$ $C_1 = \frac{C_{1k}}{m}$ $\quad C_2 = \left(\frac{m}{1 - m^2}\right)C_{1k}$ $\quad C_3 = mC_{2k}$

The value of m varies from 0 to 1, with 0.6 a commonly used value for low- and medium-frequency work and for end sections. When $m = 1$, we have the constant-k condition.

The mathematics for filter design by the modern method becomes more involved than we can handle here and is usually done by referring to tables or normalized graphs. Examples of modern filter design include Butterworth and Chebishev shapes in either the constant-k or m-derived equivalents. Computer analysis of the requirements of modern filter theory has greatly simplified design for some specific cases by producing charts of component values for various types of desired response.

Table 6-4 Shunt *m*-derived filters

Type	Diagram	Impedances	Design formulas
Low pass	Fig. 6-54*a*	$Z_{T_{2l}} = \dfrac{k^2}{Z_{\pi 1l}}$ $Z_{\pi 2l} = Z_{\pi kl}$	$k^2 = \dfrac{L_k}{C_k}$ $\omega_c^2 = \dfrac{1}{L_k C_k}$ $\quad m^2 = \dfrac{1}{(t_c \omega_\infty)^2}$ $L_1 = mL_k$ $C_1 = \left(\dfrac{1-m^2}{m}\right) C_k$ $\quad C_2 = mC_k$
High pass	Fig. 6-54*b*	$Z_{T_{2h}} = \dfrac{k^2}{Z_{\pi 1l}}$ $Z_{\pi 2h} = Z_{\pi kh}$	$k^2 = \dfrac{L_k}{C_k}$ $\omega_c^2 = \dfrac{1}{L_k C_k}$ $\quad m^2 = 1 - (t_c \omega_\infty)^2$ $L_1 = \left(\dfrac{m}{1-m^2}\right) L_k$ $\quad L_2 = \dfrac{L_k}{m}$ $C_1 = \dfrac{C_k}{m}$
Bandpass	Fig. 6-54*c*	$Z_{T_{2p}} = \dfrac{k^2}{Z_{\pi 1p}}$ $Z_{\pi 2p} = Z_{\pi kp}$	See series *m*-derived (Table 6-3) $L_1 = \dfrac{m_1 L_{2k}}{(B/\omega_o)^2 - m'}$ $\quad L_2 = \dfrac{L_{2k}}{m_2}$ $\quad L_3 = \left(\dfrac{m_2}{1-m_2^2}\right) L_{2k}$ $C_1 = \dfrac{C_{2k}}{m}\left[\left(\dfrac{B}{\omega_o}\right)^2 - m'\right]$ $\quad C_2 = m_1 C_{2k}$ $\quad C_3 = \left(\dfrac{1-m_1^2}{m_1}\right) C_{2k}$
Bandstop	Fig. 6-54*d*	$Z_{T_{2s}} = \dfrac{k^2}{Z_{\pi 1s}}$ $Z_{\pi 2s} = Z_{\pi ks}$	See series *m*-derived (Table 6-3) $L_1 = mL_{1k}$ $\quad L_2 = \dfrac{L_{2k}}{m}$ $\quad L_3 = \left(\dfrac{m}{1-m^2}\right) L_{2k}$ $C_1 = \dfrac{C_{1k}}{m}$ $\quad C_2 = mC_{2k}$ $\quad C_3 = \left(\dfrac{1-m^2}{m}\right) C_{2k}$

Example 6.15 Find the values for the constant-*k* section shown in Fig. 6.55*a* if the terminating resistance that the filter is to be connected with is 600 Ω. The desired cutoff frequency for the filter is 10 kHz.

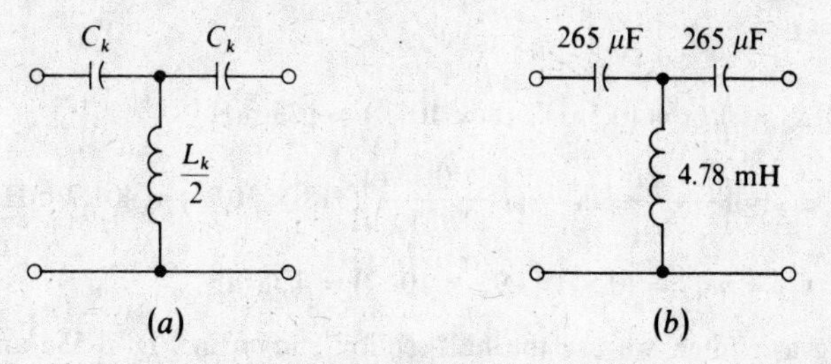

Fig. 6-55

By our definitions,

$$k = 600\ \Omega$$

and

$$\omega_c = 2\pi f_c = 2\pi(10 \times 10^3) = 6.28 \times 10^4 \text{ rad/s}$$

Figure 6-55*a* shows a high-pass T full section; so from Table 6-2 we obtain the design equations and substitute values:

$$t_c = \frac{1}{\omega_c} = \frac{1}{6.28 \times 10^4} = 1.59 \times 10^{-5} \text{ s}$$

$$L_k = kt_c = (600)(1.59 \times 10^{-5}) = 9.55 \times 10^{-3} \text{ H} = 9.55 \text{ mH}$$

$$C_k = \frac{t_c}{k} = \frac{1.59 \times 10^{-5}}{600} = 2.65 \times 10^{-8} \text{ F} = 265 \ \mu\text{F}$$

Once we have C_k and L_k, we can assign values to Fig. 6-55*a*, remembering that we need $\frac{1}{2}L_k$ for the shunt arm. The final circuit appears in Fig. 6-55*b*.

We may wish to check our results by calculating the k value using the design equation from Table 6-2:

$$k^2 = \frac{L_k}{C_k} = \frac{9.55 \times 10^{-3}}{2.65 \times 10^{-8}} = 3.60 \times 10^5$$

Taking the square root of both sides, we get

$$k \doteq 600 \ \Omega$$

Example 6.16 Design a low-pass series m-derived π filter that has an $f_c = 100$ Hz and a maximum attenuation at 120 Hz. Assume the terminating resistance is 200 Ω.

Since $f_c = 100$ Hz,

$$\omega_c = 2\pi f_c = 2\pi(100) = 628 \text{ rad/s}$$

Since $f_\infty = 120$ Hz,

$$\omega_\infty = 2\pi f_\infty = 2\pi(120) = 754 \text{ rad/s}$$

From Table 6-3,

$$t_c = \frac{1}{\omega_c} = \frac{1}{628} = 1.59 \times 10^{-3} \text{ s}$$

$$m^2 = 1 - \frac{1}{(t_c\omega_\infty)^2} = 1 - \frac{1}{[(1.59 \times 10^{-3})(754)]^2} = 1 - (0.696) = 0.304$$

Taking the square root of both sides,

$$m = 0.551$$

To get the m-derived values, we first need to find the constant-k values L_k and C_k. From Table 6-2,

$$L_k = kt_c = (200)(1.59 \times 10^{-3}) = 0.318 \text{ H} = 318 \text{ mH}$$

$$C_k = \frac{t_c}{k} = \frac{1.59 \times 10^{-3}}{200} = 7.95 \times 10^{-6} \text{ F} = 7.95 \ \mu\text{F}$$

Now, from Table 6-3,

$$L_1 = mL_k = (0.551)(318 \times 10^{-3}) = 175 \text{ mH}$$

$$L_2 = \left(\frac{1 - m^2}{m}\right)L_k = \left(\frac{1 - 0.304}{0.551}\right)(318 \times 10^{-3}) = 401.7 \text{ mH}$$

$$C_2 = mC_k = (0.551)(7.95 \times 10^{-6}) = 4.38 \ \mu\text{F}$$

Since we are designing a π filter, we use the half-section shown in Fig. 6-53*a* and combine it with another identical half-section so that the terminations on both ends are Z_π as shown in Fig. 6-56*a*.

The filter with the calculated values appears as Fig. 6-56*b* and a practical filter using rounded off values is shown in Fig. 6-56*c*.

As a check, we calculate the cutoff frequency of the practical circuit using Table 6-3 and

$$2L_1 = 350 \text{ mH}$$

and

$$C_2 = 5.0 \text{ F}$$

So

$$L_1 = 175 \text{ mH}$$

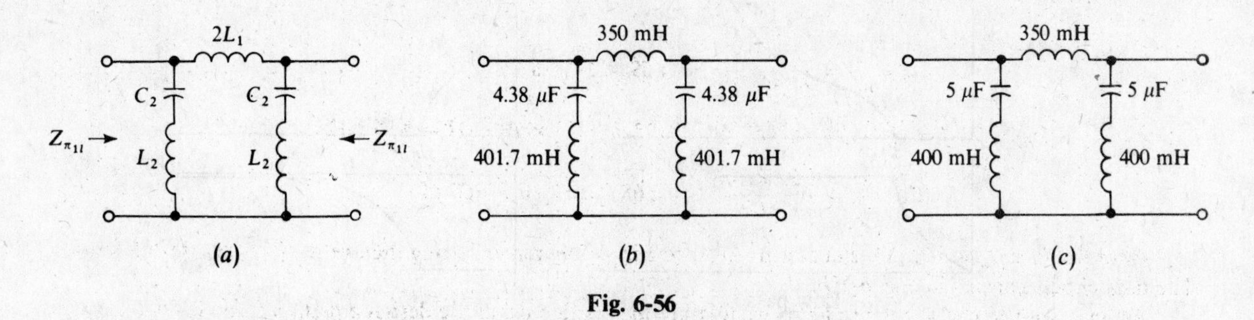

Fig. 6-56

$$\omega_c^2 = \frac{1}{L_k C_k} = \frac{m^2}{L_1 C_2} = \frac{0.304}{(175 \times 10^{-3})(5 \times 10^{-6})} = 3.47 \times 10^5$$

and

$$\omega_c = \sqrt{3.47 \times 10^5} = 589$$

This is less than 10 percent from the design requirements and for a power supply filter would be acceptable. For critical filters such as bandpass or notch types, precision components would be used to get as close to the desired result as was economically practical.

One of the most important rules about filters can be seen by studying their attenuation and phase angle graphs.

For a resistive termination, an ideal filter should have no attenuation and zero phase angle in the pass region and constant attenuation in the stop region, as shown in Fig. 6-57.

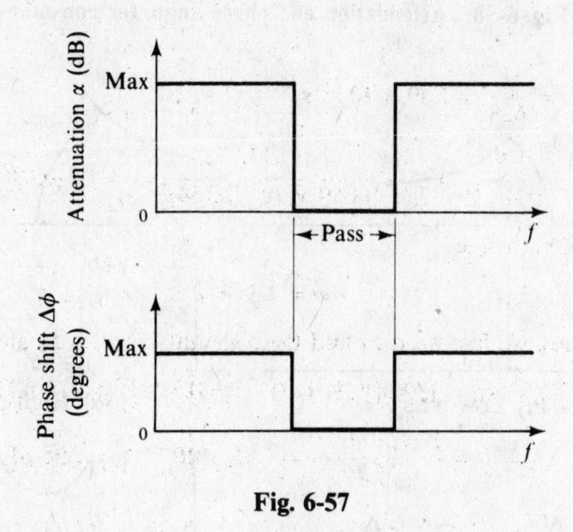

Fig. 6-57

The attenuation and phase angle graphs for constant-k filters are plotted against frequency in Fig. 6-58, while the attenuation and phase angle vs. frequency graphs for m-derived filters are shown in Fig. 6-59. Studying Figs. 6-58 and 6-59, we see that the general rules for filters may be stated as follows:

1. The phase angle varies throughout the range where the attenuation is zero.
2. The attenuation varies throughout the range where the phase angle is constant.

With modern network filter theory, it is possible to design a filter that is close to ideal in certain respects.

Active or *electronic* filters are used often in IC design. They are characterized by sharp response and high Q and with the availability of inexpensive op amps have been increasing in popularity.

There are tables and graphs for the serious filter designer who wishes to use this approach and design books are also available from integrated circuit manufacturers.

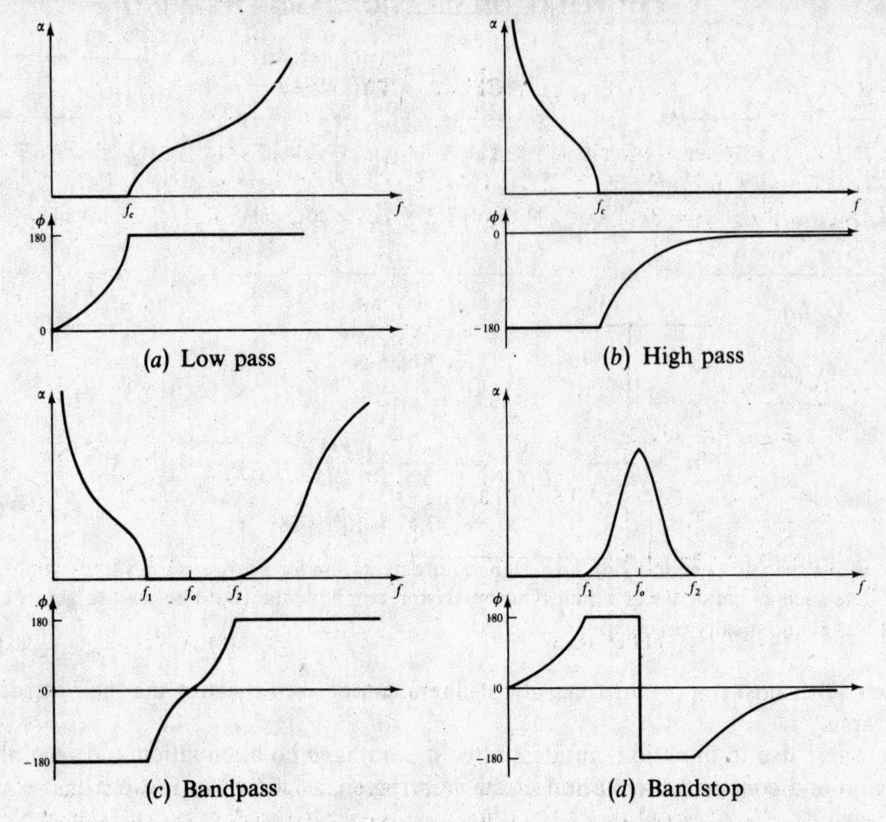

Fig. 6-58 Attenuation and phase angle for constant-k filters

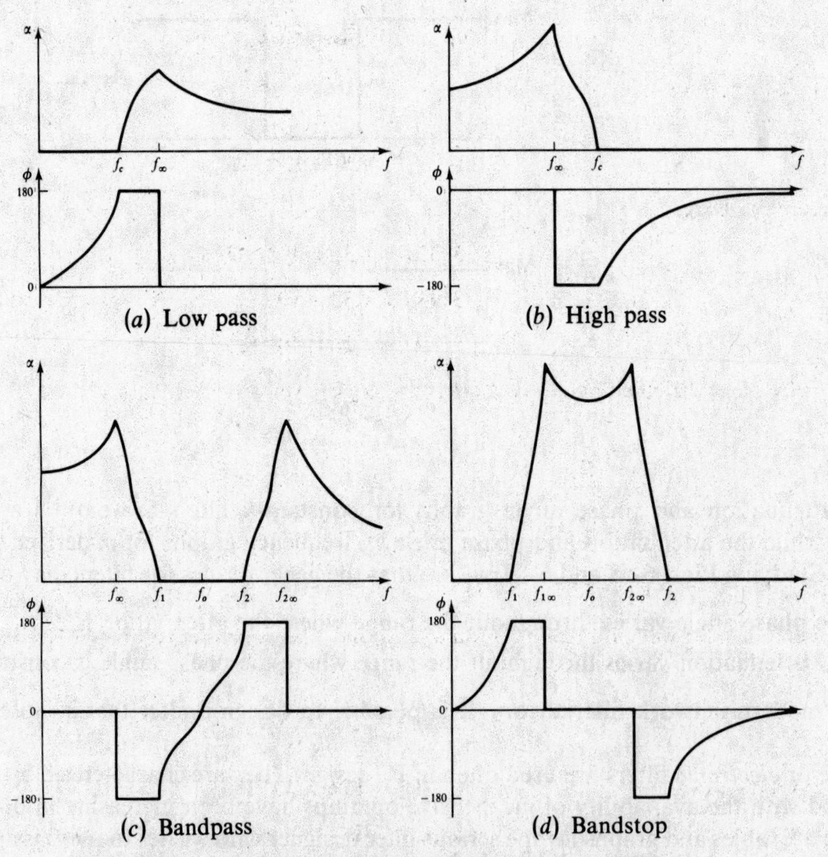

Fig. 6-59 Attenuation and phase angle for m-derived filters

Solved Problems

6.1 The FET amplifier stage shown in Fig. 6-60*a* is to have a midband gain of 36.0, a gain of 34.0 at 100 kHz, and a gain of 27.0 at 125 Hz. C_S is a perfect bypass for all frequencies under consideration, $\mu = 380$, and $r_{DS} = 120\ \text{k}\Omega$. Find the value of C_c and C_g.

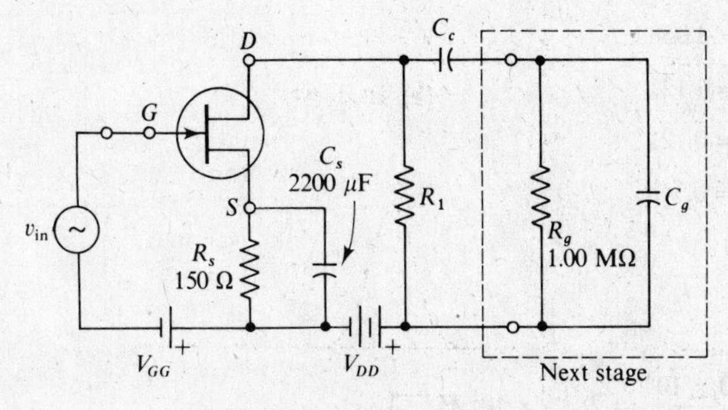

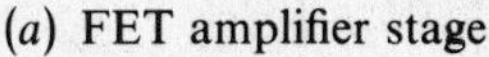

(*a*) FET amplifier stage

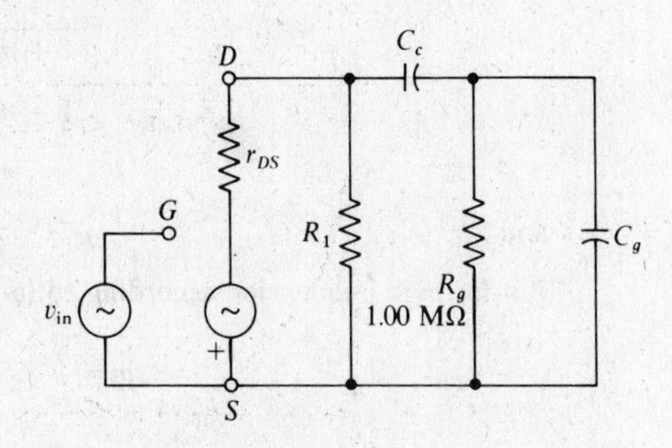

(*b*) All-frequency equivalent circuit

Fig. 6-60

First we draw the all-frequency equivalent circuit shown in Fig. 6-60*b* for this *RC* coupled amplifier noting that C_S shorts out R_S for all the frequencies under consideration as per the problem statement. Also the equivalent load resistance R_L can be seen in Fig. 6-60*b* to be R_1 in parallel with R_g. So from (6-5*a*),

$$A_{\text{mid}} = \frac{\mu R_L}{r_{DS} + R_L}$$

Substituting values,

$$36.0 = \frac{(380)R_L}{120 \times 10^3 + R_L}$$

Cross multiplying,

$$380R_L = 4.32 \times 10^6 + 36R_L$$

$$344R_L = 4.32 \times 10^6$$

$$R_L = 1.26 \times 10^4\ \Omega$$

Since R_L is the combination of R_1/R_g, and R_g is given as 1.00 MΩ in Fig. 6-60*a*,

$$\frac{1}{R_L} = \frac{1}{R_1} + \frac{1}{R_g}$$

and
$$\frac{1}{R_1} = \frac{1}{R_L} - \frac{1}{R_g} = \frac{1}{1.26 \times 10^4} - \frac{1}{1.00 \times 10^6} = 7.863 \times 10^{-5}$$

So
$$R_1 = 1.27 \times 10^4\ \Omega$$

Notice that if we had made the assumption that R_g is much bigger than R_1 (the usual case), we could have solved for R_1 directly using

$$A_{\text{mid}} = \frac{\mu R_1}{r_{DS} + R_1}$$

and from (6-5*a*)
$$R_1 = 1.26 \times 10^4\ \Omega$$

which is only a 100-Ω difference. This approach is often used in amplifier analysis, and we will use the approximate result of $1.26 \times 10^4\ \Omega$ for the remaining calculations.

For the high-frequency range, we need to find f_H. Since from (6-7)

$$A_{100\,\mathrm{k}} = \frac{A_{\mathrm{mid}}}{\sqrt{1+m^2}}$$

$$34.0 = \frac{36.0}{\sqrt{1+m^2}}$$

$$\sqrt{1+m^2} = 1.0588$$

$$1+m^2 = 1.121$$

$$m^2 = 0.121$$

and

$$m = 0.348$$

But for high frequencies, according to (*6-8b*)

$$m = \frac{f}{f_H}$$

So

$$f_H = \frac{f}{m} = \frac{100 \times 10^3}{0.348} = 287.4 \times 10^3 \text{ Hz}$$

Also, from (*6-3b*),

$$f_H = \frac{r_{DS} + R_1}{2\pi C_g r_{DS} R_1}$$

Solving for C_g,

$$C_g = \frac{r_{DS} + R_1}{2\pi f_H r_{DS} R_1} = \frac{120 \times 10^3 + 1.26 \times 10^4}{(2\pi)(287.4 \times 10^3)(120 \times 10^3)(1.26 \times 10^4)}$$

$$= 4.86 \times 10^{-11} = 48.6 \times 10^{-12} \text{ F} = 48.6 \text{ pF}$$

For the low frequencies, we need to find f_L. Since from (6-7)

$$A_{125} = \frac{A_{\mathrm{mid}}}{\sqrt{1+m^2}}$$

$$27.0 = \frac{36.0}{\sqrt{1+m^2}}$$

$$\sqrt{1+m^2} = 1.3333$$

$$1+m^2 = 1.778$$

$$m^2 = 0.778$$

and

$$m = 0.882$$

But for low frequencies, according to (6-8*a*)

$$m = \frac{f_L}{f}$$

Solving for f_L,

$$f_L = mf = (0.882)(125) = 110 \text{ Hz}$$

Also, according to (6-1*a*)

$$f_L = \frac{1}{2\pi C_c R_g}$$

Solving for C_c,

$$C_c = \frac{1}{2\pi f_L R_g} = \frac{1}{(2\pi)(110)(1 \times 10^6)} = 1.45 \times 10^{-9} = 1450 \times 10^{-12} \text{ F} = 1450 \text{ pF}$$

6.2 A common-emitter RC coupled amplifier shown in Fig. 6-61 has a phase shift of $-30°$ at a frequency of 25.0 kHz. Find the high-frequency half-power point f_H, the gain A_i at 25 kHz, and the value of C_c. The value of β is 60.0.

First we use the known phase shift to determine m at 25 kHz. From (6-9)

$$\Delta\phi_{25\text{ k}} = -\arctan m$$

$$\not-30.0° = \not-\arctan m$$

So $$\tan 30.0° = m$$

and $$m = 0.577$$

But we know for high frequencies that according to (6-8b)

$$m = \frac{f}{f_H}$$

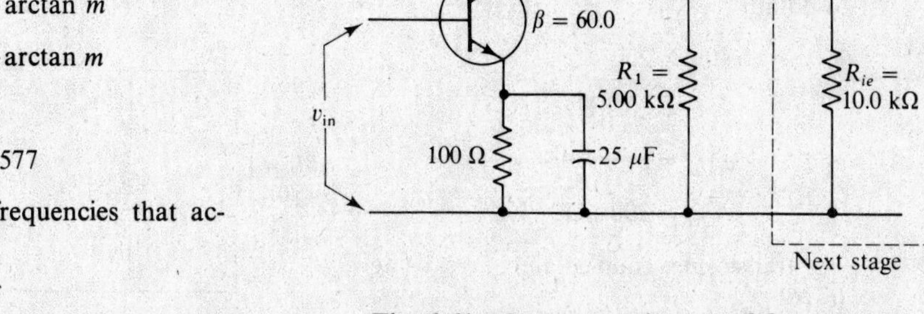

Fig. 6-61 Common-emitter amplifier stage

So $$f_H = \frac{f}{m} = \frac{25 \times 10^3}{0.577} = 43.3 \times 10^3 \text{ Hz} = 43.3 \text{ kHz}$$

We can now solve for the gain at 25 kHz using (6-7):

$$A_{25\text{ k}} = \frac{A_{\text{mid}}}{\sqrt{1 + m^2}}$$

with $A_{\text{mid}} \cong \beta$ for the common-emitter configuration. Substituting values,

$$A_{25\text{ k}} = \frac{60.0}{\sqrt{1 + (0.577)^2}} = 52.0$$

We may now get the value of C_c using (6-3b):

$$f_H = \frac{R_{ie} + R_1}{2\pi(\beta + 1)C_c R_{ie} R_1}$$

Solving for C_c, $$C_c = \frac{R_{ie} + R_1}{2\pi(\beta + 1)\, f_H R_{ie} R_1}$$

Substituting $R_{ie} = 10$ kΩ, $R_1 = 5.00$ kΩ from Fig. 6-61, and $\beta = 60.0$ and $f_H = 43.3 \times 10^3$,

$$C_c = \frac{10 \times 10^3 + 5 \times 10^3}{(2\pi)(60 + 1)(43.3 \times 10^3)(10 \times 10^3)(5 \times 10^3)} = 1.81 \times 10^{-11} = 18.1 \times 10^{-12} \text{ F} = 18.1 \text{ pF}$$

6.3 The input impedance of the Darlington amplifier shown in Fig. 6-62 is 10 MΩ. Find the midband gain of each identical transistor.

For the Darlington connection,

$$Z_{\text{in}} = A_i R_L$$

Solving for A_i and substituting values from Fig. 6-62,

$$A_i = \frac{Z_{\text{in}}}{R_L} = \frac{10 \times 10^6}{1 \times 10^3} = 1.00 \times 10^4$$

But for identical transistors in the Darlington configuration,

$$A_i = \beta^2$$

$$1.00 \times 10^4 = \beta^2$$

and $$\beta = 100$$

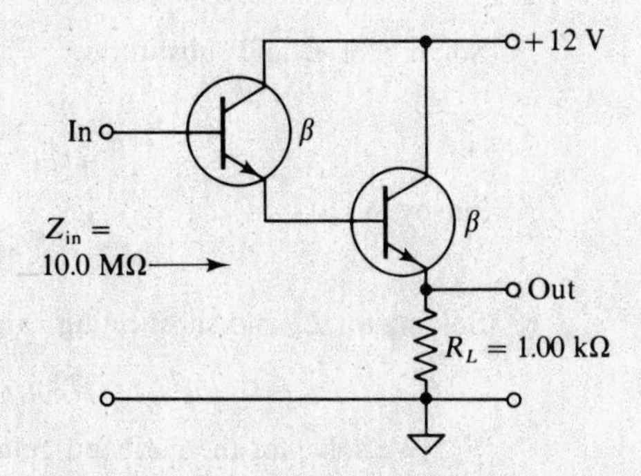

Fig. 6-62 Darlington amplifier

If the transistors did not have identical β's, 100 would be the average β of the two transistors.

6.4 The transformer coupled amplifier equivalent circuit shown in Fig. 6-63 has a midband range of 200 Hz to 20.0 kHz. Find the transformer turns ratio a, L_p, and the midband gain A_{mid}.

With a midband range given as $200\text{ Hz} < f_m < 20.0\text{ kHz}$, we substitute into (*6-10*)

$$10f_L < f_m < 0.1f_H$$

to obtain

$$10f_L = 200$$

$$f_L = 20.0\text{ Hz}$$

and

$$0.1f_H = 20.0 \times 10^3$$

$$f_H = 200 \times 10^3\text{ Hz}$$

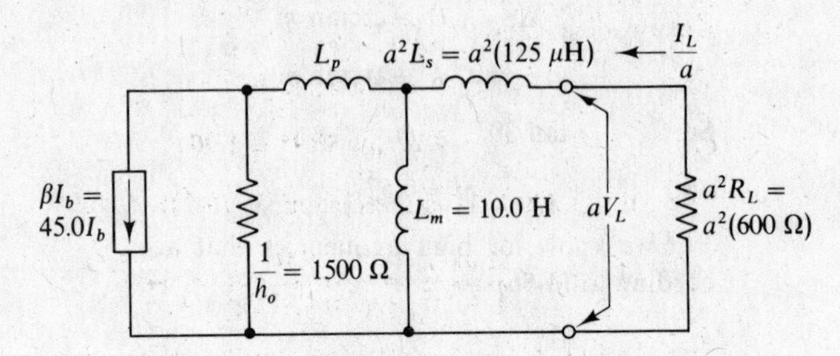

Fig. 6-63 Transformer coupled equivalent circuit

For transformer coupled amplifiers, using (*6-16*),

$$f_L = \frac{R'}{2\pi L_m}$$

Solving for R' and substituting values,

$$R' = 2\pi L_m f_L = (2\pi)(10.0)(20.0) = 1.26 \times 10^3\ \Omega$$

But according to (*6-17*),

$$R' = \frac{a^2R_L}{1 + h_o a^2 R_L}$$

Substituting $R_L = 600\ \Omega$ and $1/h_o = 1500\ \Omega$ from Fig. 6-63,

$$1.26 \times 10^3 = \frac{(a^2)(600)}{1 + (a^2)(600)/1500} = \frac{600a^2}{1 + 0.4a^2}$$

Cross multiplying,

$$1260 + 504a^2 = 600a^2$$

$$96a^2 = 1260$$

$$a^2 = 13.125$$

$$a = 3.62$$

To solve for L_p, we first find L'' and R''. Using (*6-19*),

$$R'' = \frac{1}{h_o} + a^2R_L = 1500 + (13.125)(600) = 9375\ \Omega$$

We know that according to (*6-18*),

$$f_H = \frac{R''}{2\pi L''}$$

Solving for L'' and substituting,

$$L'' = \frac{R''}{2\pi f_H} = \frac{9375}{(2\pi)(200 \times 10^3)} = 7.46 \times 10^{-3}\text{ H} = 7.46\text{ mH}$$

But by (*6-20*),

$$L'' = L_p + a^2L_s$$

Solving for L_p and substituting values,

$$L_p = L'' - a^2L_s = (7.46 \times 10^{-3}) - (13.125)(125 \times 10^{-6}) = 5.82 \times 10^{-3}\text{ H} = 5.82\text{ mH}$$

We solve for the midband gain using $\beta = 45.0$ from Fig. 6-63 and (*6-15*):

$$A_{mid} = \frac{a\beta}{1 + h_o a^2 R_L} = \frac{(3.62)(45.0)}{1 + (13.125)(600)/1500} = \frac{162.9}{1 + 5.25} = 26.1$$

6.5 The midband gain of the FET transformer coupled amplifier equivalent circuit shown in Fig. 6-64 is 100. The phase shift at 200 kHz is $-50.2°$. Find the gain at 200 kHz, the value of the load resistance R_L, and μ for the FET.

Using the given phase shift at 200 kHz, from (*6-9*)

$$\Delta\phi_{200\,\text{k}} = -\arctan m$$

Substituting values,

$$\cancel{-}50.2° = \cancel{-}\arctan m$$

or

$$m = \tan 50.2° = 1.20$$

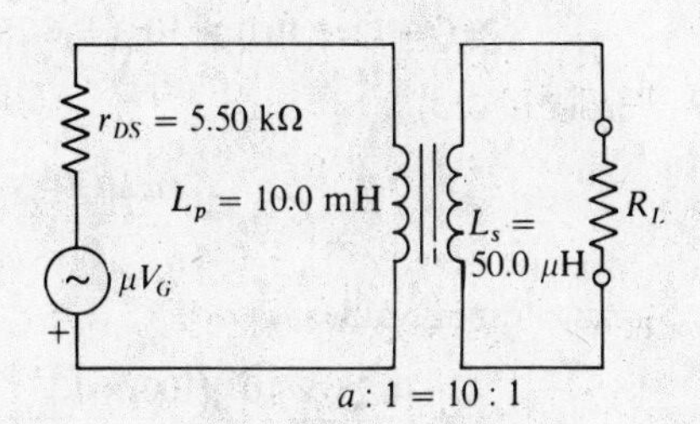

Fig. 6-64 Transformer coupled equivalent circuit

The gain at 200 kHz is then given by (*6-7*):

$$A_{200\,\text{k}} = \frac{A_{\text{mid}}}{\sqrt{1+m^2}} = \frac{100}{\sqrt{1+(1.2)^2}} = 64.0$$

To find R''_L, we need to calculate f_H. According to (*6-8b*),

$$m = \frac{f}{f_H}$$

Solving for f_H and substituting,

$$f_H = \frac{f}{m} = \frac{200 \times 10^3}{1.2} = 167 \times 10^3 \text{ Hz}$$

Since according to (*6-20*),

$$L'' = L_p + a^2 L_s$$

substituting values from Fig. 6-64,

$$L'' = (10 \times 10^{-3}) + (10)^2(50 \times 10^{-6}) = 15.0 \times 10^{-3} \text{ H}$$

Also, according to (*6-18*),

$$f_H = \frac{R''}{2\pi L''}$$

Solving for R'' and substituting values,

$$R'' = 2\pi L'' f_H = (2\pi)(15.0 \times 10^{-3})(167 \times 10^3) = 1.57 \times 10^4\ \Omega$$

For the transformer coupled FET circuit, we may use (*6-19b*):

$$R'' = r_{DS} + a^2 R_L$$

Solving for R_L and substituting values,

$$R_L = \frac{R'' - r_{DS}}{a^2} = \frac{(1.57 \times 10^4) - (5.5 \times 10^3)}{(10)^2} = 102\ \Omega$$

We now solve for μ using the FET midband gain formula (*6-15b*):

$$A_{\text{mid}} = \frac{\mu a^2 R_L}{r_{DS} + R_L}$$

Substitution yields

$$100 = \frac{\mu(10)^2(102)}{5500 + 102}$$

Cross multiplying and rearranging terms,

$$\cancel{(100)}(102\mu) = (5500 + 102)\cancel{(100)}$$

$$102\mu = 5500 + 102$$

$$\mu = 54.9$$

6.6 The low-frequency equivalent circuit for a single-tuned FET amplifier stage is shown in Fig. 6-65. If the resonant frequency is 10.0 kHz, find the value of C_1 and the Q_{eff} for the circuit.

We first calculate Q_o. Since

$$\omega_o = 2\pi f_o = (2\pi)(10.0 \times 10^3) = 6.28 \times 10^4 \text{ rad/s}$$

from *(6-21a)*

$$Q_o = \frac{\omega_o L_1}{R_1}$$

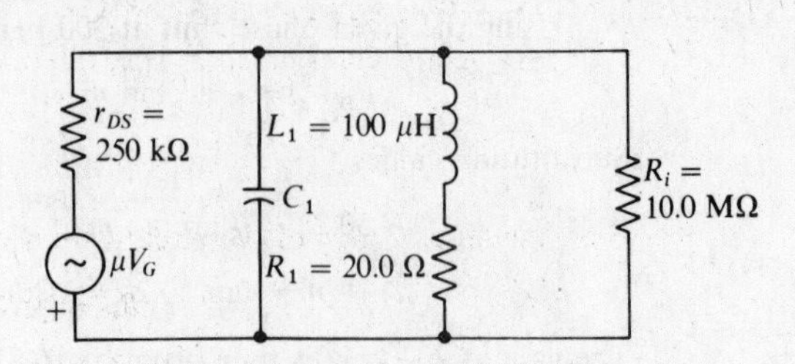

Fig. 6-65 Low-frequency equivalent circuit

Substituting values,

$$Q_o = \frac{(6.28 \times 10^4)(100 \times 10^{-3})}{(20)} = 314$$

But we also know from *(6-21a)* that

$$Q_o = \frac{1}{\omega_o R_1 C_1}$$

Solving for C_1,

$$C_1 = \frac{1}{\omega_o R_1 Q_o} = \frac{1}{(6.28 \times 10^4)(20)(314)} = 2.54 \times 10^{-9} = 2540 \times 10^{-12} \text{ F} = 2540 \text{ pF}$$

To solve for Q_{eff}, we need to calculate R_o and R_{Sh}. According to *(6-21c)*,

$$R_o = Q_o^2 R_1 = (314)^2(20.0) = 1.97 \times 10^6 \ \Omega$$

And according to *(6-26b)*,

$$\frac{1}{R_{Sh}} = \frac{1}{r_{DS}} + \frac{1}{R_o} + \frac{1}{R_i}$$

Substituting r_{DS} and R_i values from Fig. 6-65,

$$\frac{1}{R_{Sh}} = \frac{1}{250 \times 10^3} + \frac{1}{1.97 \times 10^6} + \frac{1}{10.0 \times 10^6} = 4.61 \times 10^{-6}$$

Inverting both sides,

$$R_{Sh} = 2.17 \times 10^5 \ \Omega$$

Now we solve for Q_{eff} using *(6-27)*:

$$Q_{eff} = Q_o \frac{R_{Sh}}{R_o} = \frac{(314)(2.17 \times 10^5)}{1.97 \times 10^6} = 34.6$$

6.7 Find the resonant frequency f_o, Q_{eff}, the gain A_o at resonance, and the half-power points f_L and f_H for the single-tuned transistor amplifier equivalent circuit shown in Fig. 6-66.

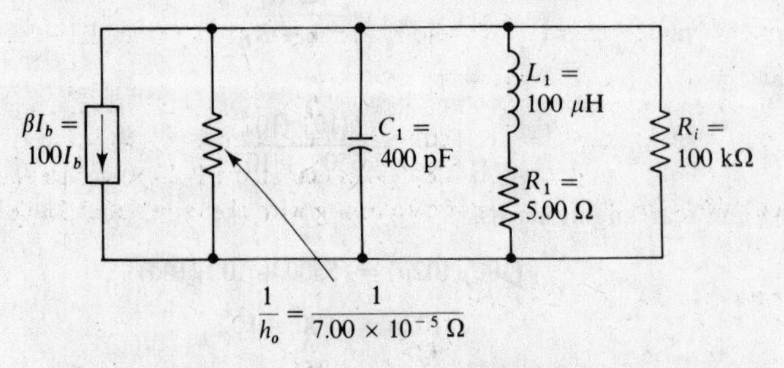

Fig. 6-66 Single-tuned transistor amplifier equivalent circuit

We may calculate the resonant frequency using the values shown in Fig. 6-66 from (*6-20b*):

$$f_o = \frac{1}{2\pi\sqrt{L_1C_1}} = \frac{1}{(2\pi)\sqrt{(100\times10^{-6})(400\times10^{-12})}} = 7.96\times10^5 = 796\times10^3\ \text{Hz} = 796\ \text{kHz}$$

Using the values given in Fig. 6-66, we now determine Q_o using (*6-21b*):

$$Q_o = \frac{1}{R_1}\sqrt{\frac{L_1}{C_1}} = \frac{1}{5.00}\sqrt{\frac{100\times10^{-6}}{400\times10^{-12}}} = 100$$

To calculate the Q_{eff} of the circuit, we need to find R_o and R_{Sh}. Using (*6-21c*),

$$R_o = Q_o^2R_1 = (100)^2(5.00) = 5.00\times10^4\ \Omega$$

and using (*6-26*),

$$\frac{1}{R_{\text{Sh}}} = h_o + \frac{1}{R_o} + \frac{1}{R_i} = (7\times10^{-5}) + \frac{1}{5\times10^4} + \frac{1}{100\times10^3} = 1.00\times10^{-4}$$

Inverting yields $$R_{\text{Sh}} = 1.00\times10^4\ \Omega$$

Now using (*6-27*),

$$Q_{\text{eff}} = Q_o\frac{R_{\text{Sh}}}{R_o} = \frac{(100)(1.00\times10^4)}{(5.00\times10^4)} = 20.0$$

The gain at resonance can be found from (*6-28*):

$$A_o = \beta\frac{R_{\text{Sh}}}{R_i} = \frac{(100)(1.00\times10^4)}{100\times10^3} = 10.0$$

The half-power points are determined from (*6-29a*),

$$f_L = f_o\left(1 - \frac{1}{2Q_{\text{eff}}}\right) = (796\times10^3)\left(1 - \frac{1}{(2)(20)}\right) = 776.1\times10^3\ \text{Hz}$$

and from (*6-29b*),

$$f_H = f_o\left(1 + \frac{1}{2Q_{\text{eff}}}\right) = (796\times10^3)\left(1 + \frac{1}{(2)(20)}\right) = 815.9\times10^3\ \text{Hz}$$

6.8 Draw the gain-frequency curve for the single-tuned transistor amplifier shown in Fig. 6-66.

We have solved this amplifier problem for $f_o, f_L, f_H, Q_{\text{eff}}$, and A_o in Prob. 6.7. Now we need to calculate the gain for frequencies above and below f_o using (*6-30*):

$$A = \frac{A_o}{\sqrt{1 + (m_oQ_{\text{eff}})^2}}$$

We find it easier when drawing a curve to first make a table of the pertinent values using (*6-31a*):

$$m_o = \frac{f}{f_o} - \frac{f_o}{f}$$

Selecting convenient values above and below 796 kHz to draw a smooth curve, we may construct Table 6-5.

From the previous problem, we have $f_L = 776.1$ kHz and $f_H = 815.9$ kHz with $A_o = 10$ and $Q_{\text{eff}} = 20$, so we may start our curve by locating the A_o, f_o point as (10, 796) and locating the half-power points at (7.07, 776.1) and (7.07, 815.9), since at the half-power frequencies the gain is $0.707A_o$.

Sample calculations for 800 kHz are shown along with the completed Table 6-5. The required curve is shown in Fig. 6-67.

Sample calculation: At $f = 800$ kHz,

$$\frac{f}{f_o} = \frac{800}{796} = 1.005$$

Table 6-5

A	$B = \dfrac{A}{796}$	$C = \dfrac{796}{A}$	$D = B - C$	$E = (D)(20.0)$	$F = 1 + E^2$	$G = \sqrt{F}$	$H = \dfrac{10.0}{G}$
f(kHz)	f/f_o	f_o/f	$m_o \times 10^{-2}$	$m_o Q_{\text{eff}}$	$1 + (m_o Q_{\text{eff}})^2$	$\sqrt{1 + (m_o Q_{\text{eff}})^2}$	A
770	0.96734	1.03377	−6.643	−1.329	2.766	1.663	6.01
$f_L \rightarrow$ 776	0.97487	1.02577	−5.090	−1.018	2.036	1.427	7.01
780	0.97990	1.02051	−4.061	−0.8122	1.650	1.288	7.76
790	0.99246	1.00759	−1.513	−0.3026	1.092	1.045	9.57
$f_o \rightarrow$ 796	1.00000	1.00000	0	0	1.000	1.000	10.00
800	1.00503	0.99500	1.003	0.2006	1.040	1.020	9.80
810	1.01759	0.98272	3.487	0.6974	1.486	1.219	8.20
$f_H \rightarrow$ 816	1.02513	0.97549	4.966	0.9928	1.986	1.409	7.10
820	1.03015	0.97073	5.942	1.1884	2.412	1.553	6.44

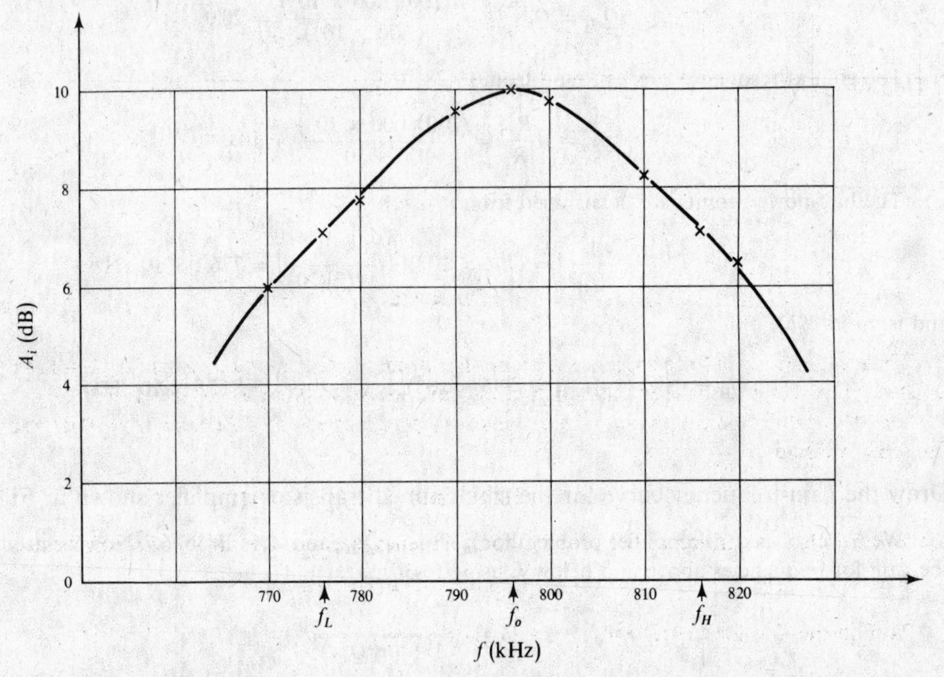

Fig. 6-67 Gain-frequency curve for the single-tuned transistor amplifier shown in Fig. 6-66

$$\frac{f_o}{f} = \frac{796}{800} = 0.995$$

$$m_o = \frac{f}{f_o} - \frac{f_o}{f} = 1.005 - 0.995 = 1.00 \times 10^{-2}$$

$$m_o Q_{\text{eff}} = (1.00 \times 10^{-2})(20) = 0.200$$

$$1 + (m_o Q_{\text{eff}})^2 = 1 + (0.200)^2 = 1.04$$

$$\sqrt{1 + (m_o Q_{\text{eff}})^2} = \sqrt{1.04} = 1.02$$

$$A = \frac{A_o}{\sqrt{1 + (m_o Q_{\text{eff}})^2}} = \frac{10.0}{1.02} = 9.80$$

The table values may vary slightly depending on round-off errors encountered in the calculations.

6.9 What are the values of the Q_{eff} and bandwidth B of a single-tuned amplifier with the gain-frequency curve shown in Fig. 6-68*a*? Assuming the Q_{eff} of the amplifier circuit were equal to 10.0 and the resonant frequency remained the same, sketch the new gain-frequency curve.

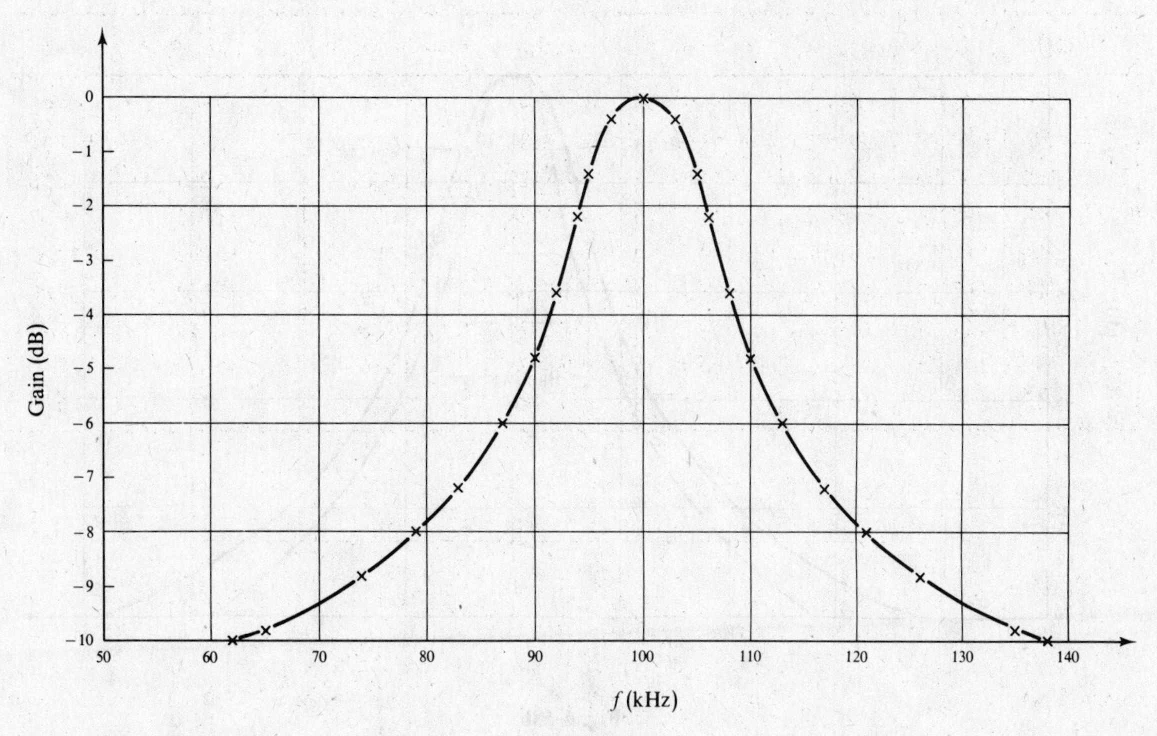

Fig. 6-68a

The maximum gain for a single-tuned amplifier occurs at the resonant frequency. Looking at the curve in Fig. 6-68*a*, we see that the maximum gain of 0 dB occurs at 100 kHz, so $f_o = 100\,\text{kHz}$.

The half-power points occur where the gain has dropped to −3 db, so from the graph shown in Fig. 6-68*a* we find

$$f_L = 93\text{ kHz} \qquad \text{and} \qquad f_H = 107\text{ kHz}$$

The bandwidth B is defined as the difference in frequency between the half-power points, so using (*6-23*),

$$B = f_H - f_L = 107 - 93 = 14\text{ kHz}$$

To find the Q_{eff} for the circuit, we use (*6-23b*):

$$B = \frac{f_o}{Q_{eff}}$$

Solving for Q_{eff} and substituting values,

$$Q_{eff} = \frac{f_o}{B} = \frac{100 \times 10^3}{14 \times 10^3} = 7.14$$

In order to sketch the $Q_{eff} = 10$ curve (which should have a smaller bandwidth), we need to calculate the new bandwidth and use it to obtain the new half-power points. From (*6-23b*),

$$B' = \frac{f_o}{Q'_{eff}} = \frac{100 \times 10^3}{10} = 10 \times 10^3\text{ Hz} = 10\text{ kHz}$$

And the new half-power points will be symmetrically located on each side of f_o at the values

$$f_L = f_o - \frac{B}{2} = 100 - 5 = 95\text{ kHz} \qquad \text{and} \qquad f_H = f_o + \frac{B}{2} = 100 + 5 = 105\text{ kHz}$$

The original curve with descriptive labels is shown in Fig. 6-68*b* and the $Q_{eff} = 10$ curve is sketched on the same set of axes with dashed lines.

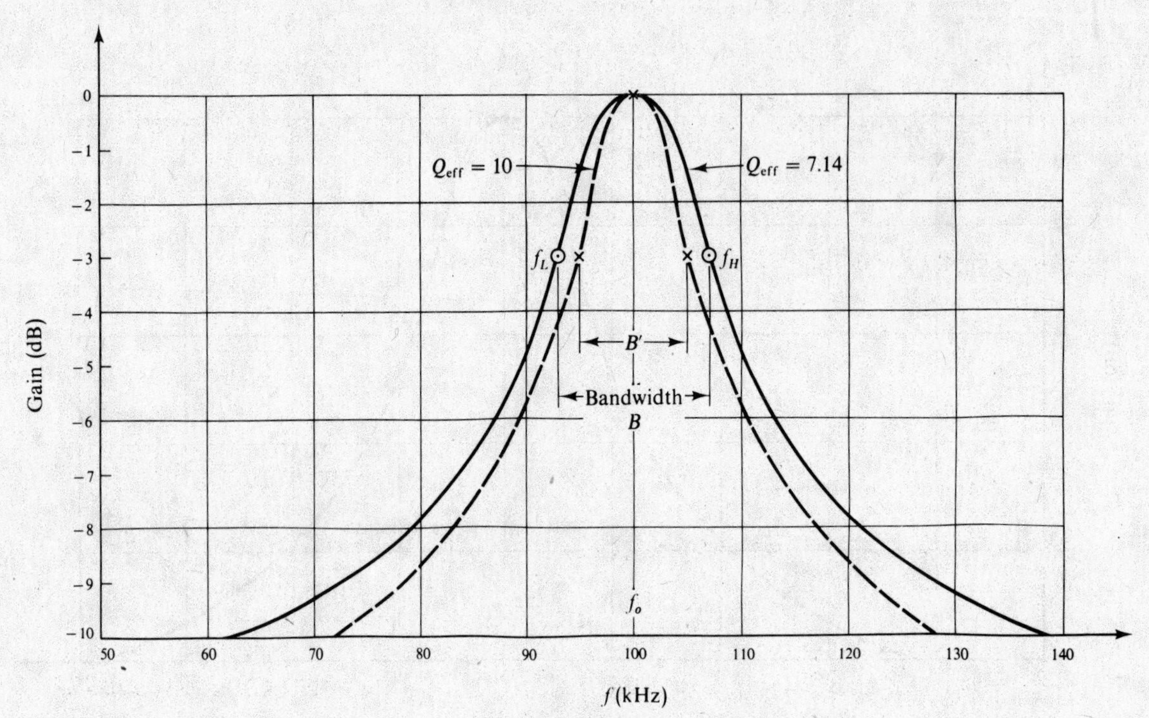

Fig. 6-68b

6.10 An amplifier has a voltage gain of 25.0 with feedback and a gain of 100 without feedback. Is this an example of positive or negative feedback? What is β_F?

Since the gain with feedback is lower than the gain without feedback (sometimes called the "open-loop gain"), this is an example of degenerative or negative feedback. Since according to (*6-35a*)

$$A_{vFB} = \frac{A_v}{1 + \beta_F A_v}$$

cross multiplication yields

$$A_{vFB} + \beta_F A_v A_{vFB} = A_v$$

Solving for β_F,

$$\beta_F = \frac{A_v - A_{vFB}}{A_v A_{vFB}}$$

Substituting values,

$$\beta_F = \frac{100 - 25}{(100)(25)} = 0.03$$

6.11 Find the operating frequency and the minimum gain for the Colpitts oscillator shown in Fig. 6-69.

Using Table 6-1, we find that for the Colpitts oscillator

$$\omega_o = 2\pi f_o = \sqrt{\frac{C_1 + C_2}{LC_1 C_2}}$$

Substituting the values found in Fig. 6-69,

$$\omega_o = 2\pi f_o = \sqrt{\frac{(2.00 \times 10^{-12}) + (18.0 \times 10^{-12})}{(14.1 \times 10^{-3})(2.00 \times 10^{-12})(18.0 \times 10^{-12})}} = 6.28 \times 10^6 \text{ rads/s}$$

So
$$f_o = \frac{6.28 \times 10^6}{2\pi} = 1.00 \times 10^6 \text{ Hz} = 1.00 \text{ MHz}$$

We also find from Table 6-1 that the minimum gain for a Colpitts oscillator is given by

$$A_v = \frac{C_2}{C_1}$$

Substituting the circuit values,
$$A_v = \frac{18.0 \times 10^{-12}}{2.00 \times 10^{-12}} = 9.0$$

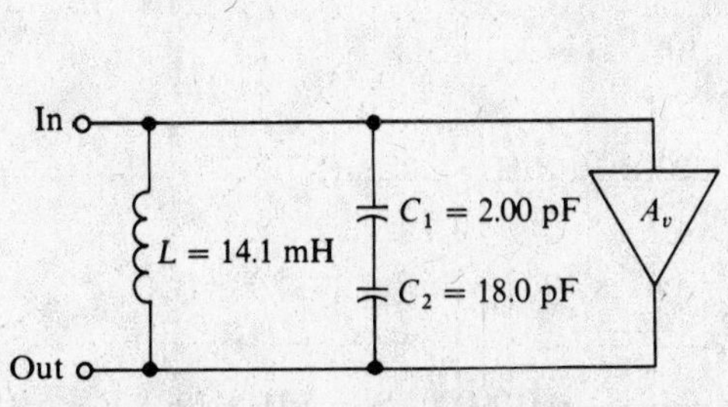

Fig. 6-69 Colpitts oscillator

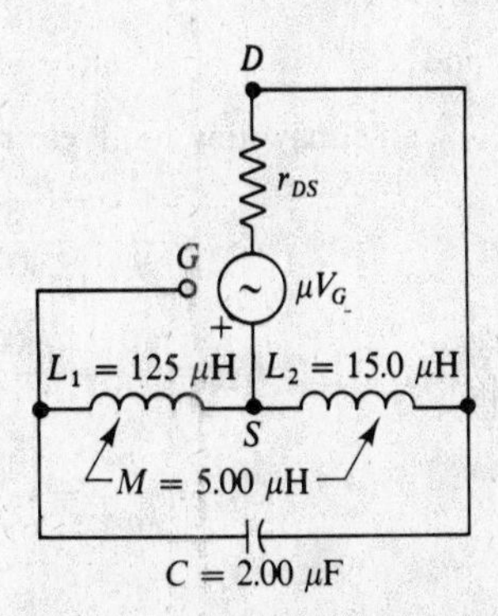

Fig. 6-70 FET Hartley oscillator

6.12 Will the FET Hartley oscillator equivalent circuit shown in Fig. 6-70 oscillate at 1 MHz if the gain of the FET amplifier used is 10.0 and the mutual inductance M is 5.00 μH?

Table 6-1 gives the Hartley gain criterion as

$$A_v \geq \frac{L_1 + L_2 + 2M}{L_2}$$

Substituting values from Fig. 6-70 with $M = 5.00\ \mu$H,

$$A_v \geq \frac{(125 \times 10^{-6}) + (15.0 \times 10^{-6}) + (2)(5.00 \times 10^{-6})}{15.0 \times 10^{-6}} \geq 10.0$$

Since the gain of the amplifier is given as 10.0, it is just equal to the minimum gain and the circuit *will* oscillate.

In practice we would use a gain that was higher than the minimum to ensure oscillation when lead and other losses were taken into account.

To check the oscillator frequency, we again use Table 6-1 and write

$$\omega_o = 2\pi f_o = \sqrt{\frac{1}{(L_1 + L_2 + 2M)C}}$$

Substituting circuit values,

$$\omega_o = 2\pi f_o = \sqrt{\frac{1}{\{(125 + 15.0) + [(2)(5.00) \times 10^{-6}]\}(2.00 \times 10^{-6})}} = 5.77 \times 10^4 \text{ rad/s}$$

and
$$f_o = \frac{5.77 \times 10^4}{2\pi} = 9.19 \times 10^3 \text{ Hz} = 9.19 \text{ kHz}$$

Although the circuit will oscillate, it will oscillate at 9.19 kHz *not* at 1.0 MHz.

For operation at 1.0 MHz, the tank capacitor could be lowered in value without affecting the minimum gain requirement for oscillation, which depends only on the inductances.

6.13 A midrange speaker has an impedance of 8 Ω throughout its operating range of 955 to 4775 Hz. Design a constant-k filter that could be used with this speaker.

Since a range of frequencies is being used, and we wish the impedance to be low throughout the range and high outside the range, the filter to be used is a bandpass constant-k type with $k = 8.00\ \Omega$.

From the problem statement, $f_1 = 955$, so

$$\omega_1 = 2\pi f_1 = (2\pi)(955) = 6.00 \times 10^3 \text{ rad/s}$$

and $f_2 = 4775$, so

$$\omega_2 = 2\pi f_2 = (2\pi)(4775) = 3.00 \times 10^4 \text{ rad/s}$$

From Table 6-2,

$$B = \omega_2 - \omega_1 = (3.00 \times 10^4) - (6.00 \times 10^3) = 2.40 \times 10^4$$

and $$\omega_o^2 = \omega_1 \omega_2 = (6.00 \times 10^3)(3.00 \times 10^4) = 18.0 \times 10^7$$

We may now use these results to calculate the component values:

$$L_{1k} = \frac{k}{B} = \frac{8.00}{2.40 \times 10^4} = 0.333 \times 10^{-3} \text{ H} = 333 \text{ mH}$$

$$L_{2k} = \frac{kB}{\omega_o^2} = \frac{(8.00)(2.40 \times 10^4)}{18.0 \times 10^7} = 1.07 \times 10^{-3} \text{ H} = 1.07 \text{ mH}$$

$$C_{1k} = \frac{B}{k\omega_o^2} = \frac{2.40 \times 10^4}{(8.00)(18.0 \times 10^7)} = 1.67 \times 10^{-5} = 16.7 \times 10^{-6} \text{ F} = 16.7\ \mu\text{F}$$

$$C_{2k} = \frac{1}{kB} = \frac{1}{(8.00)(2.40 \times 10^4)} = \frac{1}{192\,000} = 0.00000521 = 5.21 \times 10^{-6} \text{ F} = 5.21\ \mu\text{F}$$

These values are put into the form of a T equivalent filter and shown in Fig. 6-71.

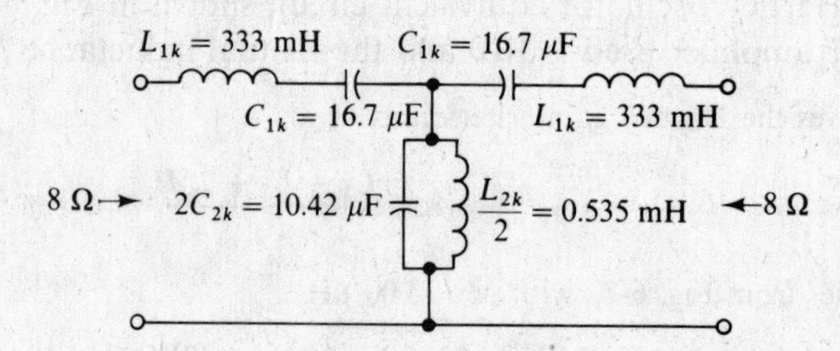

Fig. 6-71 Bandpass filter for 8-Ω midrange speaker

6.14 Find the components needed to construct a series m-derived low-pass filter with $f_\infty = 4.00$ kHz, $f_c = 1.0$ kHz, and $k = 16.0\ \Omega$.

Using Table 6-2, we first find the values for the constant-k filter:

$$\omega_c = 2\pi f_c = (2\pi)(1.00 \times 10^3) = 6.28 \times 10^3 \text{ rad/s}$$

and $$t_c = \frac{1}{\omega_c} = \frac{1}{6.28 \times 10^3} = 1.59 \times 10^{-4} \text{ s}$$

$$L_k = kt_c = (16.0)(1.59 \times 10^{-4}) = 2.55 \times 10^{-3} \text{ H}$$

$$C_k = \frac{t_c}{k} = \frac{1.59 \times 10^{-4}}{16.0} = 9.95 \times 10^{-6} \text{ F}$$

Using Table 6-3 for the series m-derived filter, we convert the constant-k values to m-derived values:

$$\omega_\infty = 2\pi f_\infty = (2\pi)(4.00 \times 10^3) = 2.51 \times 10^4 \text{ rad/s}$$

$$m^2 = 1 - \frac{1}{(t_c\omega_\infty)^2} = 1 - \frac{1}{(1.59 \times 10^{-4})(2.51 \times 10^4)} = 1 - 0.25 = 0.75$$

and $\quad m = 0.866$

$$L_1 = mL_k = (0.866)(2.55 \times 10^{-3}) = 2.21 \times 10^{-3}\ \text{H} = 2.21\ \text{mH}$$

$$L_2 = \left(\frac{1-m^2}{m}\right)L_k = \left(\frac{1-0.75}{0.866}\right)(2.55 \times 10^{-3}) = 0.736 \times 10^{-3} = 736 \times 10^{-6}\ \text{H} = 736\ \mu\text{H}$$

$$C_2 = mC_k = (0.866)(9.95 \times 10^{-6}) = 8.62 \times 10^{-6}\ \text{F} = 8.62\ \mu\text{F}$$

6.15 Find the k, m, and ω_o of the shunt m-derived bandstop half-section shown in Fig. 6-72 if

$$\omega_1 = 2.5 \times 10^3\ \text{rad/s} \qquad \omega_{1\infty} = 4.00 \times 10^3\ \text{rad/s}$$
$$\omega_2 = 17.5 \times 10^3\ \text{rad/s} \qquad \omega_{2\infty} = 16.0 \times 10^3\ \text{rad/s}$$

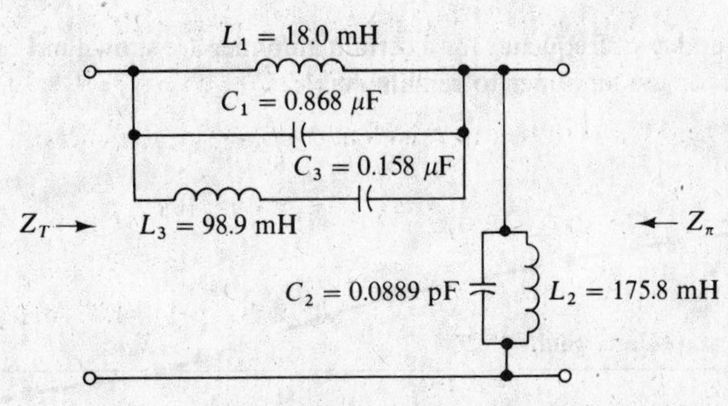

Fig. 6-72 Shunt *m*-derived bandstop half-section

We can obtain the value of m from the given information once we calculate B. From Tables 6-3 and 6-4 for a bandstop filter,

$$B = \omega_2 - \omega_1 = (17.5 \times 10^3) - (2.5 \times 10^3) = 15.0 \times 10^3$$

Now $$m^2 = 1 - \frac{(\omega_{2\infty} - \omega_{1\infty})^2}{B^2} = 1 - \frac{[(16 \times 10^3) - (4 \times 10^3)]^2}{(15 \times 10^3)^2} = 1 - (0.80)^2 = 0.36$$

and $\quad m = 0.6$

To find k, we need to know L_{1k} and C_{2k}. From Table 6-4 for the bandstop filter,

$$L_1 = mL_{1k}$$

Solving for L_{1k}, and using the value of L_1 from Fig. 6-72,

$$L_{1k} = \frac{L_1}{m} = \frac{18.0 \times 10^{-3}}{0.6} = 3.00 \times 10^{-2}\ \text{H}$$

Also $$C_2 = mC_{2k}$$

Solving for C_{2k},

$$C_{2k} = \frac{C_2}{m} = \frac{0.0889 \times 10^{-6}}{0.6} = 1.48 \times 10^{-7}\ \text{F}$$

So $$k^2 = \frac{L_{1k}}{C_{2k}} = \frac{3.00 \times 10^{-2}}{1.48 \times 10^{-7}} = 2.027 \times 10^5$$

and $$k = 450\ \Omega$$

Since Table 6-3 shows for a bandstop filter

$$\omega_o^2 = \omega_{1\infty}\omega_{2\infty}$$

we substitute the given values and obtain

$$\omega_o^2 = (4.00 \times 10^3)(16.0 \times 10^3) = 64.0 \times 10^6$$

and $$\omega_o = 8.00 \times 10^3\ \text{rad/s}$$

Supplementary Problems

6.16 Find the bandwidth and gain of a three-stage cascaded amplifier if each of the stages is identical, with a gain of 10.0 and a bandwidth of 10.0 MHz. Assume the gain-bandwidth product of the cascaded amplifier is the same as for an individual stage.

6.17 A phase-locked loop circuit has a supply voltage of 15.0 V. Find the locking range if the input frequency f_o is 300 kHz.

6.18 A filter has $Z_1 = 675\ \Omega$ and $Z_2 = 300\ \Omega$. Find Z_T, Z_π, and the characteristic impedance Z_o.

6.19 The plots of $\beta_F A$ and ϕ vs. frequency for a certain amplifier are shown in Fig. 6-73. Find the gain margin and the phase margin for the amplifier to remain stable.

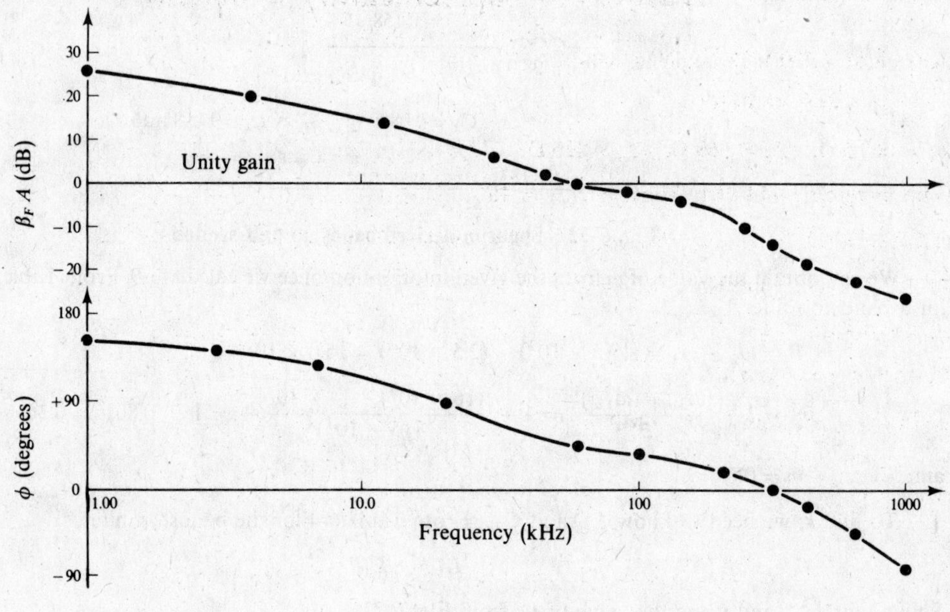

Fig. 6-73

6.20 A constant-k high-pass filter with $C_k = 2.00\ \mu F$ is converted to a shunt m-derived high-pass filter with $C_1 = 2.50\ \mu F$ and $L_2 = 158\ \mu H$. Find m, k, ω_c, and ω_∞.

6.21 An amplifier has a low-frequency cutoff of 4.00 kHz, and a high-frequency cutoff of 1.00 MHz. Find the gain and midband range if the gain at 1.20 MHz is 48.0.

6.22 The current gain A_i of an FET amplifier without feedback (open-loop gain) is -400. If the current feedback loop provides a β_F of -0.06, find the A_{iFB}.

6.23 An RC phase shift oscillator has a gain of 35 and an RC network shown in Fig. 6-74. Will the circuit actually oscillate? If so, find the frequency of oscillation f_o.

6.24 A transformer is to be used to match the output impedance of a transistor amplifier to an 8-Ω speaker. If the amplifier has an output impedance $Z_{out} = 2048\ \Omega$ and a $v_{out} = 5.12$ V, find the transformer turns ratio a needed and the speaker voltage v_2 and current i_2.

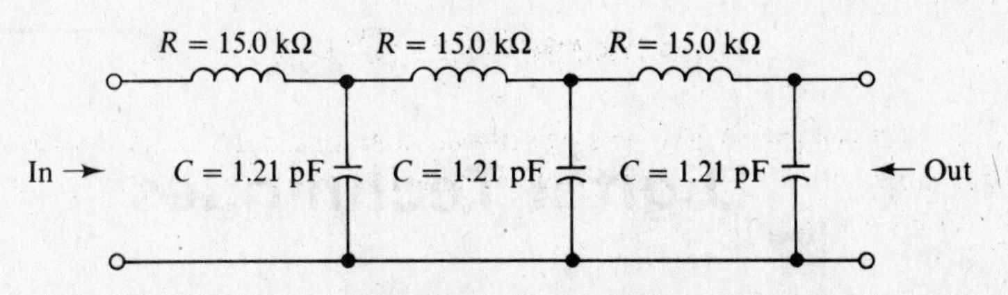

Fig. 6-74 ***RC* network for phase shift oscillator**

6.25 A parallel resonant circuit with $R_1 = 100\ \Omega$ and $Q_o = 100$ is connected to an amplifier output circuit. R_{Sh} for the combined circuit equals 500 kΩ. Find the Q_{eff} of the amplifier.

Answers to Supplementary Problems

6.16 Bandwidth = 100 kHz, gain = 1000.

6.17 f_l is from 140 to 460 kHz.

6.18 $Z_T = 562.5\ \Omega$, $Z_\pi = 360\ \Omega$, $Z_o = 450\ \Omega$.

6.19 Gain margin = -14 dB; phase margin = 45°.

6.20 $m = 0.8$, $k = 7.95\ \Omega$, $\omega_c = 6.29 \times 10^4$ rad/s, $\omega_\infty = 3.77 \times 10^4$ rad/s.

6.21 40.0 kHz $< f_m <$ 100 kHz, $A_{mid} = 75.0$.

6.22 $A_{iFB} = -16.0$.

6.23 Yes; $f_o = 3.58$ MHz.

6.24 $a = 16$, $v_2 = 320$ mV, $i_2 = 40.0$ mA.

6.25 $Q_{eff} = 50.0$.

Chapter 7

Digital Techniques

7.1 INTRODUCTION

With the realization of large-scale integrated (LSI) circuits and the smaller physical size of the newer logic families, the density of the number of circuits has increased to the point where thousands of discrete components can be replaced by a single chip.

This is giving new impetus to the use of digital techniques to handle circuit situations that were previously only handled in the analog manner.

An analog signal is the type of variation that we usually encounter in the real world, whether it is from measuring equipment, control equipment, or some other phenomenon such as sound waves.

Analog signals are characterized by the fact that they are continuously varying with time, as shown in Fig. 7-1. Because of this continuity, the bandwidth of the signal is usually restricted to a range that we can handle with conventional linear techniques such as small- and large-signal amplifiers, modulators, mixers, oscillators, filters, etc.

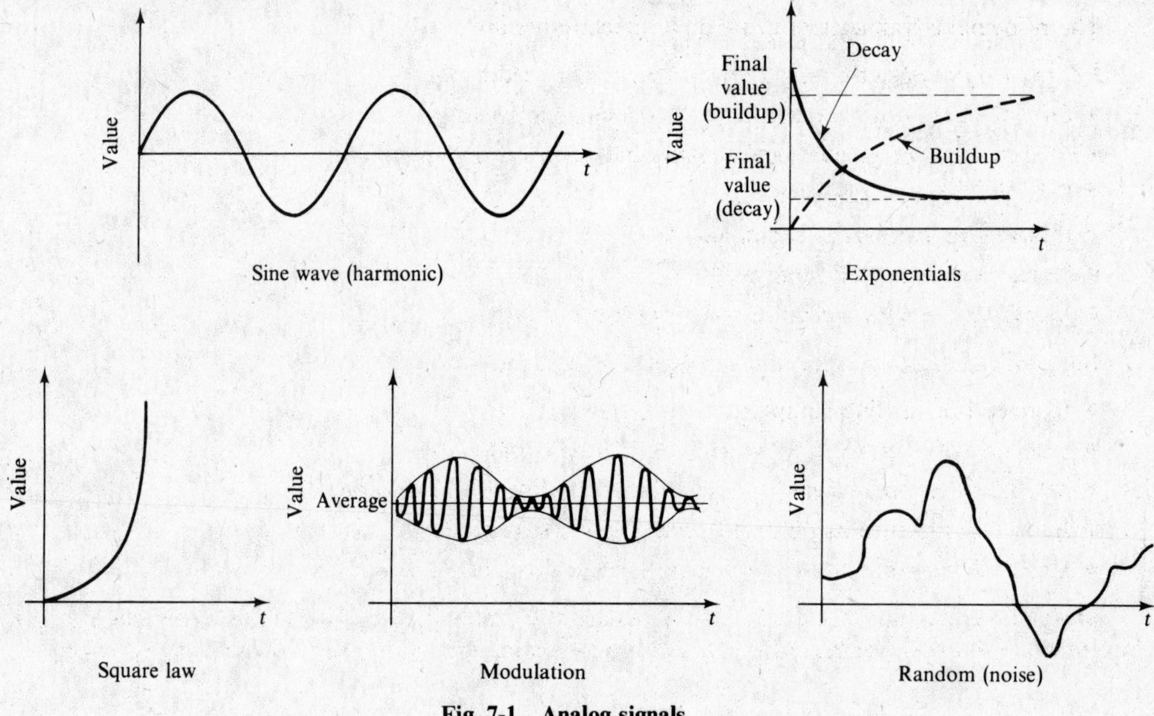

Fig. 7-1 Analog signals

In contrast, digital signals are characterized by discrete, quantized pulses arranged in a pattern or code. Some examples of digital signals are given in Fig. 7-2. Although there are many different positional and coding arrangements that can be used to represent values in digital systems, we will be concerned mostly with the approach that treats the digital signal as the result of switches.

As such, the digital system recognizes only two states, 0 and 1.

Because of the simplicity of the two-state or binary approach, complex tasks can be broken down into a series of simpler decisions that can be handled with an ON or OFF result. Although an easy analog circuit task may entail hundreds of digital decisions, they are two-state decisions and can be handled by a limited number of digital building blocks.

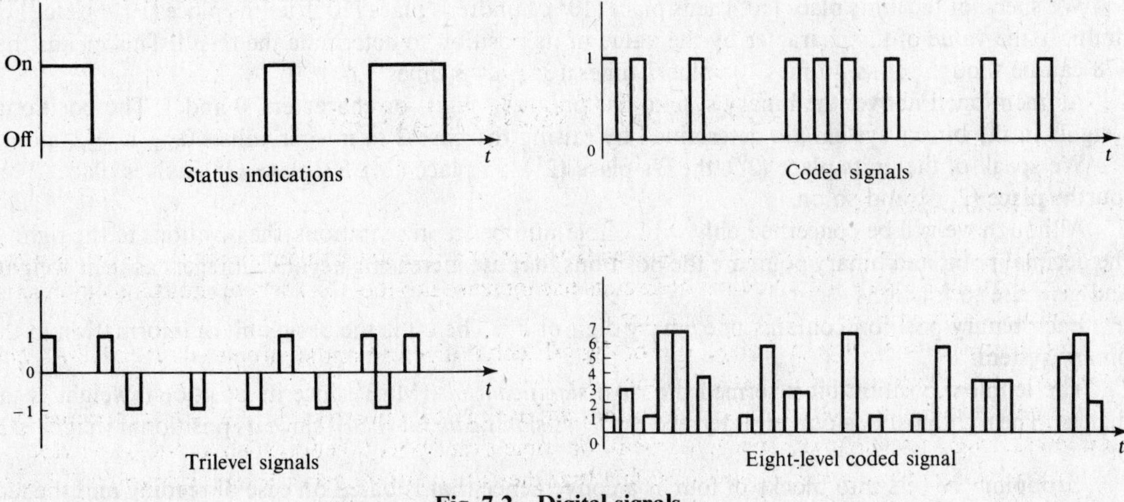

Fig. 7-2 Digital signals

Without the ability to combine hundreds of these ON-OFF building blocks into a small physical space, digital techniques would never have progressed to the present level of wide availability at reasonable cost.

The principal advantages of using digital techniques are

- Simplicity of design—only a few basic circuits need be used.
- Density of circuitry can be greater—this leads to lower costs.
- Precision can be greater—particularly in numerical manipulations.
- Processing speed may be greater.
- Information can be easily stored.
- Noise can be rejected more easily.
- Signals are easy to recognize and regenerate.

Some of the disadvantages of using digital devices include

- Increased processing complexity
- Difficulty of interfacing with the analog "real world"
- Added conversion expense—analog-digital, coding, digital-analog
- Information transmission speed usually lower than analog
- Limit on sampling rate and quantizing levels—loss of precision

Despite these limitations, digital techniques are still expanding rapidly and digital designs are being implemented in areas like broadcasting where only analog methods were viable in the past.

7.2 BINARY SYSTEMS

The word "binary" means two. Switches are a perfect application for the binary number system since switches are either off or on, that is, they can be in either one of two states.

In other words, the binary system has only two states or numbers it can recognize, 0 and 1. Another way of saying that we have a binary system is to describe it as a number system to the base 2.

Converting from one number system to another involves recognition of the character values and the relative weights of the character positions.

In the familiar decimal, or base 10 system, for example, there are 10 characters or digits which increase in value from 0 to 9. The positional weights are determined by raising the base 10 to integer values 0, ± 1, ± 2, etc.

We speak of the units place (10^0), tens place (10^1), hundreds place (10^2), tenths place (10^{-1}), etc., and multiply the value of the character by the value of its position to determine the result. This means that 478 can be thought of as 4 times 10^2 plus 7 times 10^1 plus 8 times 10^0.

As mentioned above, the binary system has only two digits or characters, 0 and 1. The positional weights in the binary system are determined by raising the base 2 to integer values 0, ± 1, ± 2, etc.

We speak of the units place (2^0), the 2's place (2^1), 4's place (2^2), 8's place (2^3), halves place (2^{-1}), fourths place (2^{-2}), and so on.

Although we will be concerned only with whole-number representations, the positions to the right of the decimal point and binary point are the positions that use increasing negative integers as their weights and give rise to fractions.

Each binary position contains one *b*inary dig*it* or *bit*. The bit is the basic unit of information in the binary system.

The leftmost position bit is termed the *most-significant bit* (MSB) since its positional weight is the highest. The rightmost position bit is termed the *least-significant bit* (LSB) since its positional weight is 2^0 or 1.

Grouping the bits into blocks of four is a convenience that is based on ease of reading and the fact that many codes use groups of four. It is somewhat analogous to the groupings in threes for decimal numbers.

A group of 4 bits is sometimes called a *nibble*.

A group of 8 bits is sometimes called a *byte*.

One bit of information per second (BPS) is called a *baud*.

Any successive arrangement of bits taken together is termed a *word*. Common values for words are 16 and 32 bits.

One of the simplest ways of converting a binary number to its decimal equivalent is to add the weighted values of the binary positions that have the binary digit 1 in them.

The weighted values of the first 25 positions of a binary number are given in Table 7-1 for binary integers.

Example 7.1 Convert the binary number 1100 1101 to its decimal equivalent.

We see that there are 1s in the first, third, fourth, seventh, and eighth positions counting from right to left. Adding the weighted values of these positions from Table 7-1 gives

Position	Weighted value
1	1
3	4
4	8
7	64
8	128
	Sum = 205

So the equivalent of binary 1100 1101 is 205 in decimal form.

Table 7-2 lists the binary and decimal equivalents for the first 31 numbers.

There are many ways of converting decimal numbers to binary, but one of the simplest is to repeatedly divide the decimal number by 2 with the remainders forming the binary number.

Example 7.2 Find the binary equivalent for the decimal number 19.

To arrive at the binary representation we successively divide by 2, recording the remainder:

$19/2 = 9$	remainder 1 (LSB)
$9/2 = 4$	remainder 1
$4/2 = 2$	remainder 0
$2/2 = 1$	remainder 0
$1/2 = 0$	remainder 1 (MSB)

Table 7-1

Position	Power of 2	Weighted value
1	2^0	1
2	2^1	2
3	2^2	4
4	2^3	8
5	2^4	16
6	2^5	32
7	2^6	64
8	2^7	128
9	2^8	256
10	2^9	512
11	2^{10}	1024
12	2^{11}	2048
13	2^{12}	4096
14	2^{13}	8192
15	2^{14}	16384
16	2^{15}	32768
17	2^{16}	65536
18	2^{17}	131072
19	2^{18}	262144
20	2^{19}	524288
21	2^{20}	1048576
22	2^{21}	2097152
23	2^{22}	4194304
24	2^{23}	8388608
25	2^{24}	16777216

Table 7-2

Decimal number	Binary number
0	0000 0000
1	0000 0001
2	0000 0010
3	0000 0011
4	0000 0100
5	0000 0101
6	0000 0110
7	0000 0111
8	0000 1000
9	0000 1001
10	0000 1010
11	0000 1011
12	0000 1100
13	0000 1101
14	0000 1110
15	0000 1111
16	0001 0000
17	0001 0001
18	0001 0010
19	0001 0011
20	0001 0100
21	0001 0101
22	0001 0110
23	0001 0111
24	0001 1000
25	0001 1001
26	0001 1010
27	0001 1011
28	0001 1100
29	0001 1101
30	0001 1110
31	0001 1111

We then order the first binary remainder at the right (next to the binary point) and the last binary digit at the right and obtain

1 0011

We then fill in the leftmost group to four places by adding 0s to obtain

0001 0011

This filling in to make even groups of four is not always done but will be the standard practice in this chapter.

The basic numerical manipulations of addition, subtraction, multiplication, and division can be carried out in binary notation and, in fact, this is how computers and calculators actually handle these operations.

Since multiplication is performed by shifting to the left and adding and division is performed by shifting to the right and subtracting, we will only show examples for addition and subtraction.

The binary rules for addition are

$$0 + 0 = 0$$
$$0 + 1 = 1$$
$$1 + 0 = 1$$
$$1 + 1 = 0 \text{ plus a 1 carry to the next position}$$

The binary rules for subtraction are

$$0 - 0 = 0$$
$$1 - 1 = 0$$
$$1 - 0 = 1$$
$$0 - 1 = 1 \text{ plus a 1 borrow from the next position}$$

A more practical method of subtraction, used extensively in computers and microprocessors, is called the *2's complement method.*

In the 2's complement method, the subtrahend (the number you are subtracting) is first converted to what is known as the "1's complement" by changing all the 1s to 0s and all the 0s to 1s. A 1 is then added to the LSB of the 1's complement to form the 2's complement. This 2's complement is then added to the minuend (the number you are subtracting from).

If the minuend is greater than the subtrahend, the result of the subtraction process is a positive number. In the 2's complement method, this is indicated by a 1 carry and the answer may be read directly from the remaining bits.

If the subtrahend is larger than the minuend, the result of the subtraction process is a negative number. In the 2's complement method, this is indicated by a 0 carry (no carry), and the remaining bits form the answer in 2's complement form.

Some methods make use of the 1's complement directly by forming the 1's complement of the subtrahend and adding it directly to the minuend. The carry, if any, is then shifted to the LSB position and added. With this procedure, a negative number will be indicated by a 0 carry, and the remaining bits will form the answer in 1's complement form.

Example 7.3 Add the binary numbers 1101 and 1001.

We start with the LSB position and apply our addition rules just as in decimal addition. For the LSB position, $1 + 1 = 0$ plus a 1 carry:

$$\begin{array}{r} 1 \\ 1101 \\ +\ \underline{1001} \\ 0 \end{array}$$

For the next position, the carry $1 + 0 = 1$ and that is added to the 0 to obtain $1 + 0 = 1$:

$$\begin{array}{r} 1101 \\ +\ \underline{1001} \\ 10 \end{array}$$

Next we have $1 + 0 = 1$:

$$\begin{array}{r} 1101 \\ +\ \underline{1001} \\ 110 \end{array}$$

And finally $1 + 1 = 0$ plus a 1 carry to the next place:

$$\begin{array}{r} 1101 \\ +\ \underline{1001} \\ 1\ 0110 \end{array}$$

Filling in the group with zeros, we have

$$\begin{array}{r} 1101 \\ +1001 \\ \hline 0001\ 0110 \end{array}$$

We may check our result by using Table 7-2 to convert to decimal numbers.

$$\begin{aligned} 1101 &= 13 \\ 1001 &= 9 \\ 0001\ 0110 &= 22 \end{aligned}$$

And $$13 + 9 = 22\checkmark$$

Example 7.4 (*a*) Subtract 0101 from 1100 in the binary system using the subtraction rules. (*b*) Subtract 0101 from 1100 using the 2's complement method. (*c*) Subtract 0101 from 1100 using the 1's complement method. (*d*) Subtract 1100 from 0101 using the 2's complement method. (*e*) Subtract 1100 from 0101 using the 1's complement method.

(*a*) We start at the LSB position and apply the subtraction rules as we would in decimal subtraction: $0 - 1 = 1$ plus a borrow 1:

$$\begin{array}{r} 1100 \\ -0101 \\ 1 \\ \hline 1 \end{array}$$

For the next position we have $0 - 0 = 0$ but the borrow must be considered. And $0 - 1 = 1$ plus a borrow 1:

$$\begin{array}{r} 1100 \\ -0101 \\ 1 \\ \hline 11 \end{array}$$

Next we have $1 - 1 = 0$, but we have to use the borrow, so $0 - 1 = 1$ plus a 1 borrow:

Finally,
$$\begin{array}{r} 1100 \\ -0101 \\ 1 \\ \hline 111 \end{array}$$

$1 - 0 = 1$ and taking the borrow into account $1 - 1 = 0$,

$$\begin{array}{r} 1100 \\ -0101 \\ \hline 0111 \end{array}$$

Again, we may check our result by using Table 7-2 to convert to decimal:

$$\begin{aligned} 1100 &= 12 \\ 0101 &= 5 \\ 0111 &= 7 \end{aligned}$$

And $$12 - 5 = 7\checkmark$$

(*b*) Performing the same subtraction using the 2's complement method is done as follows: The number we are subtracting (the subtrahend) is 0101. First form the 1's complement by changing all the 0s to 1s and all the 1s to 0s.

Therefore the 1's complement of 0101 is written 1010.

Now form the 2's complement by adding 1 to the LSB to get

$$\begin{array}{rl} 1010 & \leftarrow \text{1's complement of subtrahend} \\ +0001 & \\ \hline 1011 & \leftarrow \text{2's complement of subtrahend} \end{array}$$

The 2's complement just formed is then added to the minuend, 1100, to give

$$\begin{array}{rrl} & 1100 & \leftarrow \text{minuend} \\ & +\ 1011 & \leftarrow \text{2's complement of subtrahend} \\ \hline \text{carry} \rightarrow & 1\ 0111 & \leftarrow \text{result} \end{array}$$

The 1 carry tells us that the result is a positive number and that the number may be read directly as 0111, the same result as in part (*a*).

(*c*) In the 1's complement method, adding the 1's complement of the subtrahend to the minuend directly gives us

$$\begin{array}{rrl} & 1100 & \leftarrow \text{minuend} \\ & +\,1010 & \leftarrow \text{1's complement of subtrahend} \\ \hline \text{carry} \rightarrow & 1\;0110 & \leftarrow \text{result (partial)} \end{array}$$

The 1 carry tells us that the result is a positive number, and when we shift the carry to the LSB position and add we obtain

$$\begin{array}{rl} 0110 & \leftarrow \text{partial result} \\ +\,0001 & \leftarrow \text{shifted carry} \\ \hline 0111 & \leftarrow \text{result} \end{array}$$

This is the same result as the previous methods gave.

(*d*) Now 0101 is the minuend and 1100 is the subtrahend. Forming the 1's complement of the subtrahend gives us 0011. Adding 1 to the LSB gives the 2's complement as

$$\begin{array}{rl} 0011 & \leftarrow \text{1's complement of subtrahend} \\ +\,0001 & \\ \hline 0100 & \leftarrow \text{2's complement of subtrahend} \end{array}$$

Adding the 2's complement to the minuend, we obtain

$$\begin{array}{rl} 0101 & \leftarrow \text{minuend} \\ +\,0100 & \leftarrow \text{2's complement of subtrahend} \\ \hline 1001 & \leftarrow \text{result (in 2's complement form)} \end{array}$$

There is no carry from this operation, which indicates that the result is a negative number in 2's complement form.

If we wish to convert the 2's complement form back to binary, we may use the same operations that were done to convert a binary number to a 2's complement number. We first replace all 1s with 0s and all 0s with 1s and then add a 1 to the LSB position. In our example this procedure yields

$$\begin{array}{rl} 0110 & \leftarrow \text{1's complement of result} \\ +\,0001 & \\ \hline 0111 & \leftarrow \text{binary form of result} \end{array}$$

As we expect, this gives as the final result of our subtraction process negative 0111.

Some calculators form the two 2's complements directly by adding 1 to the 1's complement. They then ignore the MSB. In this method, the problem would appear as: 2's complement of 0101 is

$$\begin{array}{r} 1010 \\ +\,0001 \\ \hline 1011 \end{array}$$

Then

$$\begin{array}{r} 1100 \\ +\,1011 \\ \hline 1\;0111 \end{array}$$

Ignoring the MSB 1, we have 0111 as before but we need an additional test to determine whether the answer is positive or negative.

(*e*) In the 1's complement method, we add directly the 1's complement of the subtrahend to the minuend to get

$$\begin{array}{rl} 0101 & \leftarrow \text{minuend} \\ +\,0011 & \leftarrow \text{1's complement of subtrahend} \\ \hline 1000 & \leftarrow \text{result (in 1's complement form)} \end{array}$$

The absence of a carry indicates that the result is a negative number in 1's complement form.

To convert back to binary, we simply replace all the 0s with 1s and all the 1s with 0s and have 0111. This is the same negative number as calculated in part (*d*).

There are systems that use binary numbers other than the straight binary we are talking about here. These are coded systems such as binary-coded decimal (BCD), binary-coded octal (BCO), and binary-coded hexadecimal (BCH).

These systems depend on the coding of each digit into binary on a digit-by-digit basis. For example, the decimal number 57 would be coded as 0101 0111. The first group of 4 bits represents the digit 5 and the second group of 4 bits represents the digit 7. In straight binary, the same grouping of 1s and 0s would be equivalent to the decimal number 87.

With four binary positions, we can recognize 2^4 or 16 characters, which gives rise to the hexadecimal number system. Binary-coded hexadecimal makes efficient use of the 4 bits per position needed.

On the other hand, binary-coded decimal only needs to recognize 10 characters, and while it is an inefficient use of the 4 bits needed per character, BCD is useful because of the ease in making BCD-to-decimal and decimal-to-BCD conversions. As a result, BCD is commonly encountered in the digital field.

The octal system has eight characters and needs only 3 bits per character. Many computers use binary-coded octal notation in their internal operations.

A comparative table for decimal, binary, octal, hexadecimal, BCD, BCO, and BCH is shown as Table 7-3.

Table 7-3 Comparative number system representations

Decimal	Binary	Octal	Hexadecimal	BCD	BCO	BCH
0	000000	0	0	0000 0000	000 000	0000 0000
1	000001	1	1	0000 0001	000 001	0000 0001
2	000010	2	2	0000 0010	000 010	0000 0010
3	000011	3	3	0000 0011	000 011	0000 0011
4	000100	4	4	0000 0100	000 100	0000 0100
5	000101	5	5	0000 0101	000 101	0000 0101
6	000110	6	6	0000 0110	000 110	0000 0110
7	000111	7	7	0000 0111	000 111	0000 0111
8	001000	10	8	0000 1000	001 000	0000 1000
9	001001	11	9	0000 1001	001 001	0000 1001
10	001010	12	A	0001 0000	001 010	0000 1010
11	001011	13	B	0001 0001	001 011	0000 1011
12	001100	14	C	0001 0010	001 100	0000 1100
13	001101	15	D	0001 0011	001 101	0000 1101
14	001110	16	E	0001 0100	001 110	0000 1110
15	001111	17	F	0001 0101	001 111	0000 1111
16	010000	20	10	0001 0110	010 000	0001 0000
17	010001	21	11	0001 0111	010 001	0001 0001
18	010010	22	12	0001 1000	010 010	0001 0010
19	010011	23	13	0001 1001	010 011	0001 0011
20	010100	24	14	0010 0000	010 100	0001 0100
21	010101	25	15	0010 0001	010 101	0001 0101
22	010110	26	16	0010 0010	010 110	0001 0110
23	010111	27	17	0010 0011	010 111	0001 0111
24	011000	30	18	0010 0100	011 000	0001 1000
25	011001	31	19	0010 0101	011 001	0001 1001
26	011010	32	1A	0010 0110	011 010	0001 1010
27	011011	33	1B	0010 0111	011 011	0001 1011
28	011100	34	1C	0010 1000	011 100	0001 1100
29	011101	35	1D	0010 1001	011 101	0001 1101
30	011110	36	1E	0011 0000	011 110	0001 1110
31	011111	37	1F	0011 0001	011 111	0001 1111
32	100000	40	20	0011 0010	100 000	0010 0000

7.3 TRUTH TABLES AND SYMBOLIC LOGIC

In order to understand how we can use ON-OFF or binary systems to perform useful tasks, let us examine some basic switching circuits.

Figure 7-3*a* shows a normally open (N.O.) switch. When switch A is activated, the device is on. When switch A is in the open position (not activated), the device is off.

The circuit shown in Fig. 7-3*b* is a normally closed (N.C.) switch. It is the complement of the N.O. switch because when it is in the open position (not activated) the device is on, and when it is activated the device is off.

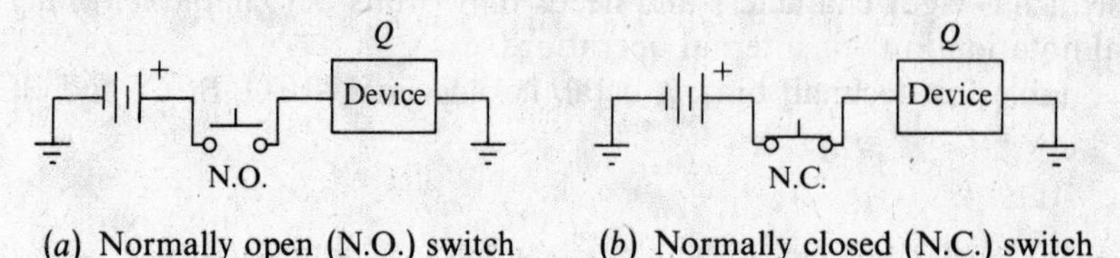

(*a*) Normally open (N.O.) switch (*b*) Normally closed (N.C.) switch

Fig. 7-3

Most mechanical switches are types of N.O. and N.C. devices.

Although the switches shown in Fig. 7-3*a* and *b* are represented by mechanical switches, they may be constructed from electrical devices such as transistors, diodes, tubes, or any other device that has two states.

In binary systems, since we have only two states, the complement of one state is the other. Another way of saying this is that the opposite state is the same as the complement.

A convenient way of symbolically representing the action of these two-state devices is to construct a chart of ON-OFF conditions for the inputs and corresponding outputs called a *truth table*.

In the binary system, the two states recognized are called 0 and 1. When the ON condition (higher voltage) means 1 (*yes* or true) and the OFF condition (lower voltage) means 0 (*no* or false), we have what is known as *positive logic*. When the ON condition (higher voltage) means 0 (*no* or false) and the OFF condition (lower voltage) means 1 (*yes* or true), we have *negative logic*. Unless otherwise indicated, in this chapter we will use positive logic.

To describe the circuit shown in Fig. 7-4*a*, which is sometimes called a *DO circuit*, we would set up a truth table as shown in Fig. 7-4*b*.

Input (cause) Output (effect)

N.O. + A Q

(*a*)

A	Q
0	0
1	1

(*b*)

Fig. 7-4

Q is used to denote the state of the output and A is used to denote the state of the switch or input. Since A can have only two states, OFF or ON (0 or 1), we write these possibilities in columnar form in the A column. The output for this circuit has been described as being off (0) when the switch A is off (0), so we place a 0 in the Q column opposite the 0 in the A column. When A is 1, Q is 1 and that information is written in by placing the 1 in the Q column opposite the 1 in the A column.

When constructing a truth table, we first list all the possible combinations of input conditions. This is the same as writing the binary numbers from 0 to $(2^n - 1)$, where n equals the number of inputs. Then, for each set of input conditions, we fill in the appropriate 1 or 0 output.

For example, with four inputs, each input may have two states (0 and 1) and the total number of combinations is 2^4 or 16. Each input is assigned a column, and the combinations are written out starting with 0000 and ending with 1111, which is the binary representation of $2^4 - 1$ or 15.

Example 7.5 Construct the truth table for the circuit shown in Fig. 7-3*b*.

The possible states for A are 0 and 1. We have previously described the output as 1 when the switch was off (0) and 0 when the switch was on (1). These conditions are filled in in the truth table as shown in Fig. 7-5.

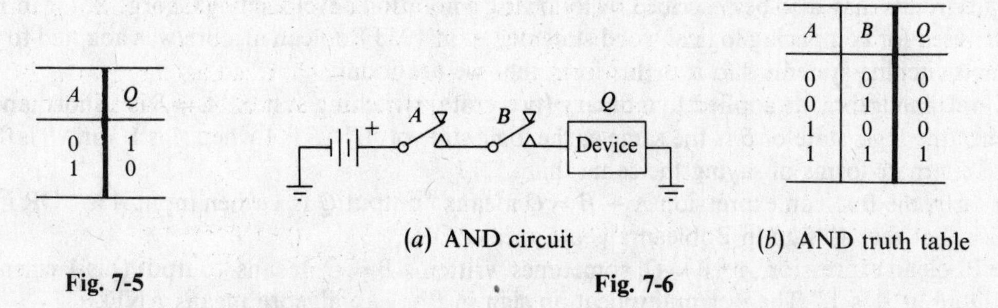

A	Q
0	1
1	0

Fig. 7-5

(a) AND circuit

A	B	Q
0	0	0
0	1	0
1	0	0
1	1	1

(b) AND truth table

Fig. 7-6

Figure 7-6*a* shows a switching circuit with an output device and two switches in series. This circuit is called the AND circuit or *coincidence* circuit because both switches have to be on at the same time for the output to be on.

The truth table for this two-input AND circuit is shown in Fig. 7-6*b*. Notice that when we have two switches, we have 2^2 or 4 possible ON-OFF combinations and that they may be thought of as being the binary equivalents to the numbers 0 to 3. The truth table tells us that the output is 1 only when both *A* and *B* are 1.

The AND circuit is characterized by the fact that, regardless of the number of switches that we place in series for this type of circuit, the output is 1 only when *all* the switches are 1 at the same time.

The dual of this arrangement is shown in Fig. 7-7*a* for two switches in parallel. This is called an OR or ANY circuit since turning either of the switches on will turn the output on. The OR circuit is also referred to as an "EITHER/OR circuit."

The truth table for the OR circuit is shown in Fig. 7-7*b*.

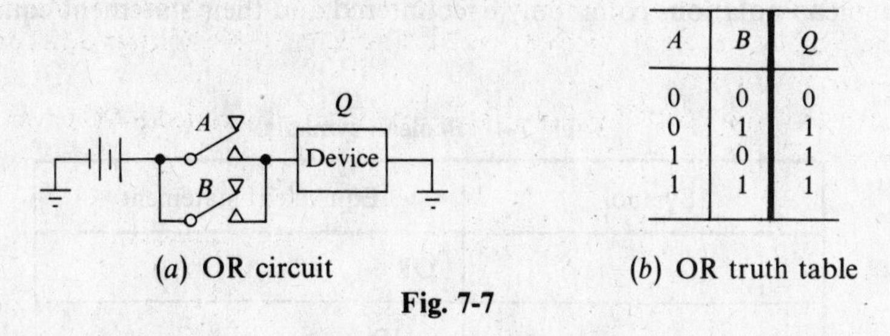

(a) OR circuit

A	B	Q
0	0	0
0	1	1
1	0	1
1	1	1

(b) OR truth table

Fig. 7-7

The OR circuit is characterized by the fact that regardless of the number of switches we place in parallel in this type of circuit, if any of the switches are 1, the output is 1. Notice that the only time the output is 0 is when both *A* and *B* are 0. This suggests that an OR for positive logic is an AND for negative logic. Similarly, the AND for positive logic is the same as the OR for negative logic. These relations between positive and negative logic are often helpful to the logic circuit designer.

A simple inverting amplifier and its truth table are shown in Fig. 7-8*a* and *b*. The inverting amplifier gives the complement at the output for the state at the input. This is the same action as the N.C. switch and is commonly called an "inverter" or "NOT circuit."

A ○—▷○—○ Q

(a) NOT circuit

A	Q
0	1
1	0

(b) NOT truth table

Fig. 7-8

A NOT circuit is characterized by the fact that whatever state is applied to the input, the output shows its complement. The NOT circuit, although extremely simple to build, is a very powerful tool in logic circuit design.

Logic circuits may also be described by following a notation developed by George Boole in 1847 and originally used for comparing logical word statements. In 1938 Boolean algebra was adapted to describe telephone switching systems and it is this form that we use today.

In Boolean algebra, as applied to a binary (two-state) switching system $A = B$ is a shorthand for the statement "the logic state of B is the same as the logic state of A." "B is 1 when A is 1" or "B is 0 when A is 0" are alternate forms of saying the same thing.

Similarly, the Boolean expression $A + B = Q$ means "output Q is 1 when input A is 1 OR input B is 1." Notice that the + sign in Boolean algebra means OR.

The Boolean expression $A \cdot B = Q$, sometimes written $AB = Q$, means "output Q is 1 when input A is 1 AND input B is 1." The $\cdot$ or multiplication sign in Boolean algebra means AND.

To indicate the complement of A, we place a bar over it ($\bar{A}$) and read it as NOT A. The statement $\bar{A} = Q$ means "output Q is 1 when input NOT A is 1."

Since A and NOT A are complements, NOT A is 1 when A is 0 and we can also say "output Q is 1 when input A is 0." This latter form is the one most commonly employed.

Since we are applying Boolean algebra to a two-state system, we can also read the previous Boolean statement as "output Q is 1 when input A is NOT 1."

Some texts indicate the complement by a prime designation, so B' would be the same as NOT B.

Example 7.6 What does the statement $\bar{A} \cdot B = Q$ mean in Boolean algebra?

Since multiplication means AND in Boolean algebra, we read the statement as "output Q is 1 if input A is NOT 1 AND input B is 1." Since we also know that NOT A is 1 if A is 0, we may substitute and read "output Q is 1 if input A is 0 AND input B is 1."

A list of the Boolean notations commonly encountered and their statement equivalence is provided in Table 7-4.

Table 7-4 Boolean symbols

Symbol	Equivalent statement
+	OR
$\cdot$	AND
$\bar{A}$ or A'	NOT A
$Q = A$	Q is true if A is true
$Q = A + B$	Q is true if A OR B is true
$Q = A \cdot B = AB = (A)(B)$	Q is true if A AND B are true

Sometimes the Boolean expressions can be simplified or put in a different form by using De Morgan's theorem or law. This theorem states that to change the form of any expression you replace each symbol with its complement, replace each + with $\cdot$ and each $\cdot$ with +. An equivalent to the original expression can then be obtained by complementing the entire result.

Remember that you can always check the validity of any Boolean statement by constructing a truth table.

Example 7.7 Use De Morgan's theorem to obtain the complement of $A \cdot B + \bar{A} = Q$ and then find an equivalent expression.

De Morgan's theorem says to first replace the symbols with their complements:

$$\bar{A} \cdot \bar{B} + A = \bar{Q}$$

Then replace the operators with their duals to obtain

$$(\bar{A} + \bar{B}) \cdot A = \bar{Q}$$

To obtain the original equivalent expression, we complement both sides of the equation to get

$$\overline{(\bar{A} + \bar{B}) \cdot A} = \bar{\bar{Q}} = Q$$

Notice that a double complement (double bar) returns the original symbols. Therefore $\bar{\bar{Q}} = Q$.

Some useful identities in Boolean algebra are listed in Table 7-5.

Table 7-5 Boolean identities

$A \cdot A = A$	(7-1a)	$A + A \cdot B = A$	(7-5a)
$A + A = A$	(7-1b)	$A(A + B) = A$	(7-5b)
$A \cdot \bar{A} = 0$	(7-2a)	$\bar{A} + \bar{B} = \overline{A \cdot B}$	(7-6)
$A + \bar{A} = 1$	(7-2b)	$\overline{A + B} = \bar{A} \cdot \bar{B}$	(7-7)
$A \cdot 1 = A$	(7-3a)	$A \cdot B + A \cdot C = A(B + C)$	(7-8)
$A + 1 = 1$	(7-3b)	$A + \bar{A} \cdot B = A + B$	(7-9)
$A \cdot 0 = 0$	(7-4a)	$A \cdot B + A \cdot \bar{B} = A$	(7-10a)
$A + 0 = A$	(7-4b)	$A \cdot B + \bar{A} \cdot B = B$	(7-10b)
		$\bar{\bar{A}} = A$	(7-11)

Example 7.8 Prove that the Boolean identity (*7-10b*) in Table 7-5 is true by using a truth table.

Identity (*7-10b*) in Table 7-5 states that

$$A \cdot B + \bar{A} \cdot B = B$$

This identity involves A, $\bar{A}$, and B, so we first set up the usual truth table for inputs A and B, as shown in Fig. 7-9*a*. We then complement the A column to get the $\bar{A}$ column, as shown in Fig. 7-9*b*. Since the identity is set up as the Boolean sum of products $A \cdot B$ and $\bar{A} \cdot B$, using

$$P_1 = A \cdot B$$

we may fill in a P_1 column using the logical AND rules and the values in columns A and B as shown in Fig. 7-9*c*. Using

$$P_2 = \bar{A} \cdot B$$

with the AND rules applied to the values in columns $\bar{A}$ and B, we may fill in a P_2 column as shown in Fig. 7-9*d*. Using

$$S_1 = P_1 + P_2$$

we can use the logical OR rules and the values in columns P_1 and P_2 to construct the S_1 column as shown in Fig. 7-9*e*.

A	B
0	0
0	1
1	0
1	1

(*a*)

A	$\bar{A}$
0	1
0	1
1	0
1	0

(*b*)

A	B	P_1
0	0	0
0	1	0
1	0	0
1	1	1

(*c*)

$\bar{A}$	B	P_2
1	0	0
1	1	1
0	0	0
0	1	0

(*d*)

P_1	P_2	S_1
0	0	0
0	1	1
0	0	0
1	0	1

(*e*)

Fig. 7-9

Comparing the S_1 column in Fig. 7-9*e* with the *B* column in Fig. 7-9*a*, we see they are identical, proving

$$S_1 = P_1 + P_2 = A \cdot B + \bar{A} \cdot B = B$$

the right-hand equation being identity (*7-10b*).

We may also employ Boolean statements to construct a switching circuit from a truth table.

Using the truth table shown in Fig. 7-7*b*, we see that the output is 1 for either of three cases: Either *A* is 0 and *B* is 1, *or A* is 1 and *B* is 0, *or A* is 1 and *B* is 1. We could then write $Q = \bar{A} \cdot B + A \cdot \bar{B} + A \cdot B$.

This can be simplified to $Q = \bar{A} \cdot B + A$ by noting that $A \cdot \bar{B} + A \cdot B = A$ identity (*7-10a*) from Table 7-5. Put another way, if *A* is 1, *Q* is 1, and it does not matter what state *B* is in. We can further simplify by using identity (*7-9*) from Table 7-5 and noting that our resulting expression can be written $\bar{A} \cdot B + A = A + B$. So the simplest realization of the truth table would be the OR circuit shown in Fig. 7-10*a*.

An important point to consider is that each of the other nonreduced Boolean expressions also could be constructed in the form of the switching circuits shown in Fig. 7-10*b* and *c*.

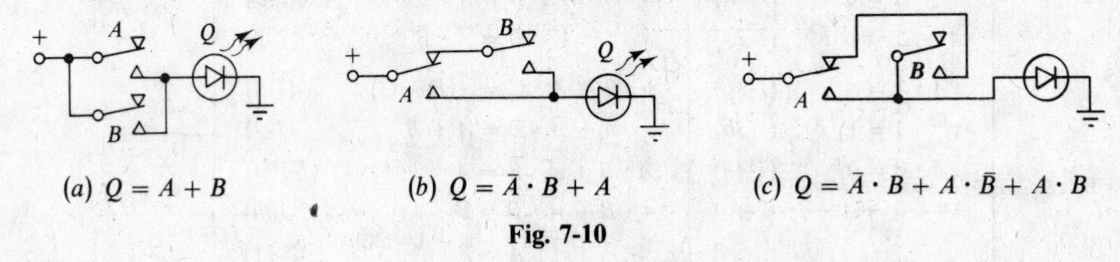

(*a*) $Q = A + B$ (*b*) $Q = \bar{A} \cdot B + A$ (*c*) $Q = \bar{A} \cdot B + A \cdot \bar{B} + A \cdot B$

Fig. 7-10

As we will see in Sec. 7.5, when we construct a switching circuit from a truth table, we have no guarantee that the circuit is the simplest one that will do the job.

While the laws of Boolean algebra form the bases for all logic circuit simplification, they are sometimes cumbersome to apply.

There are other methods, such as Karnaugh or K mapping, that may help in arriving at the most efficient switching circuit, but the methods do not always work and are themselves complicated.

In general, we can construct a truth table from a Boolean expression by breaking down the expression into a Boolean sum of products or product of sums and applying the OR and AND rules to each group of inputs (variables) formed.

To construct a Boolean expression from a truth table, we write as the Boolean sum of products all combinations of the input variables that give a true 1 output and try to use the Boolean identities to reduce the expression to its simplest form.

Naturally the fewer and simpler the switching circuits needed, the easier it becomes to build the circuit. Simple circuits have the lowest cost and are therefore a desirable design goal.

7.4 LOGIC FAMILIES

We have seen in Sec. 7.3 that switching circuits can be constructed from truth tables and that truth or function tables can describe the action of switching circuits. In this section we will describe the characteristics of some of the electronic circuits that have been developed to duplicate the switching functions of mechanical switches.

Table 7-6 lists the commonly encountered logic families and their main features. The most popular of these in current use are TTL, both regular and low-power Schottky (LS) types, ECL, MOS, CMOS, and I^2L.

Logic families are compared on the basis of

1. Switching speed
2. Power dissipation
3. Speed-power product
4. Noise margin
5. Fanout capability
6. Density
7. Cost

Table 7-6 Logic family information

Acronym	Name	Typical schematic	Comments
TTL (T^2L)	*T*ransistor-*T*ransistor *L*ogic	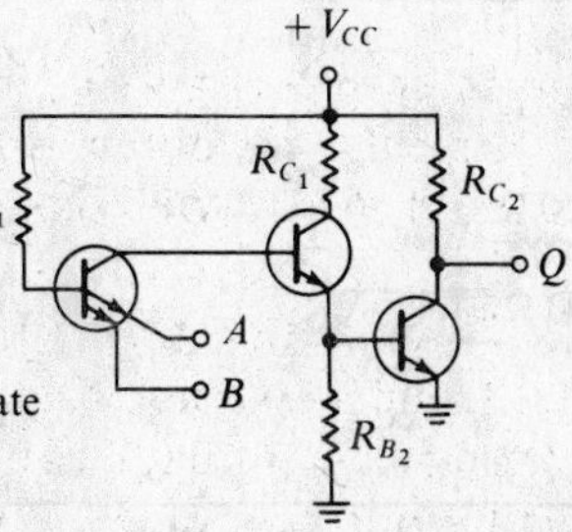 NAND gate $Q = \overline{A \cdot B}$ ($+V_{CC}$, R_{B_1}, R_{C_1}, R_{C_2}, R_{B_2}, A, B, Q)	Economical Good current capability Compatible with DTL and CMOS Good switching speed Schottky type has very high switching speed and low power consumption (built-in Schottky diodes provide clamping)
ECL	*E*mitter *C*oupled *L*ogic	NOR gate $Q = A + B$; OR gate $\bar{Q} = \overline{A + B}$ ($+V_{CC}$, R_{C_1}, R_{C_2}, R_E, V_{EE}, V_{REF}, A, B, Q, $\bar{Q}$) 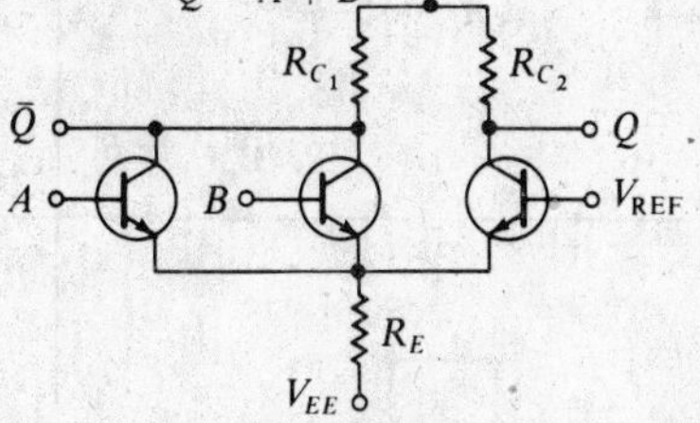	Fastest speed High power consumption Needs good heat sinking Usually provides both regular and inverted outputs
MOS	*M*etal-*O*xide *S*emiconductor	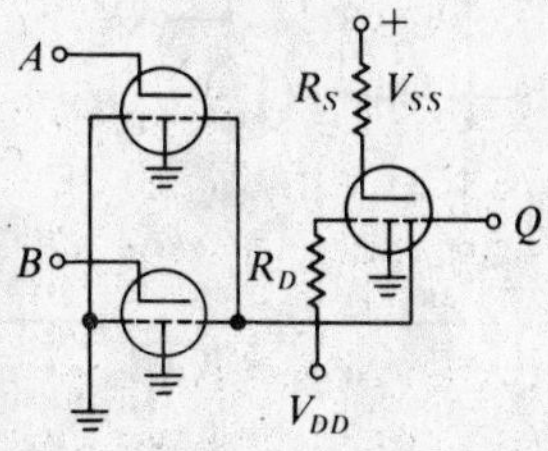 NAND gate $Q = \overline{A \cdot B}$ (+, R_S, V_{SS}, R_D, V_{DD}, A, B, Q)	This family uses negative logic Needs dual power supplies (+ and −) Easy to make large, complex chips Good economy Moderate speed
CMOS	*C*omplementary *M*etal-*O*xide *S*emiconductor	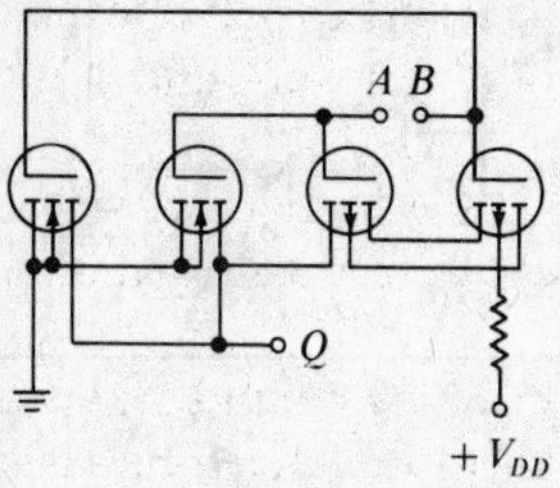 NOR gate $Q = \overline{A + B}$ (A, B, Q, $+V_{DD}$)	Uses more area than MOS More complicated to process (manufacture) High noise margin Moderate speed Very low power dissipation
IIL (I^2L)	*I*ntegrated *I*njection *L*ogic	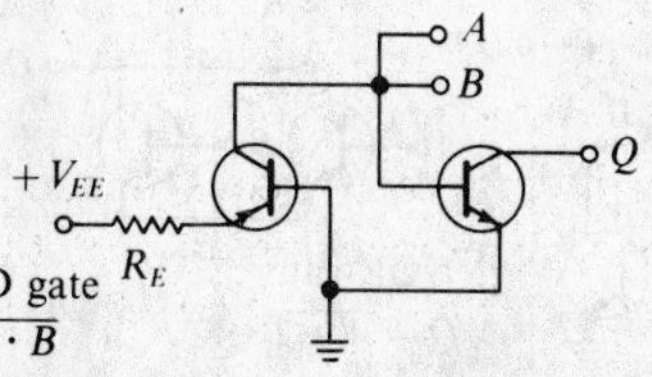 NAND gate $Q = \overline{A \cdot B}$ ($+V_{EE}$, R_E, A, B, Q)	High speed One "transistor" grown for each gate Low power consumption Good density Only one output per gate
RTL	*R*esistor-*T*ransistor *L*ogic	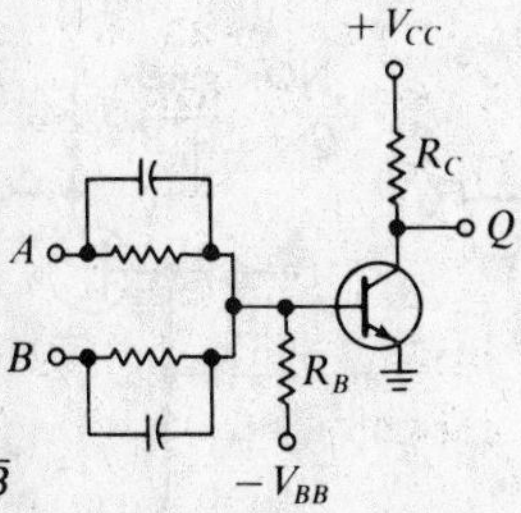 NOR gate $Q = \overline{A + B}$ ($+V_{CC}$, R_C, R_B, $-V_{BB}$, A, B, Q)	Economical and straightforward in design Easily interfaced with other common families Poor noise immunity Low threshold and fanout

Table 7-6 (*continued*)

Acronym	Name	Typical schematic	Comments
RCTL	*Resistor-Capacitor-Transistor Logic*	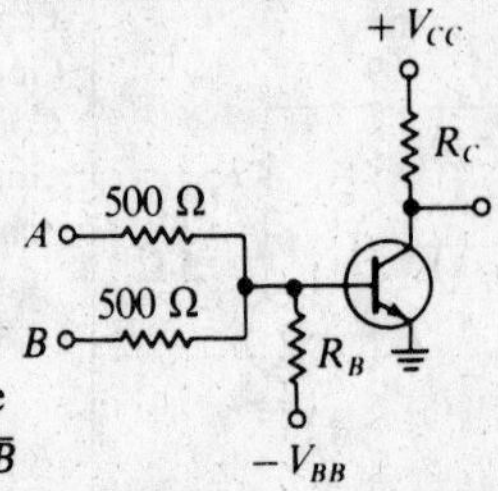 NOR gate $Q = \overline{A + B}$ ($+V_{CC}$, R_C, 500 Ω, 500 Ω, R_B, $-V_{BB}$)	Addition of capacitors increases switching speed but is more expensive than RTL Severe limitation because of stored charge requirements
DL	*Diode Logic*	OR gate 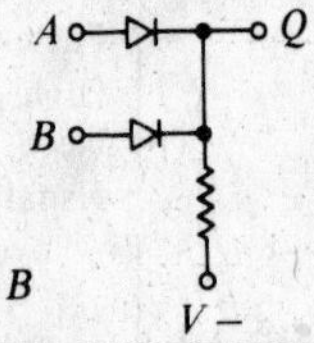 $Q = A + B$ ($V-$); AND gate 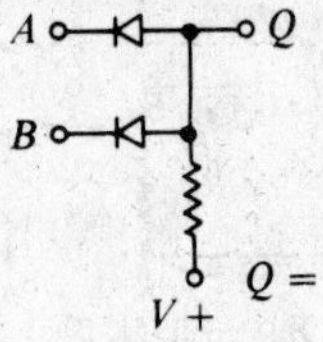 $Q = A \cdot B$ ($V+$)	Amplifiers needed for fanout above three Inexpensive components Relatively high speed
LLL (L^3) (CSDL)	*Low-Level Logic (Current Switching Diode Logic)*	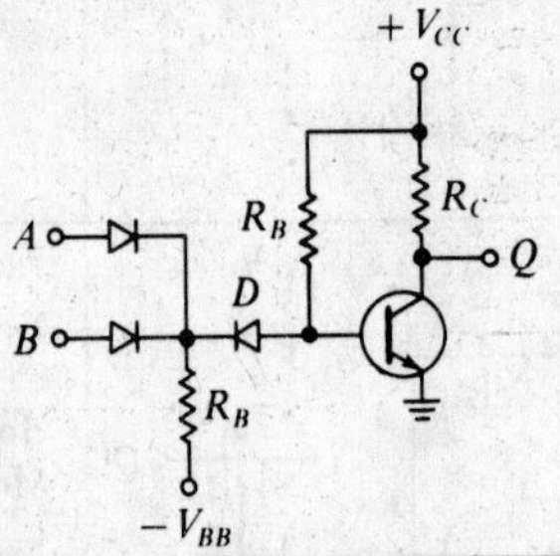 NOR gate $Q = \overline{A + B}$ ($+V_{CC}$, R_C, R_B, D, $-V_{BB}$)	D is isolating diode permitting R_B to turn on collector current Number of inputs does not affect base current Good switching speed
HTL	*High-Threshold Logic*	NAND gate $Q = \overline{A \cdot B}$ 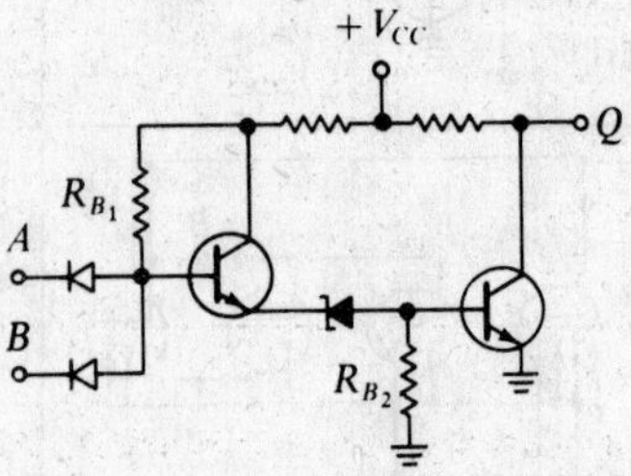 ($+V_{CC}$, R_{B_1}, R_{B_2})	Better fanout than DTL High noise immunity, typically 10 V Moderate speed
DCTL	*Direct-Coupled Transistor Logic*	NAND gate $Q = \overline{A \cdot B}$ 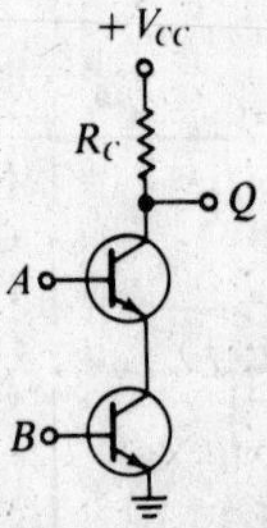 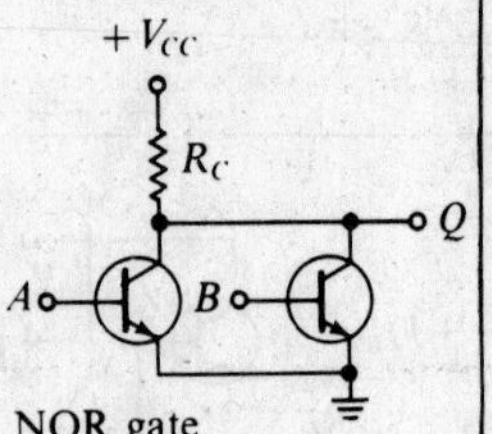 ($+V_{CC}$, R_C); NOR gate $Q = \overline{A + B}$ ($+V_{CC}$, R_C)	Moderately fast switching speeds Low power dissipation Needs uniform transistors Transistors operate in saturation region V_{BE} = logic 1; V_{CE} = logic 0 $V_{BE} \gg V_{CE}$ for stability
DTL	*Diode-Transistor Logic*	NAND gate $Q = \overline{A \cdot B}$ 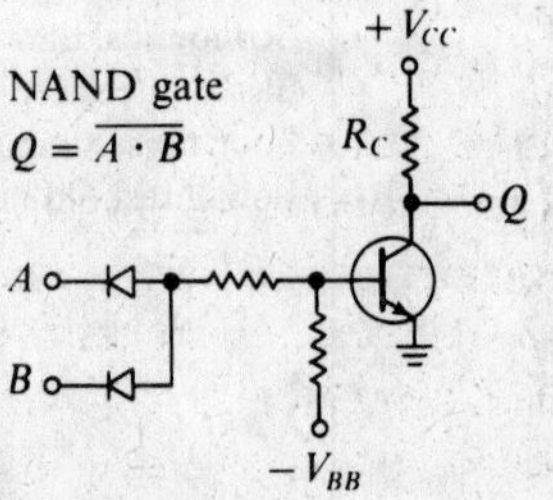 ($+V_{CC}$, R_C, $-V_{BB}$); NOR gate $Q = \overline{A + B}$ 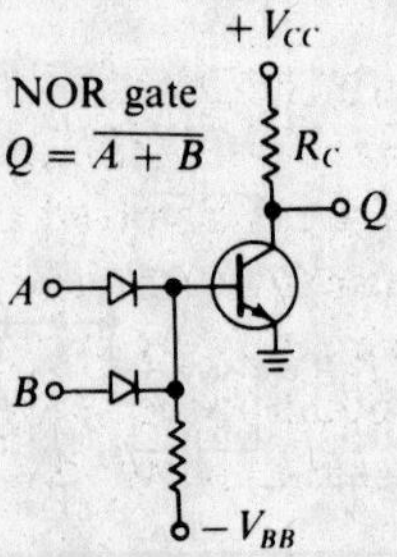 ($+V_{CC}$, R_C, $-V_{BB}$)	Logic performed by diodes Transistor acts as inverting amplifier Poor noise immunity and fanout High speeds

Table 7-6 *(continued)*

Acronym	Name	Typical schematic	Comments
CML	*C*urrent *M*ode *L*ogic	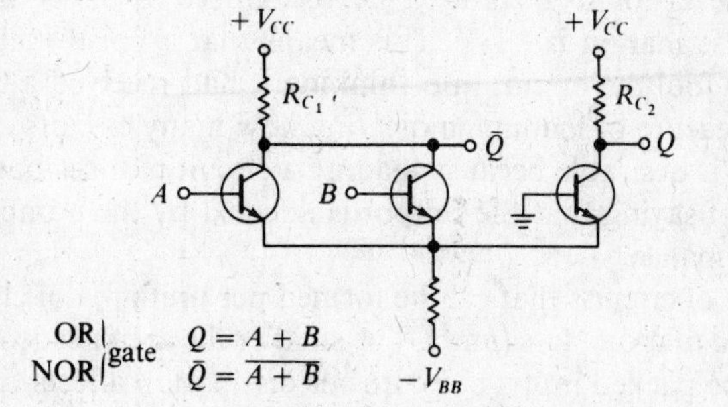$+V_{CC}$ $+V_{CC}$ $R_{C_1}{}'$ R_{C_2} $\bar{Q}$ Q A B OR } gate $Q = A + B$ NOR } $\bar{Q} = \overline{A + B}$ $-V_{BB}$	Very high switching speeds Good noise margin Transistor bias is from constant-current generator which keeps transistor far away from saturation Usually provides both regular and inverted outputs
CDL	*C*ore *D*iode *L*ogic	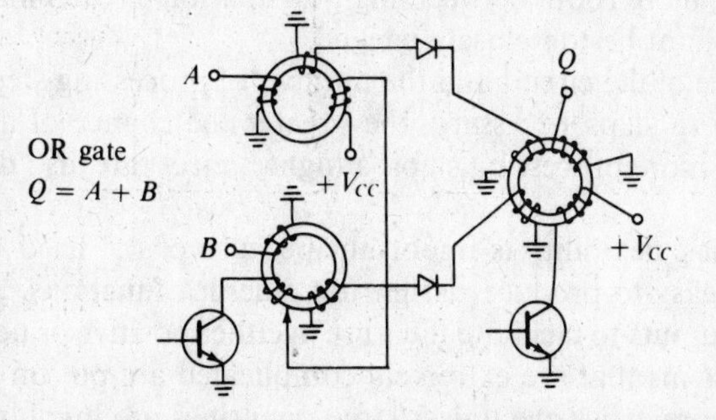OR gate $Q = A + B$ A B Q $+V_{CC}$ $+V_{CC}$	Logic performed by cores and transmitted by diodes Transistors act as drivers and can drive many cores as long as they are not successive cores in the logic line Speed limited by core switching speed
4LDL	*4-L*ayer *D*evice *L*ogic	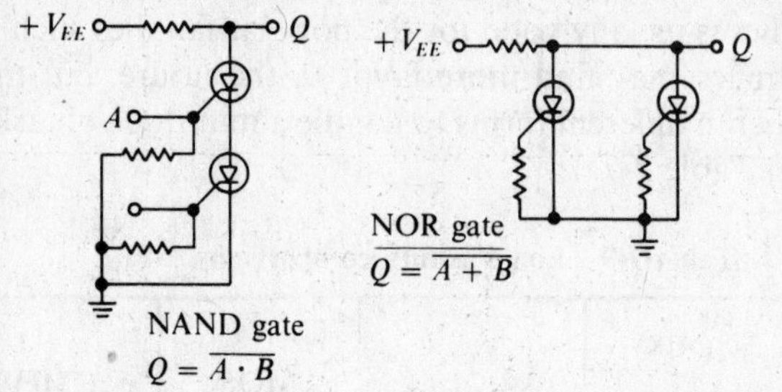$+V_{EE}$ Q A $+V_{EE}$ Q NOR gate $Q = \overline{A + B}$ NAND gate $Q = \overline{A \cdot B}$	Logic performed by silicon controlled switches Triggering by dc levels or pulses Gates have memory—must be reset Input does not have to be maintained High power output capability
TDL	*T*unnel *D*iode *L*ogic	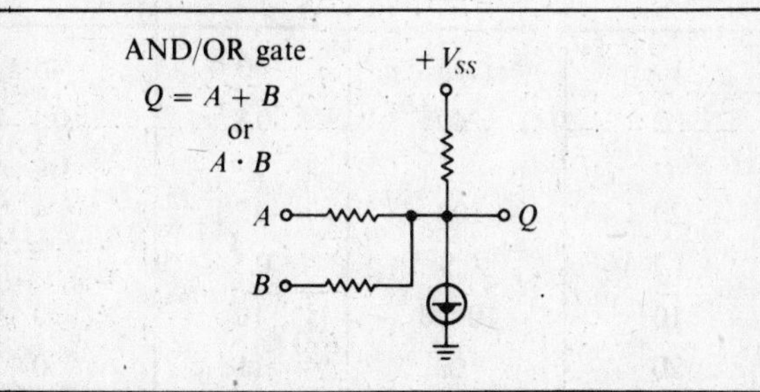AND/OR gate $Q = A + B$ or $A \cdot B$ $+V_{SS}$ A B Q	Logic obtained from tunnel diode changing states Basic gate is AND or OR depending on bias: Bias current near peak value, OR Bias current near zero, AND Simple High speed

The *switching speed* is measured in terms of propagation delay for the input change to appear at the output. The propagation delay should be as small as possible and is typically measured in nanoseconds (ns).

Power dissipation, measured in milliwatts (mW), is the amount of power dissipated in the circuit, usually in the form of heat. We want the dissipation to be as low as possible both because of the difficulty in removing heat from the small areas involved and for economy of operation.

The *speed-power product* is just what its name indicates. It is a figure of merit found by multiplying the propagation delay in nanoseconds by the power dissipation in milliwatts. This gives energy in picojoules (pJ). We want the product to be as small as possible. Usually the reduction of propagation delay is accompanied by an increase in power dissipated. The speed-power product gives us some idea of the tradeoffs involved.

Noise margin is the difference between the level transmitted for each logic state and the minimum level necessary at the input of the next circuit to correctly recognize the proper logic state. A high noise margin, measured in volts (V), is good protection against random (noise) voltages confusing the circuit. For example, if the minimum level for a 1 state to be recognized is 3.5 V and the output of the previous circuit is 4.6 V, the noise margin is 1.1 V. This means that a noise voltage would have to be greater than -1.1 V in order to fool the circuit into thinking it had received a 0.

The *fanout capability* is a measure of loading, indicating how many circuits can be driven from one output. A high fanout capability is desirable because loading a circuit reduces both its noise margin and speed. Loading is another way of saying that the output is affected by the number of circuits attached, which changes the circuit performance.

Density refers to the number of circuits that can be formed per unit area of chip surface. The area is measured in square mils or square micrometers (μm^2). The smaller the area needed per circuit, the greater the number of circuits that can be packed into a chip. Power dissipation affects the density since, even if the circuit takes little room itself, more room between high-dissipation circuits must be provided to help remove heat, and the circuits cannot be too closely packed.

Cost is dependent on the size of the circuit and the number of processing steps required to produce the circuit on the chip. The more steps necessary, the greater the chance of a defect in the chip. In addition to the increased cost of more processing steps, a higher reject rate also drives up the cost of the chip.

In designing IC logic families, flexibility is important because of the need to interface with other families and to use one basic, easy-to-produce design for different functions. Sometimes circuits are designed with complementary outputs to facilitate their use in either positive or negative logic structures.

In some cases complete systems that are extremely complicated are put on a single chip and only used for that purpose. This is more expensive unless large quantities are involved.

Compatibility between families is usually good for the popular families such as TTL and CMOS or DTL and TTL. Other logic families may find more favor in the future, but they all consist of basic building-block circuits put together in different forms to handle a multitude of tasks. A comparison of the popular logic families is given in Table 7-7.

Table 7-7 Logic family comparisons

Characteristics	TTL	Schottky TTL	ECL	MOS	CMOS	I^2L	Units
Switching speed	10	10	5	40	40	20	ns
Power dissipation	12	2	40	0.8	0.1	0.05	mW
Speed-power product	120	20	200	32	4	1.0	pJ
Noise margin	1.0	1.1	0.9	0.5	1.5	0.7	V
Fanout capability	10	10	10–20	10	50	1	...
Density	20	20	20	11	50	7	mils^2
Cost	Low	Moderate	Moderate	Very low	Moderate	High	...
Flexibility	Good	Good	Very good	Good	Very good	Fair	...

7.5 LOGIC GATES

The switching circuits we have been discussing are fundamental building blocks of digital systems and are referred to as *gates*. A gate is named for its ability to control an output with a specific combination of logic states on the inputs to the gate and for the switching functions performed and described by truth tables.

While we are primarily concerned with the function of the gate in this chapter and will not go into a description of the actual gating circuit (FET, MOS, bipolar transistor layouts, etc.), this construction and the selection of the best logic family to use for a particular gating circuit are of primary concern to the designer of the integrated circuits and to the circuit designer who uses them. Manufacturers' specification sheets can be of great help in the proper gate selection since new manufacturing techniques are continually providing the user with improved selections.

In addition to using the truth table to describe the gate's action, we use a specific graphical symbol to represent each of the common logic gate forms.

The symbol for a common amplifier with a noninverting output is shown in Fig. 7-11*a*. While this is not really a gate, the symbol is used in digital drawings to denote a buffer amplifier. A buffer amplifier usually is an op amp with unity gain and a noninverting input used to isolate one part of a circuit from another.

When an open circle is added to the output of the buffer, a NOT gate is formed, as shown in Fig. 7-11*b*. A NOT or inverting (complement) gate is usually an op amp with the inverting input used to produce an output that is the complement of the input. The use of the open circle or invert bubble at an input, as shown in Fig. 7-11*c*, is used to indicate that the input is inverted and is often used instead of drawing a separate NOT gate feeding that input. Figure 7-11*d* shows these two equivalent forms.

The symbol for an OR gate is shown in Fig. 7-12*a*, while the symbol for an AND gate is shown in Fig. 7-12*b*. The OR gate is sometimes called an "EITHER/OR BOTH gate" and the AND gate is sometimes referred to as a "COINCIDENCE gate."

When OR and AND gates have inverted outputs, they are known as "negative OR" ("NOR") and "negative AND" ("NAND") gates. Because of the ease of building NOR and NAND gates with MOS circuitry, these gates are frequently used, and their symbols are shown in Fig. 7-13*a* and *b*.

Another commonly encountered gate is the exclusive OR (XOR) gate shown in Fig. 7-14*a* and its complement the exclusive NOR (XNOR) gate shown in Fig. 7-14*b*.

Input Output (*a*) Input Output (*b*) Input Output (*c*) In Out In Out (*d*)

Fig. 7-11

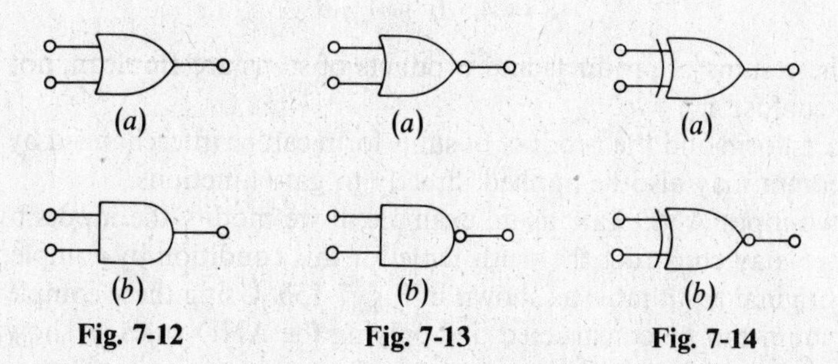

Fig. 7-12 **Fig. 7-13** **Fig. 7-14**

The EXCLUSIVE OR gate is sometimes called an "EITHER/OR BUT NOT BOTH" gate since its function duplicates the OR gate except when both inputs are 1, in which case the OR output is still 1 while the XOR output is 0.

While we have shown all the gates as having two inputs, many of the common gates have three, four, or even more inputs and are classified as three-input AND, four-input NOR, etc. It is also common practice to manufacture several gates of the same type on a single chip. This gives rise to dual (2) four-input NAND gates, quad (4) two-input OR gates, and hex (6) inverters.

As in algebra, the output may be designated with any symbol. In addition to Q, some manufacturers use Y.

Table 7-8 shows some of the commonly encountered gates along with their symbols, Boolean representation, and truth tables.

Table 7-8

DO (buffer) — $Q = A$

A	Q
0	0
1	1

NOT (inverter) — $Q = \bar{A}$

A	Q
0	1
1	0

2-Input OR — $Q = A + B$

A	B	Q
0	0	0
0	1	1
1	0	1
1	1	1

2-Input AND — $Q = A \cdot B$

A	B	Q
0	0	0
0	1	0
1	0	0
1	1	1

2-Input NOR — $Q = \overline{A + B}$

A	B	Q
0	0	1
0	1	0
1	0	0
1	1	0

2-Input NAND — $Q = \overline{A \cdot B}$

A	B	Q
0	0	1
0	1	1
1	0	1
1	1	0

EXCLUSIVE OR (XOR) — $Q = A\bar{B} + B\bar{A}$

A	B	Q
0	0	0
0	1	1
1	0	1
1	1	0

EXCLUSIVE NOR (XNOR) — $Q = \bar{A} \cdot \bar{B} + A \cdot B$

A	B	Q
0	0	1
0	1	0
1	0	0
1	1	1

With gates, as with switching circuits, the truth tables can be converted to Boolean expressions in the form of sum of products or products of sums. The sum of products of a two-input XOR gate, for example, would look like

$$S = \bar{A} \cdot B + A \cdot \bar{B}$$

Remember that these sums of products and products of sums are Boolean, not algebraic, and care must be taken not to confuse the two.

The sum of products form and the product of sums form can be interchanged by using De Morgan's theorem, and that theorem may also be applied directly to gate functions.

Looking at the two-input AND gate as an example, if we modify the inputs by inverting them as shown in Fig. 7-15*a*, we may construct the truth table for this condition by complementing each of the input columns of the original truth table as shown in Fig. 7-15*b*. Using these complemented columns ($\bar{A}$ and $\bar{B}$), the output column can be constructed by applying the AND rules as shown in Fig. 7-15*c*.

(*a*) $Q = \bar{A} \cdot \bar{B}$

(*b*)

A	B	$\bar{A}$	$\bar{B}$
0	0	1	1
0	1	1	0
1	0	0	1
1	1	0	0

(*c*)

$\bar{A}$	$\bar{B}$	Q
1	1	1
1	0	0
0	1	0
0	0	0

Fig. 7-15

Comparing this output column of the modified AND gate to the output column of the NOR gate shown in Table 7-8, we see they are equivalent.

This leads us to the fact that we may easily find equivalent gates by adapting De Morgan's law to read: If we replace all the inputs to a gate with their complements, and the gate itself with its negative dual (NAND for OR, NOR for AND, etc.), the gate function remains the same.

Example 7.9 Find the equivalent gating circuit for the diagram shown in Fig. 7-16*a* using De Morgan's theorem.

Using the modified De Morgan's theorem, we first complement each of the inputs as shown in Fig. 7-16*b*. We then convert the AND gate to a NOR gate and the NOR gate to an AND gate to obtain the circuit shown in Fig. 7-16*c*.

Since the connection in Fig. 7-16*c* shows two inversions in series, we can replace the double inversion with the positive-connection OR gate shown in Fig. 7-16*d*.

The truth table for Fig. 7-16*a* is shown in Fig. 7-16*e*, and the truth table for Fig. 7-16*d* is shown in Fig. 7-16*f*. The truth tables show that the circuit in Fig. 7-16*a* and the circuit shown in Fig. 7-16*d* perform the same logic functions.

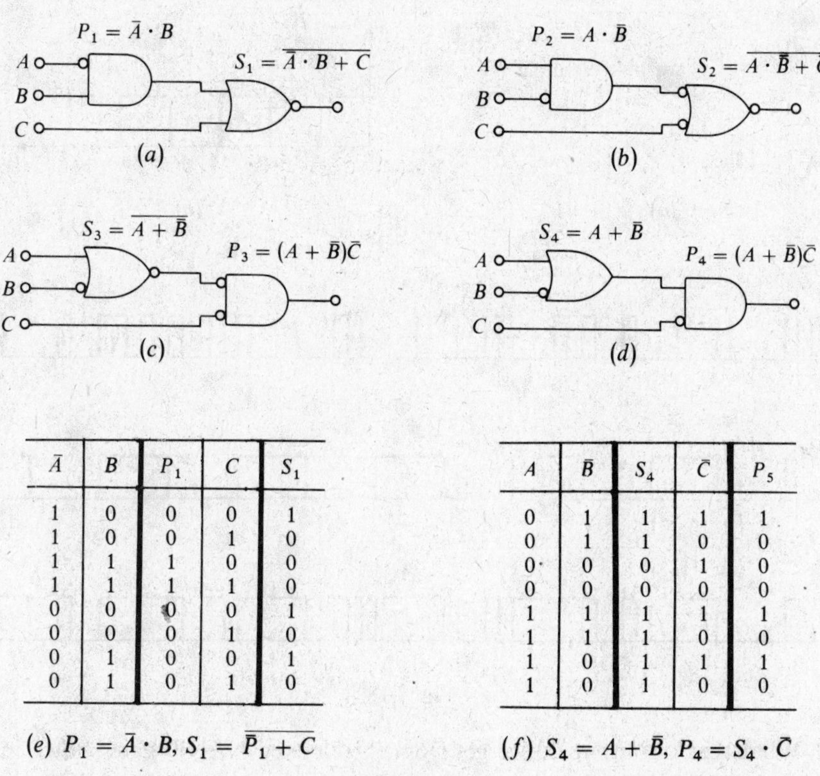

$\bar{A}$	B	P_1	C	S_1
1	0	0	0	1
1	0	0	1	0
1	1	1	0	0
1	1	1	1	0
0	0	0	0	1
0	0	0	1	0
0	1	0	0	1
0	1	0	1	0

(*e*) $P_1 = \bar{A} \cdot B$, $S_1 = \overline{P_1 + C}$

A	$\bar{B}$	S_4	$\bar{C}$	P_5
0	1	1	1	1
0	1	1	0	0
0	0	0	1	0
0	0	0	0	0
1	1	1	1	1
1	1	1	0	0
1	0	1	1	1
1	0	1	0	0

(*f*) $S_4 = A + \bar{B}$, $P_4 = S_4 \cdot \bar{C}$

Fig. 7-16

The ability to use NAND gates to perform OR functions and NOR gates to perform AND functions by inverting the inputs gives the circuit designer a great deal of flexibility in selecting gates.

By combining gates as building blocks, we can create many useful circuits performing such functions as

- Counting
- Routing
- Adding
- Multiplexing and demultiplexing
- Coding and encoding
- Sampling
- Latching
- Storing
- Detecting
- Masking
- Dividing or multiplying

Sampling is an important part of converting analog to digital signals, and sample-and-hold gates will be mentioned in Sec. 7.10.

The sampling gates pass information at a particular rate determined by a *clock*. A clock is a square wave oscillator, usually crystal controlled for accuracy, that determines the sequence of events in complicated logic circuits such as counters and microprocessors. We will show the action of counting gates in Sec. 7.7 using the clock output as a reference.

If we hold one input of a standard two input AND gate at 1, the output will exactly follow the input. This type of arrangement can be controlled by a clock, as shown in Fig. 7-17*a*. If we look at the clock output as shown in Fig. 7-17*b*, we see that, at all times when the clock is 0, the input to the AND gate will also be 0, and the AND gate will have a 0 output independent of the data input. When the clock output is 1, however, from $2T$ to $3T$ for example, the clock input to the AND gate will be held at 1, and the output will depend on the logic level present at the data input.

This same type of sampling arrangement may be achieved by tying one OR gate input to 0. The output of the OR gate follows the data input when the clock is 0. When the clock output is 1, the OR gate output is always 1, independent of the data input. This action is shown in Fig. 7-18.

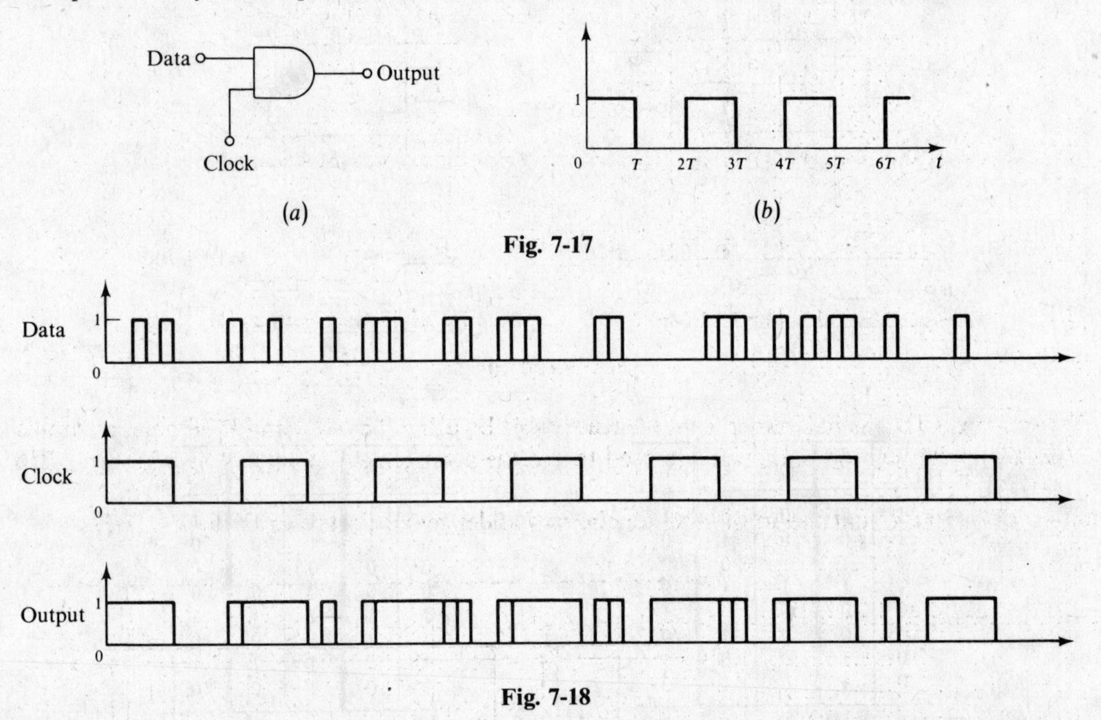

Fig. 7-17

Fig. 7-18

Example 7.10 What type of output do we get from the clocked NAND gate shown in Fig. 7-19*a*?

We may use the truth table for the two-input NAND gate with the designations CLOCK for A, DATA for B, and OUTPUT for Q.

From this truth table, shown in Fig. 7-19*b*, we see that, when the CLOCK is 0, the OUTPUT is the complement of the input (DATA) terminal. When the CLOCK is 1, the OUTPUT is 0 regardless of the logic state of the DATA terminal.

This is sometimes called a "sample and invert gate" and finds application in analog-to-digital conversion.

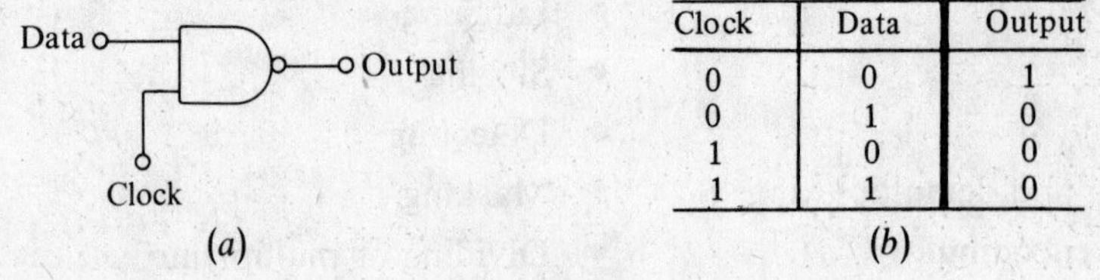

Clock	Data	Output
0	0	1
0	1	0
1	0	0
1	1	0

Fig. 7-19

For arithmetic operations, we may use a combination of gates to perform addition and, by some manipulation, the other arithmetic functions as well. As a matter of fact, almost all arithmetic calculations make use of the gate combinations known as the "full adder" and the "half-adder."

The full adder has three inputs. Two are bit inputs from the two numbers being added, while the third is the carry bit (N) from the previous stage. Also, it has two outputs: the sum (S) of the three inputs and the carry (N) bit output to the next stage.

A half-adder has two inputs; these are the bit inputs from the numbers being added. Since there is no provision for a carry input, the half-adder is used at the LSB (rightmost) position or where carries are ignored. The half-adder outputs are the same as for the full adder, with one output for the sum (S) of the bits and the other for the carry (N) to the next stage.

The truth tables for the full adder and the half-adder are shown in Fig. 7-20*a* and *b*, respectively.

A = bit from number A
B = bit from number B
N = carry bit from previous stage
C = carry bit
S = sum bit

N	B	A	S	C
0	0	0	0	0
0	0	1	1	0
0	1	0	1	0
0	1	1	0	1
1	0	0	1	0
1	0	1	0	1
1	1	0	0	1
1	1	1	1	1

(*a*) Full-adder truth table

B	A	S	C
0	0	0	0
0	1	1	0
1	0	1	0
1	1	0	1

(*b*) Half-adder truth table

Fig. 7-20

We can construct the half-adder from the truth table by using the AND and XOR gates as shown in Fig. 7-21*a*. For the full adder, though, we need to use the arrangement of gates shown in Fig. 7-21*b*.

Example 7.11 Prove that the truth table for the half-adder may be represented by Fig. 7-21*a*.

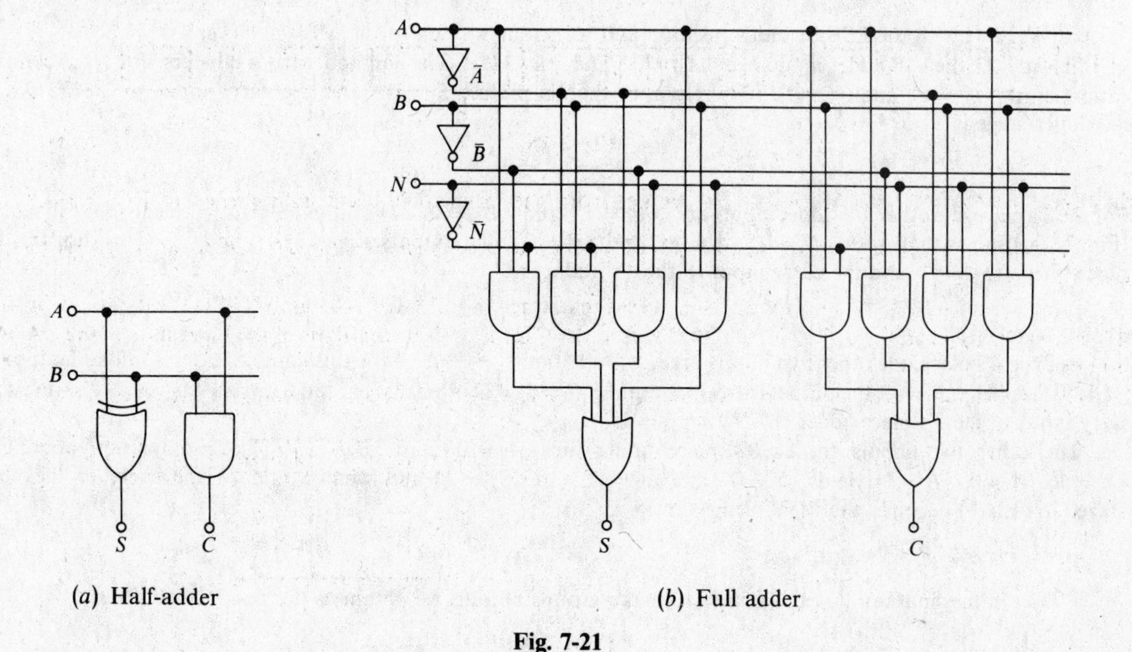

(*a*) Half-adder (*b*) Full adder

Fig. 7-21

Looking at the truth table in Fig. 7-20*b*, we see that the only time the carry bit C is 1 occurs when A AND B are 1. In Boolean form,

$$C = A \cdot B$$

This may be realized by an AND gate with inputs A and B, as shown in Fig. 7-21*a*.

For the sum S bit, we notice that the only time S is 0 is when both A and B are the same. This is the exclusive-or function and may be realized with an XOR gate with inputs A and B generating the S bit, as shown in Fig. 7-21*a*.

Adders and half-adders may be combined sequentially to obtain a 4-bit adder, as shown in Fig. 7-22, and can continue in the same manner to add any number of bits.

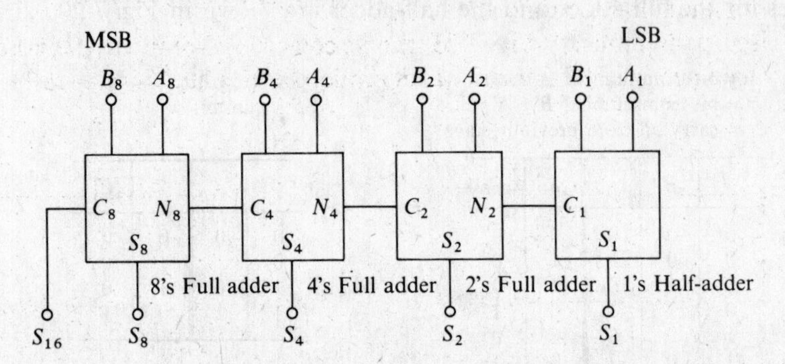

Fig. 7-22 Four-bit adder (subscripts indicate bit position weight; sum is 5 bits)

Example 7.12 Show how the 4-bit adder shown in Fig. 7-22 would add $A = 0110$ and $B = 1101$.

The first number to be added, $A = 0110$, has a 0 in the 8's (2^3) place, which we may designate as

$$A_8 = 0$$

Similarly $$A_4 = 1 \quad A_2 = 1 \quad A_1 = 0$$

For the second number, $B = 1101$, we designate

$$B_8 = 1 \quad B_4 = 1 \quad B_2 = 0 \quad B_1 = 1$$

To add, we start with the LSB or unit's position and work our way up to the MSB position.

In Fig. 7-22, the half-adder inputs are $A_1 = 0$ and $B_1 = 1$. The half-adder truth table for this input combination shows $S = 1$ and $C = 0$. Therefore, in the 1's place, $S_1 = 1$ and the carry input to the 2's place full-adder stage is

$$N_2 = C_1 = 0$$

For the 2's place, the other full-adder inputs are $A_2 = 1$ and $B_2 = 0$. The full-adder truth table for the full adder (Fig. 7-20*a*) shows that, with $N = 0$, $A = 1$, and $B = 0$, the outputs are $S = 1$ and $C = 0$. For the 2's place, then, $S_2 = 1$ and the carry input to the 4's adder is

$$N_4 = C_2 = 0$$

For the 4's place, the other two inputs are $A_4 = 1$ and $B_4 = 1$. For this combination of full-adder inputs ($N = 0$, $A = 1$, $B = 1$), the output is $S = 0$ and $C = 1$. This means that the 4's place has $S_4 = 0$ and the carry input to the 8's place adder is $N_8 = C_4 = 1$.

The other two inputs to the 8's place adder are $A_8 = 0$ and $B_8 = 1$. This combination of inputs ($N = 1$, $A = 0$, $B = 1$) yields $S = 0$ and $C = 1$, so $S_8 = 0$ and what would be the carry to the next stage (16's place) becomes the MSB of the sum:

$$S_{16} = C_8 = 1$$

Filling in the final result with extra 0s to make groups of four, we obtain as the result of addition

0001 0011

This is the decimal number 19, and checking our result with the original data by means of Table 7-2,

$$\begin{array}{r} 0110 = \ \ 6 \\ +\ 1101 = 13 \\ \hline 0001\ 0011 = 19 \end{array}$$

7.6 MULTIVIBRATORS AND FLIP-FLOPS

In dealing with digital circuits, we often need devices that will change state at certain controllable time intervals. A *multivibrator*, or MV, is the general term for a circuit that can produce a pulse, a series of pulses, or a change of state. These multivibrators are termed "monostable," "astable" or "free-running," and "bistable" or "flip-flop," respectively.

Multivibrator circuits, as shown in Fig. 7-23, can be constructed of tubes, transistors, FETs or gates, but we will be concerned here with the functional operation of the multivibrator rather than its circuitry.

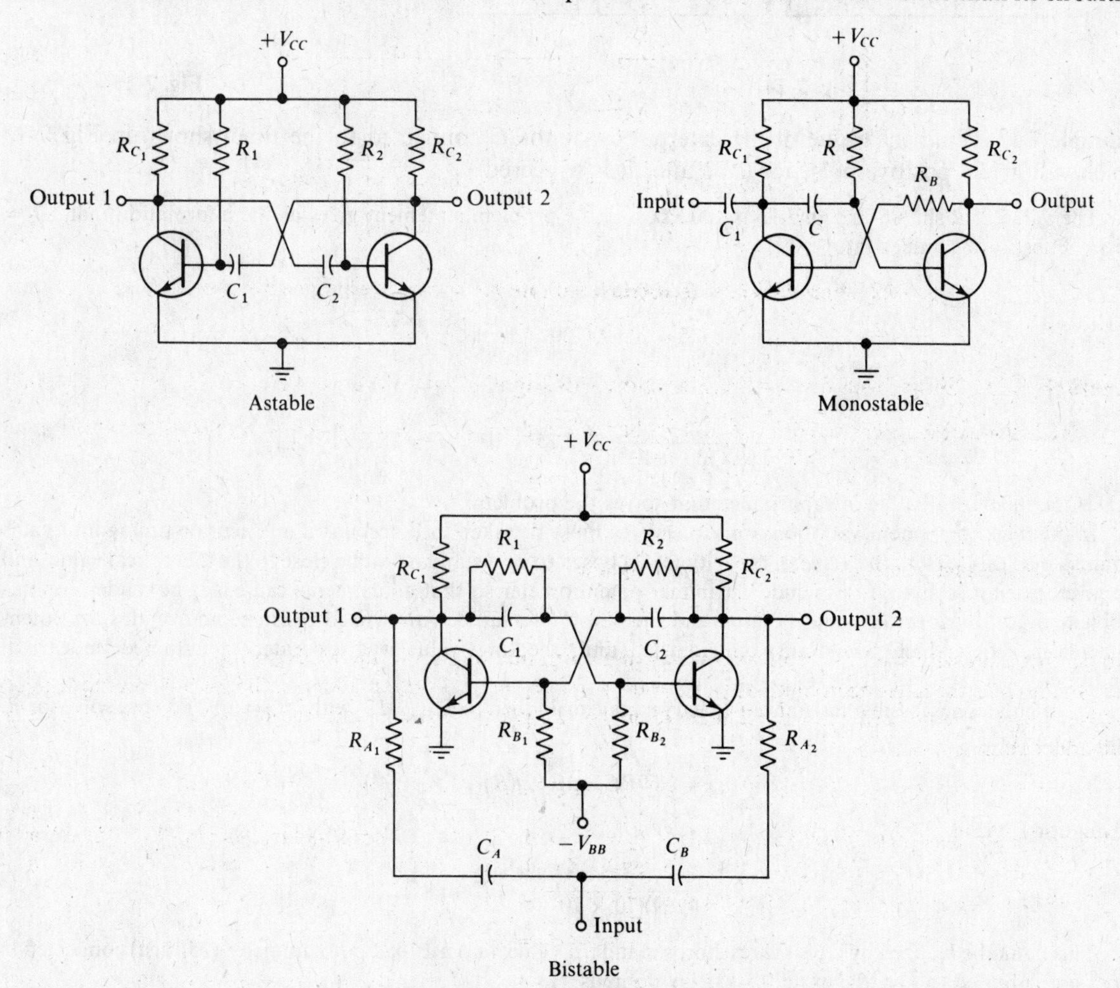

Fig. 7-23 Multivibrator circuits

A monostable or one-shot MV will produce either a positive or a negative pulse when the input is triggered by a pulse. After a length of time T, determined by an RC time constant, the output returns to its original state, as demonstrated in Fig. 7-24.

The one-shot MV can be used in conjunction with a gate to enable or inhibit a particular signal until a certain period of time has elapsed. The RC combination that determines the length of time for the pulse is usually connected externally and for most practical circuits the time T for the pulse is given by

$$T = 0.69RC \tag{7-12}$$

where T is in seconds, R is in ohms, and C is in farads.

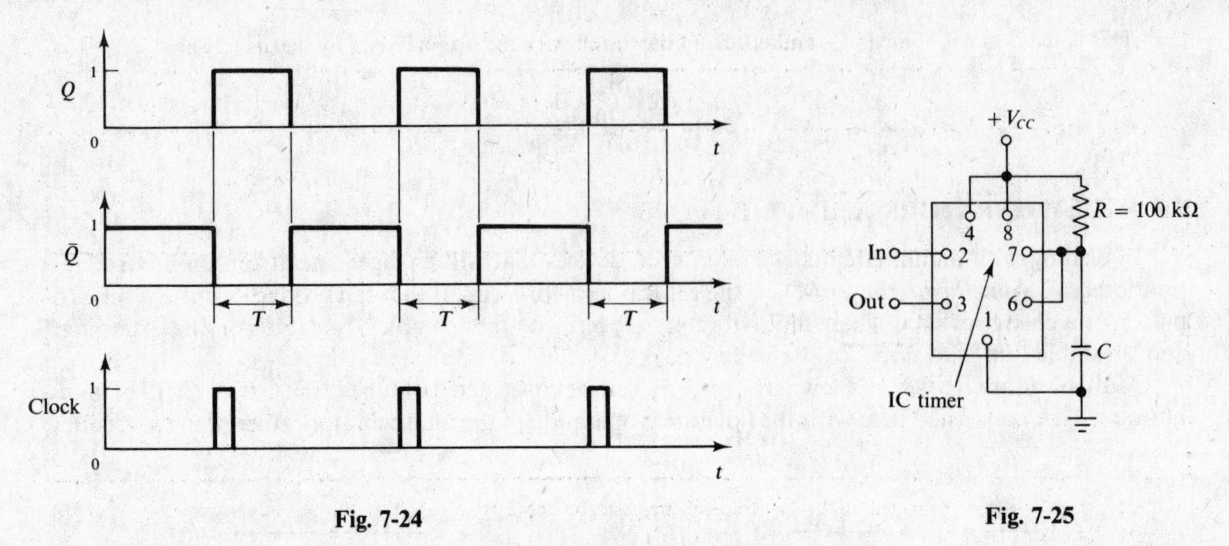

Fig. 7-24

Fig. 7-25

Example 7.13 Find the value of the external capacitor C connected to the timer shown in Fig. 7-25 which will give a positive pulse for 0.5 s after it is triggered.

The value of R shown in Fig. 7-25 is 100 kΩ, and the problem statement gives us the information that $T = 0.5$ s. Substituting values into (*7-12*),

$$T = 0.69RC$$

$$0.5 = (0.69)(100 \times 10^3)(C)$$

Solving for C,

$$C = \frac{0.5}{(0.69)(100 \times 10^3)} = 7.25 \times 10^{-6}\ \text{F} = 7.25\ \mu\text{F}$$

This is the theoretical value of capacitance that solves the problem.

In practice, component variations and tolerances must be taken into account if an accurate timing interval is required. For this reason, the value of capacitance is chosen to be a standard value close to the theoretical value, and the given resistor is altered to include a trimmer potentiometer so that adjustments can easily be made. For this problem, a 10-μF capacitor could be used and the resistor replaced with a fixed resistor and a multiturn potentiometer in series which would then be adjusted until the time value was accurately determined under real conditions.

To calculate the possible resistance values, we may again turn to Eq. (*7-12*) with $C = 10\ \mu\text{F}$ and solve for R. Since from (*7-12*)

$$T = 0.5 = 0.69RC = (0.69)(R)(10 \times 10^{-6})$$

solving for R yields

$$R = \frac{0.5}{(0.69)(10 \times 10^{-6})} = 72.46\ \text{k}\Omega$$

This could be realized by a 68-kΩ resistor, a standard value, and a 10-kΩ potentiometer (25 turn) connected in series and adjusted to give the exact 0.5-s delay desired.

As we mentioned in Sec. 7.5, a clock is a square wave generator that controls gating and other functions. Astable MVs can be used to produce a continuous string of variable-pulse-width square waves. The clock on and off times are also controlled with external RC circuits, and the relations between the RC combinations and the pulse times are listed in Table 7-9.

By far the most useful of the MV devices is the bistable MV or flip-flop. The flip-flop, which we will denote as FF, is a complementary circuit with complementary outputs and two stable states.

One state has the first transistor or other active device in the ON state and the second one OFF. The other stable state is the reverse, where the first transistor is OFF and the second one ON.

Table 7-9 Multivibrator time constants

Type	Waveform	Actual formulas (transistor)	Approximate formulas
Monostable	T	$T = RC \ln \dfrac{2V_{CC} - V_{CES} - V_{BE}}{V_{CC} - V_{BE}}$	$T = 0.69RC$
Astable	T_1, T_2, T	$T_1 = R_1 C_1 \ln \dfrac{2V_{CC} - V_{CES} - V_{BE}}{V_{CC} - V_{BE}}$ $T_2 = R_2 C_2 \ln \dfrac{2V_{CC} - V_{CES} - V_{BE}}{V_{CC} - V_{BE}}$ $T = T_1 + T_2$	$T_1 = 0.69R_1 C_1$ $T_2 = 0.69R_2 C_2$ $T = 0.69(R_1 C_1 + R_2 C_2) = T_1 + T_2$

Note: RC networks are connected at the base of each transistor in the astable case and to the base of transistor 2 for the monostable type.

What makes the circuit particularly useful is the ability of the FF to hold either of these stable conditions indefinitely until a pulse is applied to the input, at which point it flips to the other stable state.

The circuit gets its descriptive name from the action that occurs. The first pulse at the input flips the circuit to one stable state and the next pulse flops it back to the original stable state. This flip-flop action is used in counters, dividers, and memory circuits.

The FF is the first gate that we have encountered that has a memory. In other words, the state at a particular point in time depends not only on the presence of a pulse but also on the condition that the FF is in before the pulse arrives. Gates and groups of gates that "remember" their previous conditions are called *sequential logic circuits* or *sequential-type gates*.

There are different types of FFs, each of which have slightly different characteristics.

In examining gate symbols and truth tables such as the toggle or T FF shown in Fig. 7-26, we find subscript notation is sometimes used. The subscripts are used for clocked FFs where the 0 subscript indicates the state of the input or output before the clock pulse and the 1 subscript indicates the state of the input or output after the clock pulse.

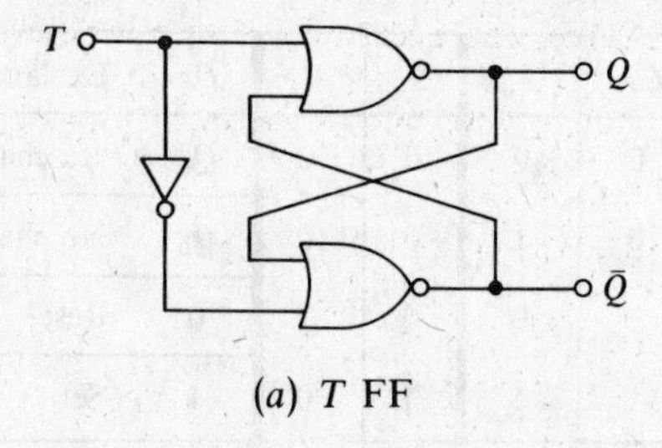

(*a*) T FF

T_0	T_1	Q_1	Result
0	0	Q_0	No change
0	1	Q_0	No change
1	0	$\bar{Q}_0$	Toggle
1	1	Q_0	No change

(*b*) T FF truth table (0 subscript means before T changes; 1 subscript means after T changes)

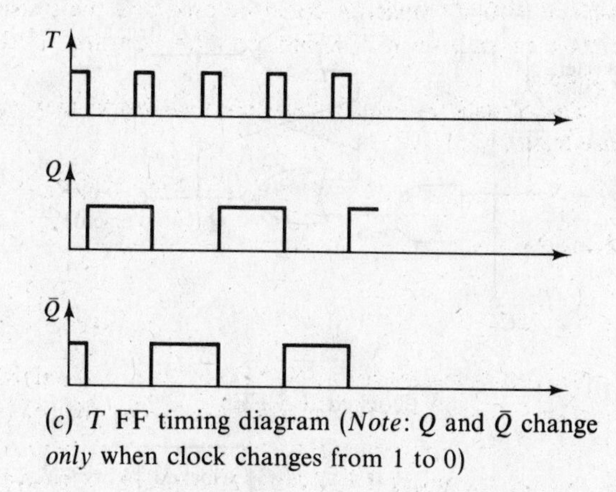

(*c*) T FF timing diagram (*Note*: Q and $\bar{Q}$ change *only* when clock changes from 1 to 0)

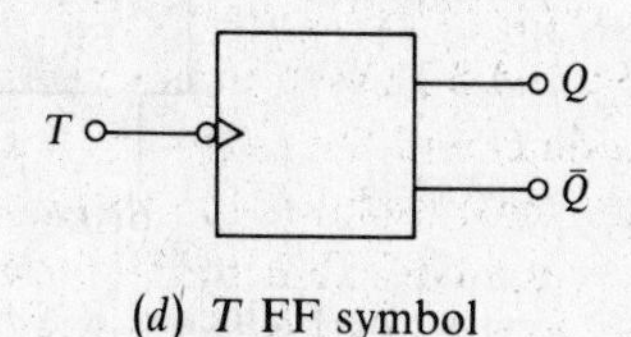

(*d*) T FF symbol

Fig. 7-26

A triangle on a FF input, usually at the *CK* input, is used to indicate that the FF is edge-triggered. "Edge triggering" means that the FF responds only to changes in the input. Another way of stating this is to say that the FF has a dynamic input.

The circle used on a dynamic input indicates that the edge triggering occurs on the trailing edge of the pulse when the input changes from 1 to 0.

Leading edge triggering, when the input changes from 0 to 1, is assumed if the input has a triangular symbol without a circle.

Another commonly encountered FF is the unclocked *R-S* FF shown in Fig. 7-27*a*. The *R* and *S* lines stand for RESET and SET and indicate the state of the output Q. When the S line is 1 and the R line is 0, the FF is SET. This means that the output $Q = 1$ and $Q_2 = 0$. Q is the true and Q_2 is the $\bar{Q}$ (complementary) output. When the R line equals 1 and $S = 0$, the Q_2 output equals 1 and the Q output equals 0. This is the RESET condition and again $Q_2 = \bar{Q}$. When $S = R = 1$, $Q = Q_2 = 0$ since a 1 on a NOR gate forces a 0 output.

When either the SET or RESET condition is followed by the high line going to 0, the *R-S* FF "remembers" its past state and Q and Q_2 are unchanged. However, if $S = R = 0$ following an $S = R = 1$ condition, an ambiguity develops and this is usually considered a prohibited mode. Some texts prefer to call the $S = R = 1$ state the "prohibited mode," although no ambiguities occur as long as a SET or RESET follows the $S = R = 1$ condition.

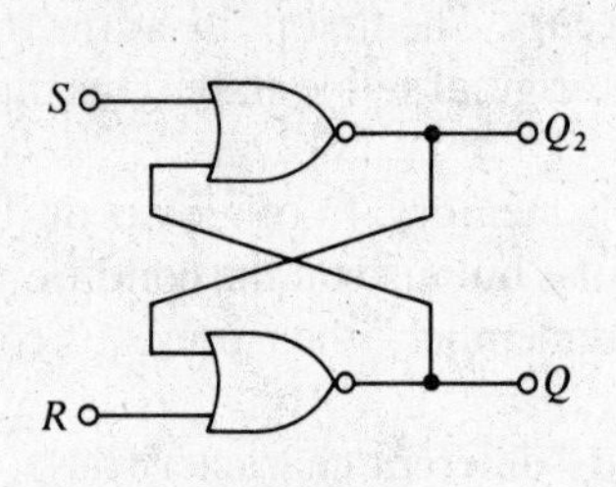

(*a*) Unclocked *R-S* FF or latch

S	R	Q	Explanation
0	0	Q_0	Previous state if preceded by set or reset
		1/0?	Ambiguity if preceded by $S = R = 1$
0	1	0	Reset
1	0	1	Set
1	1	0	"Prohibited mode"

(*b*) Unclocked $R \cdot S$ FF truth table

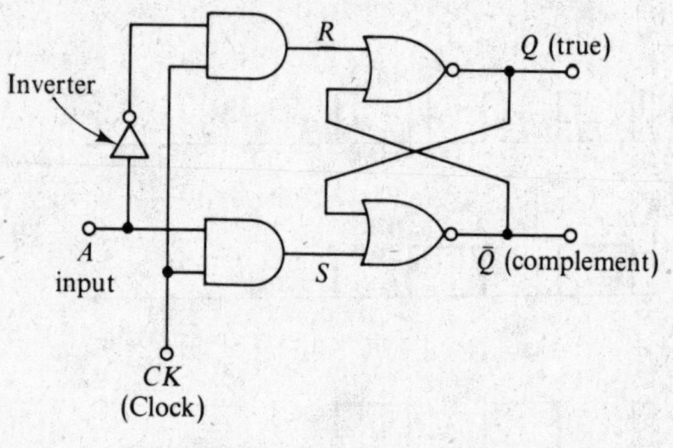

(*c*) Clocked *R-S* FF

CK	A_0	S_1	R_1	Q_1	Explanation
0	0	0	0	Q_0	No change
0	1	0	0	Q_0	No change
1	0	0	1	0	Reset
1	1	1	0	1	Set

(*d*) Clocked *R-S* FF truth table

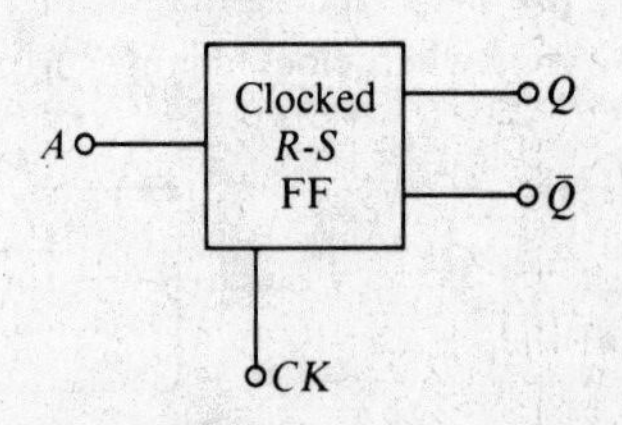

(*e*) Clocked *R-S* FF, 1 input

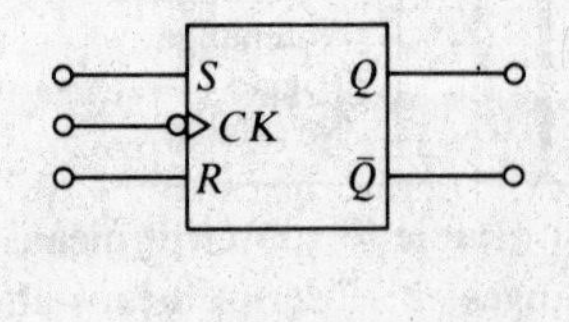

(*f*) Clocked *R-S* FF, 2 input

Fig. 7-27

The unclocked *R-S* FF truth table is shown in Fig. 7-27*b*.

Notice that in the *R-S* as well as the *T* FF, the NOR gates are cross-coupled. In other words, the output of each is connected to an input of the other. This cross coupling is typical of all FFs.

When the unclocked *R-S* FF is combined with AND gates as shown in Fig. 7-27*c*, the configuration is called a "clocked *R-S* FF" or a "gated latch." It includes a gating or "steering" network and an *R-S* FF or latch. The gating network serves to pass *A* and $\bar{A}$ through to the *R* and *S* lines whenever the clock is high (1). This is the unlatched position and the output *Q* follows the input *A* whenever the clock is 1. Notice that, in this configuration, *R* and *S* are always complementary because of the inverter on the *A* input.

When the clock is low (0), the gating network prevents the input data from affecting the *R* and *S* lines, which are both held at 0. This allows the output *Q* to remain or "remember" its original state regardless of the state of *A*.

The truth table for the clocked *R-S* FF is shown in Fig. 7-27*d*, while the symbol for the clocked *R-S* FF is shown in Fig. 7-27*e* and *f*.

The output depends on the state of the *R* and *S* inputs when the clock goes from 1 to 0 (trailing edge-triggered). If the inverter is removed and the inputs are independent, we encounter the same ambiguous situation that we did with the unclocked *R-S* FF when both inputs are 1.

If we cannot ensure that the $R = S = 1$ state will always be followed by a SET or RESET, then we should never clock the *R-S* FF when both lines are high, and again we may specify the $R = S = 1$ state as a prohibited mode.

Example 7.14 Trace the effects of the following sequence of events on the *Q* output of the clocked *R-S* FF. Assume $Q = 0$ before the first event.

Input *A*	Clock
1	1
0	1
1	0
0	0
0	1
1	1

The first two events take place with the clock high (1). In the clocked *R-S* FF, when the clock equals 1, the output follows the input. So *Q* goes from its original state of 0 to 1 for the duration of the first input pulse and back to 0 during the time *A* is 0.

The next two input changes occur when the clock is low, so the output remains at 0 and is not affected by the input. In effect, the last 0 that occurred when the clock was high is being stored.

When the clock returns to 1, the output again follows the input, and the complete timing sequence is shown in the table of Fig. 7-28*a* and the timing diagram of Fig. 7-28*b*.

CK	*A*	*Q*	Explanation
...	...	0	Original state
1	1	1	Set
1	0	0	Reset
0	1	0	No change
0	0	0	No change
1	0	0	No change (reset)
1	1	1	Set

(*a*)

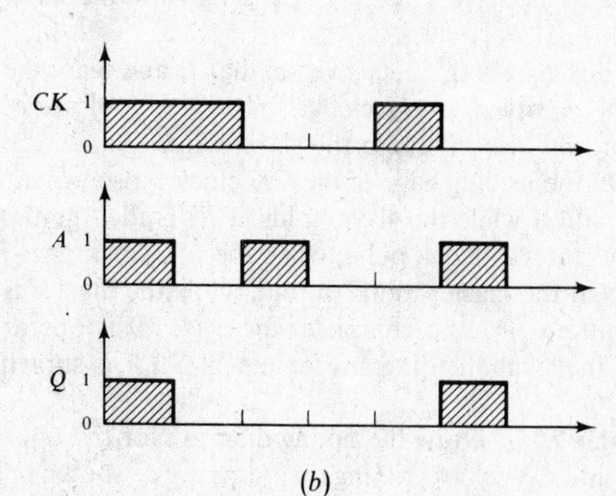

(*b*)

Fig. 7-28

An FF that transfers data when it is edge-triggered by the clock pulse is referred to as a *D* ("data" or "delay") FF. An *R-S* FF is shown in Fig. 7-29*a* wired to act as a *D* FF. There are, however, many other ways of producing a *D* FF.

Connecting the inverter between *R* and *S* assures that the inputs will always be complementary and that the problem with prohibited modes cannot occur.

The abbreviated truth table for all *D*-type FFs is shown in Fig. 7-29*b*, while the *D* FF timing diagram is shown in Fig. 7-29*c*. Note that the data is delayed two clock pulse widths from appearing at the *Q* output.

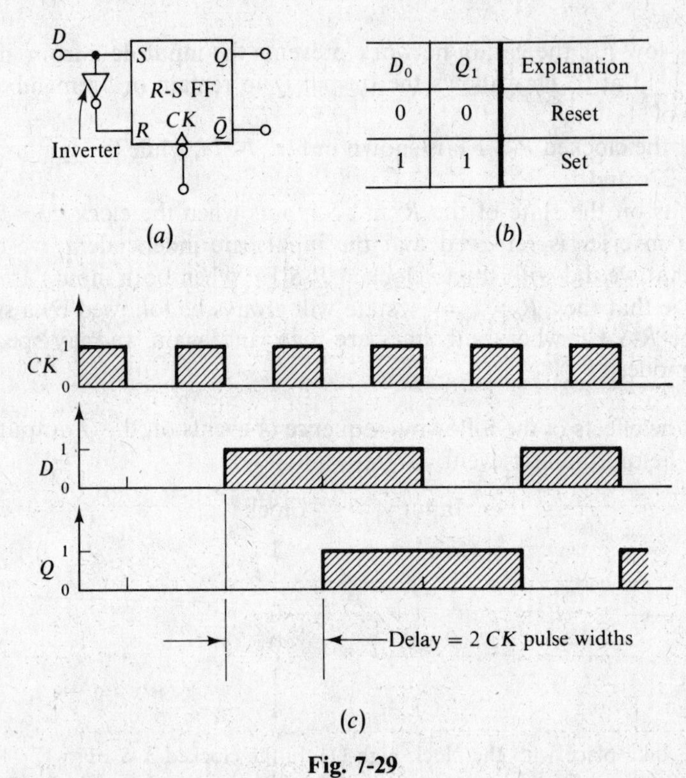

D_0	Q_1	Explanation
0	0	Reset
1	1	Set

Fig. 7-29

In many applications, data must be moved or shifted from one FF to another with all FFs triggered by the same clock signal. This shifting must be done so that no ambiguities arise and is best handled by a FF that stores the data before shifting. We can achieve this result by wiring two clocked *R-S* FFs as shown in Fig. 7-30*a*. This type of FF is termed an *M-S* or "master-slave" FF.

Since the slave clock pulse is always the complement of the master clock pulse, the master will shift its data to the slave while its clock is high and the slave will shift information to the output while its clock is high.

Looking at Fig. 7-30*a*, we see that *R* and *S* are the complementary inputs to the master, *CK* and $\overline{CK}$ are the complementary clocks for master and slave, *Y* and $\bar{Y}$ are the complementary outputs of the master and also inputs to the slave, while *Q* and $\bar{Q}$ are the complementary slave outputs.

On the leading edge of the *CK* clock pulse, when the *CK* clock goes from 0 to 1, the master receives information while the slave holds. This is the "getting ready to shift" part of the cycle. At the trailing edge of the *CK* clock pulse, when the *CK* clock goes from 1 to 0, the slave transfers the data previously stored in the master to the output, while the master is immune to any changes at the data input. This is the shifting part and completes the cycle. This operation is shown in the timing diagram of Fig. 7-30*b*, while the symbolic diagram for the *M-S* FF is shown in Fig. 7-30*c*.

Example 7.15 Draw the timing diagram for the *Q* output of the *M-S* FF shown in Fig. 7-30*a* if the *R* input and *CK* (clock) timing diagrams are as shown in Fig. 7-31*a*. Assume that $Q = 0$ at $t = 0$ and that *R* and *S* are always complementary.

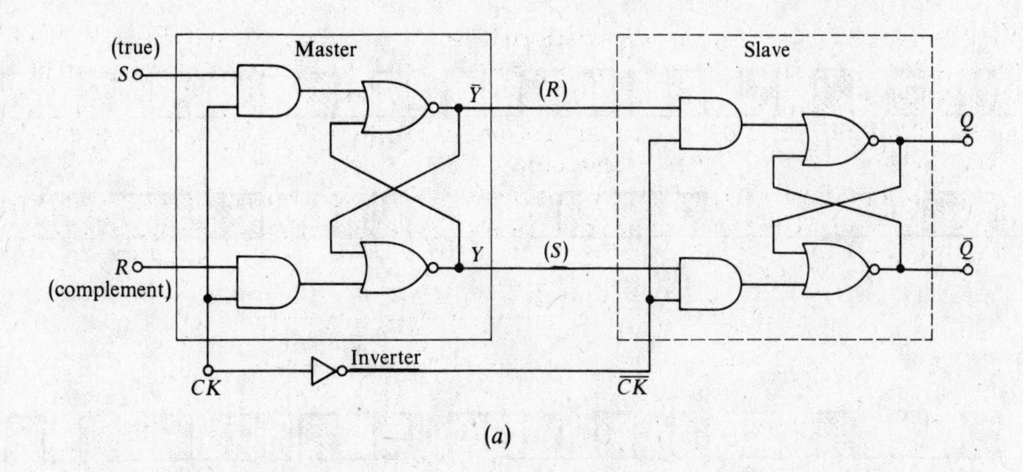

(true)
S
Master
Ȳ
(R)
Slave
Q
Y
(S)
Q̄
R
(complement)
Inverter
CK
C̄K̄

(a)

CK
C̄K̄
S
R
Y
Ȳ
Q
Q̄
1
0
t
Delay = 1 CK pulse width

(b)

(True)
Master
Slave
S
Ȳ
R
Q
R
Y
S
CK
Q̄
(Complement)
CK
(C̄K̄)
Sometimes
in single IC
package

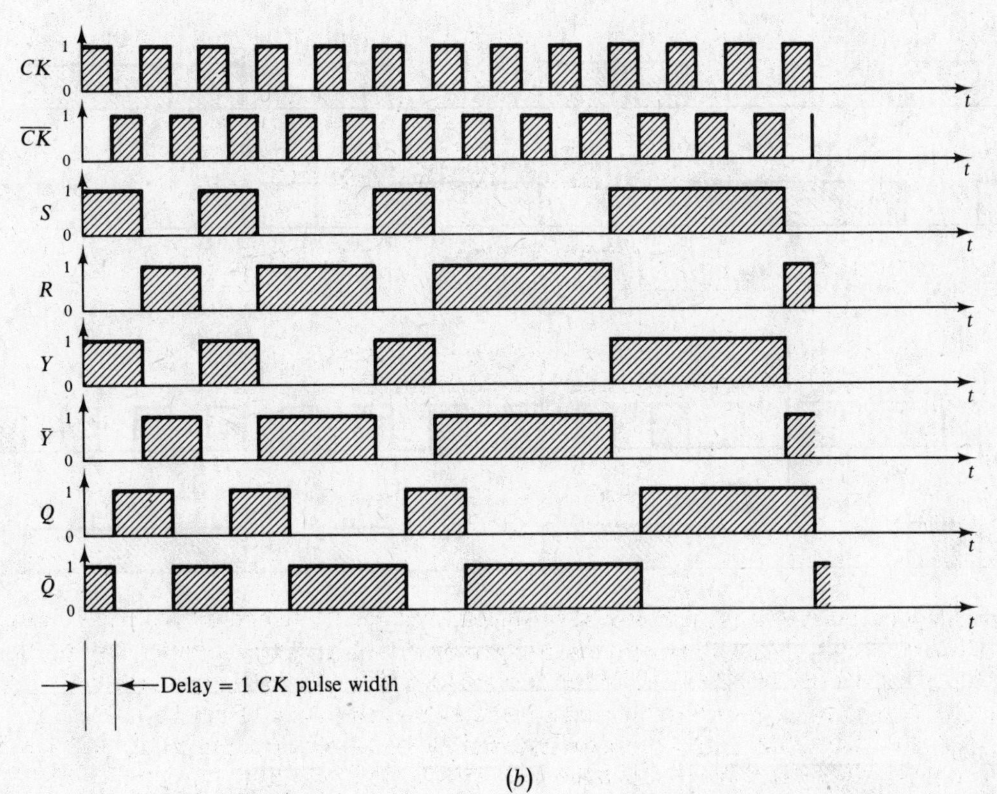

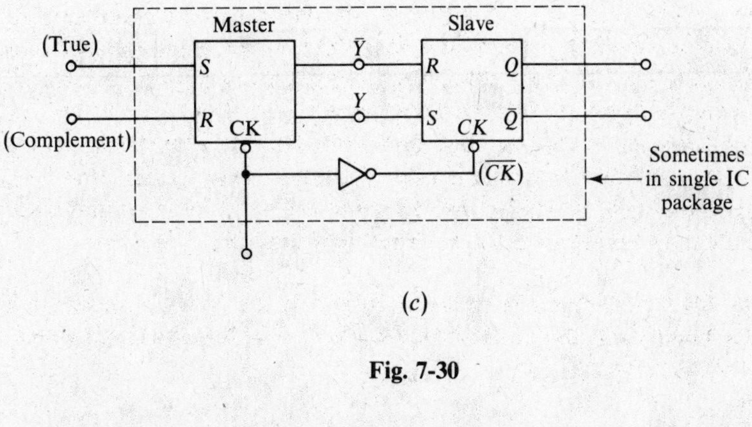

(c)

Fig. 7-30

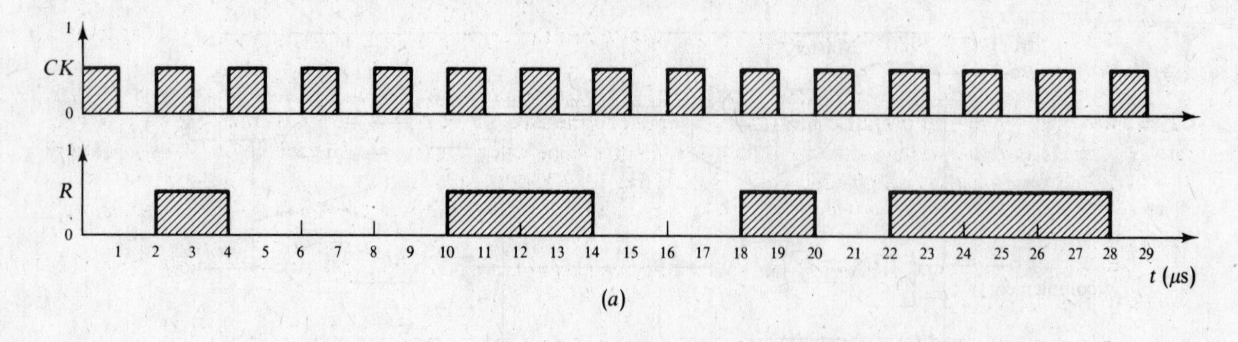
CK
R
1
0
1 2 3 4 5 6 7 8 9 10 11 12 13 14 15 16 17 18 19 20 21 22 23 24 25 26 27 28 29
t (μs)

(a)

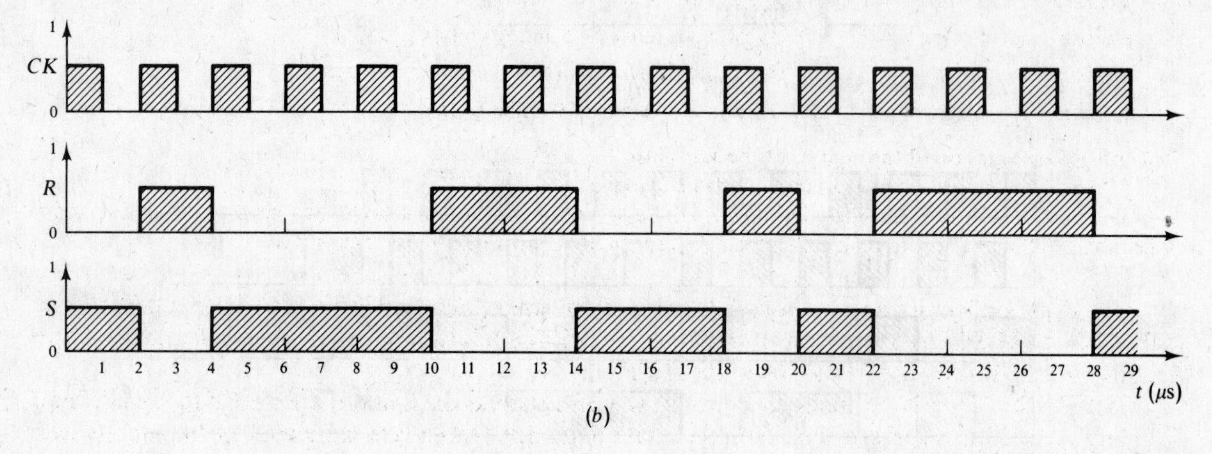
CK
R
S
1
0
1 2 3 4 5 6 7 8 9 10 11 12 13 14 15 16 17 18 19 20 21 22 23 24 25 26 27 28 29
t (μs)

(b)

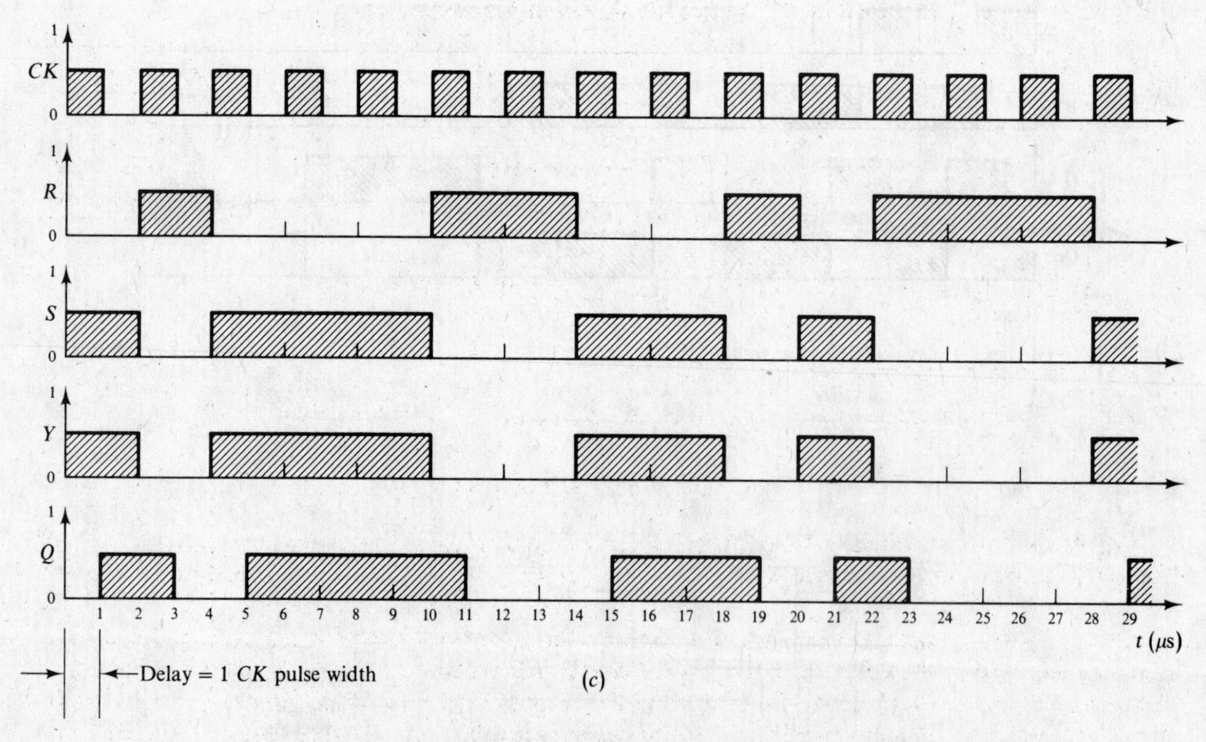
CK
R
S
Y
Q
1
0
1 2 3 4 5 6 7 8 9 10 11 12 13 14 15 16 17 18 19 20 21 22 23 24 25 26 27 28 29
t (μs)
Delay = 1 CK pulse width

(c)

Fig. 7-31

Given the fact that R and S are always complementary, we can easily draw the S input timing diagram as shown in Fig. 7-31*b*.

Looking at Fig. 7-30*a*, we see that the Y output of the master is connected to the S input of the slave. When the CK pulse is high, the master output Y follows the S input and the slave output Q remains in its previous state. When the CK pulse is low, the master output Y holds at its former value while the slave output follows Y. This is another way of saying that, on the leading edge of the clock pulse, the FF gets ready to shift and on the trailing edge of the clock pulse it actually shifts. The net effect of this action is to delay the Q output from following the S input by one clock pulse width.

Looking at the timing diagram of Fig. 7-31*a* and *b*, we may use this reasoning to draw the Y and Q outputs as shown in Fig. 7-31*c*:

- From $t = 0$ to $t = 1$, CK is high, so $Y = S = 1$ and Q stays at 0.
- From $t = 1$ to $t = 2$, CK is low, so Y stays at 1 and Q changes to 1.
- From $t = 2$ to $t = 3$, CK is high again, so $Y = S = 0$ and Q holds at 1.
- From $t = 3$ to $t = 4$, CK is low and Y stays at 0 while Q changes to 0.

This procedure is followed through until the last pulse.

The action seen here is called *pulse* or *M-S triggering*.

If the M-S FF were trailing-edge triggered, the Q output would be delayed two clock pulse widths from following the S input, rather than the one clock pulse width shown with pulse triggering.

The same difficulties encountered with the R-S FF when both inputs are 1 at the clock pulse also exist with the M-S FF. To remove the ambiguity, we cross-couple the outputs to the input gates as shown in Fig. 7-32*a*. This type of FF is called a J-K FF.

In the J-K FF, the K terminal performs the same function as the R and resets the FF to 0, while the J terminal acts as the S and sets the FF to the 1 state. The uncertain state is eliminated by flipping the state if both inputs are 1, as shown in the truth table shown in Fig. 7-32*b*.

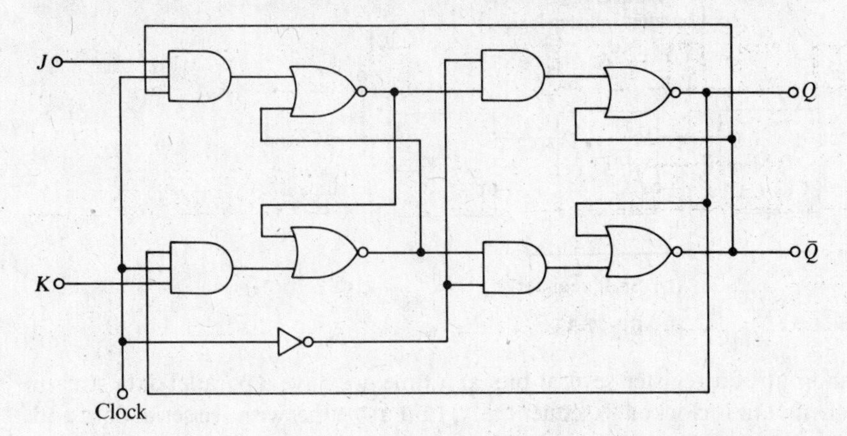

J_0	K_0	Q_1
0	0	Q_0
0	1	0
1	0	1
1	1	$\bar{Q}_0$

Fig. 7-32 *J-K* FF and truth table

Many FFs are provided with additional R- and S-type terminals termed C or "clear" and P or "preset." The clear terminal acts as a RESET to make $Q = 0$, while the preset terminal acts as a SET by setting $Q = 1$.

Normally, the preset and clear lines are connected as additional inputs to the cross-coupled NOR gates and are kept at 0 so they do not affect the regular FF operation. A short 1 pulse on the preset or clear line between clock pulses sets or resets the FF output, overriding all the other inputs. Clear and preset are useful for loading parallel data into registers and for setting or resetting all the FFs in a counter at the same time.

FFs that are dependent upon a timing signal or clock pulse are termed *synchronous FFs* and those that do not depend on a timing signal are termed *asynchronous*.

There are other variations of FF gates but they all have the same important function. FFs can store 1 bit of information until the next triggering clock pulse. As long as the power to the circuit remains on, the FF can store this bit of information indefinitely.

7.7 REGISTERS AND COUNTERS

Data may be manipulated in two forms: serial and parallel. We store data by clocking many bits of information into individual FFs in one pulse or we may send the data through a series of storage FFs 1 bit at a time. These storage units are called *registers*. A parallel register is shown in Fig. 7-33*a* while a shift register is shown in Fig. 7-33*b*.

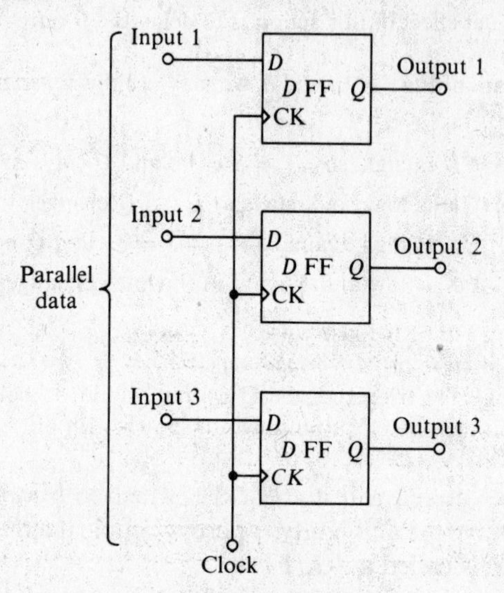

(*a*) Parallel register

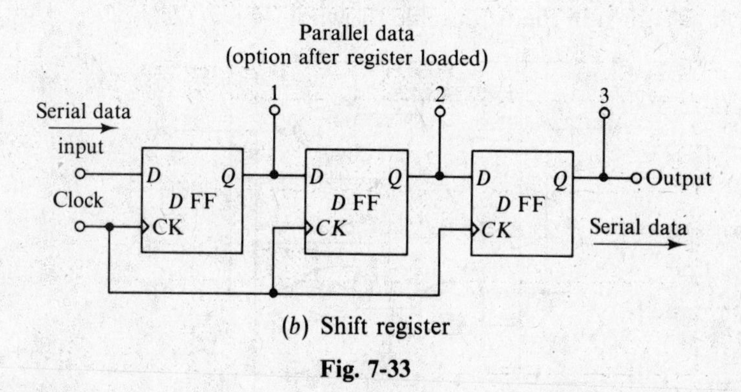

(*b*) Shift register

Fig. 7-33

When data is clocked in and out of a register several bits at a time, we have a parallel data stream and use the parallel register. When data is clocked in sequentially, 1 bit at a time, we have serial data and use a shift register. Notice that once a shift register is loaded, we have the option of clocking all the data out at once in parallel form. One of the uses of the shift register is to convert serial data to parallel form.

All units of register FFs are clocked at the same time, and timing diagrams for both serial and parallel registers are shown in Fig. 7-34*a* and *b*.

Example 7.16 Show how the serial data 1 (entered first), 0, 0, 1, 1, 0, 1, 0 (entered last) would be stored in an 8-bit shift-right register. Assume the register outputs are all initially at 0.

At the first clock pulse, the first register shows the first bit of the serial data input. For the second clock pulse, the 1 from the first register is shifted to the second register and the first accepts the second bit of the data, or 0. This process continues until the registers are filled as shown in the timing diagram of Fig. 7.34*c*.

After the eighth clock pulse all the registers are filled with the data in such a way that the first bit of data is in the eighth register and the last data bit is in the first register.

If we have serial data and wish to construct a counter, we connect the output of one FF to the input of the next, as shown in Fig. 7-35.

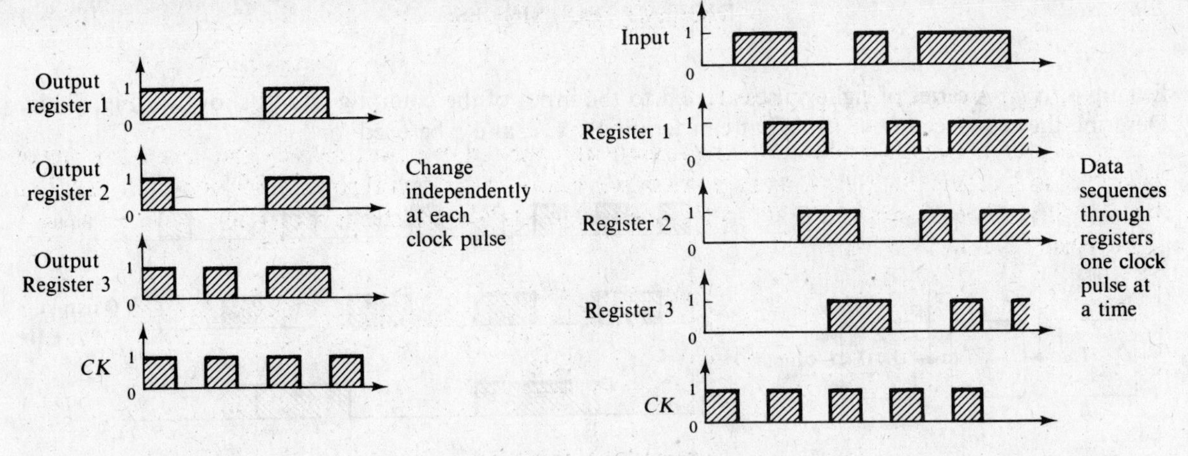

(*a*) Parallel register timing (positive edge-triggered) (*b*) Shift register timing (positive edge-triggered)

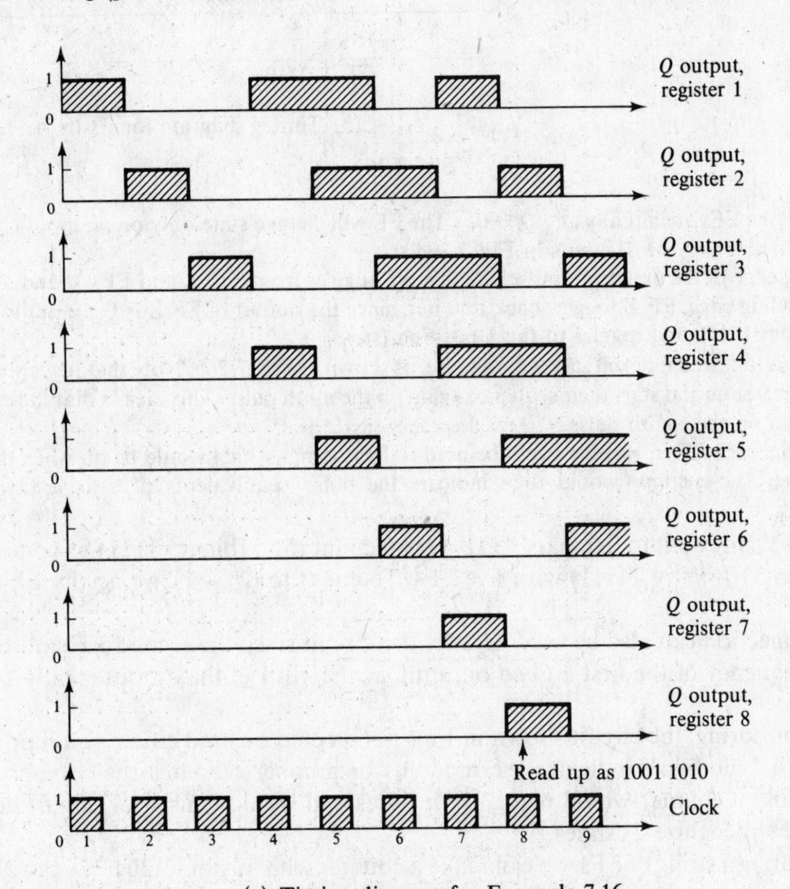

(*c*) Timing diagram for Example 7.16

Fig. 7-34

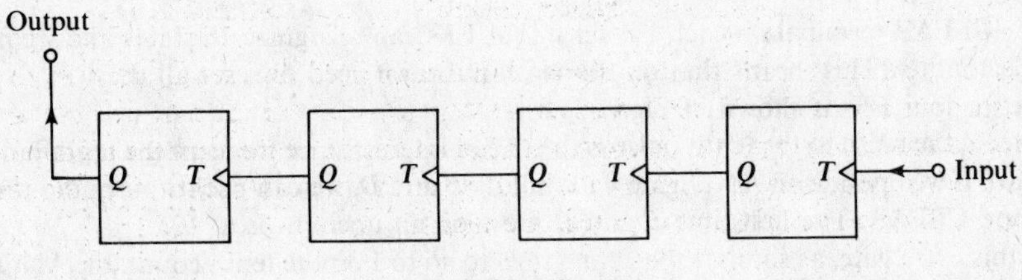

Fig. 7-35 Counter

Example 7.17 A series of eight pulses are fed to the input of the counting circuit shown in Fig. 7-36*a*. Describe the sequence of events. What can terminals *x*, *y*, and *z* be used for?

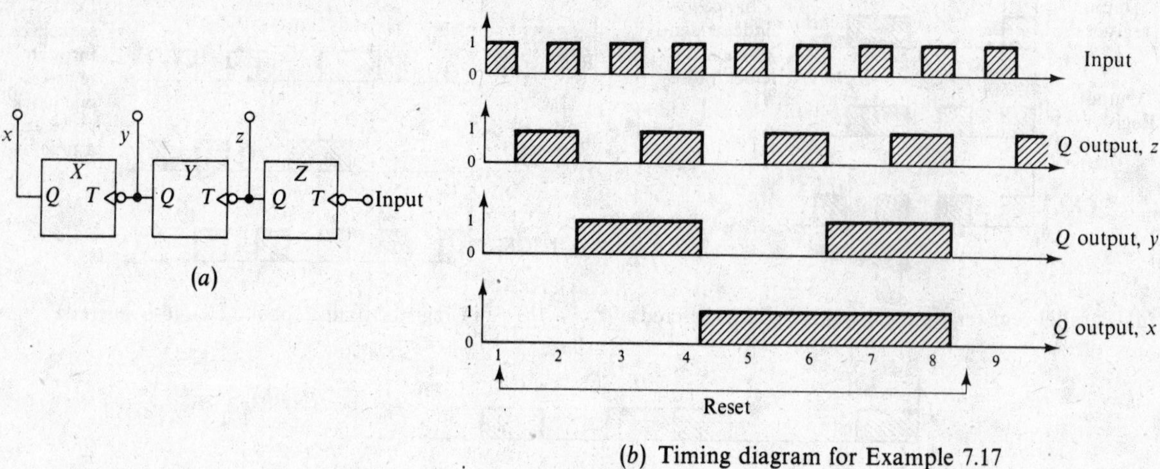

(*a*)

(*b*) Timing diagram for Example 7.17

Fig. 7-36

Assume that all the FFs are initially at $Q = 0$. The FF will change state only on an input change from 1 to 0 as denoted by the bubbles on the *T* inputs in Fig. 7.36*a*.

On the trailing edge of the first input pulse, the first FF toggles from 0 to 1, and FFs *X* and *Y* remain at 0. On the second pulse trailing edge, FF *Z* toggles back to 0 but, since the output of FF *Z* is the input to FF *Y*, FF *Y* sees an input change from 1 to 0 and toggles to the 1 position (state).

This procedure is followed for each succeeding pulse, as shown in Fig. 7-36*b*. Note that just after the eighth pulse the FFs are all in the 0 state and start their sequence again on the ninth pulse. This means that the counter will count from 0 to 7 and reset on the eighth pulse.

The outputs from each FF, *x*, *y*, and *z* could be used to light lamps which would be on when the corresponding FF was in the state. These lamps would then indicate the binary equivalents of 0 to 7. Output *z* would be the LSB.

Since three FFs can count to 7 (binary 111), we can count to 15 (binary 1111) by connecting four FFs in sequence and to 31 (binary 11111) with five FFs. To count to $(2^n - 1)$, we need *n* FFs unless there is special coding.

The same connection can also be used as a divider circuit since the second FF's output is operating at one-half the frequency of the first and so on until, at the *n*th FF, the output is only $1/2^n$ of the input frequency.

By proper monitoring, the circuit shown in Fig. 7-36*a* could be used either as a modulo 8 divider or counter. The word "modulo" indicates the modulus or number of states the counter or divider can recognize. A modulo *n* counter would recognize *n* states and would consist of $(\log n)/(\log 2)$ FFs. Put another way, *n* FFs recognize 2^n states.

By properly gating a string of FFs, we can make a counter with any modulus. We can also make an up or down counter by properly gating the signals to the FF. Up/down counting or shifting in both directions is usually accomplished with special FFs that have preset and clear terminals and a steering control to determine the direction of count.

Example 7.18 Construct a decade (modulo 10) up counter from four FFs and the appropriate gates.

We need a *C* or CLEAR terminal on each FF since four FFs can recognize 16 states and we need to recognize only 10 for a decade counter. This means that on the tenth pulse we need to reset all the FFs to 0.

The truth table for four FFs is shown in Fig. 7-37*a*.

The transition from the ninth to the tenth pulse is the one of interest since we want the tenth pulse to clear (0000) not indicate 10 (1010). If we operate an AND gate with inputs *B* and *D*, we can use its output to reset all the FFs by means of the common CLEAR. The first time *B* and *D* are 1 on an upcount is at 10.

By connecting this extra gate, as soon as the *B* line tries to go to 1 on the tenth count, the AND gate will trigger the CLEAR since the *D* is at 1 also. This provides the action we desire. The final counter is shown in Fig. 7-37*b*.

Decimal number	Binary			
	D	C	B	A
0	0	0	0	0
1	0	0	0	1
2	0	0	1	0
3	0	0	1	1
4	0	1	0	0
5	0	1	0	1
6	0	1	1	0
7	0	1	1	1
8	1	0	0	0
* 9	1	0	0	1
*10	1	0	1	0
11	1	0	1	1

(*a*) Truth table for Example 7.18

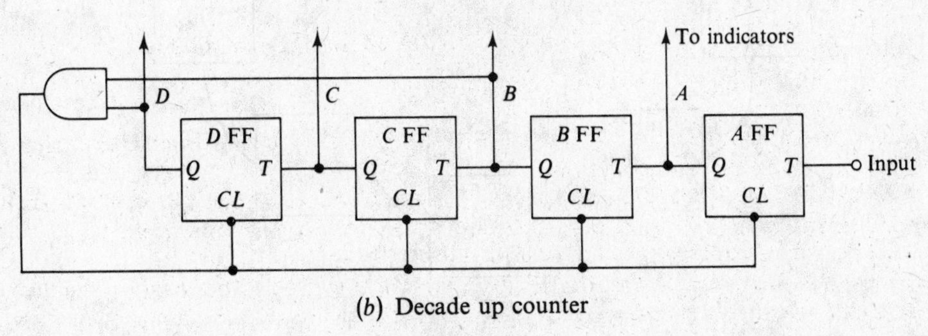

(*b*) Decade up counter

Fig. 7-37

The changes that occur in this type of counter do not take place simultaneously. The changes ripple through the counter, and while they are occurring, the outputs may not be correct. Counters which behave in this manner are called *ripple* or *asynchronous counters.*

While this is good enough for many applications such as digital clocks, there are many cases where the ripple would cause errors or become annoying. For these applications, we build clocked or *synchronous counters,* usually constructed from *J-K* FFs.

With the synchronous counter, all the FFs are triggered by the same clock pulse and all the changes occur at the same time.

The binary and decade counters are very useful devices, as they can be gated to produce different modulo counters, synchronous or asynchronous, as the application demands.

Another way of making a decade counter without gating is shown in Fig. 7-38. This type of counter is known as a *ring counter*. It has the advantage of not needing additional gates to work but the disadvantage of needing one FF for each digit. A modulo-*n* ring counter needs *n* FFs to operate. In addition, the FFs must be connected in such a manner as to allow only one FF to be in the 1 state at any one time. On the application of a pulse, the 1 state shifts to the next FF and continues in this way around the ring indefinitely.

Ring counters are very stable at high speeds and are used in many high-frequency counting circuits.

7.8 LATCHES AND MEMORIES

One of the simplest types of memories is the mechanical latching circuit shown in Fig. 7-39*a*. An example of a gated latch using NAND gates is shown in Fig. 7-39*b*.

Mechanical latches rely on a physical arrangement such as the one shown in Fig. 7-39*a* to prevent the switch from moving in the latched position. When the switch control *A* is closed, the relay physically

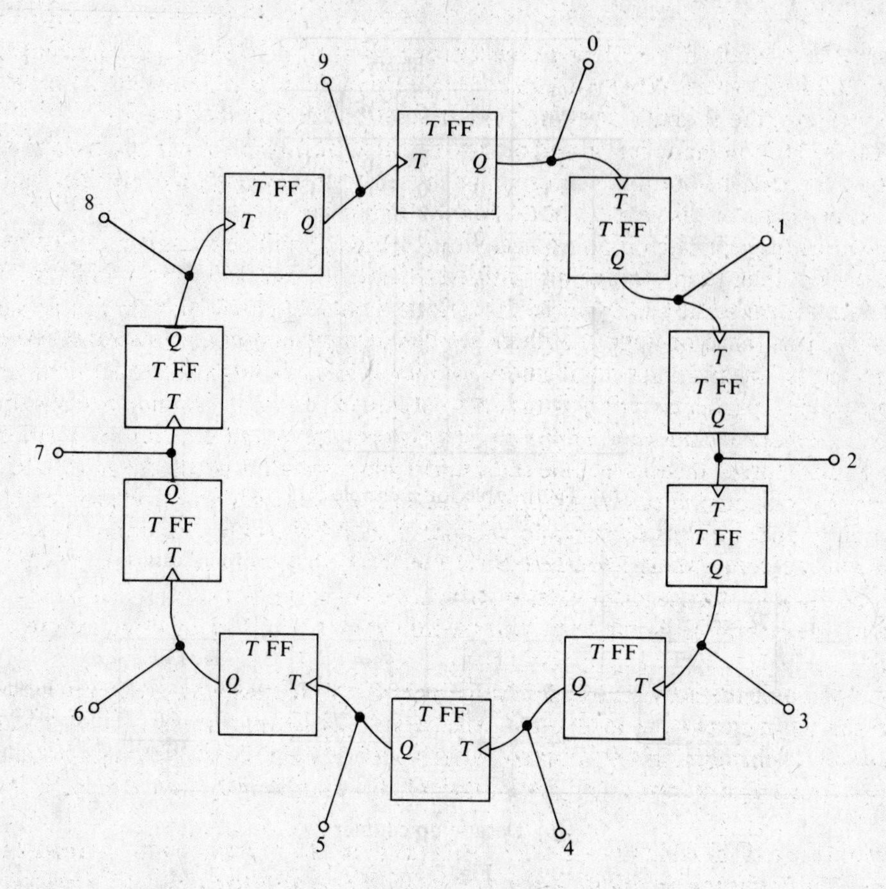

Fig. 7-38 Decade ring counter

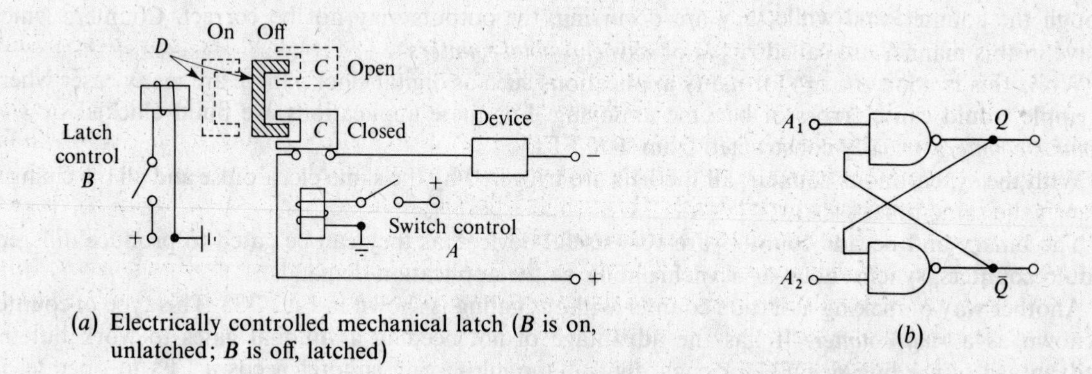

(a) Electrically controlled mechanical latch (B is on, unlatched; B is off, latched)

(b)

Fig. 7-39

pulls the metal bar C down to close the circuit. Bar C is only free to move when the latch D is pulled back by energizing the latch relay (switch B closed). If switch B is open, the D bar locks bar C in either the open or closed position and switch A has no effect on bar C. This is the latched position and "locks" or latches the circuit open or closed until D releases C.

Electronically, this same effect can be obtained by using an AND gate with the latch or steering input held at 0 to keep the gate output 0 regardless of the state of the other input line. We can also provide this action by tying the latch input of an OR gate to 1. This keeps the gate output at 1 regardless of the state of the other input.

The gated latch or clocked R-S FF mentioned in Sec. 7.6 is a typical example of latching action using clock pulses to the AND gates. When the clock is high, the input data is transferred to the R-S lines, but when the clock is low, the R and S lines are held at 0 and the output is latched.

Latches can be built in many forms and are used to hold data before a change, as in the M-S FF, or to temporarily store data before it is processed, as in a register. Latches are sometimes built into other circuits such as decoders or displays to store data for a short period of time.

For long-term storage, a latch is an expensive and space-consuming circuit, and mass memory units built specifically for long-term storage are utilized. Some of the mass memory units are designed to accept serial data such as static and dynamic shift registers; punched cards and paper tape; and magnetic tape, drums and disks (hard or floppy). Others are designed for nonserial or random access.

"Random access" means that each memory location is identified by a particular number or code, so the logic state of any particular address may be determined without examining the entire memory. Random-access memory includes the ability to write (store) and read (retrieve) information.

Random-access storage devices include static and dynamic random-access memories (RAMs), which are many memory gates combined on a chip, and read-only memories (ROMs), which do not provide for any writing ability and are thus denser and cheaper than RAMs. Other random-access mass storage devices are magnetic cores; charge-coupled devices (CCDs) that make use of the MOS construction techniques; and bubble storage memories, in which the logic state is determined by the presence or absence of a magnetic bubble in a slice of yttrium-iron garnet (YIG) sandwiched in a weakly rotating magnetic field.

The term "dynamic memory" is applied to the type of unit that depends upon movement of charge from cell to cell to maintain the logic state. The term "circulating memory" is sometimes used to describe this type of memory. If left to circulate at under 200 Hz, these memories, which depend on charge storage much like a capacitor, leak some of the charge away and the logic levels become unrecognizable.

Circulating registers, in which the last FF feeds the first, can be written into or read 1 bit at a time, with each clock pulse shifting the information to the next FF. Access to a circulating register depends on the clock rate, and you have to wait the proper number of pulses to read or write in a desired bit position. These serial registers are used to store a cathode-ray tube (CRT) pattern of dark and light pixels (picture elements) and are cleared for each scan.

Dynamic registers can store serial data at high speeds but must be continuously clocked. Static registers such as FFs can hold their state as long as the power remains on.

Dependence of the memory units on the power supply is called *volatile storage*. On the other hand, if removing the power does not alter the memory states (as is the case with magnetic tape or bubble storage), we have what is termed a *nonvolatile memory*. Most of the ROM family of devices are nonvolatile since the logic states are permanently "burned in." The logic states at the various memory locations can be detected (read) but not altered (written into).

Programmable read-only memories (PROMs) allow the user to enter his or her own set of permanent instructions as coded logic states using specialized equipment. *Erasable PROMs* or EPROMs allow the user to erase instructions using ultraviolet light and reuse the device, while electrically alterable ROMs or EAROMs can be programmed and erased while they are in a working circuit.

The PROMs, EPROMs, and EAROMs have less density than an ordinary ROM but are more flexible. They are especially useful in prototype or low-production work.

The same technique that is used to produce PROMs can also be used in what is called a *programmable logic array* or PLA, which contains information for controlling other gates to produce a desired logic result.

Magnetic cores have been the mainstay of the computer industry for many years, and despite their relatively low density and high expense continue to be used where nonvolatility and high reliability are important. Figure 7-40 shows the arrangement by which magnetic cores are written into and read. The magnetization of the cores in one direction or the other determines the particular logic state of the core.

To write into a particular location, the core drive circuitry selects the proper row and column and applies one-half the saturation current to each in the particular direction desired. The only core that receives enough magnetizing current to change state is the one where the row and column cross, as

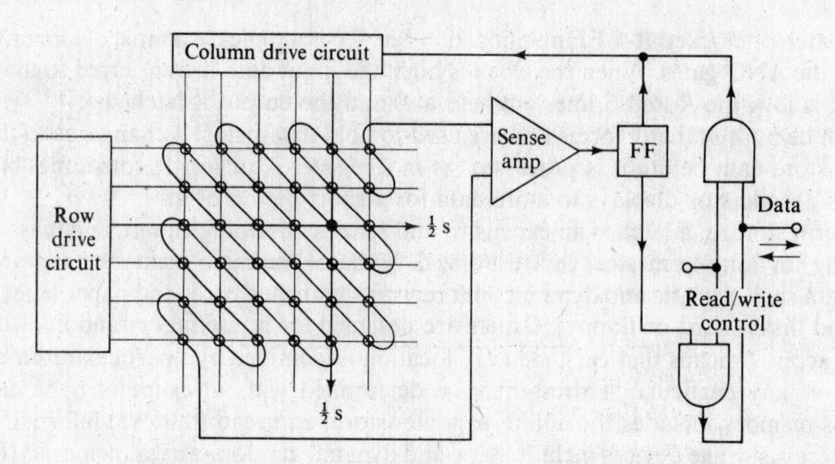

Fig. 7-40 Magnetic core memory read/write circuit

indicated by the shaded circle in Fig. 7-40. The other cores have only half the magnetizing current and remain in their original state.

The cores are read by attempting to flip the core to the 0 direction of magnetization. If the core was originally in the 0 state, no pulses are generated. If it was in the 1 state, the flip to the 0 state causes a pulse to be detected by the sense amplifier and temporarily stored in a flag register. The register is then routed back to the right circuitry and the 1 restored.

Since only one plane of cores can be accessed at a time, multiplane memories are constructed to increase the storage capacity.

A comparison of the different types of memories available is provided in Table 7-10.

Table 7-10 Memory comparisons

Type	Unit	Bits/unit	Access time	Cost/bit(¢)
Bipolar RAM	IC chip	10k	8 ns–20 ns	1
Dynamic RAM	IC chip	20k	15 ns–2 μs	0.4
MOS shift register	IC chip	2k	1 μs–100 μs	0.02
CCD Bubble device	IC chip IC wafer	100k	5 μs–2 ms	0.005
Magnetic drum	Drum	40M	0.4 ms–250 ms	0.0001
Magnetic disk	Floppy disk	5M–12M	2.0 ms–3 s	0.00004
Magnetic tape	Reel	250M	1 s–15 min	0.000002

7.9 OPERATIONAL AMPLIFIERS (OP AMPS)

The special class of amplifiers known as operational amplifiers or op amps are extremely useful in digital work. The op amp is a differential amplifier with two inputs and a single output whose magnitude is determined by the difference between the two inputs and the op amp gain. An op amp is shown symbolically in Fig. 7-41. We call the input with the − sign the inverting input and the one with the + sign the noninverting input.

Inputs − + Output

Fig. 7-41 Op amp symbol

Op amps are constructed using many different circuits and active elements, but in this chapter we are concerned with the function of the device, not its internal structure.

In analog design, op amps can be used for computational devices such as adders; subtracters; exponential, logarithmic, and antilogarithmic amplifiers; integrators; differentiators; and multipliers or scalers. For digital design, op amps are used as buffers, inverters, comparators, wave shapers, and gates. Criteria for an ideal and practical op amp are compared in Table 7-11.

Table 7-11 Op amp criteria

	Ideal	Typical
Input impedance	∞	2.0 MΩ
Input capacitance	0	1.0 pF
Open-loop gain	∞	250k
Rise time	0	0.2 μs
Overshoot	0	1.5%
Slew rate	∞	0.5 V/μs
Offset voltage	0	1.0 mV
Offset current	0	30.0 nA
Common-mode rejection ratio	∞	100 dB
Output resistance	0	50 Ω
Voltage swing	$\pm\infty$	$\pm$ 15 V
Frequency range	0–∞	0–1 MHz

Common-mode rejection ratio or CMRR is the ability of the op amp to respond to the difference component of the input voltages and to eliminate the common mode component. Most op amps have a high CMRR, which is important in instrumentation amplifiers.

Manufacturers also provide op amp terminals that can be connected to external components to provide offset corrections and frequency compensation. *Offset*, either voltage, current, or both, is the value of voltage or current at the input for a zero output. The frequency response of most op amps, even with external compensation, is only good to about 1 MHz unless special designs are employed.

Since the op amp is essentially a differential amplifier with very high open-loop gain, any difference in voltage between the noninverting (+) and inverting (−) terminals will drive the amplifier into saturation. When the + terminal is at a higher voltage than the − terminal, the amplifier is saturated on the positive side and the output voltage is theoretically equal to the positive supply voltage. When the − terminal is at a higher voltage than the + terminal, on the other hand, the amplifier is saturated on the negative side, and the output voltage is theoretically equal to the negative supply voltage. If the op amps are driven from a single-ended power supply, the negative supply voltage is 0 (ground).

Operation in this mode turns the op amp into an excellent comparator. For a typical op amp with a +5 V power supply, the saturated voltage whenever the + input is greater than the − input is about 3.5 V. For all other conditions, the output is about 0.5 V. Most gates will recognize these voltages as the 1 and 0 states, and the op amp comparator finds great use in digital circuits.

When the gain of the op amp is limited by feedback, the op amp can be used as a waveshaper, inverter, gate component, or buffer. Op amps used for gates, inverters, and buffers usually have unity gain, while those used in waveshaping may have the feedback adjusted for enough gain to produce oscillation.

Example 7.19 Use an op amp as a comparator to design a circuit that will produce the saturation output of 5 V when the input signal is greater than 3.0 V.

Since we want the op amp to produce an output pulse of 5 V only when the input is greater than 3.0 V, we connect the inverting terminal to a 3.0 V reference voltage and the input line to the noninverting terminal as shown in

Fig. 7-42. When the input line is lower than the 3.0 V reference, the op amp is in its 0 logic state since the + terminal is lower than the − terminal and the output is approximately 0 V. Only when the input is greater than the 3.0 V reference is there a positive difference voltage, which is then amplified by the gain of the op amp to produce the saturated voltage of 5 V which is recognized as a logic 1. Practically, with a +5 V supply, the positive saturation voltage would be around 3.5 V.

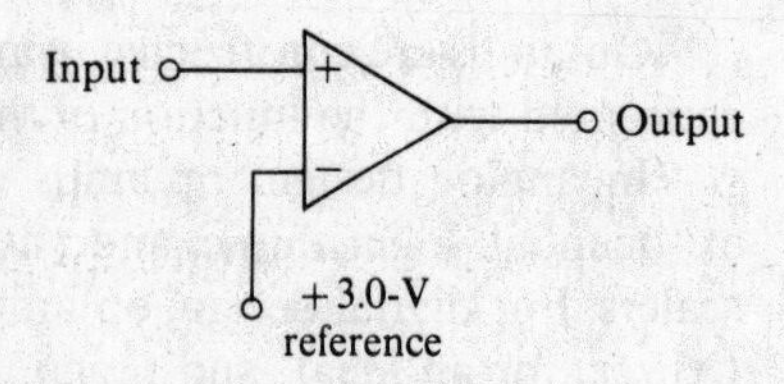

Fig. 7-42

Another useful op amp comparator circuit is the *window comparator*, which combines two comparators with an AND gate as shown in Fig. 7-43. The window comparator acts as a "window" for signals that fall within the range of the two reference voltages.

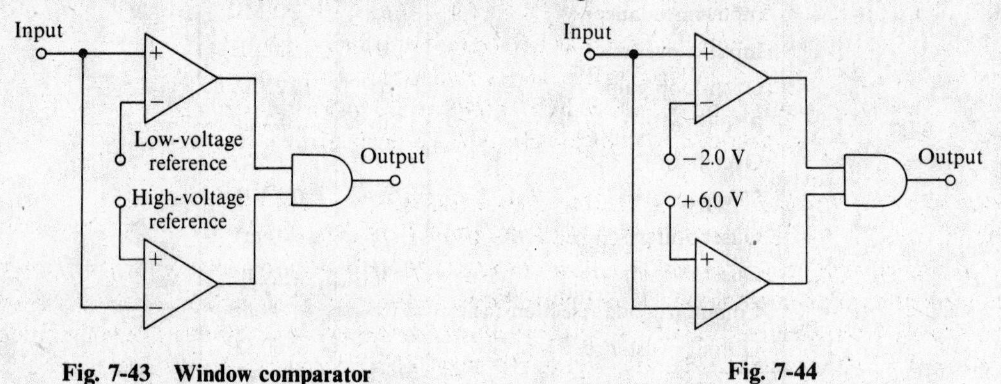

Fig. 7-43 Window comparator **Fig. 7-44**

Example 7.20 Design a window comparator that will accept voltages between +6 and −2 V and produce a logic 1 at the output.

Using the window comparator circuit shown in Fig. 7-43, we make the high-voltage reference equal to +6.0 V and the low-voltage reference equal to −2.0 V as shown in Fig. 7-44. If the signal is above 6 V, the top op amp produces a logic 1, but the bottom op amp will have the − terminal at a higher voltage than the + terminal and produce a logic 0. Since the AND gate needs both inputs to be 1 for an output to occur, and we are feeding a 1 and a 0, the output from the AND gate is 0.

If the signal is below −2 V, the top op amp produces a 0 while the bottom op amp produces a 1. To the AND gate this is still the same as before, and a 0 output from the gate results.

Only when the signal is higher than −2 V and lower than + 6 V will both op amps produce the 1 at each AND gate input necessary for the gate output to go to 1.

Op amps are also successfully used in Schmitt trigger circuits that provide excellent wave shaping characteristics. The Schmitt trigger or *regenerative comparator* is very similar to a bistable FF except it possesses a characteristic known as *hysteresis* which causes the triggering level when the voltage is increasing to be different from the triggering level when the voltage is decreasing as shown in Fig. 7-45*a*. The action of the Schmitt trigger on a distorted pulse wave train is shown in Fig. 7-45*b*.

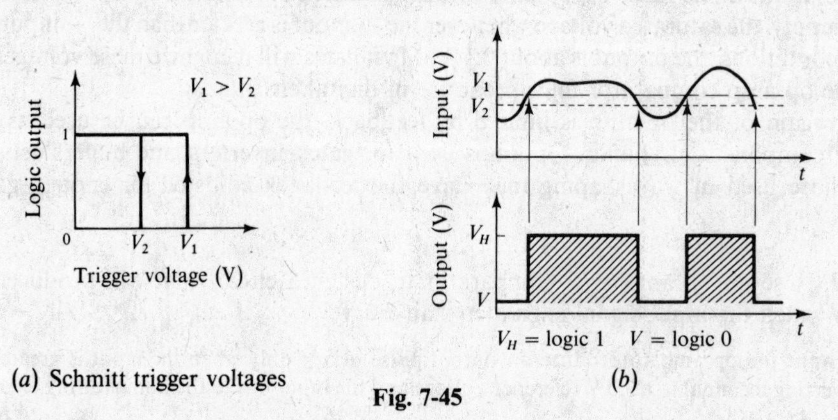

(*a*) Schmitt trigger voltages (*b*)

Fig. 7-45

It would be difficult for a gate to distinguish logic levels for the input wave shown in Fig. 7-45*b*, but the action of the Schmitt trigger can restore the pulse train to its original shape, as shown by the output in Fig. 7-45*b*.

Op amps also are used for making regular MVs and as such find great use in digital circuits.

7.10 CONVERSION DEVICES

In many cases, the digital form of binary or binary-coded signals will not properly operate a particular output device.

Output devices take the form of displays, printers, CRTs, or other indictors. Most displays use the seven-segment form shown in Fig. 7-46. While older display devices such as the Nixie tube use a separate filament for each decimal number and could easily be operated from a ring counter, the binary or BCD input to a seven-segment display must be decoded to indicate the proper decimal number on the display.

Seven-segment displays may be LED type, LCD type, or fluorescent type, but all need decoding to create the proper output.

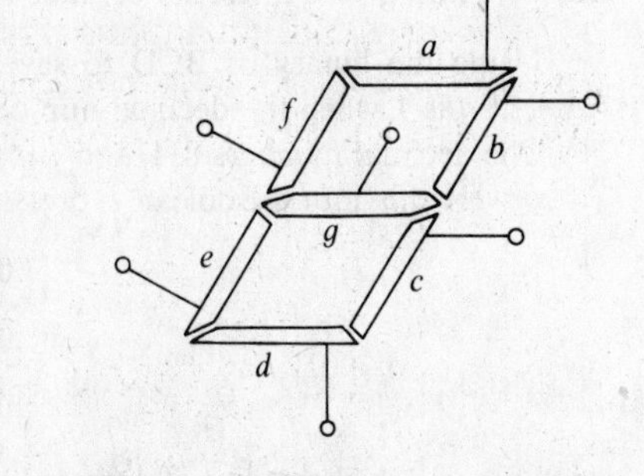

Fig. 7-46 Seven-segment display

In the typical seven-segment display shown in Fig. 7-46, each segment appears at a terminal and all segments are connected in common to a supply voltage. Segments may be connected as a common anode display, in which the positive side of the power supply is connected to the anode of each segment and a low voltage (0) at the segment cathode lights the segment; or they may be connected as a common cathode display, in which the negative side of the power supply is connected to the cathode of each segment and a high voltage (1) on the segment anode lights the segment.

Using the common anode display, we may write the seven-segment truth table as shown in

Decimal numbers	*D*	*C*	*B*	*A*
0	0	0	0	0
1	0	0	0	1
2	0	0	1	0
3	0	0	1	1
4	0	1	0	0
5	0	1	0	1
6	0	1	1	0
7	0	1	1	1
8	1	0	0	0
9	1	0	0	1

(*a*)

Decimal numbers	*a*	*b*	*c*	*d*	*e*	*f*	*g*
0	0	0	0	0	0	0	1
1	1	0	0	1	1	1	1
2	0	0	1	0	0	1	0
3	0	0	0	0	1	1	0
4	1	0	0	1	1	0	0
5	0	1	0	0	1	0	0
6	0	1	0	0	0	0	0
7	0	0	0	1	1	1	1
8	0	0	0	0	0	0	0
9	0	0	0	0	1	0	0

(*b*)

Decimal numbers	Inputs				Outputs						
	D	*C*	*B*	*A*	*a*	*b*	*c*	*d*	*e*	*f*	*g*
0	0	0	0	0	0	0	0	0	0	0	1
1	0	0	0	1	1	0	0	1	1	1	1
2	0	0	1	0	0	0	1	0	0	1	0
3	0	0	1	1	0	0	0	0	1	1	0
4	0	1	0	0	1	0	0	1	1	0	0
5	0	1	0	1	0	1	0	0	1	0	0
6	0	1	1	0	0	1	0	0	0	0	0
7	0	1	1	1	0	0	0	1	1	1	1
8	1	0	0	0	0	0	0	0	0	0	0
9	1	0	0	1	0	0	0	0	1	0	0

(*c*) Binary or BCD to common anode seven-segment display

Fig. 7-47

Fig. 7-47*b*. Comparing Fig. 7.47*b* with the truth table for the decimal numbers 0 to 9 in straight binary code as shown in Fig. 7.47*a*, we notice that the binary code for each of the decimal numbers could be treated as inputs and the seven-segment display segments for the same decimal numbers treated as outputs. This means we can combine the two tables as shown in Fig. 7-47*c* and obtain a decoding relationship between the numbers 0 to 9, lines *A*, *B*, *C*, and *D*, and the segments.

Example 7.21 Write the logic expression for segment *g* of a 4-bit output seven-segment display that shows when *g* is 1 in terms of lines *A*, *B*, *C*, and *D*. Also show your result in the form of a gating circuit.

Using the binary or BCD to seven-segment display combined truth table shown in Fig. 7-47*c*, we see that segment *g* is 1 when the decimal numbers 0, 1, and 7 are displayed.

The decimal numbers 0, 1, and 7 in terms of lines *D*, *C*, *B*, and *A* are given in Fig. 7-47*c* as 0000, 0001, and 0111. To convert this into a Boolean expression, we recall that it will be a sum of products with

$$0000 = \bar{A} \cdot \bar{B} \cdot \bar{C} \cdot \bar{D}$$

$$0001 = A \cdot \bar{B} \cdot \bar{C} \cdot \bar{D}$$

$$0111 = A \cdot B \cdot C \cdot \bar{D}$$

since Fig. 7-47*b* shows line *A* corresponding to the LSB. The Boolean sum of products for $g = 1$ is then

$$g = \bar{A} \cdot \bar{B} \cdot \bar{C} \cdot \bar{D} + A \cdot \bar{B} \cdot \bar{C} \cdot \bar{D} + A \cdot B \cdot C \cdot \bar{D}$$

Although it may be possible to simplify this expression so that fewer and less expensive gates can be used, the most straightforward way of realizing this circuit is by using four inverters to provide us with the line complements and connecting 3 four-input AND gates that feed a three-input OR gate as shown in Fig. 7-48*a*.

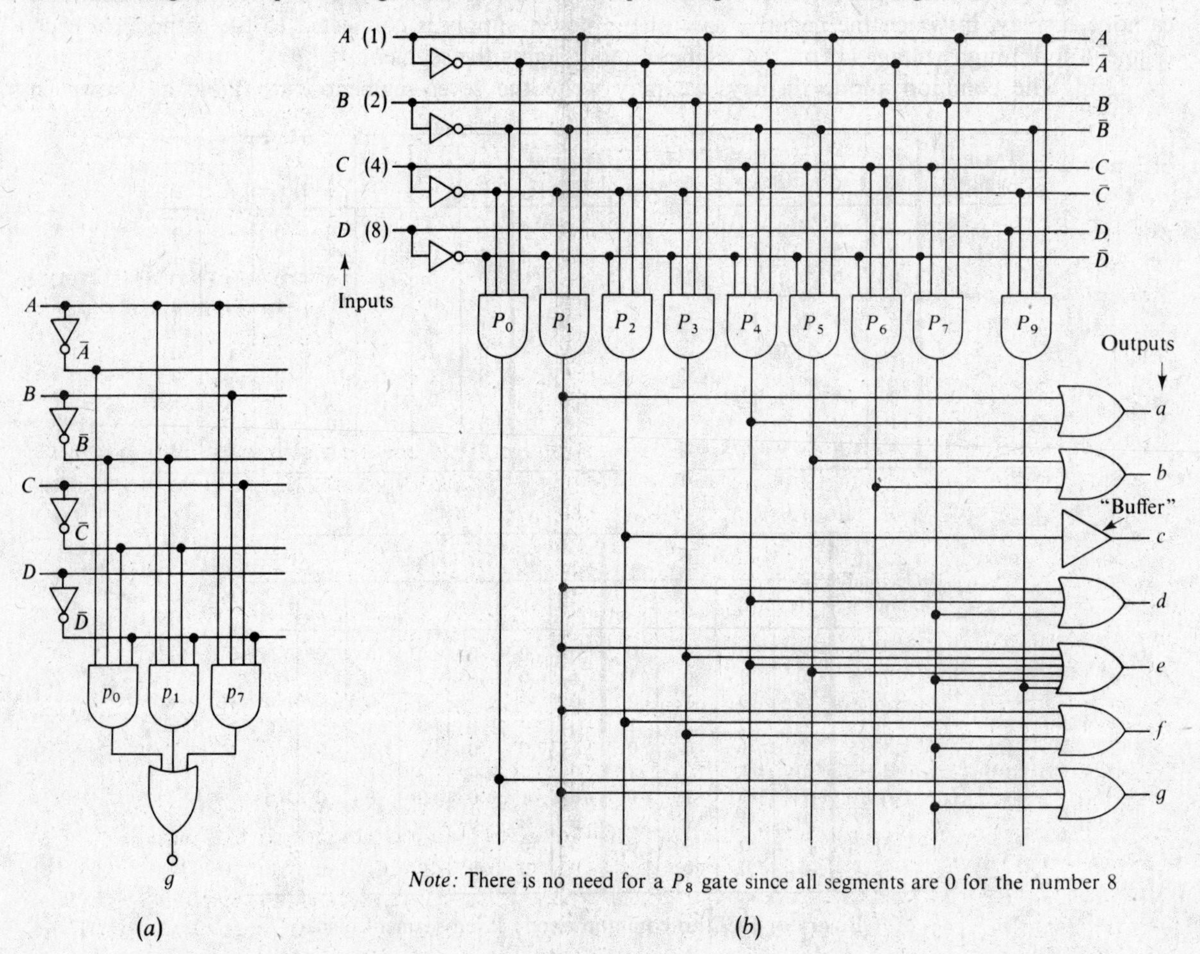

(*a*) (*b*)

Fig. 7-48

A full decoder circuit for a 4-bit output to a seven-segment display is shown in Fig. 7-48*b*.

In decoding, we take a previously coded input and convert it to its natural or uncoded form of output. The reverse procedure is called *encoding*, or *coding*, and converts a natural form of input to some type of coded output.

Coding and decoding are part of a larger conversion process called *interfacing*, in which the output of one device must be altered to properly serve as the input to a different type of device.

A common encoding circuit is a keyboard encoder such as the one shown in Fig. 7-49*a*. The logic state of the scan lines and the keyboard lines *P* and *N* provide the information for the display and for a coded input to the next stage. For example if we input the decimal number 6 by means of a keyboard, we encode the pulse from the 6 scan line and the *N* line to represent binary 6 (0110). This coded sequence of bits is now fed to the register and also to the seven-segment display, as shown in Fig. 7-49*b*. Note the decoder placed before the display in Fig. 7-49*b* to convert the signals into the proper form to display the symbols. In this case, the symbol is the number 6.

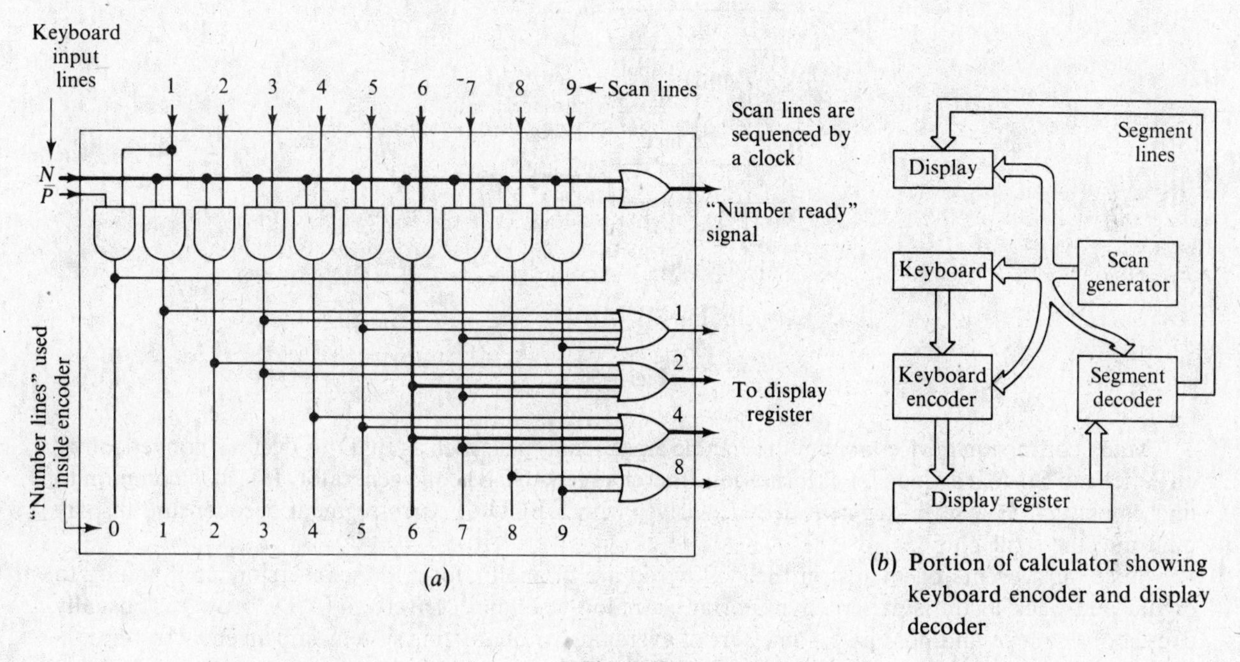

(*a*)

(*b*) Portion of calculator showing keyboard encoder and display decoder

Fig. 7-49

In encoding and decoding, it is economically advantageous to use the minimum number of gates necessary for the job. This is not always a simple task and when the use of inexpensive gates is preferred can involve a good deal of manipulation by the IC designer.

Another common conversion device used frequently in digital circuitry is the *multiplexer* shown in Fig. 7-50*a* and its dual, the *demultiplexer*, shown in Fig. 7-50*b*. Multiplexers (MUXs) and demultiplexers (DMUXs) are essentially routing switches, although there may be other features included on the same chip. A MUX converts multiple inputs to a single output, while the DMUX converts a single input to a multiple output.

When the multiplexing and demultiplexing action is done by clocking gates so that at each clock pulse a different connection is made, we term the action *time-division-multiplexing*. There is also *frequency-division-multiplexing* which is more common in analog circuits.

If a DMUX input is connected to a line which has serial data and the outputs are connected to a register, we can load serial data as if it were parallel data. If a MUX is connected so that parallel data is available on the input lines, we can obtain serial data from the output.

MUX and DMUX systems can be connected together to convert parallel data to serial data that may then be sent over a single pair of wires and converted back to parallel data at the other end as shown in Fig. 7-50*c*. Devices of this type are commercially known as MULDEMS.

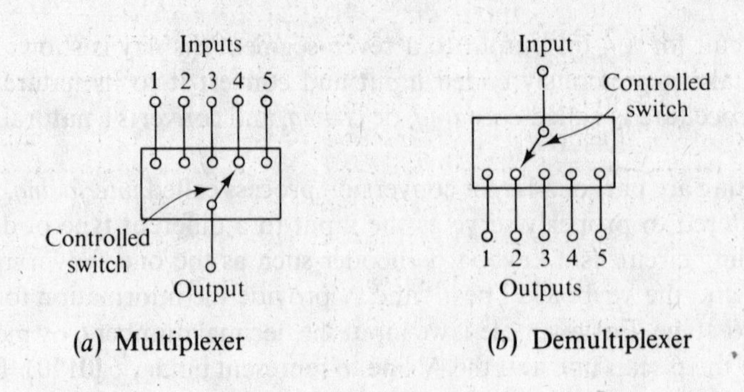

(*a*) Multiplexer (*b*) Demultiplexer

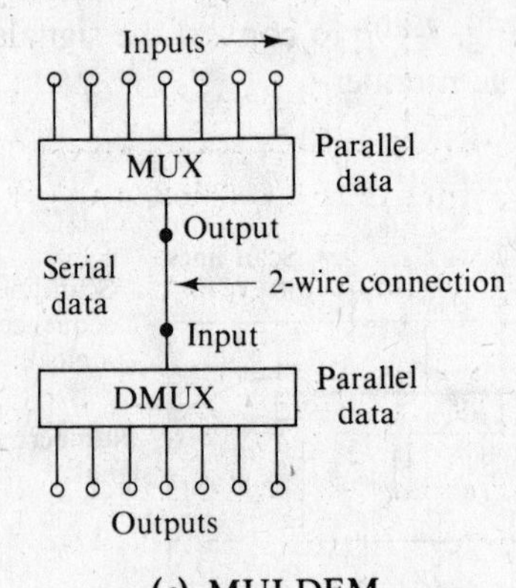

(*c*) MULDEM

Fig. 7-50

Many conversion devices are built to handle a particular task such as BCD-to-decimal conversion or an American Standard Code for Information Interchange (ASCII) code generator. It is also common to find chips such as a seven-segment decoder-driver with a BCD to seven-segment decoder and display built into the same chip.

Since most events of interest in the real world are in analog form, the conversion from analog to digital and back again is extremely important. Analog-to-digital converters (A/Ds or ADCs) usually consist of a sample-and-hold gate, some sort of averaging or quantizing system, and an encoding circuit. The two most commonly used procedures to make ADCs are the *successive approximation method* and the *integration method*.

In the successive approximation method, a digital-to-analog converter (D/A or DAC) is used along with an op amp comparator to successively latch the code bits into a register. If the DAC output is *greater* than the analog average sampled value, the bit is latched low and discarded for the next trial. If the DAC output is *less* than the analog average sampled value, that bit is latched high and kept for the next trial. The MSB of the DAC is set to 1 first and the comparison proceeds bit by bit until the LSB is latched at 0 or 1. The procedure is similar to weighing an unknown object on a precision balance, with the DAC bit positions used as decreasing measured weights.

This method can yield excellent resolution (around 12 bits), short conversion time, and easy microprocessor interfacing at the expense of higher cost per bit when compared to the integration method. Also, the successive approximation method has a poor noise rejection ratio.

Integration-type ADCs are simple, low in cost, and have high noise immunity and good IC compatibility. However, they have lower speed.

Digital-to-analog converters are less complex than ADCs and usually consist of a decoder and a series of voltage or current outputs that are related to a reference voltage by a code. A typical code and output chart is shown in Table 7-12, while a block diagram of a DAC is shown in Fig. 7-51. With DACs it is important to remember that the highest output voltage obtainable is really the full-scale value minus the LSB since the first code is for zero.

Table 7-12 DAC codes and output for eight-level code

Code	Binary fraction	Decimal fraction	% of V_{FS}	dB
000	0.000	0.0 (0)	100	0
001	0.001	0.125 ($\frac{1}{8}$)	12.5	−18.1
010	0.010	0.250 ($\frac{2}{8}$)	25.0	−12.0
011	0.011	0.375 ($\frac{3}{8}$)	37.5	−8.5
100	0.100	0.500 ($\frac{4}{8}$)	50.0	−6.0
101	0.101	0.625 ($\frac{5}{8}$)	62.5	−4.1
110	0.110	0.750 ($\frac{6}{8}$)	75.0	−2.5
111	0.111	0.875 ($\frac{7}{8}$)	87.5	−1.2

$$\text{Output voltage} = (\text{fraction value})(V_{FS})$$

$$\text{Maximum output voltage} = V_{FS}\left(1 - \frac{1}{2^n}\right)$$

$$\text{MSB} = \frac{V_{FS}}{2} \qquad \text{LSB} = \frac{V_{FS}}{2^n} \qquad \text{Error} = \tfrac{1}{2}\text{LSB}$$

where n = number of coding bits available.
For this table, $n = 3$ and

$$V_{\text{out}}\,(\text{max}) = \tfrac{7}{8}V_{FS}$$

$$\text{LSB} = \frac{V_{FS}}{2^3} = 0.125V_{FS}$$

$$\text{Error} = \tfrac{1}{2}\text{LSB} = \tfrac{1}{2}(0.125V_{FS}) = 0.0625V_{FS}$$

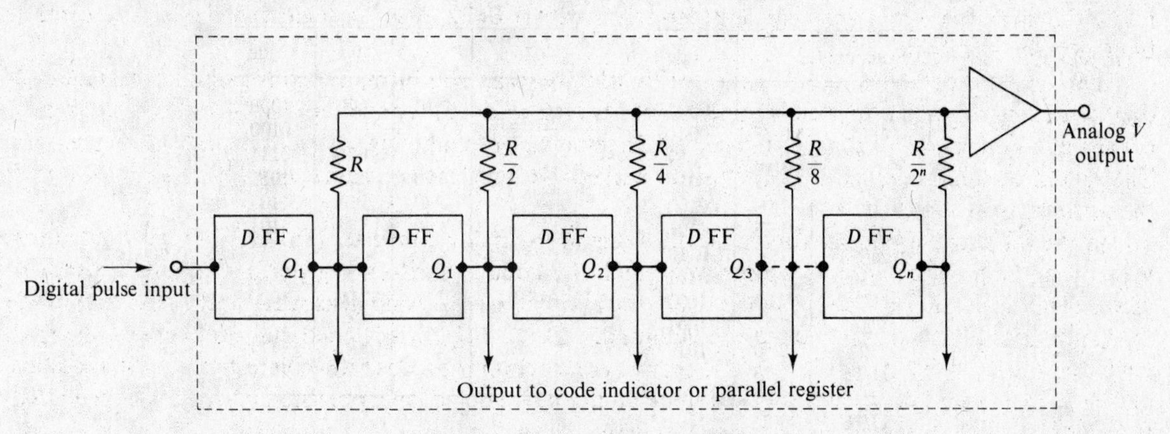

Fig. 7-51 DAC block diagram

DACs are used as component parts of ADCs, as shown in Fig. 7-52, where the counter counts the clock pulses and the DAC, op amp, and resistor configuration serve to sum the count which is then compared with the analog input. When the DAC output is the same as the analog voltage, the comparator cuts off the clock and the code in the counter is read out.

The op amp resistor network is frequently employed to produce a binary progression of currents and voltages which are then steered to the summing point to be compared with the analog current or voltage. The resistor configuration used is called an *R-2R ladder* or *network*.

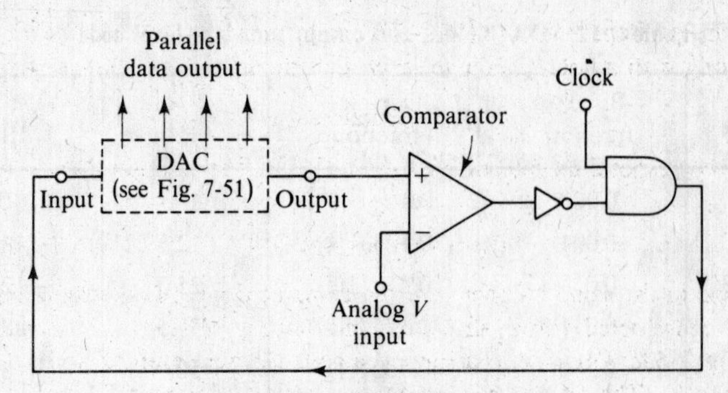

Fig. 7-52 ADC block diagram

Example 7.22 Convert the analog waveform, shown along with its sampling clock timing diagram in Fig. 7-53*a*, to a 16-level binary output code. Assume the voltage to be encoded is the highest voltage during each sampling period.

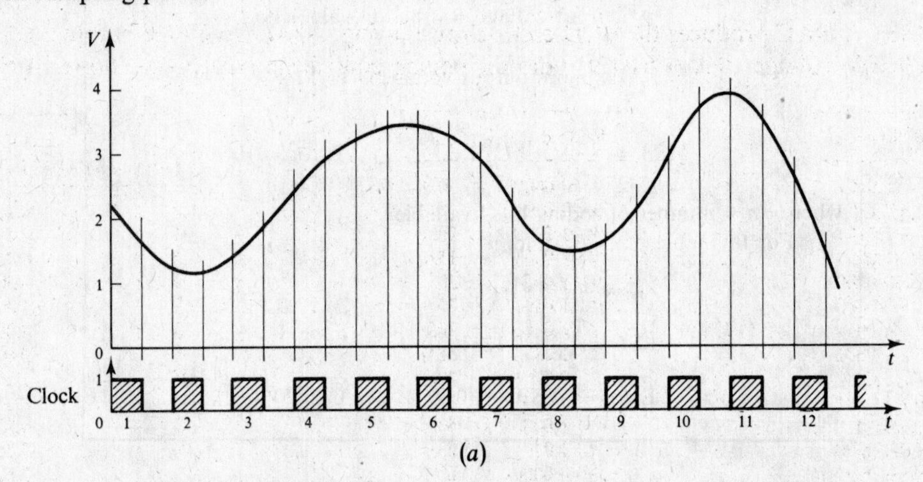

(*a*)

Fraction	V_Q	Code
0	0.0	0000
1/16	0.25	0001
1/8	0.50	0010
3/16	0.75	0011
1/4	1.00	0100
5/16	1.25	0101
3/8	1.50	0110
7/16	1.75	0111
1/2	2.00	1000
9/16	2.25	1001
5/8	2.50	1010
11/16	2.75	1011
3/4	3.00	1100
13/16	3.25	1101
7/8	3.50	1110
15/16	3.75	1111

(*b*)

Clock	V_0	V_L	Code
1	2.20	2.00	1000
2	1.20	1.00	0100
3	1.80	1.75	0111
4	2.90	2.75	1011
5	3.40	3.25	1101
6	3.40	3.25	1101
7	2.90	2.75	1011
8	1.60	1.50	0110
9	2.10	2.00	1000
10	3.70	3.50	1110
11	3.90	3.75	1111
12	2.60	2.50	1010

(*c*)

Fig. 7-53

Looking at Fig. 7-53*a*, we see that the maximum value for our sampling range is 3.90 V, so a full-scale value V_{FS} of 4.00 is selected for ease in coding. Since we wish a 16-level code, we need n bits such that

$$2^n = 16 = 2^4$$

or $n = 4$ bits. The fractional value of the LSB in a 4-bit code is

$$\text{Fraction} = \frac{1}{2^4} = \frac{1}{16}$$

A table is constructed in $\frac{1}{16}$th steps from 0 to (1 − LSB), or $\frac{15}{16}$ths in this case, and the binary fraction code written in the adjacent column along with a voltage column such that the quantized voltage step V_Q is given by

$$V_Q = (\text{fraction})V_{FS}$$

The first value is 0 with the corresponding code of 0000 and a V_Q of 0. The second value, $\frac{1}{16}$, gives a code of 0001 and a V_Q determined by

$$V_Q = \tfrac{1}{16}(4.00) = 0.25 \text{ V}$$

The rest of the table is filled in the same manner, and the completed table is shown in Fig. 7-53*b*.

A new table is then constructed which shows the maximum voltage V_p for each clock pulse and a code associated with it, as in Fig. 7-53*c*. For ease in assigning a code to each level, we add a V_L column using the next lowest quantized voltage for V_L. The maximum voltage seen during clock pulse 7, for instance, is 2.90 V. This falls between $V_Q = 2.75$ V and $V_Q = 3.00$ V, so the value for V_L is 2.75 V.

The final coding is shown in Fig. 7-53*c* and would be sent serially as

1000 (first code group) 0100 0111 1011 1101 1101 1011 0110 1000 1110 1111 1010 (last code group)

Example 7.23 A DAC produces the BCD codes shown in Fig. 7-54*a*. Graph the output analog voltage V_a if the full-scale voltage $V_{FS} = 5.00$ V and the coding equivalence table is as shown in Fig. 7-54*b*.

Clock	BCD code
1	0111 0101
2	1000 0101
3	1001 0101
4	1001 0000
5	1000 0000
6	1000 0000
7	0100 0000
8	0010 0101
9	0000 0000
10	0001 0101
11	0100 0101
12	0101 0000

(*a*)

BCD code	% V_{FS}
0000 0000	0.0
0000 0101	5.00
0001 0000	10.0
0001 0101	15.0
0010 0000	20.0
0010 0101	25.0
0011 0000	30.0
0011 0101	35.0
0100 0000	40.0
0100 0101	45.0
0101 0000	50.0
0101 0101	55.0
0110 0000	60.0
0110 0101	65.0
0111 0000	70.0
0111 0101	75.0
1000 0000	80.0
1000 0101	85.0
1001 0000	90.0
1001 0101	95.0

(*b*)

Clock	V_A output (V_{FS})(% V_{FS} from code)
1	3.75
2	4.25
3	4.75
4	4.50
5	4.00
6	4.00
7	2.00
8	1.25
9	0.00
10	0.75
11	2.25
12	2.50

(*c*)

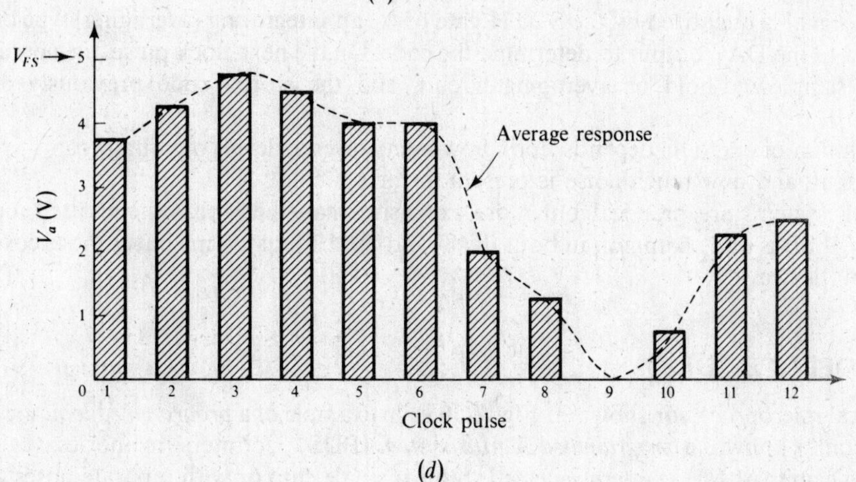

(*d*)

Fig. 7-54

To convert the code to the analog output voltage, we multiply the percentage of full-scale voltage $\%V_{FS}$ given by the code by the full-scale voltage of 5.00 V. At the sixth clock pulse, for example, the code of 1000 0000 (BCD for 0.80) is listed as $\%V_{FS} = 80.0$ in Fig. 7-54*b*. This means that the code 1000 0000 indicates 80 percent of the full-scale voltage, so

$$V_a = (0.80)(5.00) = 4.00 \text{ V}$$

The values for V_a at each clock pulse are shown in Fig. 7-54*c* and are graphed in Fig. 7-54*d*.

In practice, the pulses from the DAC would be sharp and the output would be connected across a capacitor or other low-pass filter so that the average response indicated by the dotted line would be obtained.

A sample-and-hold (S & H) gate like the one shown in Fig. 7-55 is used extensively in ADC comparator circuitry.

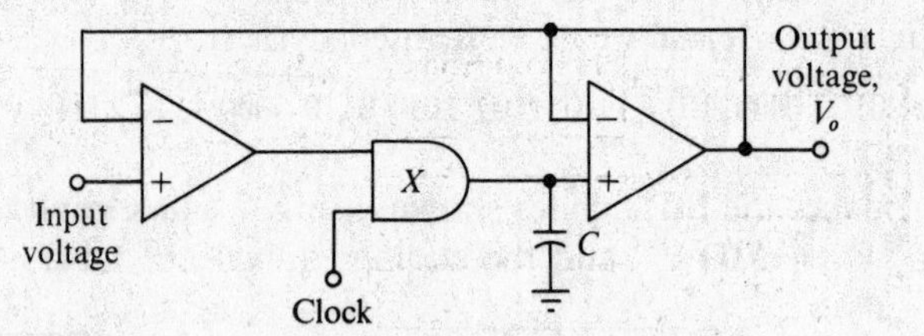

Fig. 7-55 Sample-and-hold gate

In the sample mode with the clock high, the input voltage is fed via the AND gate to the capacitor, which tries to follow the peak value of the input. This type of AND gate, sometimes designated with an X, is called a transmission gate or bilateral switch.

In the hold mode, the clock is low and the transmission gate output is cut off, allowing the voltage on the capacitor to remain at its previous value. The capacitor does not hold this voltage perfectly since it leaks charge, but the time intervals are small enough so that this effect, termed *droop*, is only a small percentage of the original voltage.

The *leakage time* is the time required for the capacitor to drop a specified percentage of its voltage. The *acquisition time* is the time necessary for the capacitor to track the signal voltage within a desired percent once the bilateral switch is closed. The length of time between the application of the part of the pulse that opens the bilateral switch and the actual switch opening is termed the *aperture time*.

Sampling theory tells us that we must filter the input and output of the S & H gate and sample at a rate at least twice the frequency of the highest frequency component in the signal wave in order to properly reconstruct the signal from the sampled pulses. Failure to sample at a high enough rate may lead to a reconstruction that has the same sampling characteristics as the input but different frequency components, usually lower. This production of false signals is called *aliasing*.

Once the signal is quantized by the S & H gate or by an integrating (averaging) type of S & H gate, it is compared to the DAC output to determine the code. On the next clock pulse, the new analog data is fed into the sample-and-hold or averaging circuit, and the stored code previously determined is transmitted.

The resolution of an ADC depends upon how many discrete levels of voltage can be recognized by the coding circuit and how much noise is present in the system.

Twelve-bit systems are practical but more expensive than the 8-bit systems used for most work. Resolution of $\frac{1}{2}$LSB is quite common in both the 8- and the 12-bit systems but is more costly to achieve with the 12-bit device.

7.11 MICROPROCESSORS

The typical microprocessor shown in Fig. 7-56 is an example of a *programmed logic system* (PLS) or, as it is sometimes known, a *programmed digital system* (PDS). Counters, memories, registers, and an arithmetic logic unit (ALU) are interconnected within a single chip or with separate buses to other chips so that different functions can be performed depending on the instructions used.

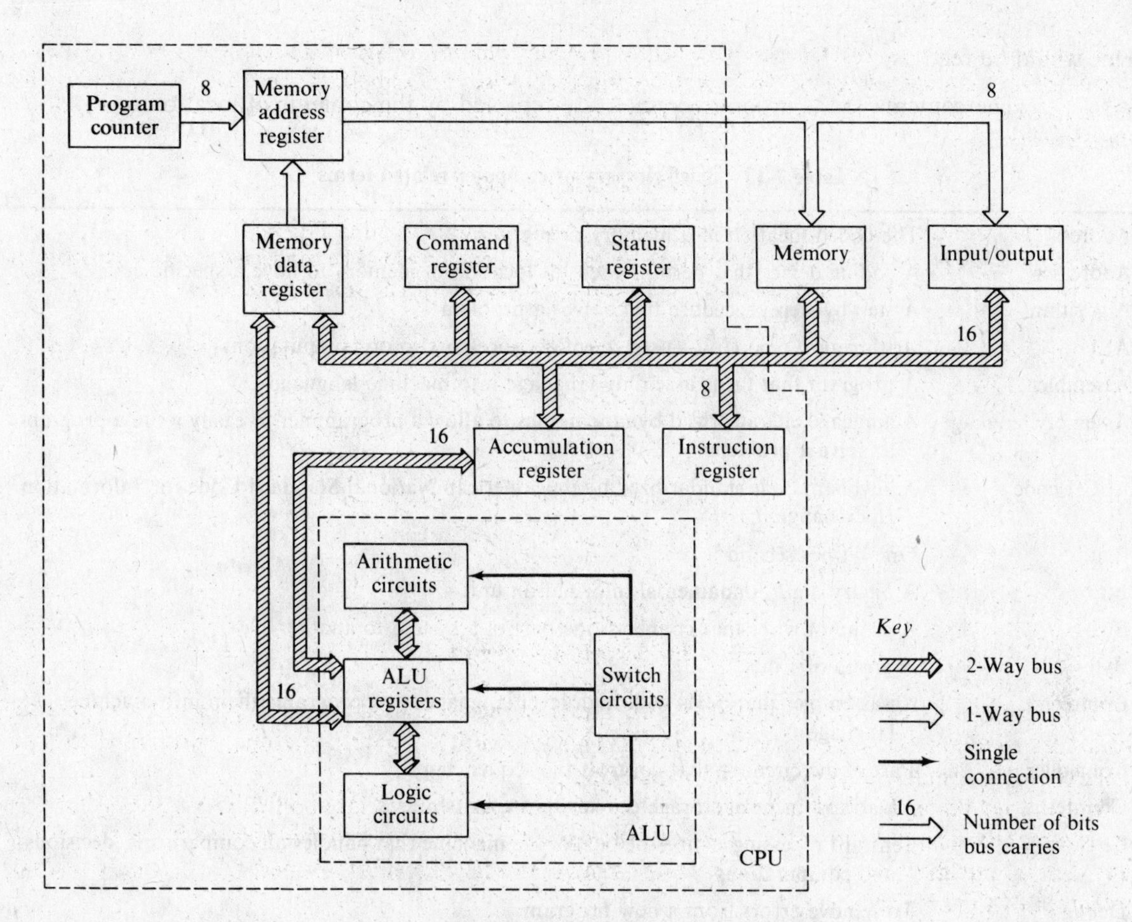

Fig. 7-56 Typical microprocessor block diagram

Since the microprocessor acts as a simple computer, many computer-related terms are used in microprocessor literature. Some of these, with brief explanations, are listed in Table 7-13.

Instructions are coded in hexadecimal (HEX) code, grouped in units of 4 bits called *nibbles*, and further grouped 2 nibbles or 8 bits at a time (called a *byte*). The instructions may be read 1, 2, or 3 bytes at a time and convention dictates that, in a 3-byte word, the first byte contains the code for the instruction, the second byte contains the low 8 bits of the operand, and the third byte contains the high 8 bits of the operand.

A list of binary to hexadecimal conversions is given in Table 7-14 and a sample of some of the 256 possible instruction codes for an 8-bit instruction byte is shown in Table 7-15.

The coded instructions are stored in an instruction register or in memory and are sequenced according to an incremented program counter. Accumulator registers are used to store numbers to be processed or new data.

The ALU performs the desired manipulation of the numbers and is usually limited to simple arithmetic operations, logic decisions, and comparisons. The ALU can be made to perform other functions, but the cost of the microprocessor increases with the versatility. The results of the ALU calculations are bussed back to an accumulator register.

The program counter is a register that holds the memory address of the next instruction in the program. The memory address register is used to store the address of the memory location we are reading from or writing to. The data is stored in a memory data register.

In describing the step-by-step functioning of a microprocessor, notation is used that looks like the following:

$$(PC) \leftarrow (LOC\ 6)$$

This would be read as

The contents of the program counter are replaced by the contents of location 6
( ) PC ← ( ) LOC 6

Table 7-13 Brief glossary of computer-related terms

Address	The coded location in a memory or register
Addresser	A coding device that permits each bit location of memory to have a specific code
Algorithm	A step-by-step procedure that solves a problem
ALU	*A*rithmetic *L*ogic *U*nit—used in microprocessors to do computations
Assembler	A program that turns assembly language into machine language
Assembly language	A language characterized by mnemonics to allow a programmer to easily write a program to solve a problem
ASCII code	A keyboard code standardized by the *A*merican National *S*tandard *C*ode for *I*nformation *I*nterchange in 1968
Baud	One bit per second
Bit	A *b*inary dig*it*; fundamental information unit
Bus	A group of wires that connects one part of a system to another
Byte	A group of 8 bits
Compiler	An assembler that deals with high-level languages and programs them into machine language
Controller	Part of the circuitry that controls the sequencing
Counter	An arrangement of storage devices that counts pulses
CPU	*C*entral *P*rocessing *U*nit—the part of the machine that handles all comparisons, decisions, and computations
Debug	To remove errors from a new program
Dedicated	Used only for one purpose
Flag	A bit that tells the status of some part of the information, sign, parity, conditional, etc.
Flowchart	A symbolic graphical representation of the steps necessary to solve a problem
Hardware	Wires, electronic circuits, and actual equipment
Hardwire	To permanently connect
Instruction	A coded statement that tells the computer what to do
LSB	*L*east *S*ignificant *B*it
Main Frame	The main memory storage circuits and CPUs
Memory	The part of the machine where information is stored
Mnemonic	A shorthand or abbreviation that indicates the program instruction
MSB	*M*ost *S*ignificant *B*it
Nibble	A group of 4 bits
Operand	A number to be used in a computation
Peripheral	Any hardware such as a terminal that is not part of the main frame
Program	Series of steps or instructions that the computer performs sequentially to solve a problem
Register	A group of memory cells
Routine	A procedure for accomplishing a particular operation
Software	Programs and instructions
Stored program	An area of memory that has been set up to perform a routine
Subroutine	A portion of a routine to handle a particular calculation or procedure
Terminal	An input or output device such as a keyboard, CRT, printer, or display
Word	Any group of bits, usually 16 or 32 bits long, which is treated as a unit

Table 7-14 Binary-to-hexadecimal conversion for 4-bit code

Decimal	Binary	Hexadecimal
0	0000	0
1	0001	1
2	0010	2
3	0011	3
4	0100	4
5	0101	5
6	0110	6
7	0111	7
8	1000	8
9	1001	9
10	1010	A
11	1011	B
12	1100	C
13	1101	D
14	1110	E
15	1111	F

Table 7-15 Assembly code and function for typical microprocessor (abbreviated list)

Code	Mnemonic	Operation	Function
03	LD3	Load via register 3	(Register 3) → (register D)
15	INC	Increment register 5	(Register 5) + 1
2B	DEC	Decrement register B	(Register B) − 1
33	BDF	Short branch if register D contents = 0	Memory location specified by byte following code → (register P) if (register D) = 0; otherwise (register P) + 1
C4	NOP	Continue (no operation)	Execution will proceed with the next instruction (time killer)
C8	LSKP	Long skip	(Register P) + 2
D6	SEP	Set register P	Register 6 becomes program counter
78	SAV	Save	(Register T) → memory location specified by register X
63	OUT	Output	M(register X) → data bus (Register X) + 1 Data bus → output #3
69	INP	Input	Input #(9-7) or 2 Input #2 → data bus Data bus → (register D) Data bus → M(register X)
73	STXD	Store in stack (PUSH)	(Register D) pushed onto stack register specified by X
FO	LDXA	Load data (POP)	(Register X) popped into register D
F6	SHR	Shift right	Shift contents of register specified in previous step 1 bit to the right
F4	ADD	Addition	Add contents of register $M1$ and register $A5$ bit by bit and store in register D

Example 7.24 Show how the contents of the registers shown in Fig. 7-57*a* change as the program counter is sequenced through a routine that adds 2 three-digit decimal numbers stored in memory and stores their four-digit result in memory. The carry, if any, is stored in register *Y* and the instructions are shown in Fig. 7-57*b*.

Register *P*	Register 1	Register 2	Register 3	Register *A*	Register *B*	Register *Y*
Program counter						
64	321	880	219	0	15	1

(*a*)

Memory

Address	Data
393	4
394	8
395	2
⋮	⋮
424	6
425	0
426	9
⋮	⋮
679	6
680	2
681	8
65	CLEAR CARRY, Y
66	3 → (REG A)
67	395 → (REG 1)
68	426 → (REG 2)
69	681 → (REG 3)
70	M(REG 1) → (REG B)
71	(REG B) + (Y) + M(REG 2) → (REG B) SET Y IF CARRY ELSE O → (Y)
72	(REG B) → M(REG 3)
73	DECREMENT REG 1, 2, 3, A
74	(A) = 0 STOP ELSE 70 → (D)

(*b*)

Fig. 7-57*a*, *b*

A microprocessor carries out the instructions in its program counter in a two-step fetch and execute sequence. During the fetch cycle, it retrieves the instruction code from memory, and during the execute cycle, it performs the operation indicated by the code.

For this routine, we will ignore the fetch cycle and concentrate on the register changes during the execute cycle when the actual computations are done. In this example, we will use an asterisk (*) to indicate a register change from the previous step.

The first instruction in the program counter is stored in memory location 65 and tells us to clear the carry register Y. The program counter is then incremented to 66 and the instruction at 66 says to place the number 3 in register A. The program counter is advanced to 67 and the instruction there followed.

This procedure continues loading the registers until the program counter is equal to 70. The instruction at location 70 says to replace the contents of register B with the contents of the memory location addressed by register 1. The contents of register 1 equal 395 and at location 395 the data equals 2. This means that 2 is placed in register B, wiping out the 15 that had been there previously.

The program counter is now advanced to 71, where the instruction tells us to take the contents of the memory location addressed by register 2 and add it to the contents of register B along with the carry bit in register Y and replace register B's contents with the sum. Implied in this instruction and stated here for convenience is the command to place a 1 in the carry register Y if the addition results in a carry and to place 0 in the Y register if it does not. Our result *does* have a carry, so Y is set to 1 and the sum 1 replaces the contents of register B.

Step 72 tells us to place the contents of register B in the memory location addressed by register 3. In step 73, registers 1, 2, 3, and A are decremented. In machine language, this would normally take several steps. Instruction 74 is a conditional test. If the contents of register A are equal to 0, then we stop. If not, we replace the contents of register P, the program counter, with the number 70 so that the next step is a repeat of the instruction found at 70. The contents of register A equal 2, so the program counter is set to 70 and the instructions 70 through 73 followed two more times until $A = 0$ and instruction 74 says to stop.

A completed list of the register changes, with * indicating a change from the previous step, is shown in Fig. 7-57*c*, while the completed memory table at the end of the routine is shown in Fig. 7-57*d*.

Register p	Register 1	Register 2	Register 3	Register A	Register B	Register Y
Program counter						
64	321	880	219	0	15	1
65	321	880	219	0	15	0*
66	321	880	219	3*	15	0
67	395*	880	219	3	15	0
68	395	426*	219	3	15	0
69	395	426	681*	3	15	0
70	395	426	681	3	2*	0
71	395	426	681	3	1*	1*
72	395	426	681	3	1	1
73	394*	425*	680*	2*	1	1
70	394	425	680	2	8*	1
71	394	425	680	2	9*	0*
72	394	425	680	2	9	0
73	393*	424*	679*	1*	9	0
70	393	424	679	1	4	0
71	393	424	679	1	0*	1*
72	393	424	679	1	0	1
73	392*	423*	678*	0*	0	1
74	STOP					

(*c*)

Memory

Address	Data
393	4
394	8
395	2
⋮	⋮
424	6
425	0
426	9
⋮	⋮
679	0*
680	9*
681	1*

(*d*)

Fig. 7-57*c*, *d*

The general approach to using microprocessors is the same as that used for computers. First, an algorithm is constructed to solve the problem under consideration. The algorithm is then converted to a flowchart and the flowchart generates a set of instructions. The memory needed and specific locations for the data necessary to solve the problem are decided upon, and the set of instructions is written out in regular form. These instructions must be coded first into assembly language, a set of mnemonics that uses four or five symbols to represent each basic instruction. The instructions can then be coded into machine language, hexadecimal coded binary in most cases, either by hand or with the use of an assembler or a compiler, depending upon the level of language used.

Some of the common high-level languages frequently used include

- Fortran
- Basic
- Algol
- Cobol
- *C* language
- PL/1
- Pascal

In programming a microprocessor, we usually have to break operations down into simple tasks that can be handled by the particular microprocessor being used. These tasks are usually repeated many times, but at high speed, so the result approaches those obtained with more flexible processing units.

Microprocessors have been successfully used for simple record keeping, metering, control functions, and data manipulation and are responsible for ushering in the era of the home computer.

Solved Problems

7.1 Add the decimal numbers 132 and 429 in straight binary code and check the result decimally.

Using the procedure of successive division by 2, we write the remainder of each step to obtain the binary representation with the LSB equal to the first remainder.

For the decimal number 132, this stepwise procedure yields

	Remainder
$132/2 = 66$	0
$66/2 = 33$	0
$33/2 = 16$	1
$16/2 = 8$	0
$8/2 = 4$	0
$4/2 = 2$	0
$2/2 = 1$	0
$1/2$	1

Therefore 132 in binary is 1000 0100.

Following the same procedure for the decimal number 429 gives us in binary 0001 1010 1101 where the leading three 0s were added in front of the MSB to make a complete group of 4 bits.

We usually add the binary numbers in groups rather than all at once to save register space in a machine, so we first add the least-significant nibble using the binary addition rules to obtain

$$\begin{array}{r} 1101 \\ +\ 0100 \\ \hline 1\ 0001 \end{array}$$

The MSB is a carry to the next nibble to produce

$$\begin{array}{rl} 1 & \text{(carry from previous step)} \\ 1000 & \\ +\ 1010 & \\ \hline 1\ 0011 & \end{array}$$

The MSB is a carry to the next group of 4 bits, which is only the most-significant nibble from 429. This gives us

$$\begin{array}{r} 1 \text{ (carry from previous step)} \\ +\ 0001 \\ \hline 0010 \end{array}$$

The final result is 0010 0011 0001.

This can be converted back to decimal form by adding the weighted values of the bit positions that have a 1 using Table 7-1. Following this procedure we obtain

$$2^9 + 2^5 + 2^4 + 2^0 =$$

$$512 + 32 + 16 + 1 = 561$$

As a check we perform the normal decimal addition with the original decimal numbers:

$$\begin{array}{r} 429 \\ +\ 132 \\ \hline 561 \end{array}$$

7.2 Subtract 0011 1001 from 0101 1100 using the 2's complement method. Check the result decimally.

In order to subtract using the 2's complement method, we first form the 1's complement of the subtrahend 0011 1001 by changing all the 1s to 0s and all the 0s to 1s. The 1's complement of 0011 1001 is therefore 1100 0110.

We form the 2's complement by adding 1 to the LSB to obtain

$$\begin{array}{r} 1100\ 0110 \\ +\qquad\quad 1 \\ \hline 1100\ 0111 \end{array}$$

The subtraction can now be carried out by adding the 2's complement just formed to the original number to be subtracted from (minuend):

$$\begin{array}{r} 0101\ 1100 \\ +\ 1100\ 0111 \\ \hline 1\ 0010\ 0011 \end{array}$$

The MSB is ignored and our result is

$$0010\ 0011$$

To check this result, we convert the original numbers into their decimal equivalents using Table 7-1 to get

$$0101\ 1100 = 92$$
$$0011\ 1001 = 57$$

Our binary answer converted to decimal form yields

$$0010\ 0011 = 35$$

$$\begin{array}{r} 92 \\ -\ 57 \\ \hline 35\checkmark \end{array}$$

7.3 Construct a truth table for the switching circuit shown in Fig. 7-58*a*. Use the completed truth table to write the Boolean equation for the output Q.

This circuit contains three input switches which can turn Q on or off from any of the three locations and is known as a four-way switch.

The truth table is constructed for the three switch inputs, A, B, and C, as shown in Fig. 7-58*b*, with a 1 indicating that the switch is in the down position and a 0 indicating the up position. The number of possible combinations with three inputs is 2^3 or 8, and this can be achieved by writing the binary equivalents of 0 to 7. Once this is set up, we need to actually trace the circuit to determine which combinations cause the LED, Q, to be on (1).

With switch A in the up position, the positive voltage is transferred to point 3. With switch B in the up position, the positive voltage is routed to point 7, and the LED, Q, can be turned on with switch C in the down position. So if A is in the up position *and* B is up, *and* C is down, $Q = 1$.

Leaving A in the up position, we can also cause Q to be on when switch B is down *and* switch C is up.

With the up position equal to 0 and the down position equal to 1, the combinations 001 and 010 will result in a 1 for Q, and we may enter the 1s into the Q column and also enter 0 for the other two possibilities with $A = 0$.

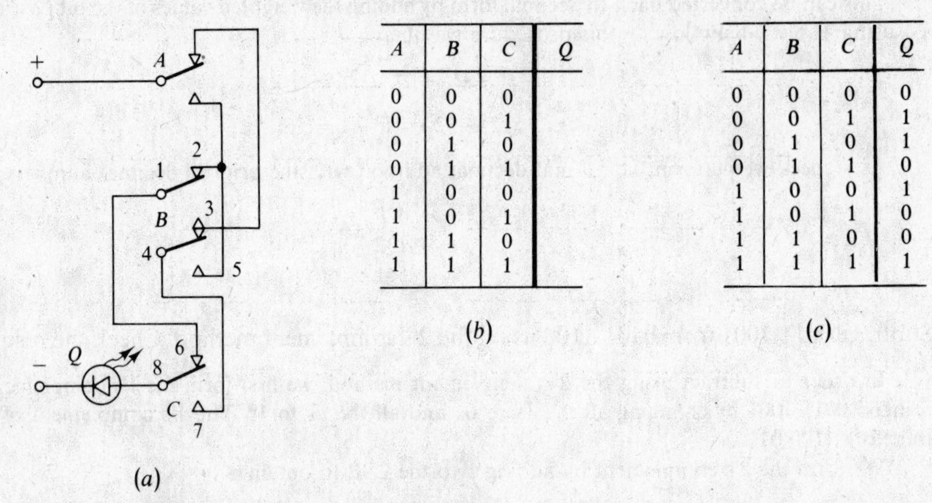

A	B	C	Q
0	0	0	
0	0	1	
0	1	0	
0	1	1	
1	0	0	
1	0	1	
1	1	0	
1	1	1	

(*b*)

A	B	C	Q
0	0	0	0
0	0	1	1
0	1	0	1
0	1	1	0
1	0	0	1
1	0	1	0
1	1	0	0
1	1	1	1

(*c*)

Fig. 7-58

The next four combinations occur with switch A in the down position (1), and we see that Q is on (1) only if *B and C* are up *or B* and *C are down.* These results are entered in the Q column to complete the truth table, as shown in Fig. 7-58*c*.

Using Fig. 7-58*c*, we write the Boolean equation for Q by listing the logic conditions for the rows where $Q = 1$. This results in the Boolean sum of products

$$Q = \bar{A}\cdot\bar{B}\cdot C + \bar{A}\cdot B\cdot\bar{C} + A\cdot\bar{B}\cdot\bar{C} + A\cdot B\cdot C$$

This can be simplified somewhat by making use of Eq. (7-8) in Table 7-5 to yield

$$Q = \bar{A}(\bar{B}\cdot C + B\cdot\bar{C}) + A(\bar{B}\cdot\bar{C} + B\cdot C)$$

It might be noted that the expressions in the parentheses can be realized with the exclusive OR and exclusive NOR functions, respectively.

7.4 **Use De Morgan's law to find out if $\bar{A}\cdot\bar{B}\cdot\bar{C} = \overline{A\cdot B\cdot C}$. Check the results with a truth table.**

Looking at the left-hand side of the equation, we apply De Morgan's law by first replacing all the variables by their complements to get

$$A\cdot B\cdot C =$$

Then we replace the $\cdot$ with a $+$ wherever it occurs to obtain

$$A + B + C =$$

We then complement the entire expression to get

$$\overline{A + B + C} =$$

This does not appear to be identical to the right-hand side of the equation. To prove the inequality, we set up the truth table as shown in Fig. 7-59*a*, with all the possible combinations of A, B, C, and their complements. Using this table, we construct additional columns for

$$S_1 = A + B + C$$
$$\bar{S}_1 = \overline{A + B + C}$$

$$P_1 = A \cdot B \cdot C$$
$$\bar{P}_1 = \overline{A \cdot B \cdot C}$$
$$P_2 = \bar{A} \cdot \bar{B} \cdot \bar{C}$$

and check the results to see if $P_2 = \bar{P}_1$. These columns are shown in Fig. 7-59*b*, and we find that $P_2 \neq \bar{P}_1$ and $P_2 = \bar{S}_1$, which checks our application of De Morgan's law. This proves that

$$\bar{A} \cdot \bar{B} \cdot \bar{C} = \overline{A + B + C}, \quad \text{not} \quad \overline{A \cdot B \cdot C}.$$

A	B	C	$\bar{A}$	$\bar{B}$	$\bar{C}$
0	0	0	1	1	1
0	0	1	1	1	0
0	1	0	1	0	1
0	1	1	1	0	0
1	0	0	0	1	1
1	0	1	0	1	0
1	1	0	0	0	1
1	1	1	0	0	0

(*a*)

S_1	$\bar{S}_1$	P_1	$\bar{P}_1$	P_2
0	1	0	1	1
1	0	0	1	0
1	0	0	1	0
1	0	0	1	0
1	0	0	1	0
1	0	0	1	0
1	0	0	1	0
1	0	1	0	0

Different ($\bar{P}_1$, P_2)

Identical ($\bar{S}_1$, P_2)

(*b*)

Fig. 7-59

7.5 Draw a logic diagram for the Boolean expression $Q = A \cdot \bar{B} + B \cdot \bar{C} + \bar{A} \cdot B \cdot C$ using only NAND gates.

Without the limitation on the type of gates to be used, we could realize Q by connecting three inverters, three AND gates, and an OR gate, as shown in Fig. 7-60*a*. To use *only* NAND gates, however, we have to modify the gating diagram of Fig. 7-60*a* by use of De Morgan's theorem for gates.

If we invert the inputs to a gate and replace that gate with its negative dual, we preserve the function. In order to perform inversion with a NAND gate, we merely have to tie the inputs together. Then a NAND with a NAND inverter following it could be utilized to give the AND function, and a NAND with its inputs inverted would perform the OR function.

Making these substitutions, we arrive at the diagram shown in Fig. 7-60*b*.

Wherever we see two NAND inverters in series, we may eliminate both. This results in the simplified NAND gate diagram shown in Fig. 7-60*c*.

This problem demonstrates that any gate may be duplicated in function using only NAND gates.

In the MOS family, the NAND is the fundamental gate structure, as it is inexpensive to manufacture. All gate functions in the MOS family are configured from only NAND gates.

7.6 Construct a gating circuit using only NOR gates to perform the XOR function.

Inversion is possible with the NOR gate by tying both inputs together as we did in the case of the NAND inverter.

The XOR truth table is shown in Fig. 7-61*a*, and the straightforward implementation of the XOR function $Q = A \cdot \bar{B} + \bar{A} \cdot B$ is shown in Fig. 7-61*b*.

Using De Morgan's gate law, we may duplicate the AND function by inverting the inputs to a NOR. The OR function can be duplicated by inverting the NOR output with a NOR inverter. These substitutions are shown in Fig. 7-61*c*.

Inverters 1 and 5 are in series feeding one NOR gate, and inverters 2 and 4 are in series feeding the other NOR gate. By removing the double inverters (two inverters in series), we can simplify the gating diagram, as is shown in Fig. 7-61*d*.

This demonstrates the fact that any gating function can be duplicated by using only NOR gates.

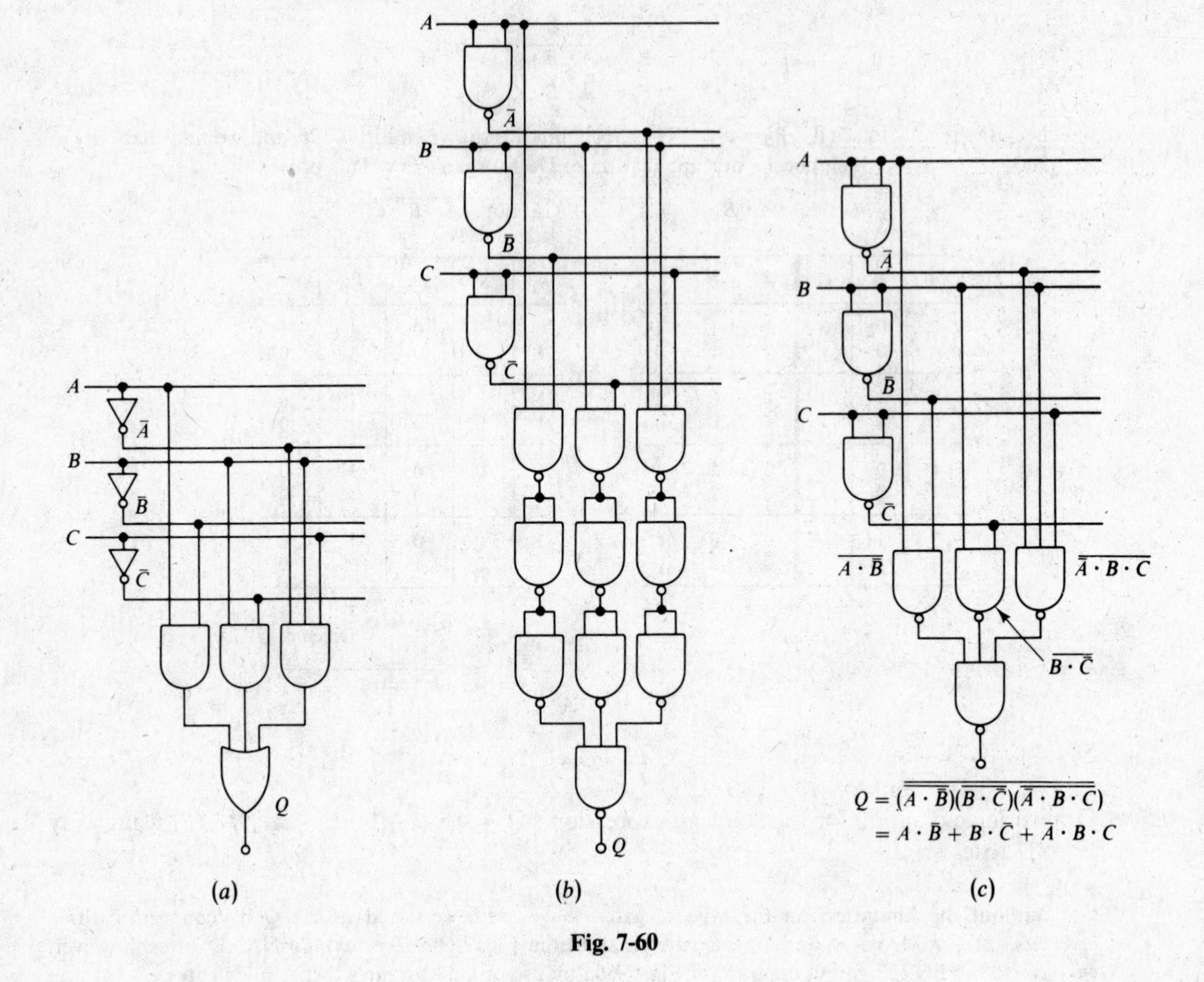

Fig. 7-60

A	B	Q
0	0	0
0	1	1
1	0	1
1	1	0

(a)

$Q = A \cdot \bar{B} + \bar{A} \cdot B$

(b)

Fig. 7-61*a*, *b*

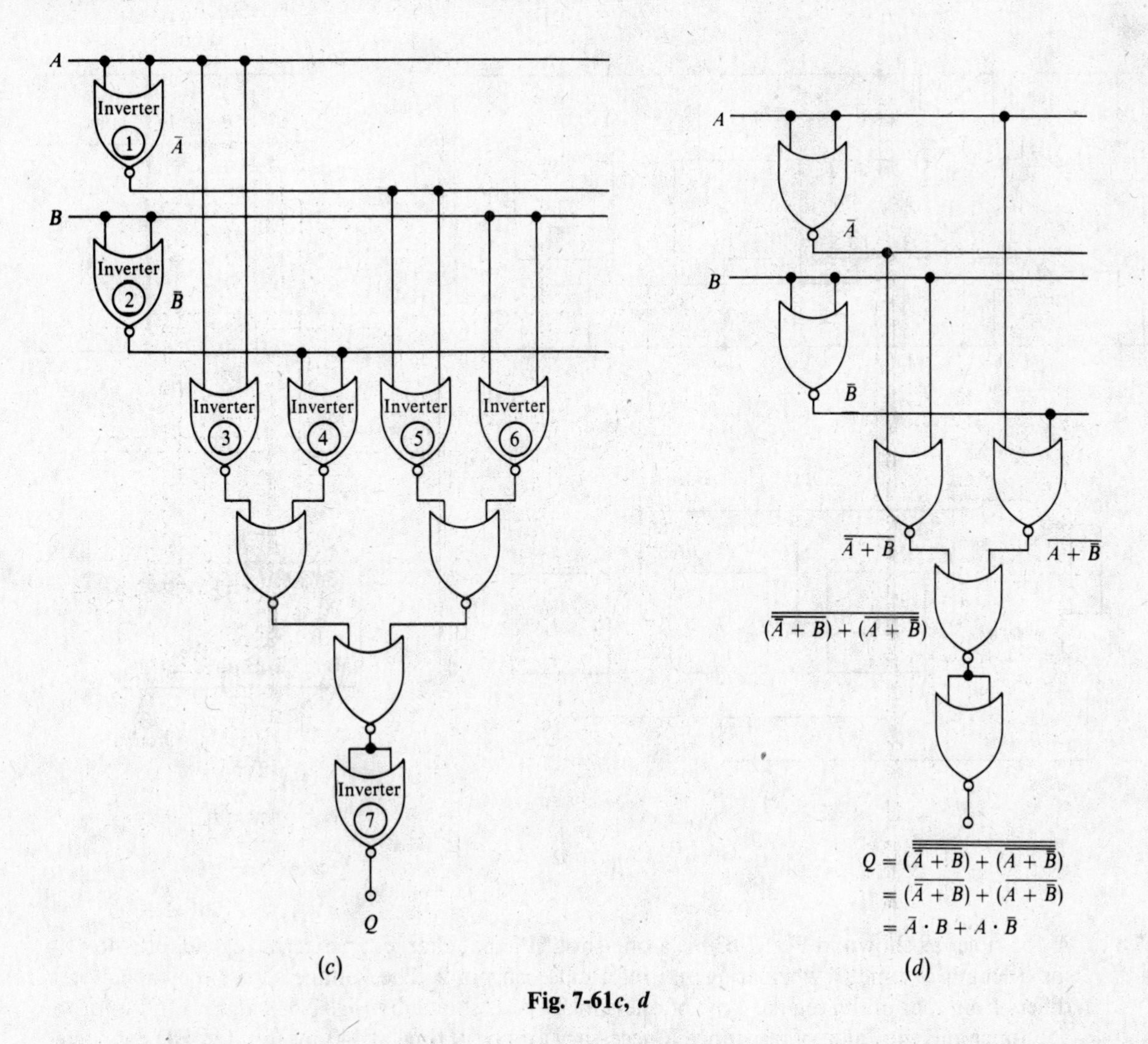

Fig. 7-61*c*, *d*

7.7 A manufacturer has provided a dual two-input NAND gate structure on the same IC chip as three *D*-type FFs, as shown in Fig. 7-62*a*. Show by a diagram how you would connect the components of this chip to obtain a modulo 5 counter.

To make a counter, we connect the input *D* terminal of the first FF (pin 7) to the input data stream. This is the LSB (*A*) FF and we connect its output Q_A (pin 15) to the input of the next FF (pin 14). The Q_B output (pin 2) is then connected to the input of the *C* FF (pin 4). Output pins 12, 2, and 15 are then the output data lines *C*, *B* and *A* and may be connected to the display decoder.

This connection would normally be a modulo 8 counter since $2^3 = 8$. We wish to modify the count sequence to a modulo 5 counter.

By constructing the truth table for the full counter as shown in Fig. 7-62*b*, we see that the desired effect could be achieved if the transition from 4 to 5 (100 to 101) activates a common clear (reset) terminal on all the FFs.

We need a logic circuit that will operate the clear terminals when *C* is 1 *and A* is 1. On the next count after 4, the FFs will all reset to 000 since the AND gate will be enabled.

We may use the dual NAND gate to perform the AND function by tying the two inputs of the second gate together as an inverter. Feeding the NAND output of the first to the common terminals of the second will cause the required AND function to occur.

Therefore, we connect pin 9 of the second NAND gate to the common clear terminal, 13; pins 10, 11, and 8 are tied together, feeding the output of the first NAND to the input of the second, and the *A* line and *C* line are connected to the inputs of the first NAND by wiring pin 15 to pin 6 and pin 12 to 7.

The counter is drawn in block form in Fig. 7-62*c*, while a possible printed circuit configuration for the output socket is shown in Fig. 7-62*d*.

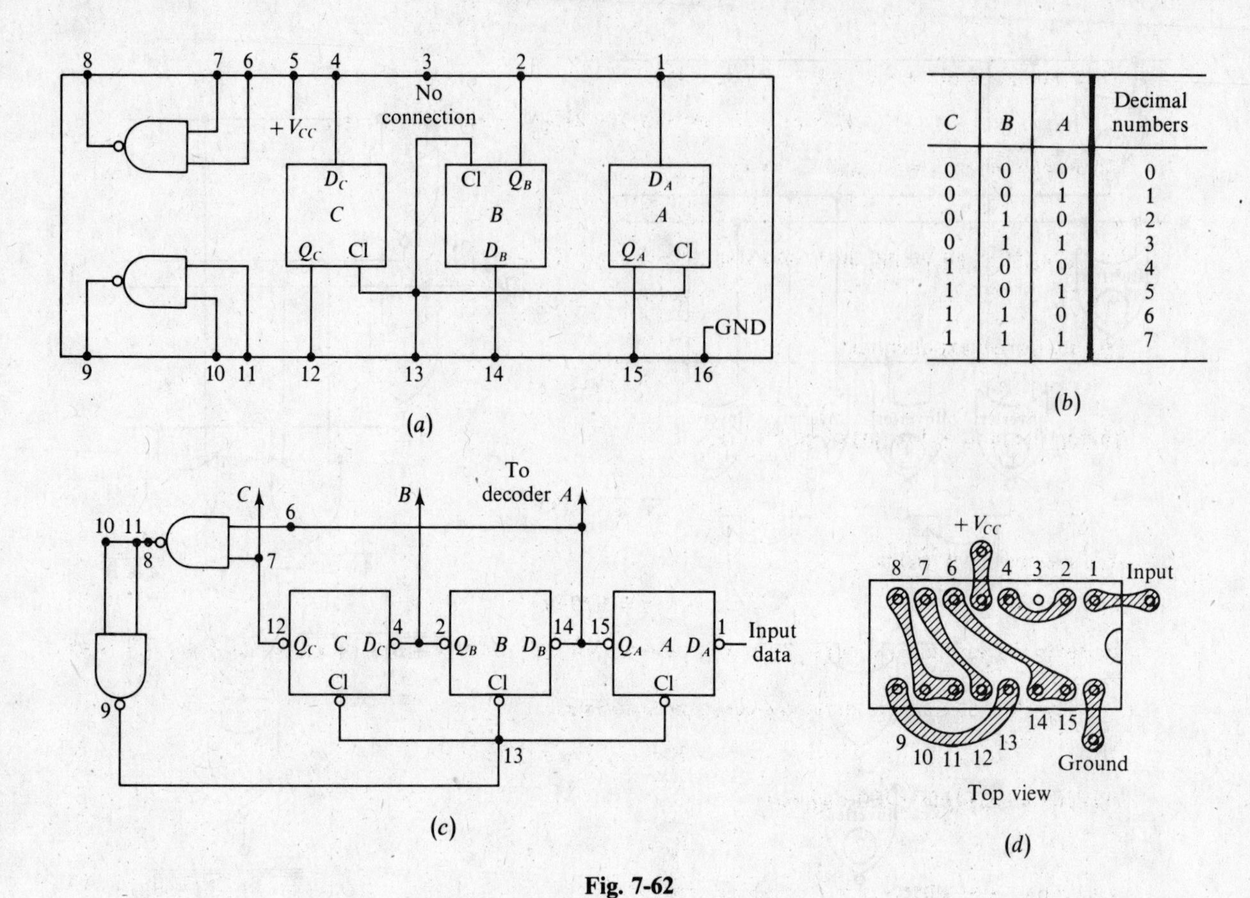

C	B	A	Decimal numbers
0	0	0	0
0	0	1	1
0	1	0	2
0	1	1	3
1	0	0	4
1	0	1	5
1	1	0	6
1	1	1	7

(b)

Fig. 7-62

7.8 A 555 timer as shown in Fig. 7-63*a* is a one-shot MV that charges an external capacitor C to $\frac{2}{3}V_{CC}$ for a length of time T when triggered by a pulse on pin 2. The voltage across the capacitor V_c differs from that of the regular type of one-shot MV, as shown by Fig. 7-63*b* and *c*. Find T for the 555 timer and the value of resistance R necessary for an ON time of 440 ms for the LED connected to the output.

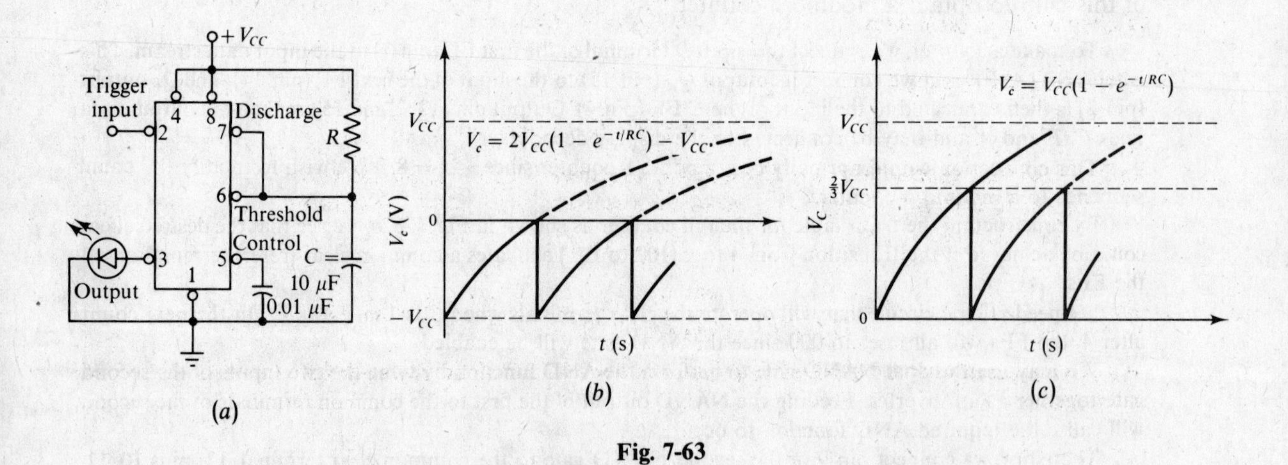

Fig. 7-63

For the regular MV shown in Fig. 7-63*b*, V_c is given by

$$V_c = 2V_{CC}(1 - e^{-t/RC}) - V_{CC}$$

The curve starts at $t = 0$ where $V_c = -V_{CC}$:

$$V_c = 2V_{CC}(1 - e^0) - V_{CC} = 2V_{CC}(1 - 1) - V_{CC} = -V_{CC}$$

and stops where $t = T$ and $V_c = 0$:

$$V_c = 2V_{CC}(1 - e^{-T/RC}) - V_{CC} = 0$$

Canceling V_{CC} and taking the exponential to the other side,

$$2e^{-T/RC} = 2 - 1 = 1$$

Solving for the exponential,

$$e^{-T/RC} = \tfrac{1}{2}$$

Taking the natural log (ln) of both sides,

$$-\frac{T}{RC} = 0.693$$

And solving for T yields

$$T = 0.693RC$$

which is the formula we normally use for a one-shot MV time constant, as shown in Table 7-9, when $V_{CC} \gg V_{BE} + V_{CES}$.

For the 555 timer V_c curve, the equation is given as

$$V_c = V_{CC}(1 - e^{-t/RC})$$

The curve starts at $t = 0$ where

$$V_c = V_{CC}(1 - e^0) = 0$$

The curve stops where $t = T$ and $V_c = \frac{2}{3}V_{CC}$:

$$V_c = V_{CC}(1 - e^{-T/RC}) = \tfrac{2}{3}V_{CC}$$

Canceling V_{CC} and removing parentheses,

$$\tfrac{2}{3} = 1 - e^{-T/RC}$$

Solving for the exponential,

$$e^{-T/RC} = 1 - \tfrac{2}{3} = \tfrac{1}{3}$$

Taking the ln of both sides, $$-\frac{T}{RC} = -1.098$$

And solving for T yields

$$T = 1.098RC = 1.1RC$$

which is the time constant formula provided by most manufacturers of the 555 timer.

Figure 7-63*a* shows $C = 10\ \mu\text{F}$, so we need to solve for R when T is 440 ms:

$$T = 1.1RC$$

So $$R = \frac{T}{(1.1)(C)} = \frac{440 \times 10^{-3}}{(1.1)(10 \times 10^{-6})} = 40.0 \times 10^3\ \Omega = 40.0\ \text{k}\Omega$$

We could probably use the standard 10 percent resistor value of 39 kΩ unless the time the LED is to remain on is critical.

7.9 The circuit shown in Fig. 7-64*a* consists of *T* FFs. Show what happens to the output data lines *C*, *B*, and *A* for the first 12 input pulses. Draw the timing diagram with the assumption that the *T* FFs will only change state on an input change from 1 to 0 and that the FFs are all in the 1 state before the first input pulse is applied.

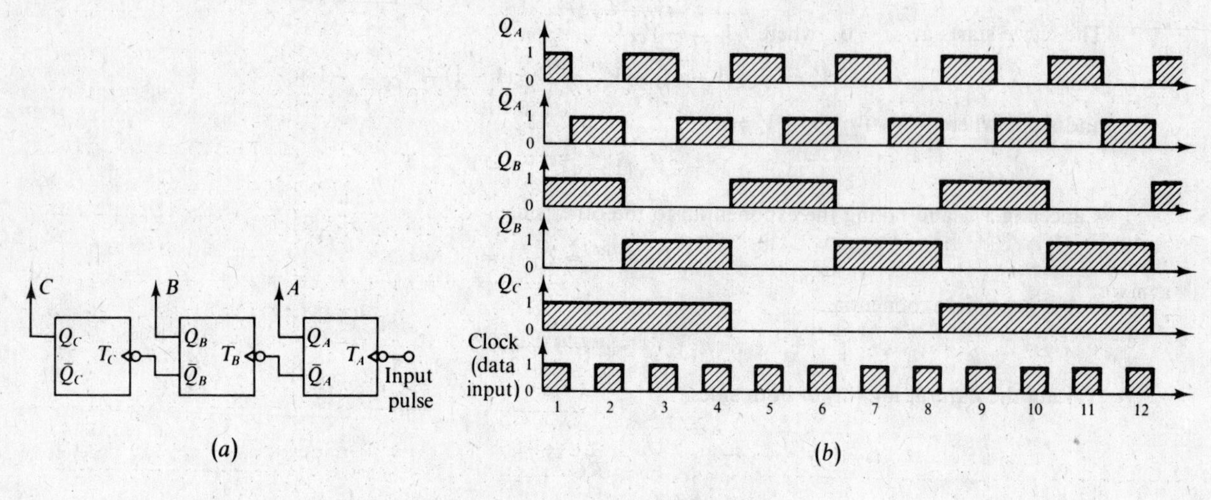

(After) Pulse number	C	B	A	Output Data Decimal numbers
1	1	1	0	6
2	1	0	1	5
3	1	0	0	4
4	0	1	1	3
5	0	1	0	2
6	0	0	1	1
7	0	0	0	0
8	1	1	1	7
9	1	1	0	6
10	1	0	1	5
11	1	0	0	4
12	0	1	1	3

(c)

Fig. 7-64

Looking at the Q_A and $\bar{Q}_A$ outputs, before the first input pulse Q_A is 1 and $\bar{Q}_A$ is 0. On the trailing edge of the first pulse, FF A changes state so that $\bar{Q}_A$ is 1 and Q_A is 0. The input to FF B is connected to $\bar{Q}_A$, so it does not see a change from 1 to 0 and remains in its initial state of 1 until after the second pulse when Q_A changes back to 1 and $\bar{Q}_A$ flops to 0.

The input to FF C is connected to $\bar{Q}_B$ and so it does not see a transition from 1 to 0 until after the fourth pulse when Q_B goes back to 1 and $\bar{Q}_B$ flops to 0.

After the seventh pulse all the FF outputs are in the 0 state and, on the trailing edge of the eighth pulse, the FF outputs return to 1 and begin the count again.

The timing diagram is shown in Fig. 7-64*b* and the truth table for this sequence is shown in Fig. 7-64*c*. Looking at the output data line in the truth table, we see that this circuit functions as a modulo 8 downcounter recognizing the numbers 7 to 0 (111 to 000).

7.10 Show the portion of a BCH decoder to a seven-segment display for the decimal numbers 10 to 15. Assume that a display segment is on when its logic state is 0 and off when its logic state is 1.

The truth table for the BCH and HEX equivalents is shown in Fig. 7-65*a* and the seven-segment truth table for the HEX numbers of interest is shown in Fig. 7-65*b* along with the segment displays.

Decimal numbers	w_8	w_4	w_2	w_1	Hexadecimal numbers
10	1	0	1	0	*A*
11	1	0	1	1	*B*
12	1	1	0	0	*C*
13	1	1	0	1	*D*
14	1	1	1	0	*E*
15	1	1	1	1	*F*

(*a*)

Decimal numbers	Display	*a*	*b*	*c*	*d*	*e*	*f*	*g*
10		0	0	0	1	0	0	0
11		1	1	0	0	0	0	0
12		0	1	1	0	0	0	1
13		1	0	0	0	0	1	0
14		0	1	1	0	0	0	0
15		0	1	1	1	0	0	0

(*b*)

Decimal numbers	Display (Hexadecimal numbers)	Inputs				Outputs						
		w_8	w_4	w_2	w_1	*a*	*b*	*c*	*d*	*e*	*f*	*g*
10	*A*	1	0	1	0	0	0	0	1	0	0	0
11	*B*	1	0	1	1	1	1	0	0	0	0	0
12	*C*	1	1	0	0	0	1	1	0	0	0	1
13	*D*	1	1	0	1	1	0	0	0	0	1	0
14	*E*	1	1	1	0	0	1	1	0	0	0	0
15	*F*	1	1	1	1	0	1	1	1	0	0	0

(*c*)

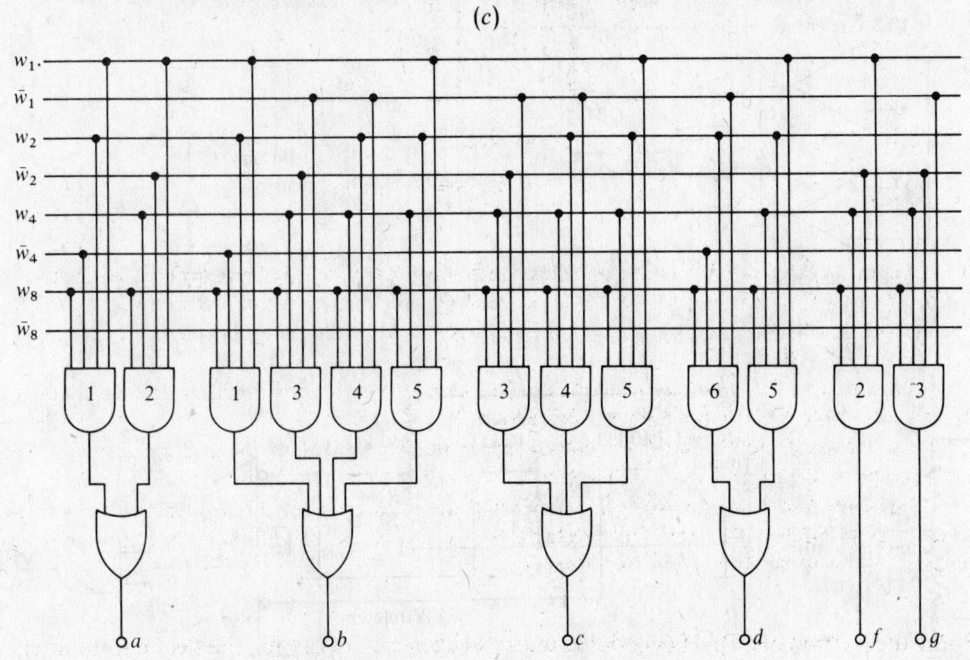

Gates with same number could be fanned out

(*d*)

Fig. 7-65

Looking at the truth table constructed in Fig. 7-65*b* we see that segment *e* is on at all times for the numbers 10 through 15 (*A* through *F*) and will not need a gate. However, segment *e* would need a buffer amplifier to handle the current drawn when the segment is on.

Combining the two truth tables as shown in Fig. 7-65*c* to show the relationship between the segments and lines w_8, w_4, w_2, and w_1, we find, for example, that segment *a* is 1 for decimal numbers 11 and 13. For 11 we have 1011 and for 13 we have 1101 as the states of the *w* lines. This means that the Boolean sum of products form for segment $a = 1$ is

$$a = w_8 \cdot \bar{w}_4 \cdot w_2 \cdot w_1 + w_8 \cdot w_4 \cdot \bar{w}_2 \cdot w_1$$

$$b = w_8 \cdot \bar{w}_4 \cdot w_2 \cdot w_1 + w_8 \cdot w_4 \cdot \bar{w}_2 \cdot \bar{w}_1 + w_8 \cdot w_4 \cdot w_2 \cdot \bar{w}_1 + w_8 \cdot w_4 \cdot w_2 \cdot w_1$$

$$c = w_8 \cdot w_4 \cdot \bar{w}_2 \cdot \bar{w}_1 + w_8 \cdot w_4 \cdot w_2 \cdot \bar{w}_1 + w_8 \cdot w_4 \cdot w_2 \cdot w_1$$

$$d = w_8 \cdot \bar{w}_4 \cdot w_2 \cdot \bar{w}_1 + w_8 \cdot w_4 \cdot w_2 \cdot w_1$$

$$e = \text{ON at all times, never 1}$$

$$f = w_8 \cdot w_4 \cdot \bar{w}_2 \cdot w_1$$

$$g = w_8 \cdot w_4 \cdot \bar{w}_2 \cdot \bar{w}_1$$

Although some of the expressions can be simplified by the use of the Boolean identities listed in Table 7-5, it is simpler to construct the gating diagram from the expressions just written. This diagram is shown in Fig. 7-65*d*.

Segment *d* could be realized with an XNOR gate with inputs w_1 and w_4 feeding an AND gate with w_2 and w_8, but this saves only one gate from the arrangement shown in Fig. 7-65*d* for segment *d* and is not usually worth the trouble unless there is a constraint in the number or type of gates available.

7.11 A battery alarm system is to be built using op amp comparators. The system should light an LED when the battery terminal voltage goes above 13.6 V or below 10.2 V. Draw the circuit.

This is a practical application of the window comparator, but the reference voltages may not be readily available. If they are, the circuit shown in Fig. 7-66*a* will do the job. If no reference voltages are available, we can create our own by using zener diodes and voltage dividers.

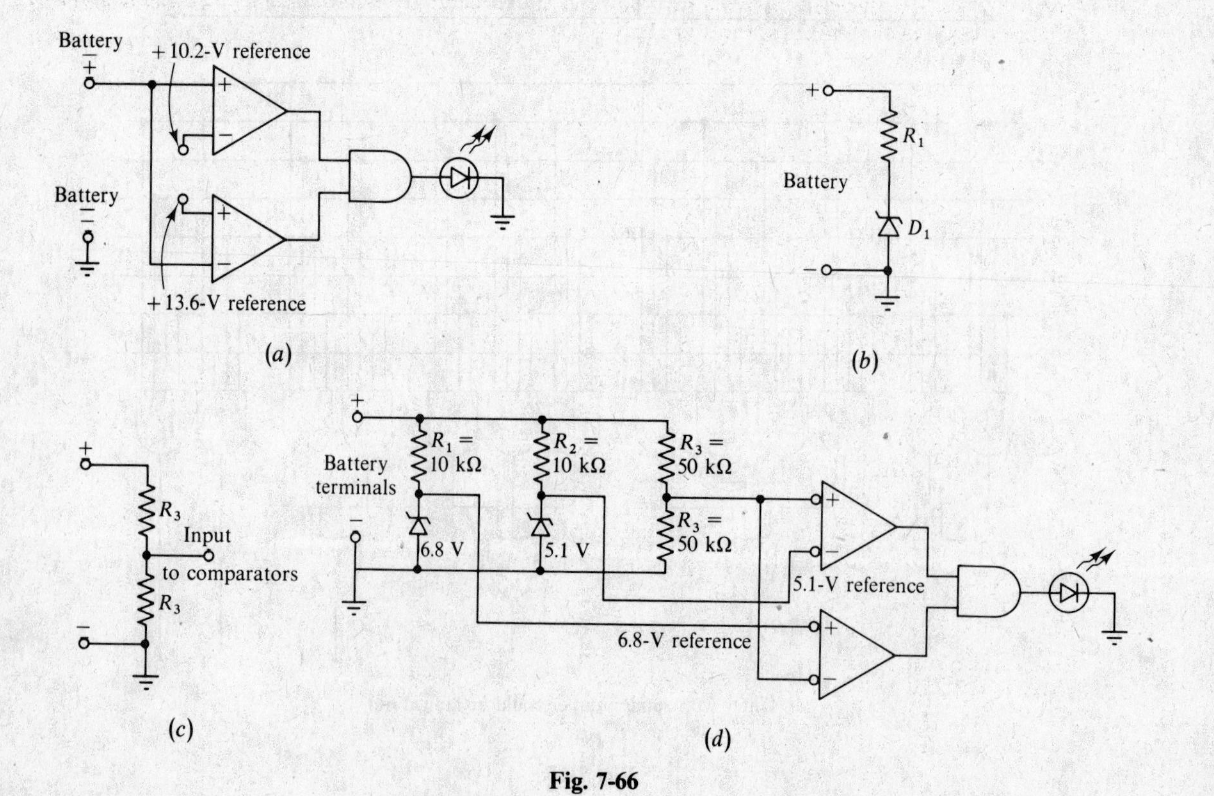

Fig. 7-66

The zener diode in series with a resistor across the battery terminals as shown in Fig. 7-66*b* will provide a good reference as long as the series resistor is capable of handling the zener breakdown current. The references must be lower than the terminal voltage of the battery in order to achieve zener action, so we need to use a voltage divider to provide the input for comparison with the reference.

If we use two precision resistors in series across the battery, as shown in Fig. 7-66*c*, we will provide the input of our comparator with one-half the terminal voltage even when that voltage varies.

When the battery is at its upper limit, the input to the comparator will be $\frac{1}{2}(13.6) = 6.8$ V, so we need a 6.8-V reference for the high reference value. When the battery is at its lower limit, the input to the comparator is $\frac{1}{2}(10.2) = 5.1$ V, so we need a 5.1-V reference for the low reference value.

The voltage divider in this case may consist of any two equal resistances or a potentiometer, as long as the total resistance across the battery is high enough to prevent loading and low enough to provide an input to the op amp. In this case, 50 kΩ has been selected as a representative value.

The desired circuit is shown in Fig. 7-66*d* using the values discussed.

7.12 A 100-Hz sine wave, 16.0 V peak to peak, is sampled 1600 times per second as shown in Fig. 7-67*a*. If this wave is digitally coded with a 16-level ADC and then converted back to an analog voltage using the same coding scheme, compare the input and output waveforms. The coding is in offset binary so that the 0 level is coded as 1 0000 as shown in Fig. 7-67*b*. Suggest an improvement to this scheme.

The actual voltage values at the sampling points are charted in Fig. 7-67*c* using the value $V_1 = 8.00 \sin \theta$, where θ is a whole number multiple of 22.5°. For example, at sampling pulse 14,

$$\theta = 13(22.5) = 292.5° \quad \text{or} \quad -67.5°$$

and
$$V_1 = 8.00 \sin(-67.5) = (8.00)(-0.924) = -7.39 \text{ V}$$

The digitally quantized voltage V_L is equal to the actual voltage V_1, or is the next lowest step for positive voltages or the next highest step for negative voltages.

At sampling pulse 14, $V_1 = -7.39$. Since this is a negative voltage between quantized steps -7.00 and -7.50, V_L is the next highest step or -7.00 V.

Notice that in the offset binary code, $-V_{FS}$ has a coded value while V_{FS} does not. The highest possible value is V_{FS} − LSB. This gives rise to a nonsymmetrical output for the symmetrical input, and the graph of the output levels of the DAC is shown in Fig. 7-67*d*.

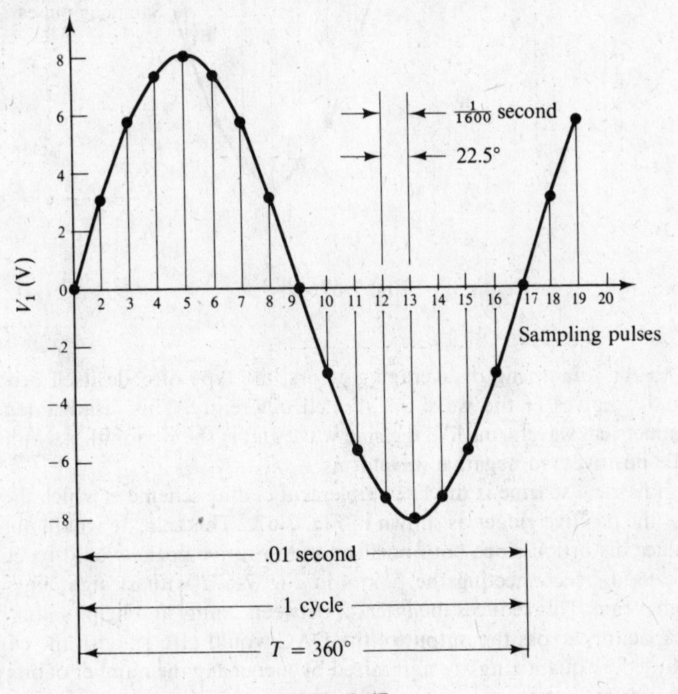

Fig. 7-67*a*

Offset binary code	Fraction	V_Q (V_{FS})(fraction)	Offset binary code	Fraction	V_Q (V_{FS})(fraction)
1 1111	$+\frac{15}{16}$	+7.5	0 1111	$-\frac{1}{16}$	−0.5
1 1110	$+\frac{7}{8}$	+7.0	0 1110	$-\frac{1}{8}$	−1.0
1 1101	$+\frac{13}{16}$	+6.5	0 1101	$-\frac{3}{16}$	−1.5
1 1100	$+\frac{3}{4}$	+6.0	0 1100	$-\frac{1}{4}$	−2.0
1 1011	$+\frac{11}{16}$	+5.5	0 1011	$-\frac{5}{16}$	−2.5
1 1010	$+\frac{5}{8}$	+5.0	0 1010	$-\frac{3}{8}$	−3.0
1 1001	$+\frac{9}{16}$	+4.5	0 1001	$-\frac{7}{16}$	−3.5
1 1000	$+\frac{1}{2}$	+4.0	0 1000	$-\frac{1}{2}$	−4.0
1 0111	$+\frac{7}{16}$	+3.5	0 0111	$-\frac{9}{16}$	−4.5
1 0110	$+\frac{3}{8}$	+3.0	0 0110	$-\frac{5}{8}$	−5.0
1 0101	$+\frac{5}{16}$	+2.5	0 0101	$-\frac{11}{16}$	−5.5
1 0100	$+\frac{1}{4}$	+2.0	0 0100	$-\frac{3}{4}$	−6.0
1 0011	$+\frac{3}{16}$	+1.5	0 0011	$-\frac{13}{16}$	−6.5
1 0010	$+\frac{1}{8}$	+1.0	0 0010	$-\frac{7}{8}$	−7.0
1 0001	$+\frac{1}{16}$	+0.5	0 0001	$-\frac{15}{16}$	−7.5
1 0000	0	±0	0 0000	$-\frac{16}{16}$	−8.0

(*b*)

Sampling point	V_1 (volts) θ	V_1 (volts)	V_L (volts)
1	0	0	0
2	22.5	3.06	3.0
3	45	5.66	5.5
4	67.5	7.39	7.0
5	90	8.00	7.5
6	112.5	7.39	7.0
7	135	5.66	5.5
8	157.5	3.06	3.0
9	180	0	0
10	202.5	−3.06	−3.0
11	225	−5.66	−5.5
12	247.5	−7.39	−7.0
13	270	−8.00	−8.0
14	292.5	−7.39	−7.0
15	315	−5.66	−5.5
16	337.5	−3.06	−3.0
17	360	0	0

(*c*)

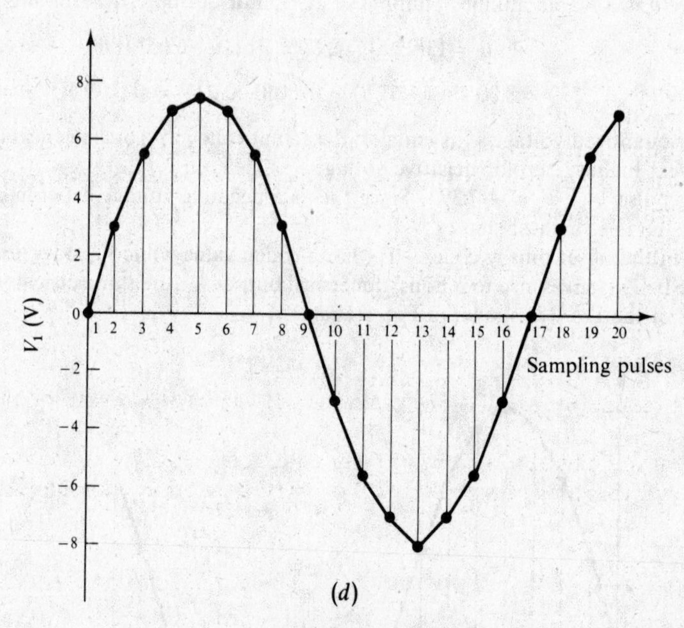

(*d*)

Fig. 7-67*b-d*

In addition to the quantizing or averaging errors, the type of code itself produces distortion as the positive and negative halves of the wave are treated differently. This disadvantage introduces harmonic distortion for symmetrical waveforms like the sine wave and is the reason that symmetrical coding schemes are used to handle positive and negative waveforms.

One such symmetrical scheme is the 1's complement coding scheme in which the negative values are the 1's complement of the positive values as shown in Fig. 7-67*e*. This scheme would improve the symmetry but would still introduce distortion since both positive and negative peaks would be clipped at V_{FS} − LSB.

The errors caused by connecting the points in Fig. 7-67*d* with straight lines can be minimized by sampling at a higher rate. This reduces the interval between points and helps smooth the curve. A low-pass filter, such as a capacitor, across the output of the DAC would also smooth the curve.

The errors caused by quantizing are minimized by increasing the number of bits used. This would allow more levels and better resolution.

The MSB used to indicate polarity in bipolar systems does decrease the resolution with respect to the same number of bits in a unipolar system.

1's Complement code	Fraction	1's Complement code	Fraction
0 1111	$+\frac{15}{16}$	1 1111	-0
0 1110	$+\frac{7}{8}$	1 1110	$-\frac{1}{16}$
0 1101	$+\frac{13}{16}$	1 1101	$-\frac{1}{8}$
0 1100	$+\frac{3}{4}$	1 1100	$-\frac{3}{16}$
0 1011	$+\frac{11}{16}$	1 1011	$-\frac{1}{4}$
0 1010	$+\frac{5}{8}$	1 1010	$-\frac{5}{16}$
0 1001	$+\frac{9}{16}$	1 1001	$-\frac{3}{8}$
0 1000	$+\frac{1}{2}$	1 1000	$-\frac{7}{16}$
0 0111	$+\frac{7}{16}$	1 0111	$-\frac{1}{2}$
0 0110	$+\frac{3}{8}$	1 0110	$-\frac{9}{16}$
0 0101	$+\frac{5}{16}$	1 0101	$-\frac{5}{8}$
0 0100	$+\frac{1}{4}$	1 0100	$-\frac{11}{16}$
0 0011	$+\frac{3}{16}$	1 0011	$-\frac{3}{4}$
0 0010	$+\frac{1}{8}$	1 0010	$-\frac{13}{16}$
0 0001	$+\frac{1}{16}$	1 0001	$-\frac{7}{8}$
0 0000	$+0$	1 0000	$-\frac{15}{16}$

Fig. 7-67*e*

7.13 Design an algorithm that will allow a microprocessor to add two numbers in BCD form and store the result. Set up as many registers as you need. Draw the flowchart for the algorithm and use the decimal numbers 837 and 414 to demonstrate the procedure.

If we try to add two BCD numbers by using the binary rules for addition, we encounter some difficulty. For example, when we add 8 and 3 in BCD form we have

$$\begin{array}{r} 1000 \\ +\,0011 \\ \hline 1011 \end{array}$$

The result, 1011, although correct in HEX, has no meaning in BCD, which is limited to 9 or 1001 as its highest character.

To surmount this difficulty, we may add 6 or 0110 to any result over 9 to throw the result into a HEX carry that will give the correct result. Since 1011 is greater than 9 in our example, we add 0110 to obtain

$$\begin{array}{r} 1011 \\ +\,0110 \\ \hline 1\ 0001 \end{array}$$

This is 11 in BCD and the correct result.

The algorithm will make use of this fact and add each BCD digit separately and set a carry flag for the next digit.

The algorithm first needs to be stated procedurally and then converted to flowchart form. The first step is to start the algorithms by taking care of any initializing that needs to be done such as clearing registers and setting flags. Next, we need to convert the given numbers to their BCD form. We may then add the least-significant nibbles and save the carry for the next addition. The next nibbles are then added, taking into account the previous carry, and this procedure loops until the final nibble is encountered. The last carry, if any, becomes the MSB of the result, and all results are then stored in memory.

If we were actually performing this routine for a microprocessor, we would need to set aside the proper amount of memory to be able to process this algorithm. In general, we would allocate one more nibble in the storage memory than the maximum nibbles for the original numbers.

The flowchart is shown in Fig. 7-68*a*, where each block is a subroutine and would need to be expanded if an actual assembly language program were to be written. The results of the flowchart process are listed in Fig. 7-68*b* using 837 and 414 for the two input numbers.

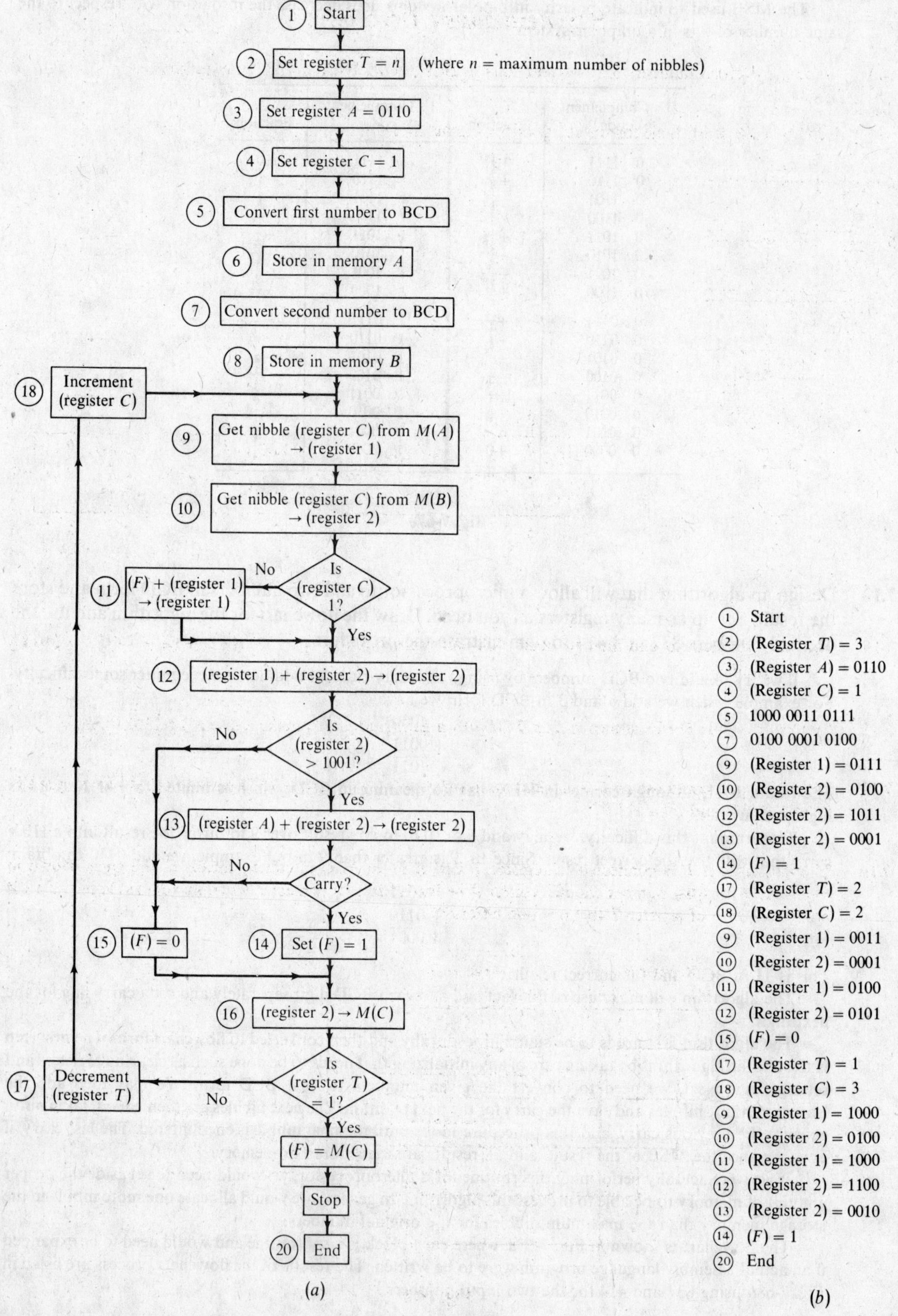

1 Start
2 Set register $T = n$ (where n = maximum number of nibbles)
3 Set register $A = 0110$
4 Set register $C = 1$
5 Convert first number to BCD
6 Store in memory A
7 Convert second number to BCD
8 Store in memory B
18 Increment (register C)
9 Get nibble (register C) from $M(A)$ → (register 1)
10 Get nibble (register C) from $M(B)$ → (register 2)
Is (register C) 1?
No
Yes
11 (F) + (register 1) → (register 1)
12 (register 1) + (register 2) → (register 2)
Is (register 2) > 1001?
No
Yes
13 (register A) + (register 2) → (register 2)
Carry?
No
Yes
15 $(F) = 0$
14 Set $(F) = 1$
16 (register 2) → $M(C)$
Is (register T) = 1?
No
Yes
17 Decrement (register T)
$(F) = M(C)$
Stop
20 End
(a)
1 Start
2 (Register T) = 3
3 (Register A) = 0110
4 (Register C) = 1
5 1000 0011 0111
7 0100 0001 0100
9 (Register 1) = 0111
10 (Register 2) = 0100
12 (Register 2) = 1011
13 (Register 2) = 0001
14 $(F) = 1$
17 (Register T) = 2
18 (Register C) = 2
9 (Register 1) = 0011
10 (Register 2) = 0001
11 (Register 1) = 0100
12 (Register 2) = 0101
15 $(F) = 0$
17 (Register T) = 1
18 (Register C) = 3
9 (Register 1) = 1000
10 (Register 2) = 0100
11 (Register 1) = 1000
12 (Register 2) = 1100
13 (Register 2) = 0010
14 $(F) = 1$
20 End
(b)

Fig. 7-68

Supplementary Problems

7.14 Write the decimal numbers 327, 94, and 1039 in BCD, octal, HEX, and BCH notation.

7.15 Construct the truth table for output Q of the gating circuit shown in Fig. 7-69.

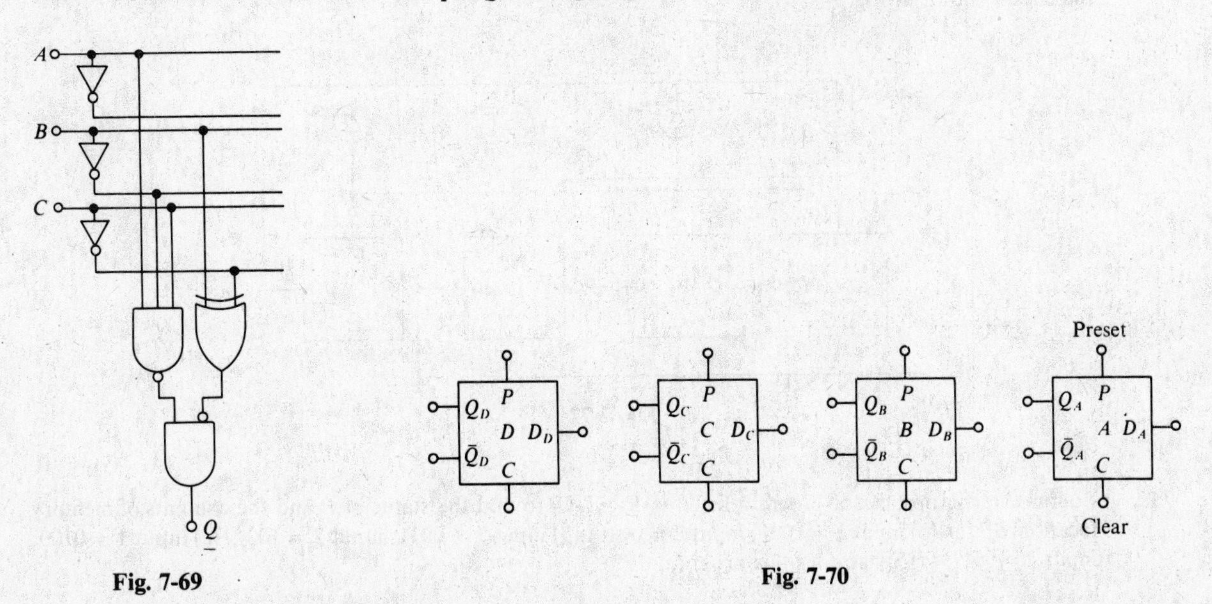

Fig. 7-69 **Fig. 7-70**

7.16 Find the complement and the equivalent Boolean expression for $Q = A \cdot B \cdot \bar{C} + (A + \bar{B})(C + \bar{A})$ using De Morgan's law.

7.17 Connect the four D FFs shown in Fig. 7-70 with a gating circuit to make a divide by 12 counter.

7.18 Show how you could use an R-S clocked master-slave (M-S) FF and NAND gates to construct (a) a D FF, (b) a J-K FF, and (c) a T FF.

7.19 The flag register F is connected to register T and R in an ALU circuit by means of the gates shown in Fig. 7-71. With what number should register R be loaded so that register F will only be set at logic 1 if a 1 is present at bit 3 of register T (bit position circled)?

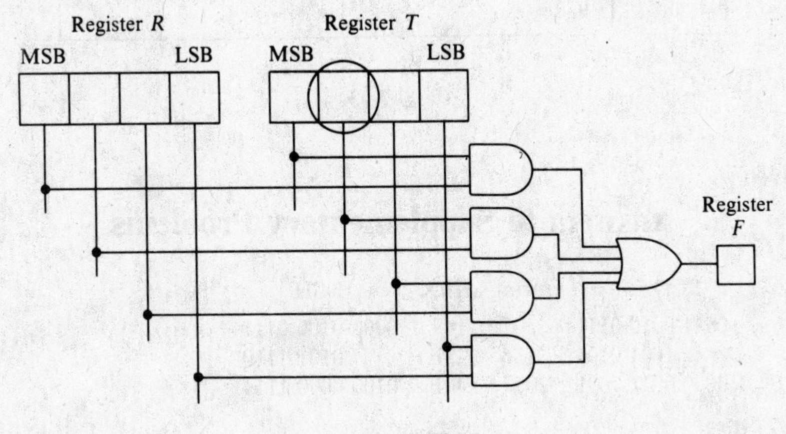

Fig. 7-71

7.20 Draw the diagram of a gating circuit that compares two bits and will light an LED when the bits are not the same.

7.21 The LEDs shown in Fig. 7-72 are on when their data lines are in the 1 state. Assume the FFs were all in the 0 state before the first input pulse. (*a*) Which LEDs are on after the twelfth pulse? (*b*) What is the function of the circuit shown in Fig. 7-72?

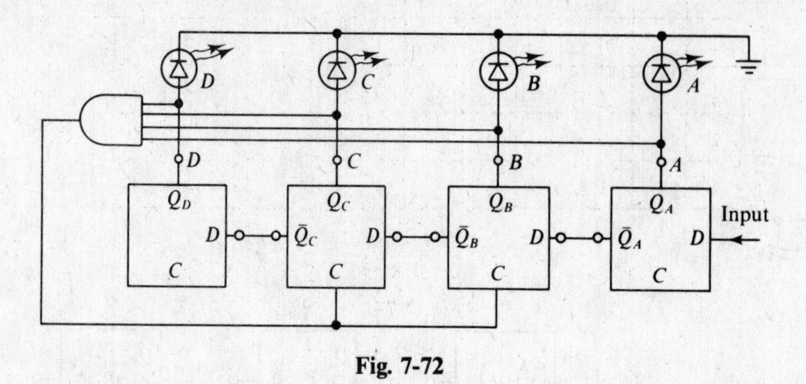

Fig. 7-72

7.22 Follow the microprocessor program listed in Fig. 7-73 to find the status of *F* and the contents of memory location *BB* for (*a*) input 1 = 1001; input 2 = 0010; (*b*) input 1 = 0011; input 2 = 1001; (*c*) input 1 = 0100; input 2 = 0111. (*d*) What does this program do?

Program counter	Instructions
2000	Load input No. 1 into register 1
2001	Load input No. 2 into register 2
2002	Load 1111 into register 3
2003	If (Register 1) > (Register 2) go to program counter 2008
2004	(Register 1) → (Register *T*)
2005	(Register 2) → (Register 1)
2006	(Register *T*) → (Register 2)
2007	Set register *F* = 1
2008	(Register 2) XOR (Register 3) → (Register 2)
2009	Increment register 2
2010	(Register 1) + (Register 2) → (Register 1)
2011	(Register 1) → *M*(*BB*)
2012	Stop
2013	End

Fig. 7-73

Answers to Supplementary Problems

7.14

Decimal	BCD	Octal	HEX	BCH
327	0011 0010 0111	507	147	0001 0100 0111
94	1001 0100	136	5E	0101 1110
1039	0001 0000 0011 1001	2017	40F	0100 0000 1111

7.15 The resulting truth table is shown in Fig. 7-74.

A	B	C	$\bar{A}$	$\bar{B}$	$\bar{C}$	P_1	$\bar{P}_1$	S_1	$\bar{S}_1$	Q
0	0	0	1	1	1	0	1	1	0	0
0	0	1	1	1	0	0	1	0	1	1
0	1	0	1	0	1	0	1	0	1	1
0	1	1	1	0	0	0	1	1	0	0
1	0	0	0	1	1	0	1	1	0	0
1	0	1	0	1	0	1	0	0	1	0
1	1	0	0	0	1	0	1	0	1	1
1	1	1	0	0	0	0	1	1	0	0

$P_1 = A \cdot \bar{B} \cdot C \quad S_1 = B \text{ XOR } \bar{C} \quad Q = \bar{P}_1 \cdot \bar{S}_1$

Fig. 7-74

7.16

$$\bar{Q} = (\bar{A} + \bar{B} + C)(\bar{A} \cdot B + \bar{C} \cdot A)$$
$$Q = \overline{(\bar{A} + \bar{B} + C)(\bar{A} \cdot B + \bar{C} \cdot A)}$$

7.17 The required diagram is shown in Fig. 7-75.

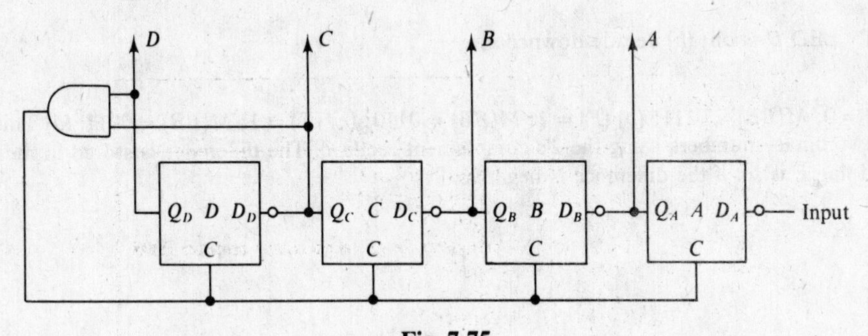

Fig. 7-75

7.18 The connections are shown in Fig. 7-76*a*, *b*, and *c*.

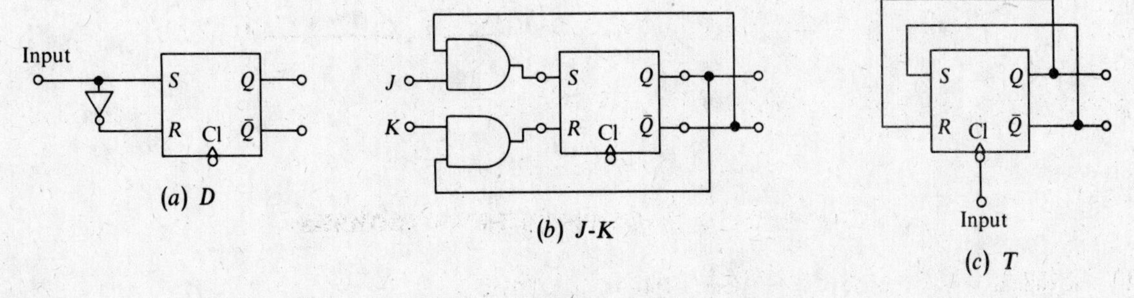

(*a*) *D* (*b*) *J-K* (*c*) *T*

Fig. 7-76

7.19 (Register R) = 0100.

7.20 The preferred arrangement is shown in Fig. 7-77*a*, with alternates shown in Fig. 7-77*b* and *c*.

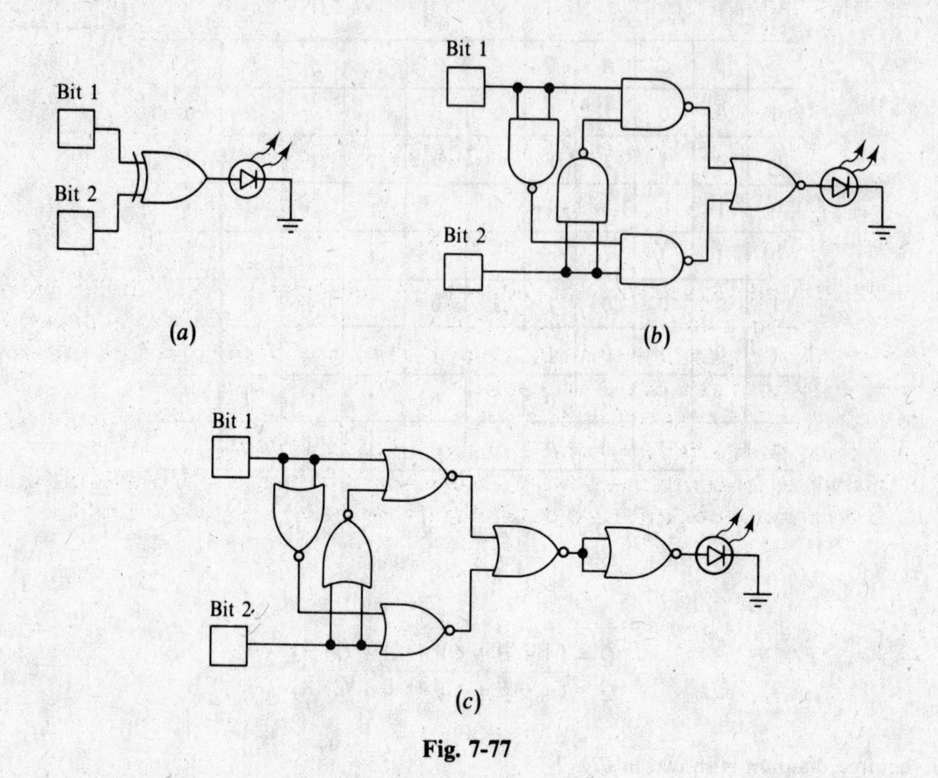

Fig. 7-77

7.21 (*a*) Only LED *D* is on; (*b*) decade downcounter.

7.22 (*a*) $(F) = 0$; $M(BB) = 0111$; (*b*) $(F) = 1$; $M(BB) = 0110$; (*c*) $(F) = 1$; $M(BB) = 0011$. (*d*) This routine subtracts two binary numbers using the 2's complement method. The difference is stored in memory location *BB* and flag *F* is set if the difference is negative.

Chapter 8

Power Supplies

8.1 INTRODUCTION

Electronic amplifying devices require a dc voltage to get them into operation. The required operating voltage usually comes from a dc *power supply.* This is an unfortunate term because it does not actually supply the power in most applications. Instead, it modifies existing power in some way. (An exception to this would be a battery used as a dc power supply.)

There are ac power supplies as well as dc supplies. They also convert existing power for use in some particular application. An *inverter,* for example, converts dc into ac power.

The four possibilities for converting power from one form to another are illustrated in Fig. 8-1. Each has applications in electronics, but the ac-to-dc rectifier type is the most popular and will be emphasized in this chapter.

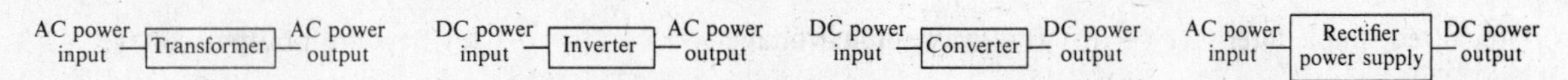

Fig. 8-1

8.2 UNREGULATED RECTIFIER POWER SUPPLIES

Figure 8-2 shows a block diagram of an unregulated supply. The power source is usually the power company that supplies 60-Hz ac power to homes and industry. However, mobile systems that use ac power—like much of the larger aircraft—may supply the power from an alternator.

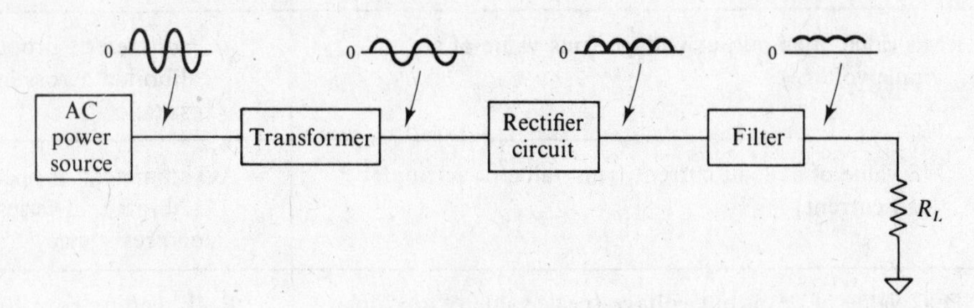

Fig. 8-2

The power transformer steps the input voltage up or down as needed to obtain the desired output voltage. It also serves to isolate the load from the ac power line. Not all rectifier power supplies have transformers. Transformerless types are sometimes called *ac-dc supplies* because they will operate with either type of input power.

The rectifier section converts the ac to a pulsating dc and may consist of tubes, semiconductor diodes, or SCRs (silicon controlled rectifiers).

The filter shown in Fig. 8-2 is a low-pass type used for removing the voltage variations in the rectifier output. Both passive and active types are used.

The load resistance of the supply may consist of tubes, transistors, or FETs. Do not confuse the terms " load " and " load resistance." The *load* of a power supply is the amount of current that it must deliver. The load resistance is the opposition to the power-supply current.

8.3 SPECIFICATIONS FOR RECTIFIER SUPPLIES

The output voltage waveform shown across R_L in Fig. 8-2 is not pure direct current. It can be considered to be a dc and ac voltage combined. Manufacturer's specifications describe the amount of dc voltage and current provided by the supply, and the maximum ac allowed in the output. The definitions for the dc and ac voltages and currents usually encountered are listed in Table 8-1 and shown graphically for an idealized filter circuit in Fig. 8-3.

The dc and ac components are related by the equations listed in Table 8-2.

Most rectifier circuits are designed for operation with a sinusoidal input. The average and effective values of the output waveform are useful in determining rectifier characteristics. Table 8-3 lists the average and effective values and form factor for common rectifier outputs.

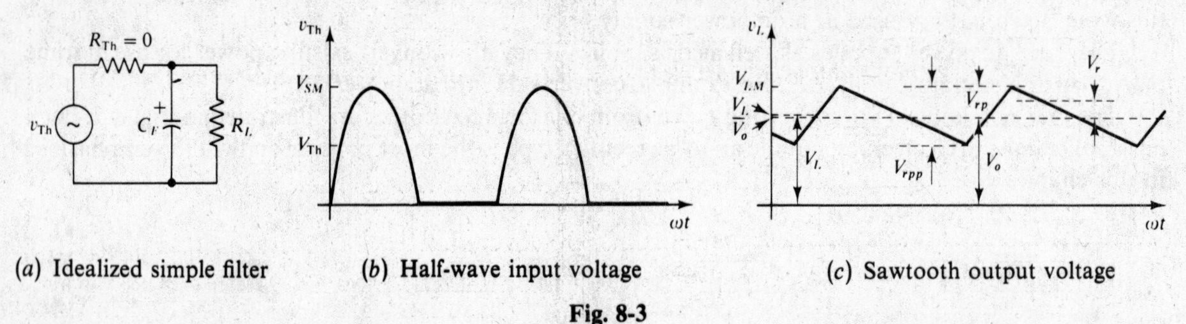

(*a*) Idealized simple filter (*b*) Half-wave input voltage (*c*) Sawtooth output voltage

Fig. 8-3

Table 8-1

Term	Meaning	Measurement
V_o	Average value of output voltage (dc value of output voltage)	DC voltmeter across load resistance
I_o	Average value of load current (dc value of load current)	DC ammeter in series with load resistance
V_r (V_a) (V_{ac})	RMS value of ac output voltage (rms value of ac ripple voltage)	AC voltmeter,* properly calibrated across load resistance
I_r (I_a) (I_{ac})	RMS value of ac load current (rms value of ac ripple load current)	AC ammeter,* properly calibrated, in series with load resistance
V_{rp}	Peak value of ac output voltage (peak value of ac ripple voltage)	Peak reading ac voltmeter* across load resistance
I_{rp}	Peak value of ac load current (peak value of ac ripple load current)	Peak reading ac ammeter* in series with load resistance
V_{rpp}	Peak-to-peak value of ac ripple voltage	Scaled from oscilloscope trace
I_{rpp}	Peak-to-peak value of ac ripple load current	Scaled from oscilloscope trace calibrated to read current
V_L	Total rms output voltage	True rms voltmeter across load resistance
I_L	Total rms load current	True rms ammeter in series with load resistance

Table 8-1 (continued)

Term	Meaning	Measurement
v_L	Actual output voltage	Oscilloscope trace
i_L	Actual load current	Oscilloscope trace calibrated to read current
V_{LM}	Peak value of output voltage (maximum possible output voltage)	Peak reading voltmeter or oscilloscope*
I_{LM}	Peak value of output load current (maximum current that rectifier can handle)	Peak reading ammeter or oscilloscope scaled to read current*
V_{sm}	Peak value of transformer secondary voltage	Peak reading voltmeter or oscilloscope*
v_{Th}	Instantaneous Thevenin voltage	Theoretical
V_{Th}	RMS Thevenin voltage	Theoretical

* The ac meters should incorporate a blocking capacitor, while the peak-reading meters should read the peak of the entire waveform.

Table 8-2

Term	Symbol	Equation	Comment
Ripple factor	$FR =$	$\dfrac{V_r}{V_0} = \dfrac{I_r}{I_o}$ (*8-1*)	0 for ideal case (pure dc)
Peak ripple factor	$FR_p =$	$\dfrac{V_{rp}}{V_o} = \dfrac{I_{rp}}{I_o}$ (*8-2*)	
	$FR^2 =$	$FF^2 - 1$ (*8-3*)	
Percent ripple		$FR \times 100$ (*8-4*)	
Percent peak ripple		$FR_p \times 100$ (*8-5*)	
(ac) Ripple voltage, peak-to-peak	$V_{rpp} =$	$2V_{rp}$ (*8-6a*)	Symmetrical wave-forms only
(ac) Ripple current, peak-to-peak	$I_{rpp} =$	$2I_{rp}$ (*8-6b*)	
Form factor	$FF =$	$\dfrac{V_L}{V_o} = \dfrac{I_L}{I_o}$ (*8-7*)	1 for ideal case (pure dc)
	$FF^2 =$	$FR^2 + 1$ (*8-8*)	
Total output voltage, peak	$V_{LM} =$	$V_o + V_{rp}$ (*8-9a*)	
Total output current, peak	$I_{LM} =$	$I_o + I_{rp}$ (*8-9b*)	

Table 8-2 (*continued*)

Term	Symbol		Equation		Comment
Efficiency	η	=	$\left(\frac{P_o}{P_a}\right)(100) = \left(\frac{P_o}{P_{in}}\right)(100)$	(*8-10a*)	100% for ideal case (pure dc)
		=	$RR \times 100$	(*8-10b*)	
DC output power	P_o	=	$V_o I_o$	(*8-11*)	
AC input power	P_a, P_{in}	=	$V_{in} I_{in}$	(*8-12*)	
Rectification ratio	RR	=	$\frac{P_o}{P_a} = \frac{P_o}{P_{in}}$	(*8-13*)	1 for ideal case (pure dc)
	V_L^2	=	$V_o^2 + V_r^2$	(*8-14a*)	
	I_L^2	=	$I_o^2 + I_r^2$	(*8-14b*)	

Table 8-3

Type (see Fig. 8-4)	RMS value (I or V)	Average value (I_o or V_o)	Form factor FF
Unrectified sine wave	$\frac{\sqrt{2}}{2} I_m = 0.707 I_m$ $\frac{\sqrt{2}}{2} V_m = 0.707 V_m$	$\frac{2}{\pi} I_m = 0.636 I_m$ $\frac{2}{\pi} V_m = 0.636 V_m$ (Half-cycle average)	
Half-wave rectified sine wave	$0.500 I_m$ $0.500 V_m$	$\frac{1}{\pi} I_m = 0.318 I_m$ $\frac{1}{\pi} V_m = 0.318 V_m$	1.57
Full-wave rectified sine wave	$\frac{\sqrt{2}}{2} I_m = 0.707 I_m$ $\frac{\sqrt{2}}{2} V_m = 0.707 V_m$	$\frac{2}{\pi} I_m = 0.636 I_m$ $\frac{2}{\pi} V_m = 0.636 V_m$	1.11
Sawtooth wave	$\frac{\sqrt{3}}{3} I_m = 0.577 I_m$ $\frac{\sqrt{3}}{3} V_m = 0.577 V_m$	$0.500 I_m$ $0.500 V_m$	1.15

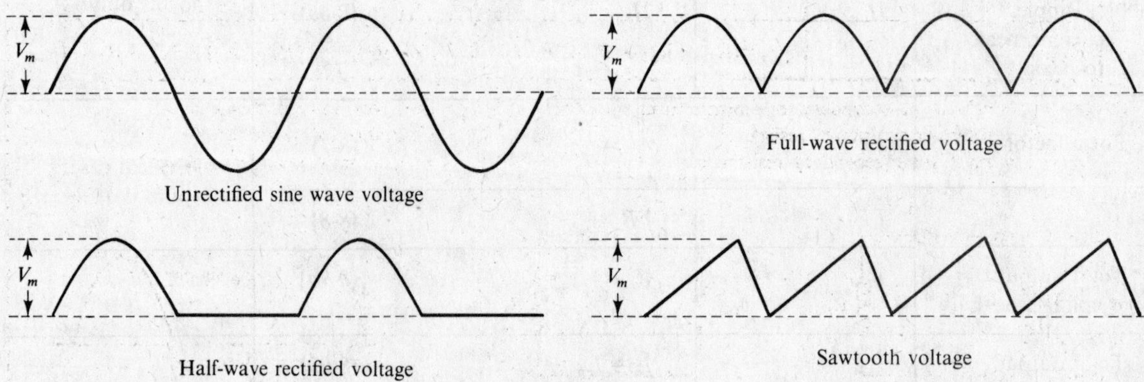

Fig. 8-4

8.4 SINGLE-PHASE RECTIFIER CONFIGURATIONS

Rectifiers are classified as being *half-wave* types if they produce an output voltage or current during one-half of the ac input cycle. If they produce an output during *both* halves of the cycle, they are *full-wave* types.

A *bridge* rectifier is a full-wave type that does not require a transformer if the ac input does not need to be stepped up or down.

Table 8-4 compares single-phase rectifiers and includes the effect of 50 percent filtering.

Table 8-4 Single-phase rectifier comparisons

Parameter	Half-wave	Full wave	Bridge	Filtered (50%) half-wave
$V_{Th} = V_{sm}$ times	0.500	0.707	0.707	0.764
$V_o = \dfrac{V_{sm}R_L}{R_{Th}+R_L}$ times	0.318	0.636	0.636	0.750
$V_L = \dfrac{V_{sm}R_L}{R_{Th}+R_L}$ times	0.500	0.707	0.707	0.764
$V_r = \dfrac{V_{sm}R_L}{R_{Th}+R_L}$ times	0.386	0.308	0.308	0.144
$FF = V_L/V_o =$	1.57	1.11	1.11	1.02
Ripple factor $= V_r/V_o =$	1.21	0.483	0.483	0.192
$P_o = \dfrac{V_{sm}^2 R_L}{(R_{Th}+R_L)^2}$ times	0.101	0.405	0.405	0.563
$P_a = \dfrac{V_{sm}^2}{R_{Th}+R_L}$ times	0.250	0.500	0.500	0.584
$\eta = \dfrac{R_L}{R_{Th}+R_L}$ times	40.5%	81.1%	81.1%	96.4%
PIV per diode $= V_{sm}$ times	1.0	2.0	1.0	2.0
Ripple frequency = input frequency f times	1.0	2.0	2.0	1.0
Diagrams	Fig. 8-5	Fig. 8-6	Fig. 8-7	Fig. 8-8
$R_{Th} =$	$n^2(R_p + R_g) + R_s + R_{FD}$		$R_g + 2R_{FD}$	$n^2(R_p + R_g) + R_s + R_{FD}$

Note: n = primary-to-secondary transformer turns ratio ($n:h$)
R_p = transformer primary resistance
R_s = transformer secondary resistance
R_g = generator internal resistance
R_{FD} = diode forward resistance
V_{sm} = peak value of secondary voltage

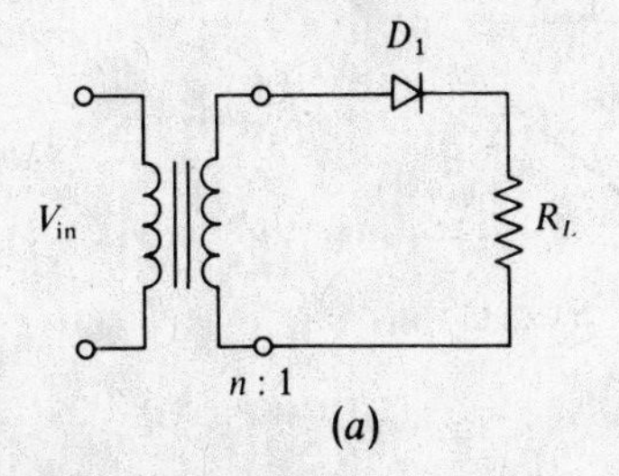

(a)

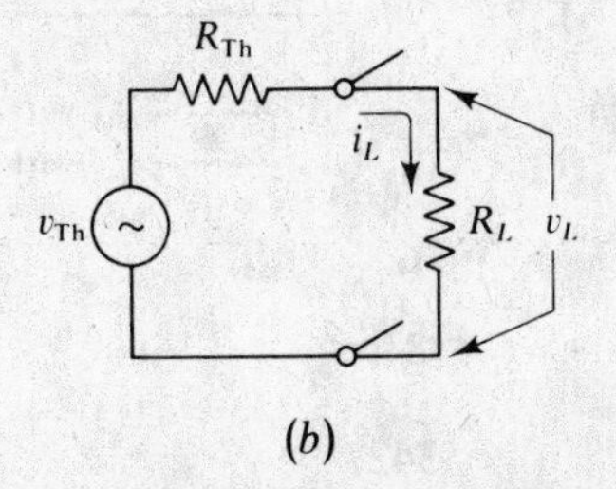

(b)

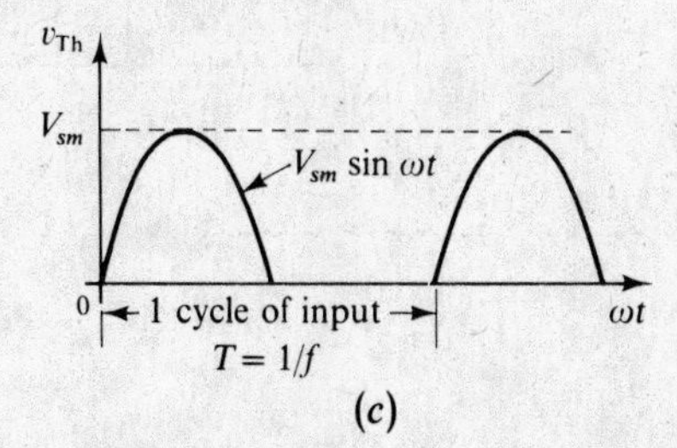

(c)

Fig. 8-5

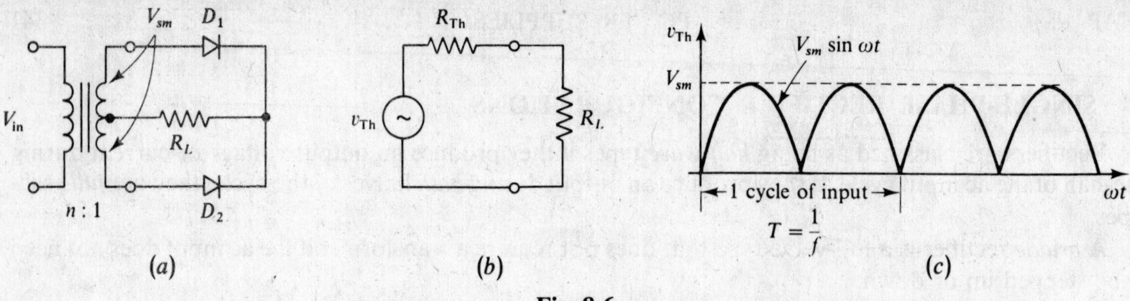

Fig. 8-6

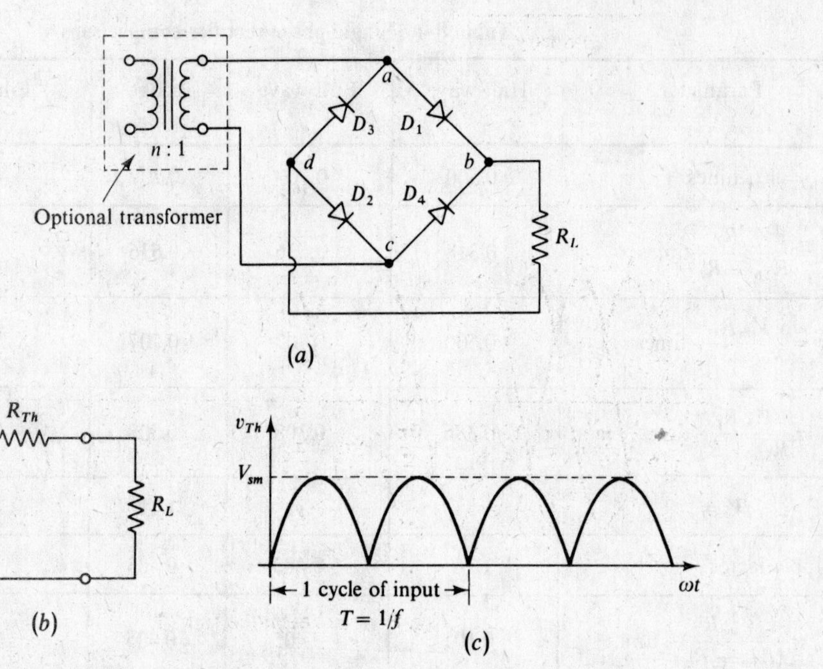

Fig. 8-7

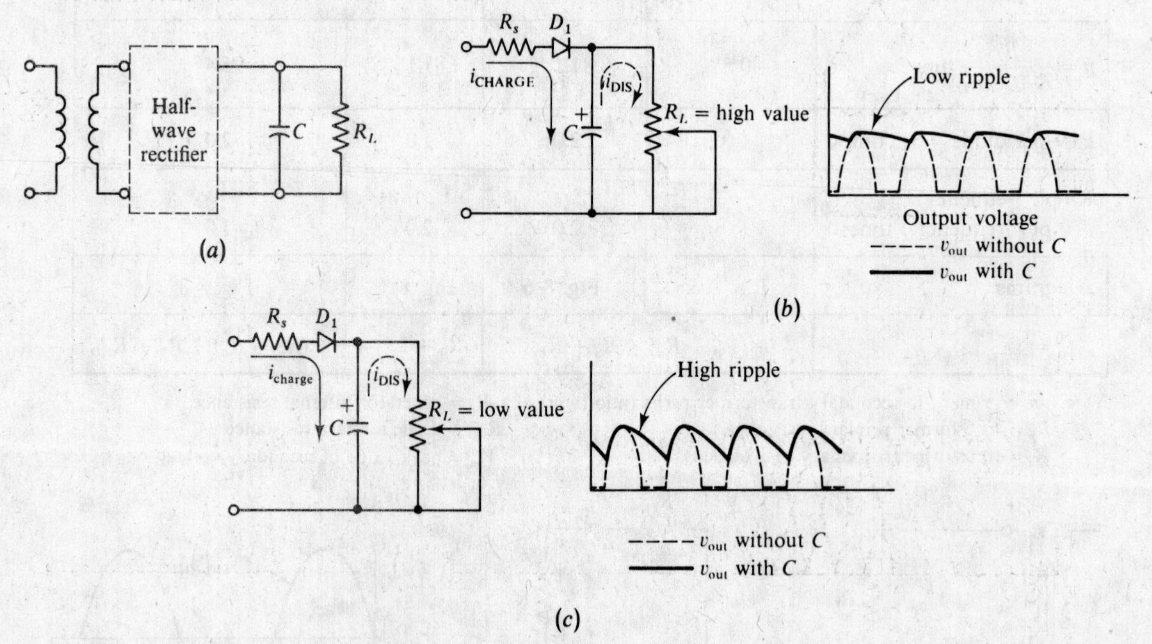

Fig. 8-8

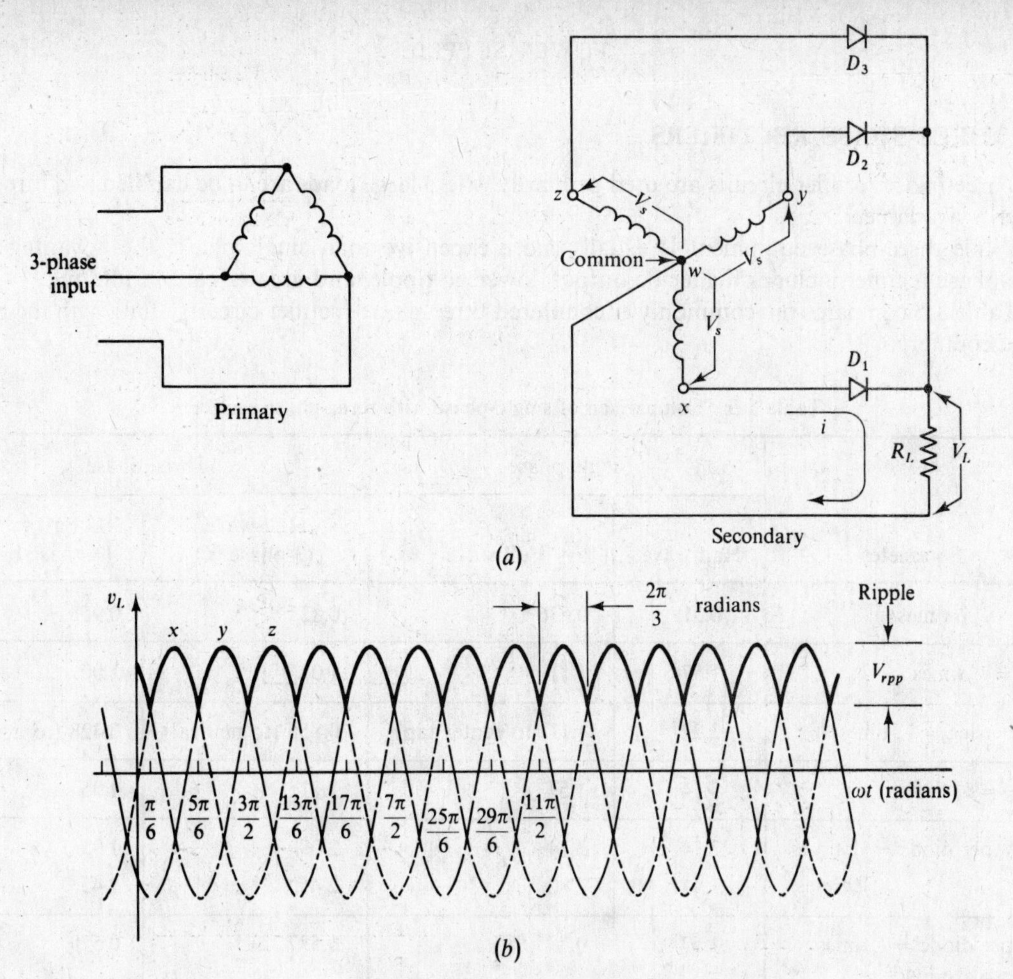

Fig. 8-9 Three-phase star (Y)

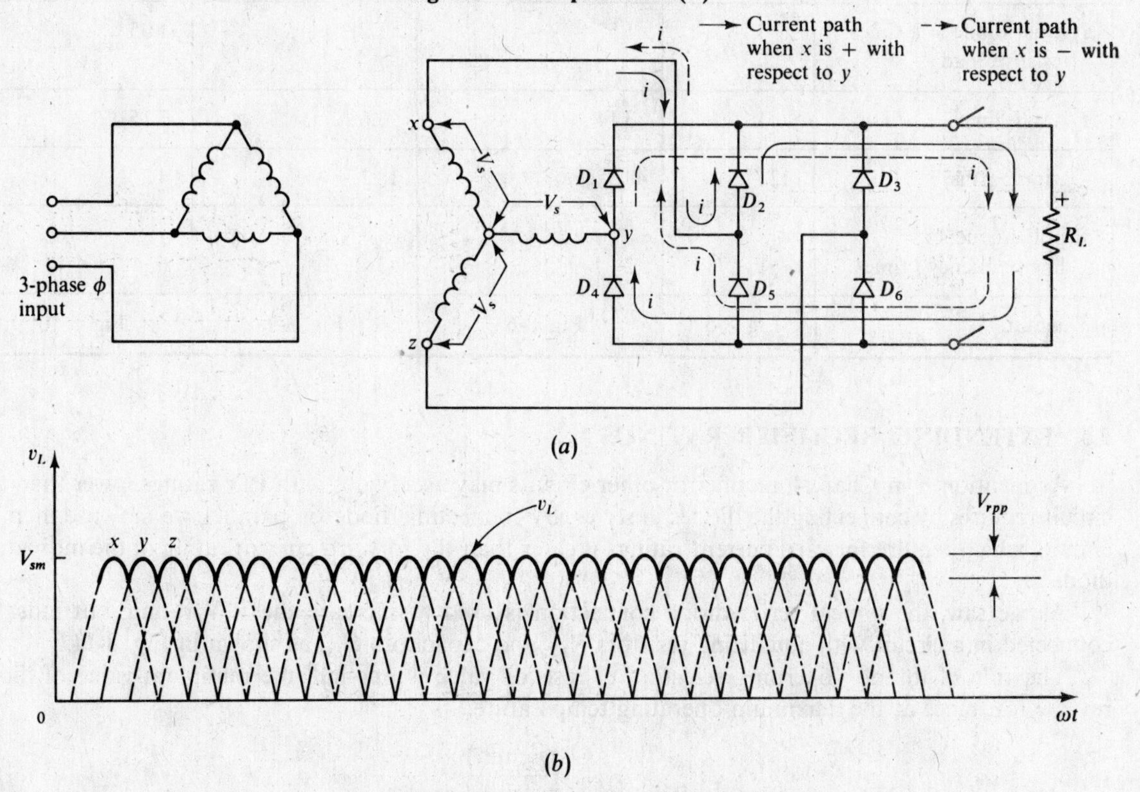

Fig. 8-10 Three-phase bridge (full wave)

8.5 THREE-PHASE RECTIFIERS

Three-phase rectifier circuits are used primarily where large loads are to be handled and three-phase power is available.

While three-phase equipment is usually more expensive than single-phase, the advantage of the three-phase rectifier includes higher dc output, lower ac ripple, and greater ease in filtering.

Table 8-5 compares the commonly encountered three-phase rectifier circuits along with their single-phase counterparts.

Table 8-5 Comparison of single-phase with three-phase rectifiers

	Single phase		Three phase	
Parameter	Half-wave	Full wave	Half-wave (3-phase Y)	Full wave (3-phase bridge)
$V_o = V_{LM}$ times	0.318	0.636	0.827	0.955
$V_L = V_o$ times	1.57	1.11	1.02	1.00
V_s per leg $= V_o$ times	2.22	1.11 (to center tap)	0.855 (to neutral)	0.428 (to neutral)
$V_{LM} = V_o$ times	3.14	1.57	1.21	1.05
PIV per diode $= V_o$ times $= V_s$ times	3.14 1.41	3.14 2.82	2.09 2.45	1.05 2.45
I_L per diode $= I_o$ times (resistive load)	1.57	0.785	0.587	0.579
I_{LM} per diode $= I_o$ times (resistive load)	3.14	1.57	1.21	1.05
$I_{LM} = I_o$ times	3.14	3.14	3.63	3.15
% ripple (rms)	121%	48.3%	18.3%	4.2%
Ripple frequency = line frequency f times	1	2	3	6
Diagram	Fig. 8-5	Fig. 8-6	Fig. 8-9	Fig. 8-10

8.6 EXTENDING RECTIFIER RATINGS

As mentioned in Chap. 4, rectifier or other circuits may use diodes with PIV ratings lower than the circuit requires by connecting the diodes in series. By connecting diodes in parallel, we may use them in circuits which require forward current ratings greater than the forward current rating of the individual diode.

As we saw, the typical series diode connection is used to increase the PIV rating, but must be connected in a circuit with equalizing resistors R_{ED} and capacitors C_{ED} as shown in Fig. 8-11.

The rule of thumb governing selection of resistor value is one-half the minimum value of diode reverse resistance at the maximum operating temperature.

$$R_{ED} = \frac{r_d(\text{min})}{2} \tag{8-15}$$

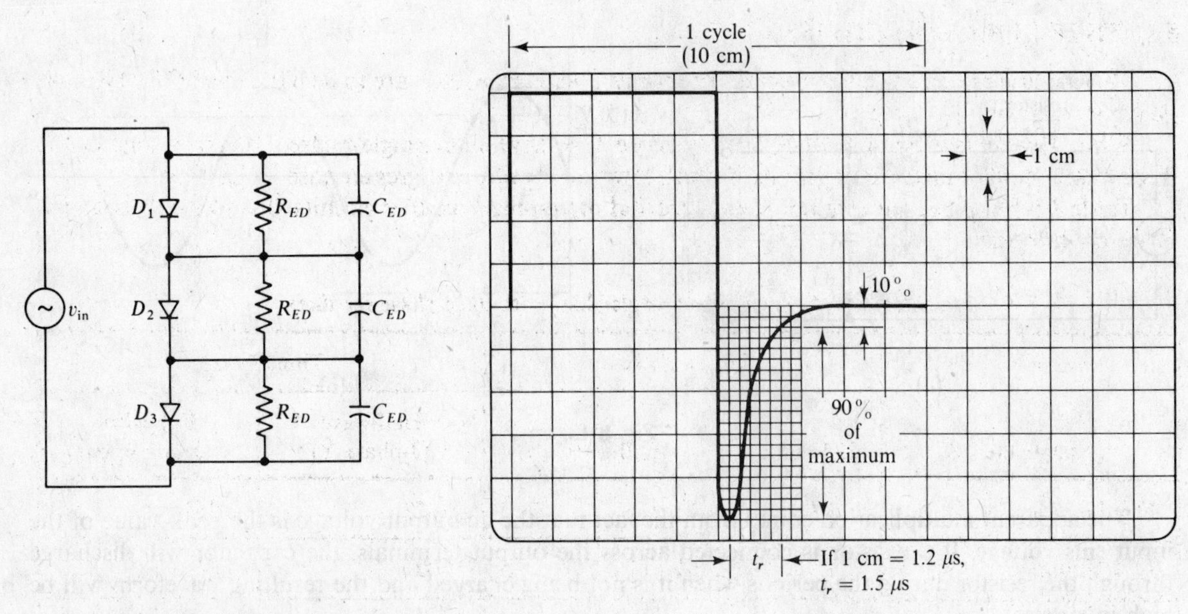

Fig. 8-11

Fig. 8-12

The rule governing selection of capacitor value depends on the diode reverse recovery time t_r and is given by the formula

$$C_{ED} = \frac{nt_r}{R_T} \tag{8-16}$$

where n = number of diodes connected in series
t_r = diode recovery time, s
R_T = internal generator resistance plus load resistance, Ω
C_{ED} = equalizing capacitance value across each diode, F

Diode recovery time t_r is measured as the time necessary for the diode to reach 90 percent of its maximum negative value, as shown in Fig. 8-12.

8.7 VOLTAGE MULTIPLIER CIRCUITS

Diodes and capacitors can be connected to ac power lines in such a way as to increase the dc output voltage to a value which is higher than that obtainable with simple rectification. There is an important tradeoff to be remembered here: The higher the dc output voltage, the less able the supply is to sustain a load. In other words, voltage multipliers with high dc output voltages cannot deliver high load currents over a long period of time.

Figure 8-13*a* shows a simple half-wave rectifier with a capacitor across the output. This circuit is not usually considered to be a voltage multiplier but is introduced here to demonstrate the effect of the capacitor.

The arrows in Fig. 8-13*a* show the current path when point *a* is positive with respect to point *b*. The capacitor is a large-valued electrolytic type, as indicated by the + symbol. Neglecting the small voltage drops caused by R_S and the diode, the capacitor charges to the peak value of the input voltage after a few cycles.

In the case of Fig. 8-13*b*, which shows an unrectified sine wave voltage with the rms value V given as 120 V, this peak value will be given by [from (*1-12b*)]

$$V_m = \sqrt{2}(V) = \sqrt{2}(120) = 170 \text{ V}$$

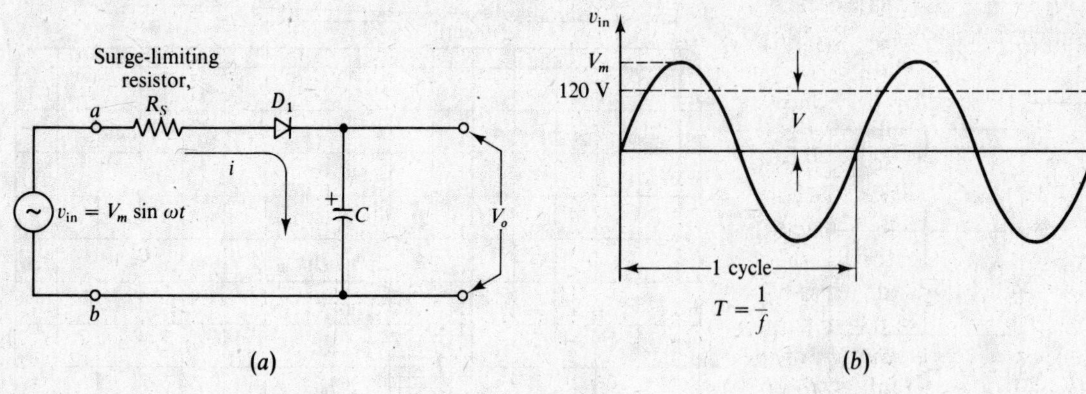

Fig. 8-13

The apparent multiplication comes from the fact that the dc output voltage is the peak value of the input rms voltage. If a resistor is connected across the output terminals, the capacitor will discharge through the resistor during the periods when it is not being charged and the resulting waveform will be high in ripple.

Figure 8-14*a* shows a circuit for a half-wave voltage doubler. As shown in the illustration, the input signal is a pure sine wave having the equation

$$V_{in} = V_m \sin \omega t$$

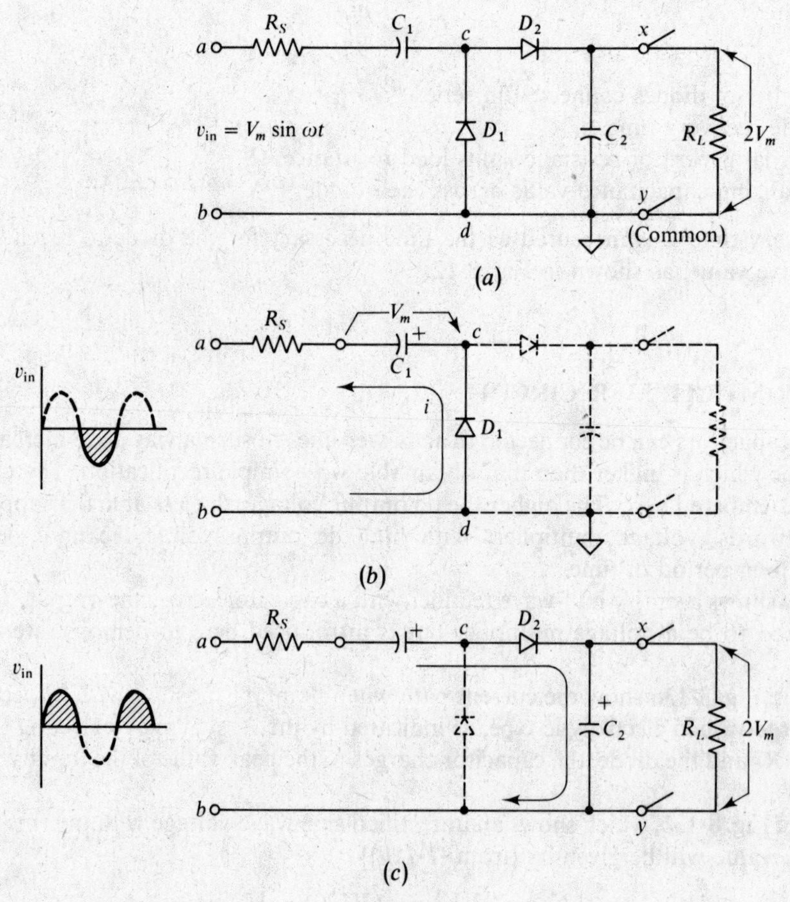

Fig. 8-14

During the negative half-cycle of input power as shown in Fig. 8-14*b*, capacitor C_1 is charged to the peak value of input voltage, V_m. Note that the voltage across capacitor C_1 has a polarity that is positive toward the cathode of diode D_1.

In the next half-cycle, the input voltage is positive at *a* with respect to the voltage at *b*. This voltage is in series with the peak voltage stored across capacitor C_1. The two voltages add in a manner similar to voltages of series-connected batteries adding.

The effect is to put a positive voltage of 2 V_m at point *c*. Diode D_2 now conducts, and the conduction path is shown in Fig. 8-14*c*. Capacitor C_2 will charge to the peak voltage of 2 V_m and this voltage appears across C_2 and load resistor R_L.

When R_L is connected across terminals *x* and *y*, capacitor C_2 will discharge through it on alternate half-cycles and cause the power supply regulation to be poor and the ripple output voltage high. This is one of the major disadvantages of the voltage doubler circuit.

Figure 8-15 shows a full-wave voltage doubler circuit. When capacitors C_1 and C_2 are both charged to the peak value of input voltage V_m, the output voltage across terminals *x* and *y* will be 2 V_m. The full-wave doubler has somewhat better regulation than the half-wave doubler because the capacitor discharge time is less, but it still has relatively high ripple content when low values of load resistance are used.

Some useful design information for voltage doublers is contained in Table 8-6.

Although only voltage doublers have been discussed in this section, the same procedures can be used to create voltage multipliers of 3, 4, 5, and more. Voltage multipliers are usually called for in circuits requiring high voltages with low current drain, as in TV circuits.

A typical TV 5 times multiplier is shown in Fig. 8-16.

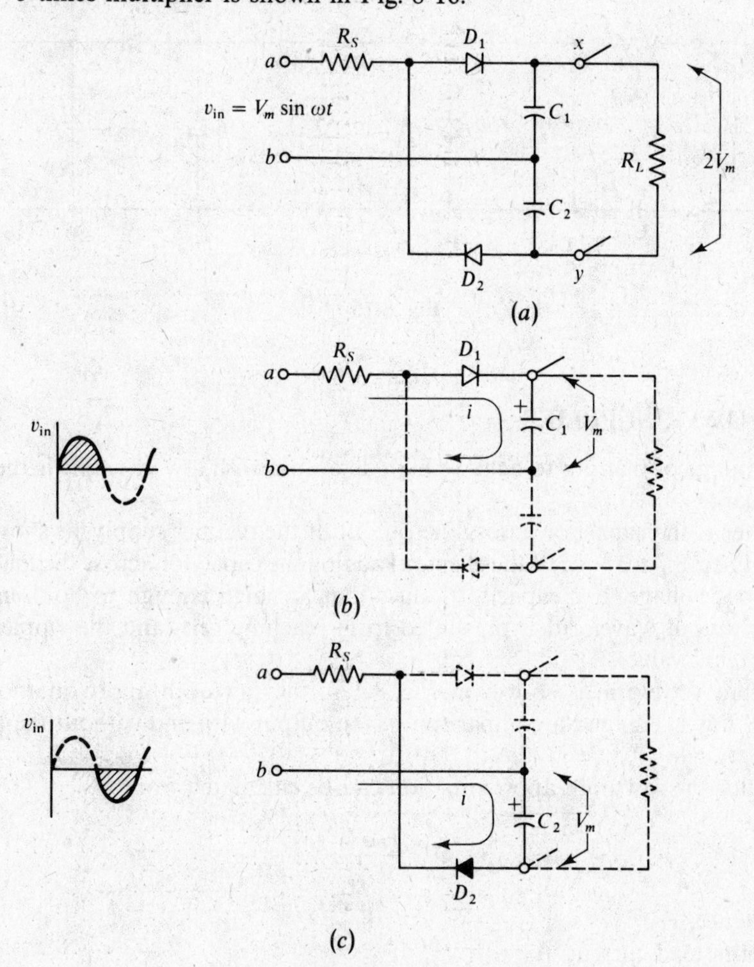

Fig. 8-15

Table 8-6

Minimum load resistance	$R'_L = \dfrac{V_o}{I_{LM}}$		
Surge-limiting resistance	$R_S = \dfrac{R'_L}{50} = \dfrac{V_o}{50 I_{LM}}$		(8-17)
Capacitive reactance	$X_C = \dfrac{R'_L}{12} = 0.083 R'_L$		(8-18)
Capacitor values	$C_1 = C_2 = \dfrac{12}{2\pi f R'_L} = \dfrac{1.91}{f R'_L}$		(8-19)
Voltage rating	C_1	$\geq V_m$	
		Half-wave	Full wave
	C_2	$\geq 2V_m$	$\geq V_m$
PIV rating	$D_1 = D_2$	$\geq 2V_m$	

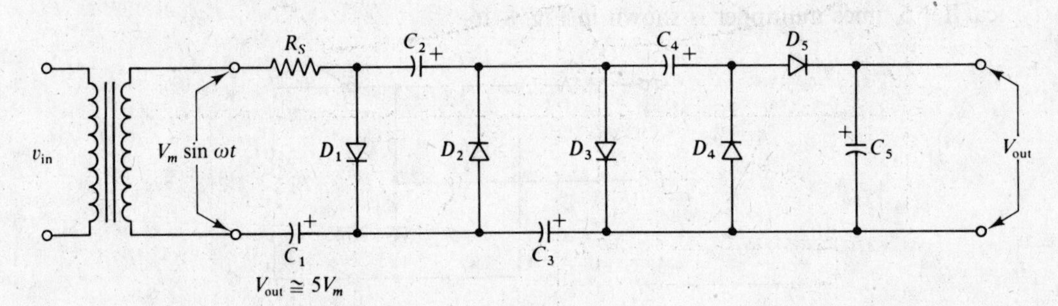

Fig. 8-16

8.8 FILTERED POWER SUPPLIES

The major advantage of filtering to achieve high dc output with low ac ripple is the wide variety of low-cost filters available.

The simplest filter is the capacitor across the output of the rectifier supply, as shown in Fig. 8-17*a*. We have seen in Sec. 8.7 that one of the effects of adding the capacitor across the load resistance is to increase the dc output voltage. For capacitor values that are high enough to hold energy during the discharge cycle, the output waveform is prevented from reaching zero and the ripple factor becomes lower than the unfiltered value.

The actual output waveform is shown in Fig. 8-17*b*, the sawtooth approximation used in most analysis is shown in Fig. 8-17*c*, and a comparison of the output with and without the filter is shown in Fig. 8-17*d*.

The value of C for the sawtooth approximation can be calculated from

$$C = \frac{I_{LM}\, t_d}{2V_{rp}} \tag{8-20}$$

where C = capacitance, F
I_{LM} = maximum load current, A
t_d = discharge time, s

and

$$V_{rp} = \frac{V_{max} - V_{min}}{2} \tag{8-21a}$$

as shown in Fig. 8-17*c*.

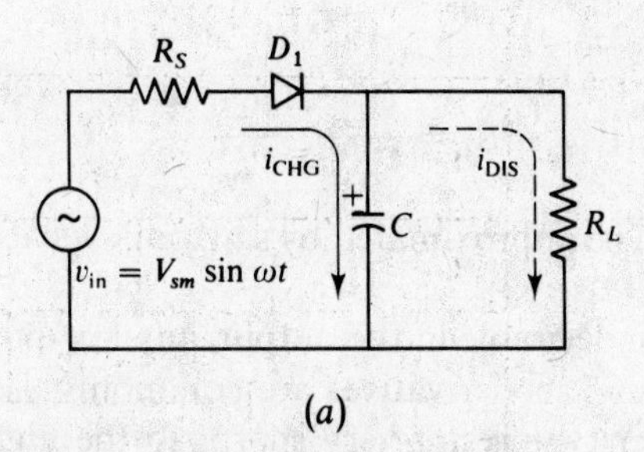

(*a*)

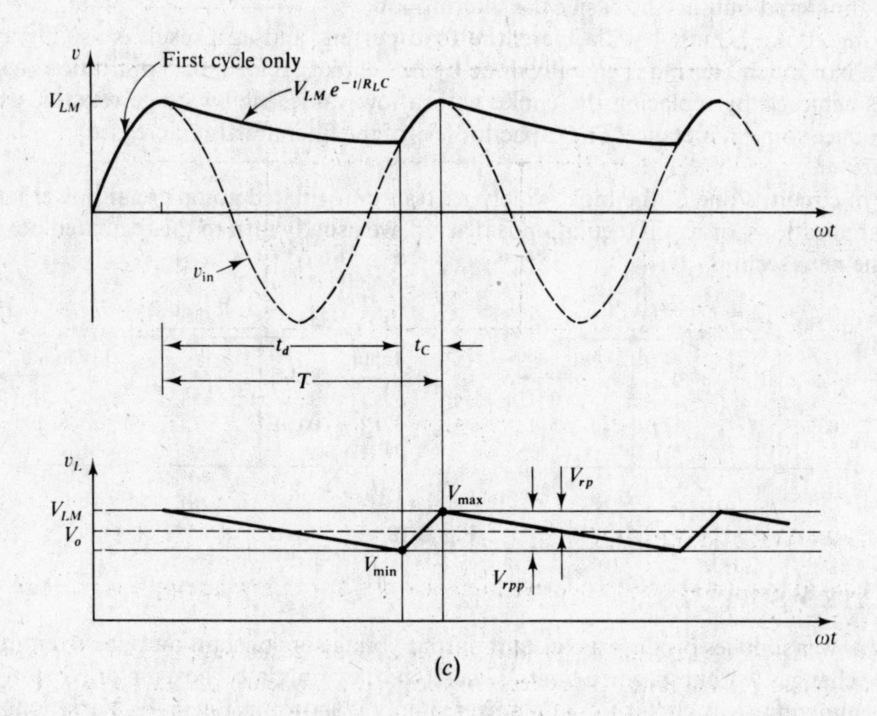

(*c*)

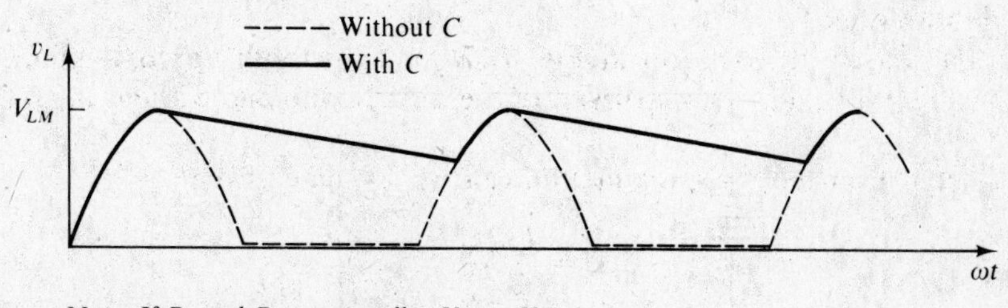

Note: If R_S and R_{FD} are small, $V_{LM} = V_{sm}$

(*d*)

Fig. 8-17

Since the average value V_o of the sawtooth wave (shown in Fig. 8-17c) is

$$V_o = \frac{V_{max} + V_{min}}{2} \tag{8-21b}$$

and the peak ripple factor is given by Eq. (8-3a) in Table 8-2 as

$$FR_p = \frac{V_{rp}}{V_o}$$

substitution yields

$$FR_p = \frac{V_{max} - V_{min}}{V_{max} + V_{min}} \tag{8-22}$$

Because most filtered waveforms can be approximated by sawtooth waves riding on a dc value, Eq. (8-22) is quite useful.

Since all alternating current is undesirable in the output, any low-pass filter is acceptable and the π low-pass filter shown in Fig. 8-18a and its derivatives are commonly used.

The cutoff frequency of this filter type is not very sharp, so the higher in frequency the harmonic content of the unfiltered output, the easier the filtering job.

The blocking choke L must handle the entire load current and as a result is usually a bulky and expensive item. Not much filtering is actually done by this choke, so for most applications a satisfactory compromise is achieved by replacing the choke with a low-value, high-wattage resistor, usually wire-wound to introduce some inductance. The capacitor legs of the filter are also increased in value, as shown in Fig. 8-18b.

This type of circuit is one of the most widely used in unregulated commercial power supplies. For more critical applications, or if stiff regulation is desired, we usually turn to the regulated power supplies discussed in the next section.

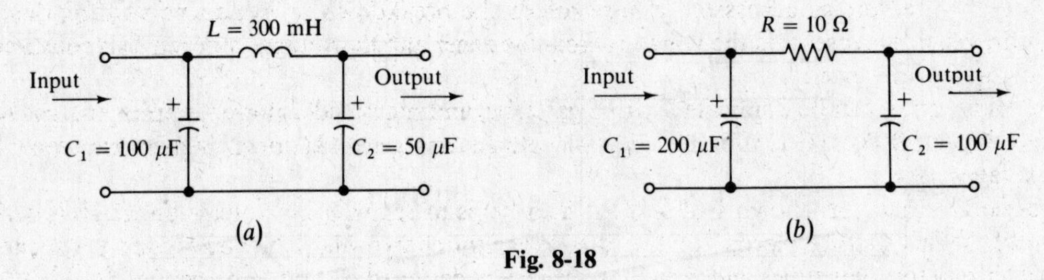

Fig. 8-18

8.9 REGULATED SUPPLIES

Regulated power supplies produce a constant output voltage or constant output current regardless of load resistance changes (within specified limits). Therefore they simulate Thevenin or Norton generators.

The Thevenin equivalent circuit for a dc power supply is shown in Fig. 8-19. Variations in the load resistance R_L cause changes in the current through R_{Th} and changes in the voltage V_{ab} across the supply terminals.

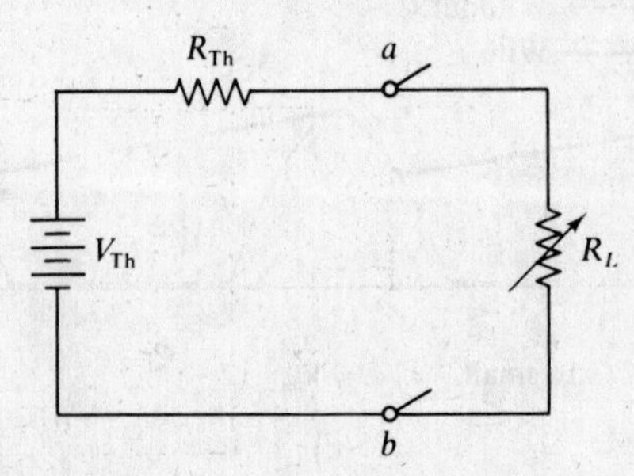

Fig. 8-19

The Thevenin resistance R_{Th} is given by

$$R_{\text{Th}} = n^2(R_p + R_g) + R_s + mR_{FD} + R_F + R_w$$

where n = primary to secondary transformer turns ratio (n:1)
R_p = transformer primary resistance
R_s = transformer secondary resistance
R_g = generator internal resistance
m = number of diodes in current path
R_{FD} = diode forward resistance
R_F = filter resistance
R_w = wiring resistance

when all values for resistance are given in ohms.

The regulation of a supply is a measure of how well it maintains a dc voltage when the load current changes. It is often expressed as a percent value:

$$\%\ \text{Regulation} = \left(\frac{V_{NL} - V_{FL}}{V_{NL}}\right)(100) \qquad (8\text{-}23)$$

where V_{NL} = no-load output voltage, V
V_{FL} = full-load output voltage, V

The *lower* the percent regulation, the more stable the output voltage for changes in load current. Also, the internal resistance of an unregulated power supply should be kept low in order to get a low (more desirable) percent regulation.

8.10 SHUNT REGULATORS

Figure 8-20 shows a simple shunt regulator using a reverse-biased zener diode across the load resistor. When the voltage across the zener exceeds the breakdown or avalanche voltage, the current through the zener increases and the voltage across the zener and the load resistor remains constant at the rated voltage for the zener.

The same action can be duplicated with a gas tube, and many older power supplies use gas tubes to regulate critical voltages. Gas-tube breakdown voltages are generally much higher than zener breakdown voltages.

A bleeder resistor, as shown in Fig. 8-21, also helps to regulate the voltage although it derives its name from the fact that it "bleeds" off the filter capacitor charge when the power supply is deenergized.

Bleeder resistors, gas tubes and zeners are generally designed so that 10 percent of the total full-load current flows through the regulating device:

$$I_{\text{Sh}} = 0.1 I_L \qquad (8\text{-}24)$$

The design of shunt regulators involves calculating the value and rating of R_S and the shunt regulator (zener diode, etc.) given the input voltage, the expected variations in either input or load, and the desired output regulation.

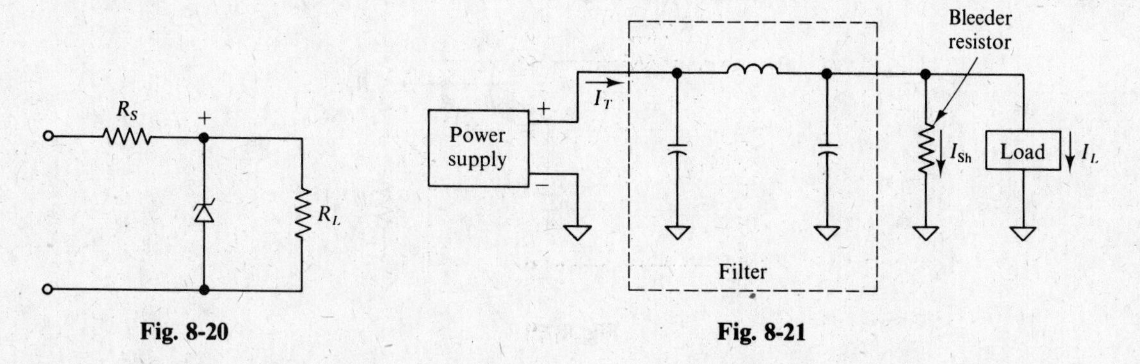

Fig. 8-20

Fig. 8-21

8.11 OPEN-LOOP SERIES REGULATORS

An example of an open-loop transistor series regulator is shown in Fig. 8-22. The base voltage of the transistor is maintained at a fixed voltage by the voltage drop across the zener diode. If the load resistance decreases, the voltage at point *a* will start to decrease. This produces an increase in voltage across Q_1. The transistor conducts harder, increasing the drop across the load resistance. The overall result is that the output voltage is maintained at a constant value.

An increase in the resistance of R_L decreases the drop across Q_1 and lowers the current through Q_1 and R_L. The decrease in current lowers the voltage across R_L to the desired value. So, regardless of whether R_L increases or decreases, the output voltage is stabilized.

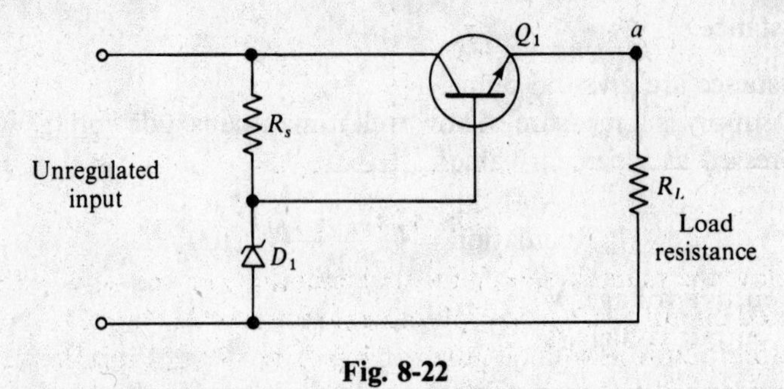

Fig. 8-22

8.12 CLOSED-LOOP SERIES REGULATORS

Figure 8-23 shows a closed-loop series regulator. The reference is the voltage across zener D_1. The output voltage of the supply is sensed at the arm of R_2. Transistor Q_2 conducts in an amount dependent upon the difference in voltage between the emitter and base.

If the output voltage starts to increase, Q_2 conducts harder. The base of Q_1 becomes less positive, and Q_1 decreases its conduction. That lowers the output voltage to the required value. If the output voltage decreases, Q_2 decreases its conduction. That makes the base of Q_1 more positive, which causes Q_1 to conduct harder. This increases the output voltage to the required value.

Closed-loop circuits like the one shown in Fig. 8-23 present an especially difficult servicing problem. If the output voltage is not correct, it may be due to a malfunction of any of the components in the loop. The voltage at each point depends upon the voltage at other points in the closed loop. The best way to troubleshoot this type of circuit is to open the closed loop. An ideal point would be the arm of R_2. A power-supply voltage can be substituted for the normal voltage at point *a*. With the loop open, standard troubleshooting procedures can be used.

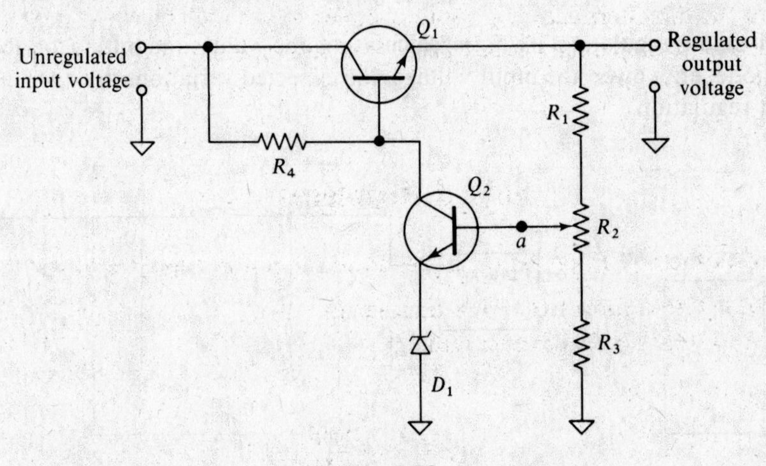

Fig. 8-23

8.13 IC REGULATORS

With the advent of low-cost LSI circuits, even inexpensive power supplies in the common voltage ranges with under 5 A loads can now have the advantage of low ripple and good (stiff) regulation.

A typical three-terminal IC regulator circuit is shown in Fig. 8-24.

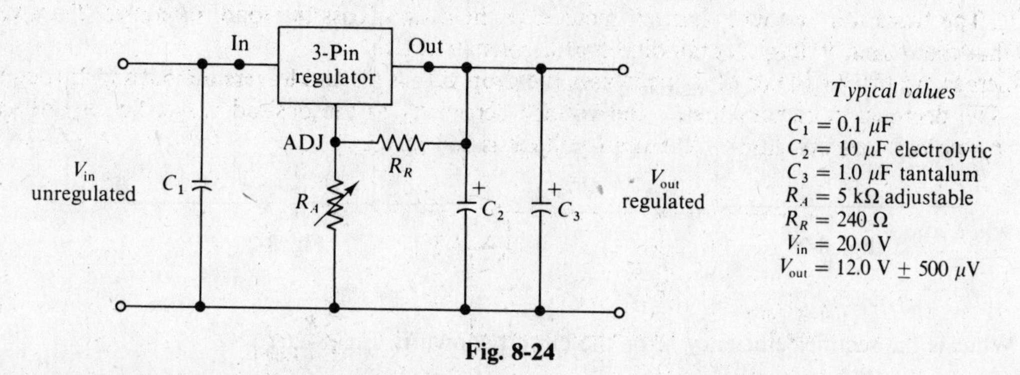

Fig. 8-24

It is important to follow the manufacturer's recommendations in positioning the input and output filter capacitors in the wired circuit as improper placement may result in unwanted oscillations set up by the transistor feedback circuit built into the regulator chip. The output capacitor serves to improve the stability and transient response of the regulator.

Typical values of output ripple are from 0.005 to 0.01 percent and uses for the three-terminal regulator include

- Constant-voltage (Thevenin) generators
- Constant-current (Norton) generators
- Battery chargers
- Logic supplies
- Adjustable power supplies

Most regulator IC chips contain provision for current limiting so that accidentally short circuiting the output will not destroy the regulator chip. A power supply that is built to provide current limiting under actual load conditions is called a *crowbar* or *foldback* supply.

The crowbar or current limiting capability comes from a shunt path to ground controlled by a high current transistor or Darlington pair. When the current limit is sensed by the reference circuit, the shunt transistor path is activated and allows the excess current to pass harmlessly to ground. Care must be taken to properly dissipate the heat caused by this type of operation or damage to the regulator will occur.

The IC regulator, in three-terminal and more complex forms such as the switching regulator, has become quite common in commercial circuits because of its flexibility in application and its ability to provide stiff regulation at low cost.

Solved Problems

8.1 What is the form factor FF for the load current in the power supply shown in Fig. 8-25?

Given: $I_L = 0.85$ A (from true rms meter reading)
$I_o = 0.61$ A (from dc meter reading)

From (8-7a),

$$FF = \frac{I_L}{I_o} = \frac{0.85}{0.61} = 1.39$$

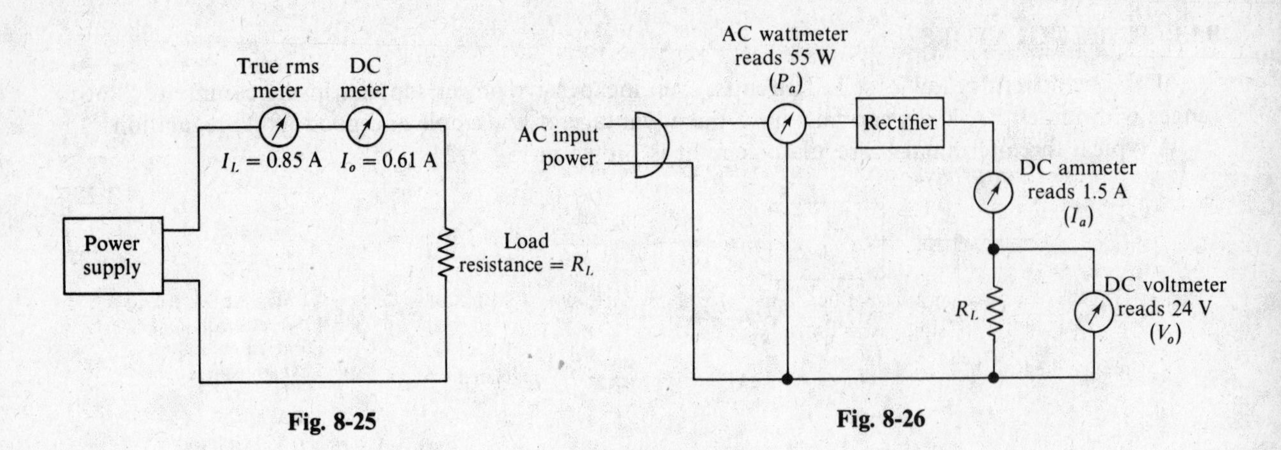

Fig. 8-25

Fig. 8-26

8.2 What is the rectifier efficiency η for the circuit shown in Fig. 8-26?

The dc output power P_o can be found from the given values of V_o and I_o:

$$P_o = V_o I_o = (24)(1.5) = 36 \text{ W}$$

The ac input power P_a is given as 55 W in Fig. 8-26. And from Eq. (*8-10a*) we know that

$$\eta = \left(\frac{P_o}{P_a}\right)(100) = \left(\frac{36}{55}\right)(100) = 65.5\%$$

8.3 In the circuit shown in Fig. 8-27, find (*a*) V_L (total rms output voltage); (*b*) V_o (dc output voltage); (*c*) V_r (rms ac ripple voltage); and (*d*) ripple frequency. Neglect R_g and the diode forward resistance R_{FD}.

Since the circuit produces an output during only one-half of the input cycle, it is a half-wave rectifier with $R_{Th} = R_g + R_{FD} = 0$ (from the problem statement).

$$v_{Th} = 8.90 \sin 377t$$

which is in the form

$$v_{Th} = V_{sm} \sin \omega t$$

So

$$V_{sm} = 8.90 \text{ V} \quad \text{and} \quad \omega = 377 \text{ rad/s}$$

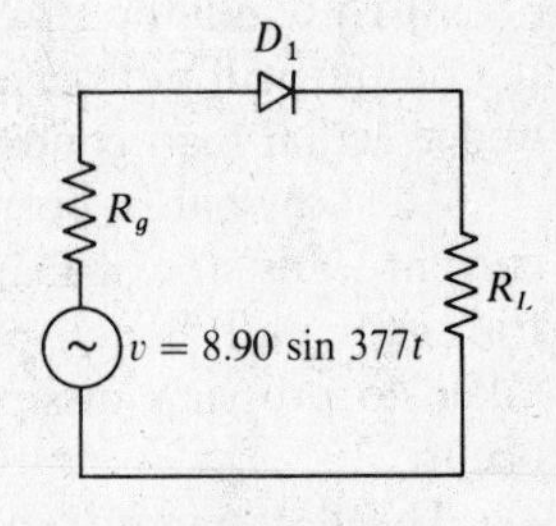

Fig. 8-27

(*a*) From Table 8-4,

$$V_L = \frac{0.5 V_{sm} R_L}{R_{Th} + R_L} \quad \text{with } R_{Th} = 0$$

Substituting,

$$V_L = \frac{(0.5)(8.90)(R_L)}{0 + R_L} = 4.45 \text{ V}$$

(*b*) From Table 8-4,

$$V_o = \frac{0.318 V_{sm} R_L}{R_{Th} + R_L}$$

Substitution yields

$$V_o = \frac{(0.318)(8.90)(R_L)}{0 + R_L} = 2.83 \text{ V}$$

(*c*) Using Table 8-2,

$$V_r = \frac{0.386 V_{sm} R_L}{R_{Th} + R_L}$$

Substitution gives us

$$V_r = \frac{(0.386)(8.90)(\cancel{R_L})}{0 + \cancel{R_L}} = 3.41 \text{ V}$$

(*d*) This is a half-wave rectifier, so the ripple frequency is the same as the generator (line) frequency (see Table 8-2). Using $\omega = 2\pi f$,

$$f = \frac{\omega}{2\pi} = \frac{377}{2\pi} = 60 \text{ Hz}$$

8.4 **Find the rms value V_r of the ac ripple voltage in the circuit shown in Fig. 8-28.**

According to (*8-14a*),

$$V_L^2 = V_o^2 + V_r^2$$

Substituting values of $V_o = 12.0$ and $V_L = 12.6$ from Fig. 8-28,

$$(12.6)^2 = (12.0)^2 + V_r^2$$
$$158.76 = 144.0 + V_r^2$$

Solving for V_r^2,

$$V_r^2 = 158.76 - 144.0 = 14.76$$

and $V_r = \sqrt{14.76} = 3.84$ V

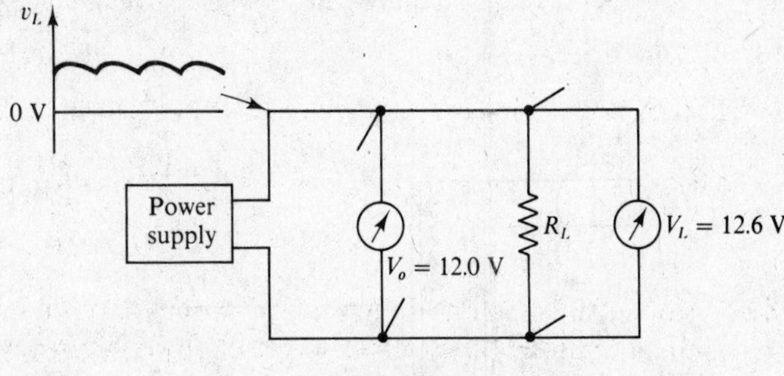

Fig. 8-28

8.5 **Two different unfiltered power supplies (A and B) are being compared. Both require 17.5 W of power from the ac input. The output of power supply A is 12.5 W, and the output power supply B is 10.0 W. Compare the rectifier efficiency η of the two supplies and state which is more desirable.**

For power supply A: $P_a = 17.5$ W, $P_o = 12.5$ W. According to Eq. (*8-10a*),

$$\eta = \left(\frac{P_o}{P_a}\right)(100) = \left(\frac{12.5}{17.5}\right)(100) = 71.4\%$$

For power supply B: $P_a = 17.5$ W, $P_o = 10.0$ W. And according to Eq. (*8-10a*) again,

$$\eta = \left(\frac{P_o}{P_a}\right)(100) = \left(\frac{10.0}{17.5}\right)(100) = 57.1\%$$

Clearly, power supply A is more efficient, and more desirable.

8.6 **The rms ripple V_r of the output of a certain filtered half-wave rectifier is measured and found to be 2 V. The average value of output voltage V_o is 100 V. What is the ripple factor and the percent ripple for this supply?**

According to (*8-1*),

$$FR = \frac{V_r}{V_o} = \frac{2}{100} = 0.02$$

According to (*8-4*),

$$\% \text{ Ripple} = (FR)(100) = (0.02)(100) = 2\%$$

8.7 For the rectifier circuit shown in Fig. 8-29, determine the following: (*a*) The peak value of output voltage V_{LM} assuming a forward voltage drop V_1 of 0.7 V across the diode. (*b*) The average value of output voltage V_o.

(*a*) The transformer secondary voltage is given as 6.3 V. Transformer voltages are normally given as rms values. The peak value of voltage, V_m, can be determined from Table 8-3:

$$V = \frac{V_m}{\sqrt{2}}$$

Transposing:

$$V_m = V\sqrt{2} = 6.3\sqrt{2} = 8.9 \text{ V}$$

From Kirchhoff's voltage rule, the peak output voltage V_{LM} is equal to the peak input voltage V_m minus the voltage drop V_1 across the diode:

$$V_{LM} = V_m - V_1 = 8.9 - 0.7 = 8.2 \text{ V}$$

(*b*) The average value V_o of a half-wave rectified sine wave is given by

$$V_o = \frac{V_m}{\pi} = \frac{8.2}{\pi} = 2.6 \text{ V}$$

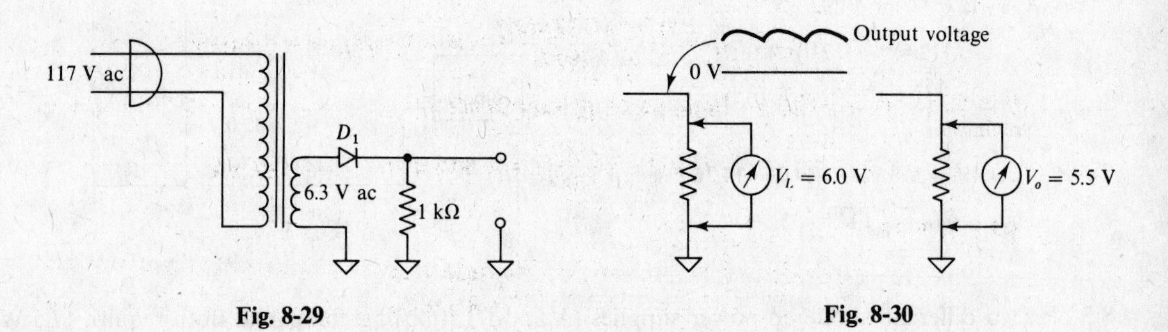

Fig. 8-29 **Fig. 8-30**

8.8 Figure 8-30 shows the different voltage measurements across a power-supply load resistor. The true rms voltage V_L (dc and ac ripple combined) is found to be 6.0 V and the average value of dc output V_o is 5.5 V. What is the rms value V_r of ripple voltage?

From (*8-14a*),

$$V_L^2 = V_o^2 + V_r^2$$

Transposing and taking the square root of both sides,

$$V_r = \sqrt{V_L^2 - V_o^2} = \sqrt{(6.0)^2 - (5.5)^2} = 2.4 \text{ V}$$

8.9 Neglecting the forward voltage drop across the diodes in the circuit shown in Fig. 8.31, calculate (*a*) the rms voltage V_L from *a* to ground; (*b*) I_L, measured from *a* to ground; (*c*) V_o, measured from *a* to ground; (*d*) I_o, measured from *a* to ground; (*e*) P_o; and (*f*) PIV across each diode.

Looking at Fig. 8-31, we identify the circuit as a bridge rectifier with $V_{\text{Th}} = 80.0$ V, $R_{\text{Th}} = 0$, and $R_L = 1000\ \Omega$. Therefore we may substitute these values in Table 8-2 and calculate the required answers.

Notice that the output is grounded at the high end of the bridge, so our voltages from *a* to ground will be negative and the currents from *a* to ground will also be negative (actual current is in the direction shown in Fig. 8-31).

Table 8-4 lists the required parameters in terms of V_{sm}, so we use the listed relationship to convert V_{Th}:

$$V_{\text{Th}} = 0.707\ V_{sm}$$

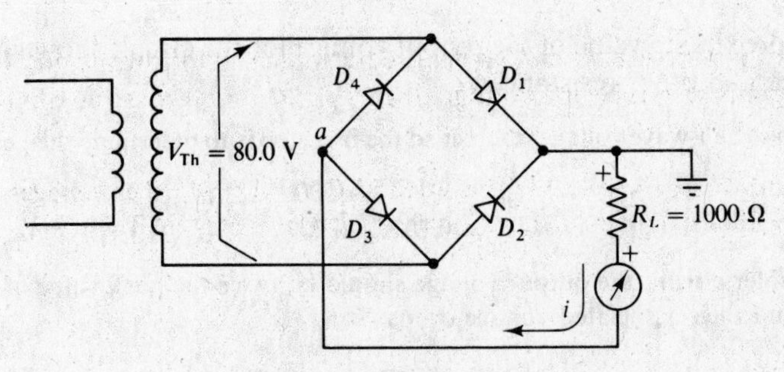

Fig. 8-31

Transposing and substituting,

$$V_{sm} = \frac{V_{Th}}{0.707} = \frac{80.0}{0.707} = 113 \text{ V}$$

(*a*) We have

$$V_L = \frac{0.707\ V_{sm}}{R_{Th} + R_L}$$

Substituting,

$$V_L = \frac{(0.707)(113)(1000)}{(0 + 1000)} = 80.0 \text{ V negative} = -80.0 \text{ V}$$

(*b*) Since $V_L = -80.0$ V [from part (*a*)], from Ohm's law

$$I_L = \frac{V_L}{R_L} = \frac{-80.0}{1000} = -80.0 \times 10^{-3} \text{ A} = -80 \text{ mA}$$

(*c*) We have

$$V_o = \frac{0.636\ V_{sm} R_L}{R_{Th} + R_L}$$

Substituting,

$$V_o = \frac{(0.636)(113)(1000)}{0 + 1000} = 71.9 \text{ V negative} = -71.9 \text{ V}$$

(*d*) Since $V_o = -71.9$ V [from part (*c*)], from Ohm's law

$$I_o = \frac{V_o}{R_L}$$

Substituting,

$$I_o = \frac{-71.9}{1000} = -71.9 \times 10^{-3} \text{ A} = -71.9 \text{ mA}$$

(*e*) We have

$$P_o = \frac{0.405 V_{sm} R_L}{(R_{Th} + R_L)^2}$$

So

$$P_o = \frac{(0.405)(113)^2(1000)}{(0 + 1000)^2} = 5.17 \text{ W}$$

We may check this result using $V_o = -71.9$ V from part (*c*), $I_o = -71.9 \times 10^{-3}$ A from part (*d*), and

$$P_o = V_o I_o = (-71.9)(-71.9 \times 10^{-3}) = 5.17 \text{ W} \checkmark$$

(*f*) Table 8-4 also shows the PIV per diode = V_{sm}. So the PIV per diode = 113 V.

8.10 What is the approximate value of the output voltage for the circuit shown in Fig. 8-32? Disregard diode resistance and other resistances.

This circuit is a half-wave voltage doubler so the first step is to determine the peak value of input voltage:

$$V_m = \frac{V}{0.707} = \frac{500}{0.707} = 707 \text{ V}$$

In a voltage doubler circuit, the output voltage should be twice the peak value of input voltage, neglecting the resistances and their associated voltage drops. So

$$V_{out} = 2V_{sm} = (2)(707) = 1414 \text{ V}$$

In practice the value will be somewhat less due to circuit losses.

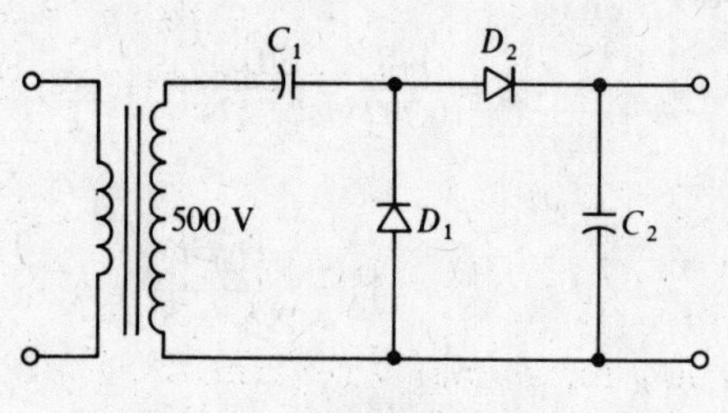

Fig. 8-32

8.11 In order to get an output voltage V_o of 75 V, what value of rms voltage V_s is required for each secondary leg of a three-phase transformer in a full-wave circuit?

V_o is given as 75 V. From Table 8-5:

$$V_s = 0.428V_o = (0.428)(75) = 32.1 \text{ V}$$

8.12 Diodes with PIV ratings of 50 V, reverse recovery times averaging 6 μs, and minimum reverse leakage resistance of 1.5 MΩ are to be used in a half-wave rectifier with a maximum applied voltage of 180 V. $R_T = 480\ \Omega$. Find the values of the equalizing resistors and the shunt capacitors and draw the circuit.

For a maximum applied voltage of 180 V and a PIV rating per diode of 50 V, the number of diodes needed is

$$\frac{180}{50} = 3.6$$

and rounding to the next highest integer gives four diodes in series.

Since the minimum reverse leakage resistance r_d(min) is given as 1.5 MΩ, using Eq. (*8-15*) we find

$$R_{ED} = \frac{r_d(\text{min})}{2} = \frac{1.5 \times 10^6}{2} = 0.75 \times 10^6 = 750 \times 10^3\ \Omega = 750 \text{ k}\Omega$$

Utilizing Eq. (*8-16*) to find the equalizing capacitance that will protect against different recovery times, we have

$$C_{ED} = \frac{nt_r}{R_T}$$

From the problem statement $t_r = 6\ \mu$s and $R_T = 480\ \Omega$. We have also determined that $n = 4$, so

$$C_{ED} = \frac{(4)(6 \times 10^{-6})}{480} = 0.05 \times 10^{-6} = 0.05\ \mu\text{F}$$

The circuit is shown in Fig. 8-33.

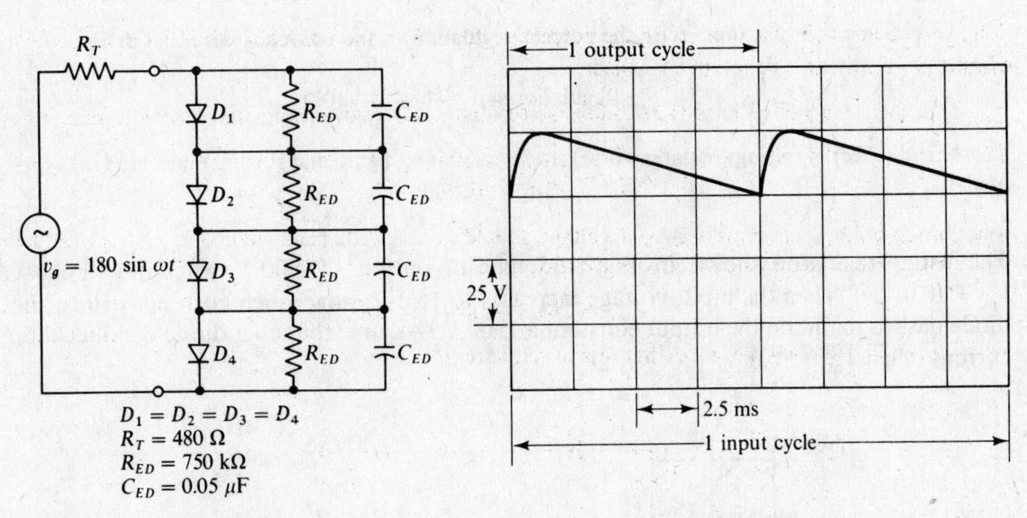

Fig. 8-33

Fig. 8-34

8.13 A filtered power supply has the output waveform shown in Fig. 8-34. (*a*) Find the output frequency f_{out} and (*b*) find the peak ripple factor. (*c*) What is the rated output voltage V_o?

(*a*) Figure 8-34 shows one horizontal division equal to 2.5 ms, so

$$T_{out} = 4 \text{ divisions} = (4)(2.5 \times 10^{-3}) = 10.0 \times 10^{-3} \text{ s}$$

But since the period and frequency are reciprocals of each other,

$$T_{out} = \frac{1}{f_{out}}$$

Transposing and substituting,

$$f_{out} = \frac{1}{T_{out}} = \frac{1}{10.0 \times 10^{-3}} = 100 \text{ Hz}$$

Note that the output undergoes two cycles for each input cycle, indicating a full-wave supply.

(*b*) Figure 8-34 shows one vertical division equal to 25 V, so

$$V_{max} = 4 \text{ divisions} = (4)(25) = 100 \text{ V}$$

$$V_{min} = 3 \text{ divisions} = (3)(25) = 75 \text{ V}$$

The waveform is approximately a sawtooth, so we may use (*8-22*):

$$FR_p = \frac{V_{max} - V_{min}}{V_{max} + V_{min}} = \frac{100 - 75}{100 + 75} = 0.143$$

(*c*) V_o of the approximate sawtooth would be the average value between V_{max} and V_{min}. Using (*8-21b*):

$$V_o = \frac{V_{max} + V_{min}}{2} = \frac{100 + 75}{2} = 87.5 \text{ V}$$

8.14 A series regulator has a measured dc output voltage of 12.10 V with the output terminals open-circuited. When the regulator delivers its rated power, the output voltage falls to 12.06 V. Find the percent regulation.

The problem statement gives $V_{NL} = 12.10$ V and $V_{FL} = 12.06$ V. Using Eq. (*8-23*):

$$\%\ \text{Regulation} = \left(\frac{V_{NL} - V_{FL}}{V_{NL}}\right)(100) = \left(\frac{12.10 - 12.06}{12.10}\right)(100) = 0.331\%$$

Note: Some manufacturers refer the percent regulation to the full-load value and define

$$\% \text{ Regulation} = \left(\frac{V_{NL} - V_{FL}}{V_{FL}}\right)(100)$$

This has the effect of raising (making worse) the % regulation figure and is sometimes used in "worst-case" analysis.

8.15 The shunt regulator shown in Fig. 8-35 has an output of 5.00 V when the input voltage $V_{in} = 10.0$ V. When the input voltage increases to 15.0 V, how much current I_Z must the zener diode pass to maintain the output constant at 5.00 V? Assume the zener diode conducts minimum current when $V_{in} = 10.0$ V.

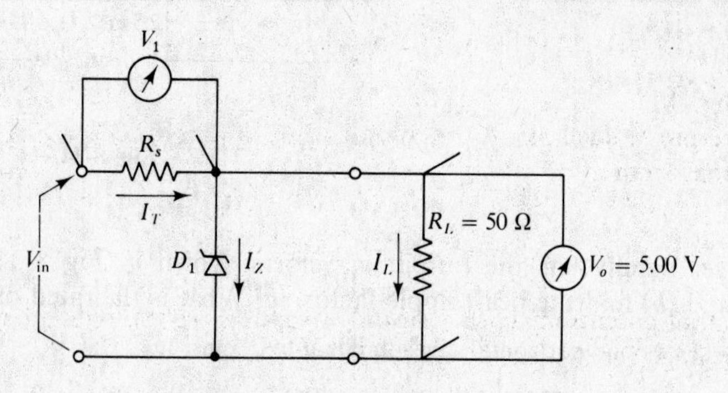

Fig. 8-35

With the voltages $V_{in} = 10.0$ V and $V_o = 5.00$ V, we first find R_S. By Kirchhoff's voltage rule,

$$V_{in} = V_1 + V_o$$

$$10.0 = V_1 + 5.00$$

Transposition yields

$$V_1 = 10.0 - 5.00 = 5.00 \text{ V}$$

Now by Ohm's law,

$$I_L = \frac{V_o}{R_L} = \frac{5.00}{50} = 0.100 \text{ A}$$

Since $I_Z(\text{min})$ occurs at $V_{in} = 10.0$ V, we use our rule of thumb:

$$I_Z(\text{min}) = 0.1 I_L = (0.1)(0.100) = 0.010 \text{ A}$$

By Kirchhoff's current rule,

$$I_T = I_Z + I_L$$

Substituting,

$$I_T = 0.010 + 0.100 = 0.110 \text{ A}$$

We find R_s from Ohm's law,

$$R_s = \frac{V_1}{I_T} = \frac{5.00}{0.110} = 45.5 \ \Omega$$

With R_s known, we can solve for the new I_T when the input voltage increases to 15.0 V. Again using Kirchhoff's voltage rule with $V_o = 5.00$ V and $V_{in} = 15.0$ V,

$$V_{in} = V_1 + V_o$$

$$15.0 = V_1 + 5.00$$

So

$$V_1 = 10.0 \text{ V}$$

And by Ohm's law,

$$I_T = \frac{V_1}{R_s} = \frac{10.0}{45.5} = 0.220 \text{ A}$$

I_L is still 0.100 A since V_o does not change. By Kirchhoff's current rule,

$$I_T = I_Z + I_L$$

$$0.220 = I_Z + 0.100$$

Transposing,

$$I_Z = 0.220 - 0.100 = 0.120 \text{ A} = 120 \text{ mA}$$

8.16 A typical three-pin adjustable IC regulator shown in Fig. 8-36 comes with manufacturer's specifications that state

$$V_{out} \cong V_{REF}\left(1 + \frac{R_2}{R_1}\right)$$

Find the value that potentiometer R_2 should have for an output voltage of 13.75 V. Assume that the reference voltage for this chip is 1.25 V.

Substituting values in the specification formula, we have

$$V_{out} = V_{REF}\left(1 + \frac{R_2}{R_1}\right)$$

$$13.75 = 1.25\left(1 + \frac{R_2}{R_1}\right)$$

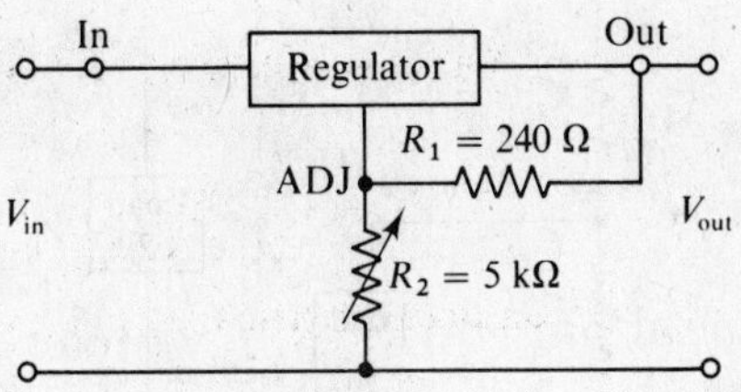

Solving for R_2/R_1,

$$\frac{R_2}{R_1} = \frac{13.75}{1.25} - 1 = 10$$

Fig. 8-36

With R_1 shown in Fig. 8-36 as 240 Ω,

$$R_2 = 10R_1 = (10)(240) = 2400\ \Omega = 2.4\ \text{k}\Omega$$

So the potentiometer should be set at $2.4/5.0 = 0.48$ of its range. If it is a precision 10-turn potentiometer, this means it should be set at 0.48 times 10, or 4.8 turns.

Supplementary Problems

8.17 Compare the waveforms shown in Fig. 8-37*a* and *b*. Which is more desirable for a power supply? Why?

8.18 If the *FF* of a filtered power-supply output waveform is given as 1.001, find the rms % ripple.

8.19 What is the *FF* of an unrectified (pure) sine wave voltage?

8.20 The dc output power of a certain rectifier power supply is 8.0 W. It is operated with an ac input power of 11.3 W. What is the rectification % efficiency η?

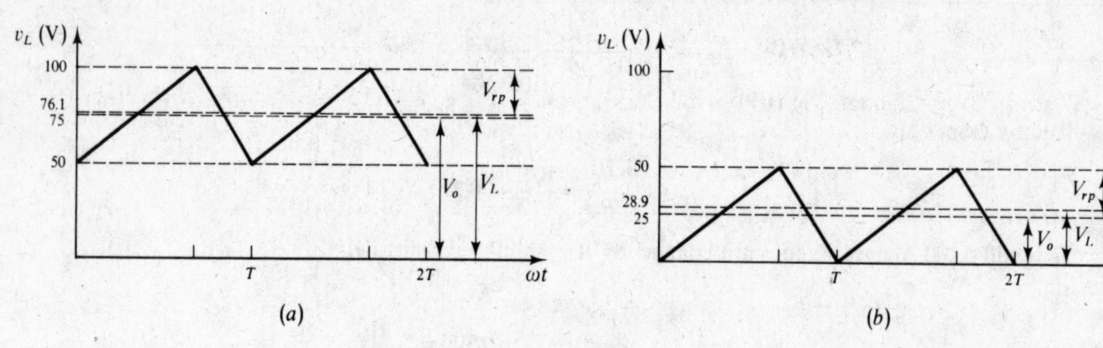

Fig. 8-37

8.21 The average value of a filtered voltage waveform from a power supply is 37 V. The rms value of the ac ripple is 31.7 mV. What is the rms value V_L of the output voltage?

8.22 The circuit shown in Fig. 8-38 is typical of resistor-voltage dividers used in the output circuit of some dc power supplies. Determine the values of R_1, R_2, and R_3 in order to obtain the values of voltages and currents shown.

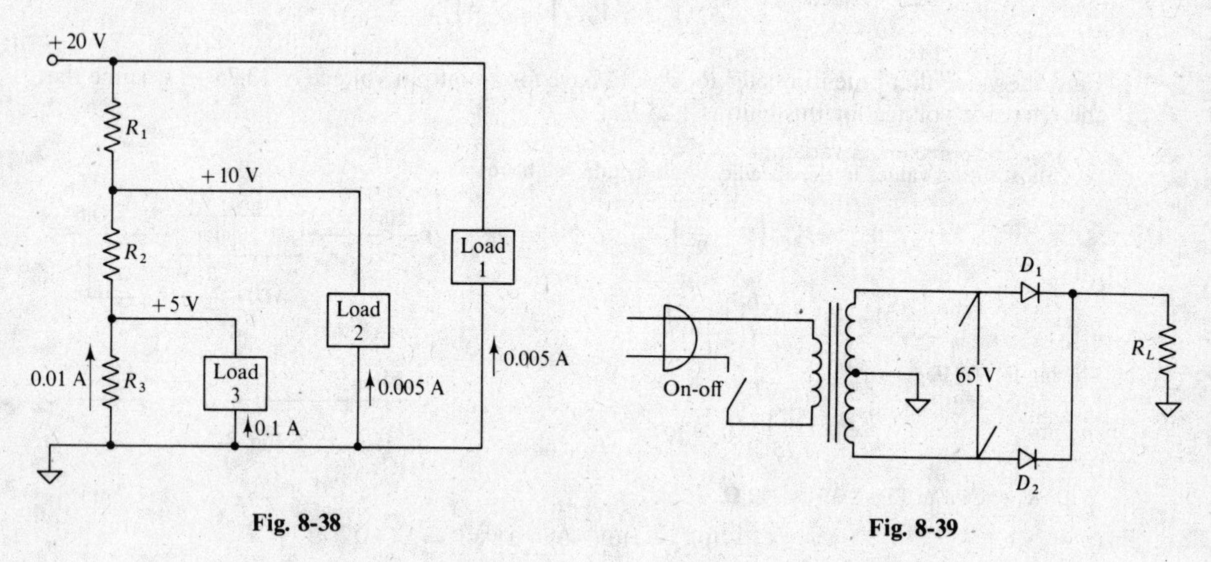

Fig. 8-38 **Fig. 8-39**

8.23 For the circuit shown in Fig. 8-39 calculate the following: (*a*) peak reverse voltage across each diode, (*b*) rms value of output voltage, and (*c*) average value of output voltage. *Note*: Disregard the forward voltage drop across each diode.

8.24 The average load current for a certain three-phase unfiltered half-wave rectifier is 250 mA. What is the peak value of forward current per diode?

8.25 The specifications for a series closed-loop IC regulator state that the output voltage is 5.00 V with 0.005% peak ripple. What is the maximum variation of the output voltage?

Answers to Supplementary Problems

8.17 Figure 8-37*a*. Although V_{rp} is identical for both parts, the V_o is higher and the ripple factor lower for Fig. 8-37*a*.

8.18 0.200%.

8.19 1.11.

8.20 70.8%.

8.21 37.6 V (ripple amplitude is too low to seriously affect rms value).

8.22 $R_1 = 87.0\ \Omega$, $R_2 = 45.5\ \Omega$, $R_3 = 500\ \Omega$.

8.23 (*a*) 91.9 V, (*b*) 32.5 V, and (c) 29.3 V.

8.24 302.5 mA.

8.25 $\pm 250\ \mu V$ or a maximum variation of 500 μV (4.99975 to 5.00025 V).

Index